Phenomenology of Space and Time

ANALECTA HUSSERLIANA

THE YEARBOOK OF PHENOMENOLOGICAL RESEARCH

VOLUME CXVI

Published under the auspices of
The World Institute for Advanced Phenomenological Research and Learning
A-T. Tymieniecka, President

For further volumes:
http://www.springer.com/series/5621

Anna-Teresa Tymieniecka
Editor

Phenomenology of Space and Time

The Forces of the Cosmos and the Ontopoietic Genesis of Life: Book One

Springer

Editor
Anna-Teresa Tymieniecka
The World Phenomenology Institute
Hanover, NH, USA

ISBN 978-3-319-02014-3 ISBN 978-3-319-02015-0 (eBook)
DOI 10.1007/978-3-319-02015-0
Springer Cham Heidelberg New York Dordrecht London

Library of Congress Control Number: 2014930023

Printed on acid-free paper

Springer is part of Springer Science+Business Media (www.springer.com)

Acknowledgements

In presenting to the scholarly public this collection, the first of two volumes gathering work from the World Phenomenology Institute's 62nd International Congress, "The Forces of the Cosmos and the Ontopoietic Genesis of Life," held in Paris, August 8–10, 2012, at Lucernaire Centre National d'Art et d'Essai, I wish to thank all who made this marvelous congress possible. Most importantly, I thank the Institute's Vice President, Daniela Verducci, who ran the conference expertly in my absence. Alessia Vitale, Mina Sehdev, Stefano Polenta, and Claire Hill deserve thanks for attending to the many essential details of a large conference. We all thank the staff of Lucernaire for making sure our rooms were comfortable and suitable to our needs. And, of course, I thank the authors, whose enthusiasm and expertise made the congress, and now this volume, a great success.

I am also grateful to Jeffrey T. Hurlburt and Louis Tymieniecki Houthakker for their assistance in preparing and editing this volume.

Anna-Teresa Tymieniecka

Contents

Part I

Toward a New Enlightenment: Metaphysics as Philosophy of Life 3
Nicoletta Ghigi

Moral Excellence as Cosmicization of Human Beingness in the *Ontopoietic* Perspective 11
Carmen Cozma

The Inseparable Link Between "Cosmology" and the "World of Life" in the Philosophy of Anna-Teresa Tymieniecka: The Originality of a New Perspective on the "Real Individual and Autonomous Being" and a Comparison with the "Phenomenological Realism" of Hedwig Conrad-Martius 21
Francesco Alfieri

The Forces of the Cosmos Before Genesis and Before Life: Some Remarks on Eugen Fink's Philosophy of the World 37
Simona Bertolini

Part II

Ontopoiesis in Ben Okri's Poetic Oeuvre and *A Time for New Dreams* (2011) 49
Rosemary Gray

Cosmic Order and Exoneration of the Beautiful: Visions of the Problem in Contemporary Philosophy 59
Ella Buceniece

The Law of Opposites in the Ontopoiesis of Life and in Language 71
Zaiga Ikere

The Forces of Darkness and the Forces of Goodness: Jerzy Nowosielski's Concept 79
Katarzyna Stark

Part III

Anthropological Regression in the Modern World Versus Anna-Teresa Tymieniecka's Metaphysics of Ontopoiesis of Life 93
Jan Szmyd

Biologically Organized Quantum Vacuum and the Cosmic Origin of Cellular Life 107
Attila Grandpierre

The Cosmos of Yolanthe: Knowing Without Seeing 135
Detlev Quintern

Philosophical Hermeneutics Confronted by That Which Is Different 147
Aleksandra Pawliszyn

Anna-Teresa Tymieniecka's and Max Scheler's Phenomenology as the *Ontopoietic Genesis of a Manager's Life* 155
Bronisław Bombała

Part IV

Comparative Phenomenology of Singing and Dance as Involving Artistic "Instruments" Incorporated Into the Body of Their Performer 169
Alessia Rita Vitale

Phenomenology and Archeology: Methodological Insights and Thematic Inspirations 195
Jaroslava Vydrová

Plotinus' "Enneads" and Self-creation 209
Ineta Kivle

Directing Anatoly Vasilyev, from Individual Creative Manner to the Method 219
Valery Kolenova

Part V

Teleology in Nature and Life-Transforming Art 227
Vladimir L. Marchenkov

The Creative Potential of Humor 237
Anna Malecka

Educational Paradigm Shift Towards Phenomenological Pedagogy 245
Kiymet Selvi

Human Soul, Body and Life Horizons 259
Maija Kūle

Part VI

The Unity of Eastern and Western Thought Traditions in Anna-Teresa Tymieniecka's Phenomenology of Life 273
Salahaddin Khalilov

Transcendental Morphology: A Phenomenological Interpretation of Human and Non-Human Cosmos 285
Bence Peter Marosan

The Outside's Inside: The Phenomenology of the External World in Hedwig Conrad-Martius' Thought 327
Ronny Miron

Kant and the Starry Heavens or the Splendor and Misery of Speculative Rationalism 359
Rihards Kūlis

Part VII

How to Approach Heideggerian Gods 369
Jani Vanhala

Meaning in the Forthcoming Sciences of Life: From Nietzsche and Husserl to Embodiment and Biosemiotics 381
Ammar Zeifa

Motion in Crisis: Why the Analytic Principles of Thought Destroy Motion and Life in the Cosmos 419
Ion Soteropoulos

"Heraclitus/Nietzsche/Heidegger in Πόλεμς" - "τὰ δὲ Πάντα οἰακίζει Κεραυνός" – Heraclitus 425
Kimiyo Murata-Soraci

***Lebenswelt* and Operational Methodology in the Philosophical and Epistemological Reflections of Hugo Dingler** 435
Dario Sacchi

The Permanent Creativity of Self 455
Stefano Polenta

Part VIII

Cognition and Emotion: From Dichotomy to Ambiguity 473
Claus Halberg and Simen Andersen Øyen

The Meeting of Man with Man 485
Leszek Pyra

Humour, an Enlightening and Restorative Force of the Inner *Cosmos*: A Phenomenological Approach 493
Tereza-Brindusa Palade

Program from The 62nd International Congress of Phenomenology: *The Forces of the Cosmos and the Ontopoietic Genesis of Life*, held in Paris, France, 2012 501

Part I

Toward a New Enlightenment: Metaphysics as Philosophy of Life

Nicoletta Ghigi

Abstract In Anna-Teresa Tymieniecka's various studies, one finds a specific focus to her work: she brings to light a new concept of the transcendental, understood no longer as an abstract concept of traditional phenomenology, but as a characteristic of living in which the human being actively participates through its "intellectual spirit" that we call the Transcendental of humanity. In our opinion, based on the creative activity of this spirit, it becomes possible to restore to philosophy a new vitality; it need no longer be seen as abstract or ideological, but as human, social. Philosophy becomes the human ground for a metaphysics of life, for a way of thinking that deals with the "great and ultimate" questions, to borrow from Husserl's turn of phrase (Edmund Husserl, *Cartesianische Meditationen und Pariser Vorträge* in *Husserliana* I, ed. B. Strasser [Den Haag: Martinus Nijhoff, 1950], p. 299). One can examine these questions in a straightforward human fashion, from the vantage point of the human being in her concrete and particular reality, who lives in the world. In its critical capacity, the "intellectual spirit" becomes the precondition for the constitution of a metaphysics that no longer has the abstract character of a science of being, understood as the intangible; rather, it configures itself as a personal, concrete human understanding of the universe with a particular focus on meaning for everyday life.

Nicoletta Ghigi (tr. Antonio Calcagno)

N. Ghigi (✉)
CIRF (Roma), Via G. Carducci, 62 Gubbio (PG), Italy
e-mail: nicoletta.ghigi@unipg.it

A.-T. Tymieniecka (ed.), *Phenomenology of Space and Time: The Forces of the Cosmos and the Ontopoietic Genesis of Life: Book One*, Analecta Husserliana 116,
DOI 10.1007/978-3-319-02015-0_1, © Springer International Publishing Switzerland 2014

"The Human Being and Its "Livingness"

Life and the Human

If philosophy has some vested interest in saying something that may be useful for humanity, for its very living, for its existential problems, it has a duty to seek to comprehend that humanity is living, rather than reflecting on itself or its variegated subfields that signify in "empty formal structures."[1] Philosophy must necessarily abandon its odious intellectualising activity, the very "calculative thinking" that Heidegger[2] spoke about. It must cease to operate in a logically calculating or critical fashion—a philosophy that works on valid, formal concepts that are, in the end, useless for living.

Philosophy must return to its fundamental task: to illumine conscience as to its own values, literally in the sense demanded by Enlightenment thought, which reflects the very thought of Anna-Teresa Tymieniecka. In other words, philosophy has the task of re-appropriating the role, drawing from psychology and the other sciences (i.e., anthropology, psychoanalysis, etc.) of being the "indicator of the sense or meaning" of living or, as Aristotle said, to teach one to reach that which is generally called wisdom.

Even the question of being, viewed from the perspective of the ontological difference, with which Heidegger thought he had definitively resolved the ontological problem, thereby defeating classical, abstract and dogmatic metaphysics, must remain primary and must serve as the principal existential argument of metaphysics. Here, we have to ask: What kind of metaphysics? Certainly not the kind that Heidegger wanted to overthrow for it is completely empty and useless when it comes to the question of the ground and of existence.

The problem of a new ground for metaphysics has been much discussed and does not simply concern the forgetting of being, which, in its unpredictable "happening", configures itself again as an abstraction. The problem, rather, concerns the forgetting of the telos and sense of "being there". In our opinion, humanity did not forget being, as Heidegger says, but what is forgotten is its true significance, the sense of every single living being in the world that surrounds being. The problem of being, therefore, remains a metaphysical problem, but not as one stemming form an ancient or dogmatic metaphysics that turns emptily upon concepts, nor as a metaphysics

[1] Edmund Husserl, *Der Encyclopaedia Britannica Artikel. Erster Entwurf* in *Phänomenologische Vorlesungen Sommersemester 1925*, in Husserliana IX, hrsg. V. W. Biemel, (Den Hagg: M. Nijhoff, 1962), 253. It is for this reason, in our opinion, that the phenomenology of Edmund Husserl, which already laments this problem in the early 1900s, is the science that "rejects every metaphysics that moves from empty formal structures. Like all authentic philosophical problems, all metaphysical problems refer back to the phenomenological terrain and find there their form and authentic, transcendental method, forged by intuition." *Ibid.*

[2] One sees the relation between calculative thinking and technique in the collection of essays and talks as found in Martin Heidegger, *Vorträge und Aufsätze* (Pfullingen: Günther Neske, 1957).

that has no bearing on practical life. Here, rather, we are speaking about a new metaphysics that *directly talks to and about life*. If we wish to avoid falling into that inauthentic metaphysics that Heidegger denounced, we must turn to speak about neither being nor its sense, understood in an abstract sense, but of the concreteness of life.

In this regard, Anna-Teresa Tymieniecka gives us the possibility that in the human being there exists a creative power, an ontopoetic logos that is at the basis for our comprehension of the universe. Here, we must begin in order to be able to conceive a possible renewal of philosophy as well as a "new Enlightenment" of human reason in relation to its being in the world.[3]

Anguish and Dehumanisation

The problem of dehumanisation, understood as the forgetting of the person and not being, is, in our age, the general character of our existence. "One" lives, we live carrying out the most varied activities while abandoning the interior dialogue and the significance of the non-repeatable individuality that is our own. One does not follow one's own telos, that which the singular and absolutely unique personality displays from "within" each of us; rather, we follow that which attracts us "from the outside", which does not call into question our true and deep individuality. In our view, this is the process of dehumanisation that lies at the base of anguish or dis-ease, both of which characterise our generation.

The question of anguish is hardly a problem abstracted from or foreign to everyday life, which must only be considered by psychologists or philosophers as something that signals the limit of "non-normality." On the contrary, here we are dealing with a *wholly real* problem that involves humanity at its very *constitutive level*. It actually emerges with great force in its characterisation as a malady that must be cured or resolved, as something that impedes the person from effectively living in the actual present, according to "externally" imposed rhythms. And, so, daily anguish, imposed more by external time than by internal solicitations, becomes a pathology.

It should be remarked that, in the past, anguish also characterised humans, but it was "lived" in a natural fashion, as a normal fact of life, as something that signalled the growth of a person. The person would follow her own inclinations (from within); one encountered others and the situations of life. Today, this natural process does not seem to proceed in the same, spontaneous manner; rather, anguish is largely characterised as a malady that needs to be cured in the quickest way possible.

[3] Tymieniecka's investigations of this theme are well known throughout the world. For a comprehensive understanding of her thematic inquiry please see the study of J.S. Smith, "The Cosmo-Tymieniecka's New Enlightenment", in *Phenomenological Inquiry*, vol. 35 (October 2011), pp. 17–24.

What has changed from not so long ago? First, as humans have resolved key problems concerning bare survival, and certainly this is the case in wealthier countries (where, in fact, anguish is recognised as a pathology[4]), anguish has been displaced from the natural level to an unnatural one. We no longer have to worry about the necessities of life insofar as these are already given. Now, we face the problem of superabundance. In fact, that which is no longer a necessity for life has become one, much like a neurosis, namely, to be "in synch" with others. Anguish refers to that which is lacking but which is not *effectively* necessary. Momentarily, turning away from the psychological and social implications, it is interesting to see how this lack of the non-necessary leads us to a non-natural anguish and pathology by which a human being loses meaning. Anguish becomes the substrate of the human being because each thing is lived as frustrating, given that the natural plane of attribution of meaning has been displaced onto things. Since one gives value to that which is not necessary (insofar as that is already given is in need of no other searching), the attribution of value is conferred on to that which is not necessary. And this is wholly unnatural, contrary to human nature. The relation to things and with others is consequently lived in a negative way (unnaturally) with much anguish because one expects from life a gift that is nothing natural, but artificial. Artificiality, then, becomes the focus of an anguish-ridden living. One seeks pleasure, objects, realities, situations that are not necessary for positive living (read natural) and one forgets singular, unique human destiny, one forgets one's own belonging to being animated by a naturalness that seeks integrated parts from nature.

Another cause of pathological anguish and, therefore, dehumanisation, is connected to the forgetting of one's own "interior voice" in favour of listening to the various solicitations that come from the outside. This attitude contains a displacement of personal equilibrium that renders almost impossible the achievement of a syntony between the self and the consciousness that flows from one's own personality and one's own will. Anguish, the actual state that once was considered a momentary dis-ease that characterised processes of settlement in the surrounding world and the growth of personality, has become an important and insistent dis-ease that is difficult to manage. This is the case because its root is not within, but is, instead, lodged in a wanting to adapt to external claims, ultimately neglecting following the telos indicated by one's own personality. This attitude leads to that which we define as the process of dehumanisation.

Natural "Livingness" and Unnatural Madness

Life has become an unnatural race toward that which does not count and we have completely lost the message that comes form our human tradition of living

[4] In many "Third World" countries, that which is understood by us as a psychotic subject is, on the contrary, honoured as belonging to the province of the shaman.

naturally according to our own telos that guides us from within or, in other words, our common sense. Pathological anguish, then, characterises our existence. How do we get back to "normal", if at all possible, or to natural living?

Living abnormally, following that which calls us from the outside, has become normal, whereas normal life, natural life, seems to have become abnormal. The person who lives in an anguished state, following outside rhythms and full of stress, is recognised as a true living being, as one who intensely lives her life and who knows how to live actively, whereas the individual person, who lives free from stress in a natural fashion, is recognised as a fish out of water, one who does not know how to live and who has "slowed down", who is abnormal. The dialectic of abnormality and normality has turned upside down to such an extent that there has been an inversion of the meaning of related emotional states.

Treatment for our psychic dis-ease (anxiety, anguish, depression, etc.), things that everyone defines as normal today, are already integrated parts of our living. No one really worries very much if one sees in oneself the symptoms of a nervous disorder. Psychologists have taught us, in fact, that it is normal that we suffer from one of these disorders in today's world. Psychotherapy teaches us that such nervous disorders are a reaction to the stress of our daily collective lives. And even if we accept this normality, it still does not give us any possibility of positing any remedy to an illness that does not have pathological roots, for the malady is only an effect, although cultural and, above all, socio-economic. One could even say that this "problem", the "dis-ease of our civilisation", is not only due to our present-day culture and to the values of our contemporary society but also to human nature, which have all led us to a certain level of development through adapting to the environment, reacting to behaviours that with respect to the past can be considered pathological.

Theories may continue to develop and give rise to further interpretations of the problem, but the situation does not change. In fact, what remains inexplicable is the level of suffering that humanity imposes on itself in order to live in a world that it itself built bit by bit. Given that we are living unnaturally in our world, we no longer ask questions except those that seem to satisfy the exigencies of the moment. Long-term, however, this does not help us, nor does it help when we "normalise" an abnormality, the anomaly of everyday living.

If a mad gene surreptitiously overtook the world,[5] transforming the cosmos into pure chaos such that if an extraterrestrial were able to observe it from on high, one could suppose that we constantly live under the influence of a self-destructive drive. And such a "transvaluation" of values, including suicide, which is usually held to be a negative value because it is a negation of life, changes and becomes a positive self-determination of a will of life that imposes itself and wishes the negation of an unnatural life contrary to natural living.

[5] This is a theory postulated by recent scholars of memetics. They draw from the work of Richard Dawkins as elaborated in his 1976 work *The Selfish Gene*.

But does it make sense to want to live in an "abnormal" manner in order to be able to feel oneself normal? If we were to become aware of this reversal of sense, how is it possible to stop this perverse logic and overcome the anguish-filled encounter with the "must live" outside of our human nature?

The Proposal of a New Enlightenment

Metaphysics as the Philosophy of Life

It is incumbent upon philosophy to respond to these demands. To construct a metaphysical response is to build a new metaphysics, a science that is capable of offering a conceptualisation of the concrete objects of living: *metaphysics as the philosophy of life*.

What use is philosophy if not to help the human being live? It is inconceivable to think philosophy otherwise. It does not make sense to think that philosophy can be other than learning how to think in order to live well. But what is life? In what does this existence that accompanies each living being consist?

Life is the whole of the conditions by which an individual is active. Its components, including reason, will, instinct, culture and sensibility are in "movement" or "functioning". Every human being acts in relation to her rationality, will, instinct and principal needs. Sensation and internal force push the human being according to seemingly non-manifested movements subjected to the human being's very existence. There are also hidden components that act from within.

Life, therefore, is the satisfaction of these demands. But if we reflect on living itself or if we complete a metaphysics of life, we will see or, better still, *feel from within* that life is also something other. Life is not only the satisfaction of our needs or what is demanded of us but also a continuous journey that lets us see all of the active operators or unconscious supports of life. We "feel from within", one used to say. But what is feeling? How can we arrive at such a situation of abstraction such that we are able to understand the meaning of our personal life? Is this feeling a feeling that characterises everyone or can only a few access it?

Feeling, understood as that sensation of one's own interiority, occurs only when one is conscious that one is no longer conscious of feeling, when that which lies before the anguish of living "decides" to let being be. Feeling, then, means to record a weakness on the part of the human being and to understand that there is a part of us that is not containable by reason. It is only in this way that we can see that this part can come under the gaze of reason. Are we, then, giving a space to the irrational, the unconscious, to that which the psychiatrists call hallucinatory fantasy?

The flow of life is signalled by an infinity of moments that do not only not follow one another in a chronological fashion but that also do not even have the rigour of an internal dialectical logic. A drive or a sensation can discharge absolutely unforeseen voluntary actions, given determined conditions. But all of this is not exempt from "internal" being, that is, rationality. There is reason even in the most unreal fantasies. But this is not exclusively the case. Rationality has its own support and its own source from that which is most properly irrational in the individual: emotional feeling.

This field, this absolute domain of drive and emotion, has a structure that is always in formation but which also has a basic structure. There is a certain form that contains all the sectors of emotion and which moves to fill them, much like any liquid in a container. Sensation is, we said, the forceful feeling of the presence of a similar territory in oneself and in others. It means empathy, understood as the force that from within us calls us to recognise the other and, as Edith Stein says, through the other we recognise ourselves.[6]

The Transcendental as the Critical and Poietic Tool of the Human Being

The personal telos of life is recognisable only by a philosophy that is constituted as a science that is attentive to a singular interiority, but that is not confused for the investigations of contemporary psychology. This philosophy, founded on a new metaphysics or on this system focussed on the sphere of interior sensibility, must have its a prioris and, in our opinion, these are based on the recognition of an intuitive force, the "intellectual spirit"; these a prioris penetrate the human structure or, even better, offer tools to everyone to reach one's own I and to listen to one's own interiority such as to render the self syntonic with the world.

On the basis of phenomenological science, which is, in our opinion, most capable of giving us tools for the constitution of a similar metaphysics, we call this "intellectual spirit" the transcendental of humanity. In light of this constitutive structure of each single individual, which everyone must learn to make function in order to understand each single part in itself, it becomes possible to think of "re-humanisation" and, therefore, of a new re-appropriation on the part of every single personality of one's own life and one's own telos.

To constitute a metaphysics as a science that makes this telos its own object or to think of a philosophical reflection that is completely turned toward life and its meaning offers us the possibility of rethinking the human and to rethink her existence as a true return to authentic existence—to an existence that does not speak of being as a stranger that encounters us and that "happens" (*Ereignis*), but as the being to which we are and in which we participate insofar as we are single personalities endowed with our own interiority and, above all, our own telos that gives form to life. Following our own telos and listening to the voice that calls us from within to become conscious of that which we are and to which we want and must reach, this means contributing to the re-humanisation of the human and the reconstitution

[6] Edith Stein, *Zum Problem der Einfühlung* (Halle: Niemeyer, 1927).

of meaning and the natural rhythm that the human being herself is called to make evident and discipline through her reason.[7] In this regard, following the reflections of Anna-Teresa Tymieniecka, we can speak of a new Enlightenment, understood as a new light, a great gust of wind of coherence that brings the human being to self-understanding within the horizon of meaning in which the human being participates and for which this human being is constitutively responsible.

[7] Tymieniecka speaks of a "critique of reason" and explains its characteristics and the forms it must have. See her *The Passions of the Skies* in *Analecta Husserliana*, vol. CVII (Dordrecht: Springer), p. xii.

Moral Excellence as Cosmicization of Human Beingness in the *Ontopoietic* Perspective

Carmen Cozma

Abstract Designing the core of Phenomenology of Life founded by Anna-Teresa Tymieniecka, "Ontopoiesis of Life" reveals a holistic and dynamical philosophy about the dialectical unity of *logos-ethos-cosmos* in the great plan of life. Grounded in the idea of firstness in existential formation by the continuous creative process of human becoming, the ontopoietic perspective opens to the understanding of moral excellence under the auspices of order and beauty that simultaneously define the cosmicization of human condition. It leads to structuring our endeavor of rising in the horizon of participation in the universal harmony, by appropriating the inward-outward oriented self-individualization through the workings of the "logos of life" in its multiple manifestations. We try to emphasize some articulations of the "Ontopoiesis of Life" as significant marks in developing our moral affirmation by following the ideal axis around which everything is harmonizing within the single whole: the cosmos.

In the context of serious shattering and overturning the valuable reference points of human existence, that of manifesting the peril of chaotic dominants deepening the alienation process from a healthy creative meaning of life, the ideal of moral excellence becomes a priority to be considered in its plentiful force of re-structuring our attitude towards the cosmos we are part of.

More than ever, the human beingness is touched by the captivities of an artificial environment made by the sophisticated information and communication technologies, by the invader digital network; it is an environment that carries new risks into increasing even the distance from the human well-being. What does it mean, briefly?

C. Cozma (✉)
Alexandru Ioan Cuza University of Iasi, Bd. Copou, 11, Iasi 700506, Romania
e-mail: carmen.cozma@uaic.ro

A.-T. Tymieniecka (ed.), *Phenomenology of Space and Time: The Forces of the Cosmos and the Ontopoietic Genesis of Life: Book One*, Analecta Husserliana 116,
DOI 10.1007/978-3-319-02015-0_2, © Springer International Publishing Switzerland 2014

It is the purpose followed along the history of moral philosophy, "the central problem" of the "best life for all".[1] In these terms, our interest for the moral excellence finds support in the necessity of a continuous insertion of the ethical dominant for a way to live well. It leads us to the significance of cultivating the human quality of doing "what is right and just", by "having received a proper upbringing in moral conduct", respectively by having some sense of good(ness)/virtue (virtuousness).[2]

Centered on the aretaic becoming, the *moral excellence* represents an eternal ideal of human living as happiness. Gradually, it can be achieved by high-minded man within the effort of inscribing and developing his individuality in distinction, but also in sameness with the whole network of life. It supposes the attainment of a harmonious combination of a plurality of inner and outer coordinates, by conquering the value of just measure – "the golden measure" in all – keeping the existential equilibrium (personal, societal, and cosmic). Finally, it is an outcome of the spiritual and cosmic perspective, by following the "music of the spheres" (from Pythagoras' theory of consonant intervals) as inspiring a sustainable living, and by integrating in it as much as possible.

Aiming to and working for the *moral excellence* show a certain process similar to that of bringing cosmos from the chaos, or of transforming chaos into *cosmos* – that means order and beauty in an inherent connection with the good, as virtue's experience.

The problem of virtue needs revaluation(s) as an essential instrument to assure a *human* content for a life in progress, in the tradition opened by Aristotle's theory of mediation. As "the midst way between excess and deficiency", a "mean between a too much and a too little", virtue remains the greatest value for the "excellence of man" as that "makes a man good and able to perform his proper function well",[3] his *àreté*. It remains the proper force, *the vehicle of moral conduct*, so much needed for the psychosomatic, cultural-societal, spiritual-natural order of equilibrium in life.

In the original *ontopoietic* perspective of Anna-Teresa Tymieniecka, the question of virtue is tackled like one of the most prominent values for understanding the progressive course of life, generally. It "lies at the heart of the life strategies of the Logos".[4]

The "Ontopoiesis of Life" – defining the nucleus of Tymienieckan phenomenology – articulates the manifestations of the *logos of life* in multifaceted inventive rationalities that it "projects in the course of carrying our lives and our world-in-transformation".[5] At the same time, the entire *ontopoietic* design engages the crucial function of *Imaginatio Creatrix* – "the fulgurating force within the human

---

[1] E.J. Bond, *Ethics and Human Well-Being. An Introduction to Moral Philosophy*, Blackwell Publishers, Oxford, 1996, pp. 208–209.

[2] Aristotle, *Nicomachean Ethics*, Romanian translation: Etica Nicomahică, Scientific and Encyclopedic Publishing House, Bucharest, 1988, 1095b.

[3] Ibid., 1106a20–1107a.

[4] Anna-Teresa Tymieniecka, *Logos and Life*, Book 4: *Impetus and Equipoise in the Life-Strategies of Reason*, Kluwer Academic Publishers, Dordrecht/Boston/London, 2000, p. 598.

[5] Anna-Teresa Tymieniecka, "The Triumph of Imagination in the Critique of Reason", in *Analecta Husserliana*, Volume LXXXIII, Kluwer Academic Publishers, Dordrecht/Boston/London, 2004, p. xviii.

creative experience", *the prime force inspiring human endeavors.*[6] With such outlined particularities, the problem of virtue can be disclosed in a larger significance.

According to the author of phenomenology of life, the issue of virtue is situated "at the primogenital human plane where reason with its faculties, on the one hand, and the vital forces, on the other, emerge as partners in the creative orchestration of human functioning that forms the crucibles of intelligibility that is specifically human and that accounts for the emergence of the human universe, that is, the human expansion of the schema of Nature".[7] (We would extend the last term, by *cosmos*). Thus, man's mandate of morality gets a chance of realizing as cosmicization, too; that claims an increased sensitivity toward a kind of syntony: an agreement/accord to be found and highlighted as concerns the embedment of human(ness) in the tangible cosmic context.

As part of the whole, man is connected to all aspects of nature, which is essentially even for his very survival. Reviving the cosmological philosophy of Pythagoras and Plato, for example, in an era full of contradictions, man can (re)discover the experience of kinship and affiliation to the cosmic order and beauty rooted in harmony – the "fitting together" of opposites –, and to reconsider natural structures as embodying virtue, after the model of "divine proportion", in functioning as moral agent in the best possible way.[8]

In searching paths toward harmony and concord, toward proportion, symmetry and measure as marks of a cosmic moral understanding of human life in the universe, the *ontopoietic* metaphysics deploys significant themes, grasping a constructive direction of philosophizing.

Frequently, Tymieniecka underlines the necessity of surmounting the multiple tensions and conflicts, of keeping a perpetual quest for equilibrium that, in a complex process of harmonizing, the *logos of life* manifests "its most powerful engines of creative advance"[9] in assuring the moral order.

In the framework of the *ontopoietic* phenomenology, the fundamental category of "Human Condition" implies – among other things – the seal of harmony. Oscillating between similarity and distinction, proceeding from natural generic roots to the cultural-spiritual situation, moving between without and within, and always in the circuits of flux and stasis, human condition reveals itself as a progressive experience into harmony, by a creative moral transformation. This is happening because the "Human Creative Condition" is fulfilling in the horizon of self-individualizing and integrating not merely in the social order, but in the cosmic one, too.

---

[6] Anna-Teresa Tymieniecka, "*Imaginatio Creatrix*, the Creative versus the Constitutive Function of Man and the Possible Worlds", in *Analecta Husserliana*, Volume III, D. Reidel, Dordrecht, 1974, pp. 3–41.

[7] Anna-Teresa Tymieniecka, *Logos and Life*, Book 4: *Impetus and Equipoise in the Life-Strategies of Reason*, op.cit., p. 598.

[8] Plato, *Timaeus*, Hackett Publishing Company, Inc., Indianapolis IN, 2000. 53a–54b.

[9] Anna-Teresa Tymieniecka, *Logos and Life*, Book 4: *Impetus and Equipoise in the Life-Strategies of Reason*, op.cit., p. 372.

An intrinsic part of the "ontopoiesis of life" perspective is the question of understanding the moral lifestyle in connection with the cosmos as steering man to what it is worthy to get as a model of becoming, to avoid the decay and destruction, to overcome the danger of breakdown in barbarity. Beyond the cogent benefits of the scientific and technological advances, we have to recognize, no less, the threatening of human essence – which Michel Henry has signaled as being "the barbarism of the monstrous objectivity of techno-science" menacing and even rejecting human subjectivity.[10] Thus, much more concern for the approach of moral-human-becoming-in-cosmos is entitled by necessity.

The issue of *moral excellence as cosmicization* has arisen in a metaphoric manner, as a process of shaping the moral character under the ideal dimension of our knowledge of cosmos, as an orderly, harmonious, beautiful and perfect living system. An ideal of governing the moral creative condition is at stake: to endeavor to enhancing, to rising in the horizon of cosmic characteristics showing a holistic entity in creative process, in continuous revival, positive lastingness and constructive coherence. Moral excellence as cosmicization covers a telos toward which man is able to work in ordering, measuring, improving and embellishing his own life, especially by living in conformity with the *harmony* – as the dominant note of cosmos. It means that man can find and appropriate the value of harmonization in the most inner self – in himself and with himself – and, no less, beyond the strict individuality – as being for the world, in and with the world. This represents the core synthesized by Anna-Teresa Tymieniecka through her thesis about "the human soul in the cosmos and the cosmos in the human soul". Insisting on the creative virtualities of man, her insight is built on "a new formulation of the concept of nature life, one open to the cosmos and to culture".[11]

Creativity – as the "Archimedean point" for phenomenology of life, as the "quintessential faculty" of man for Tymienieckan "new critique of reason" – receives an ethical resonance. It appears like a virtue of man, marking the process of his self-creation "with respect to the laws of Nature and of the Cosmos, but especially with respect to the specific ultimate significance of his existence".[12]

Owing to creativity, man can reach the plenitude of affirmation in the great plane of life by activating an orderly behavior, somehow subordinated to a vision of strength, equilibrium and harmony, of meaningful order, of cosmicization – enrooted in the Greek *kosmiotēs*, with the idea about the unity of *ethos-lógos-kósmos*.

A pillar of Anna-Teresa Tymieniecka's phenomenological inquiry is that about the role of "moral sense", which is even "a harmonizing logoic principle".[13] The authentic human life is conceived as emerging from the basis of moral values,

---

[10] Michel Henry, *La barbarie*, Editions Grasset, Paris, 1987.

[11] See Anna-Teresa Tymieniecka, *The Fullness of the Logos in the Key of Life*, Book I: *The Case of God in the New Enlightenment*, Springer, Dordrecht, 2009, pp. 181–211.

[12] Anna-Teresa Tymieniecka, "The Creative Self and the Other in Man's Self-Interpretation", in *Analecta Husserliana*, Volume VI, D. Reidel, Dordrecht, 1977, p. 161.

[13] Anna-Teresa Tymieniecka, *Logos and Life*, Book 4: *Impetus and Equipoise in the Life-Strategies of Reason*, op.cit., p. 373.

"a universal type of configuration of the life context" originating "from the universally valid moral sense".[14] Together with the "aesthetic" and "intellective" senses, it is a "giving/valuating factor of sense" to be disclosed "in the projection of individualizing circuits forging them within the inward networks of an *intimate zone* projected into a vast network of *outward* interlinkage with other living beings within a common world of life".[15]

In the territory of the dynamic and holistic philosophizing upon the "self-individualizing ontopoietic schema" applied to the total life expanse, human condition is conceived in its "knot position", namely that of the unique responsibility man has toward "everything-there-is-alive". It is the function of "moral sense" to introduce "the Sentiment of Benevolence toward other living creatures, toward oneself, and toward life in general", moving to a higher order of significance under the auspices of "the moral measure for life". This supposes the complex process of "sharing-in-life" on the ground of moral virtues of human being, eventually, in accordance with the cosmic laws – especially, of harmonious resonance, of synergy and equilibrium beyond any conflicts and contradictory forces/energies –, into a creative direction, for the common good of life. Examining the functionality of "moral sense", Tymieniecka "draws the conclusion that life manifests benevolence at the level of humanity for the well-being of all living beings"; seeing that, "Introduced by the Benevolent Sentiment, the axis of right/wrong balances out the conflict".[16]

The *moral sense* with the *benevolent sentiment* is an essential component of the "phenomenological attitude" – one given "in the seeing and experiencing act itself", a crucial one as "spiritual posture" of man in the cosmos, which marks precisely the human becoming as raising to the "openness toward the world", so speaking in the language of Max Scheler.[17]

The German philosopher has emphasized the role of "love-determined movement of the inmost personal self of a finite being toward participation in the essential reality of all possibles".[18] In her turn, Anna-Teresa Tymieniecka stresses the valences of benevolence and sympathy, and no less of commitment and responsibility to define the peculiar status of man, with respect to the totality of life,

[14] Anna-Teresa Tymieniecka, "The Moral Sense. A Discourse on the Phenomenological Foundation of the Social World and the Ethics", in *Analecta Husserliana*, Volume XV, D.Reidel, Dordrecht, 1983, p. 40.

[15] Anna-Teresa Tymieniecka, *Logos and Life*, Book 4: *Impetus and Equipoise in the Life-Strategies of Reason*, op.cit., p. 634.

[16] Gary Backhaus, "Anna-Teresa Tymieniecka: The Trajectory of her Thought from Eidetic Phenomenology to the Phenomenology of Life", in *Phenomenological Inquiry*, Volume 25, Belmont, Massachusetts, The World Institute for Advanced Phenomenological Research and Learning, 2001, pp. 42; 41.

[17] Max Scheler, *Die Stellung des Menschen im Kosmos* (1927)*; Romanian translation: Poziţia omului în cosmos*, Paralela 45 Publishing House, Bucharest, 2001, pp. 39–41.

[18] Max Scheler, *Vom Ewigen im Menschen* (1921); English translation: *On the Eternal in Man*, Harper & Brothers, New York, 1960, p. 74.

that is the "Custodian of everything there is alive". Such a key place deep in the midst of "all-alive-unity" unveils man in the singular position "to estimate the life-situation within the living kingdom as well as his own",[19] striving for the safety and flourishing of life in its plenitude.

On the ground of comprehending the significance of "Custodian of everything there is alive", the Tymienieckan phenomenology reveals a wise avenue in re-thinking and assuming an urgent duty for contemporary man as regards the protection of the health, integrity and sustainability for the natural capital, for the recovery of global ecosystem.

Accounting for the critical moral and ecological situation mankind is passing through for the last decades, respectively the sketched "*anatomy of bewilderment* – of the disarray humanity now finds itself in",[20] the syntagm of "Custodian-of-the-unity-of-everything-there-is-alive" is very important for the moral thinking, for an environmental ethics particularly. In a phenomenological insight, it proves a major function to guiding the human attitude to face the natural circumambient forces, by impelling the necessity of working for the moral excellence tying the formative role of "the mobile cosmic architectonics" with its earthly and celestial channels in the complex process of "individualizing beingness".[21]

This peculiar status of man requires thoughtful deliberation in the valuation process of concerning for the good of each present form of life and for future generations, in a wider culture of environmental stewardship. Heralding "a New Enlightenment", Anna-Teresa Tymieniecka notes that "We are challenged to enter into our depths in order to achieve a new understanding of our place in the cosmos and the web of life, to find new wisdom for charting our paths together and fresh inspiration to animate our personal conduct".[22]

The lucid awareness concerning the dual games of continuity and disruption, success and failure, advancement and destruction manifested in the universal life-system – as much as we get access –, in the mutual conversion nature-human, does sustain Tymieniecka to manifest trust in man's discernment in using the inventive faculties on the side of affirmative and constructive way.

In the light of the spirit and especially "in the moral sense", human being realizes itself a higher birth. It is a "second birth" or "a completion of the Human Condition, which was begun by the entrance into the game of life of the human drama" with "pain and suffering", "birth and death"; but, eventually, finding the force to evaluate the creative function in the "*kairic* timing" of "freedom and accomplishment" on the

[19] Anna-Teresa Tymieniecka, "Phenomenology of Life and the New Critique of Reason: From Husserl's Philosophy to the Phenomenology of Life and the Human Condition", in *Analecta Husserliana*, Volume XXIX, Kluwer Academic Publishers, Dordrecht, 1990, p. 16.

[20] Anna-Teresa Tymieniecka, "Measure and the Ontopoietic Self-Individualization of Life", in *Phenomenological Inquiry*, Volume 19, Belmont, Massachusetts, The World Institute for Advanced Phenomenological Research and Learning, 1995, pp. 26–27.

[21] Anna-Teresa Tymieniecka, "Transcendentalism Overturned: Life's Geo-Cosmic Positioning of Beingness", in *Analecta Husserliana*, Volume CVIII, Springer, Dordrecht/Heidelberg/London/New York, 2011, pp. 6–9.

[22] Anna-Teresa Tymieniecka, "Measure and the Ontopoietic Self-Individualization of Life", op.cit., p. 26.

inward/outward advance of humanness. According to Tymieniecka, "along the path of human creative self-individualizing, kairos is concurrently the timing of the propitious circumstances and forces leading toward the realization of constructive projects, their accomplishment, and concurrently and finally is their measure".[23]

We face a vision upon the human moral authority as self-accomplishment, sharing-in-life and solidarity not just at a social level of existence, but within the unity of the Deity and the cosmos throughout the inner workings of the *logos of life* in its "constructive *impetus and equipoise*". This is what Anna-Teresa Tymieniecka has acknowledged as "the ontopoietic unfolding of the logos of life" that "manifests itself in the spectacle of the All, cosmos, world, nature, life, the works of human spirit",[24] engaging various modalities of "vital", "Dionysian", "Promethean" and "sacral" logoi on the horizon of life, by a perpetual "exaltation of the ideal" of creative development.

In the architectonics of phenomenology of life, the key-concept is that of *logos of life*, as the reason of reasons, the primordial principle, and the all-pervading presence of life. Investigating the "logos of life", Tymieniecka actually establishes some harmonized modalities of progression for this universal reason and "riser" of life. So, she follows its movements from the "vital/entelechial" logos, through the "affective, sentient/emotive, sympathic sharing-in life" that she called "Dionysian" logos; also, through the "creative", "prompting new forms, qualities, hints of life" and expanding in freedom of the human spirit, with the transcendental experience, that is the "Promethean" logos. This "ontopoietic" unfolding is completed by which the author calls the "sacral/Divine" rationality as the ultimate, accomplished sense of the "logos of life" in a "logo-theic horizon".[25]

To a certain extent, phenomenology of life encompasses the fact that it is time to re-explore the Heraclitean reflection upon the *logos* embracing man, earth, and cosmos; at the same time, to re-think about the Pythagorean message about the human soul brought into harmony with the natural order. It conveys toward the importance of forming a cosmicized interior as a condition of moral excellence, by experiencing a variety of trials and selecting on the side of what really matters for individual and community, for social and natural existence, alike. In this sense, our phenomenologist of life envisions the "individualizing-ontopoietic process of life" as one of "sense and ordering".[26] Within it, man is the subject of constructive inventive evolution, by cultivating the royal path of virtue closed to all the vital forces, gathered and organized "in proper channels of growth and subsistence within the

---

[23] Anna-Teresa Tymieniecka, *The Fullness of the Logos in the Key of Life*, Book I: *The Case of God in the New Enlightenment*, op.cit., pp. 197–199; 209.

[24] Anna-Teresa Tymieniecka, *Logos and Life*, Book 4: *Impetus and Equipoise in the Life-Strategies of Reason*, op.cit., p. 10.

[25] Ibid., pp.320–321. See also Anna-Teresa Tymieniecka, *The Fullness of the Logos in the Key of Life*, Book I: *The Case of God in the New Enlightenment*, op.cit., pp. 231–255.

[26] Anna-Teresa Tymieniecka, "Inspirations of Heraclitus from Ephesus Fulfilled in Our New Enlightenment", in *Analecta Husserliana*, Volume CX/Part I, Springer, Dordrecht/Heidelberg/London/New York, 2011, p. 9.

stream of becoming", by "assuming the transcendental role formerly accorded to consciousness" for what Tymieniecka considers to be the "cosmic positioning" of human being.[27]

To the grave question: "how should one live?" a response is certainly inspired by the great learning we get from the cosmological philosophizing implied by phenomenology of the "ontopoiesis of life". It offers a scrutiny of moral virtue in its profound link with the "logos of life", as well as the emphasis of the intimacy between man and universal rhythm of life (from mineral to cosmic levels). In discussion is "the rhythm of creative process", a "precondition" of unity and stability without which neither "the values common to all mankind", nor a "world order and a meaningful life" could exist.[28] Thus, phenomenology of life gives us the opportunity to circumscribe a picture of harmonizing law that conducts the process of humanization-cosmicization in the arteries of spontaneity and hazard on the one hand, and deliberation and necessity on the other.

We have tried to deal with the concept of cosmicization as a path for human self-fulfilling simultaneously in its uniqueness and its universality, by reaching the balance between differentiation and integration in a more and more estranged from Nature world, a quite technical and material-consumerist one. We could bring part of the ideal of cosmicization in our life, by sharpening an enlightened understanding of its role in registering ourselves on the trajectory of creative moral becoming. It could help us to surpass limitations of an artificial world, by finding the sources to fostering an authentic health and prosperity for humans and nonhumans, both as individuals and communities in the unity of life.

To conclude, the *ontopoietic* phenomenology of life elaborated by Anna-Teresa Tymieniecka makes us remember the truth contained in an old Latin phrase, as a directory orienting our care about the moral living: "Naturam si sequemur ducem, nunquam aberrabimus"/"If we take nature for our guide, we shall never go astray".[29]

References

Aristotle. 1988. *Nicomachean ethics*. Romanian translation: *Etica Nicomahică*. Bucharest: Scientific and Encyclopedic Publishing House.

Backhaus, Gary. 2001. Anna-Teresa Tymieniecka: The trajectory of her thought from eidetic phenomenology to the phenomenology of life. In *Phenomenological inquiry*, vol. 25. Belmont: The World Institute for Advanced Phenomenological Research and Learning.

Bond, E.J. 1996. *Ethics and human well-being. An Introduction to moral philosophy*. Oxford: Blackwell.

Cicero, Marcus Tullius. 1913. *De Officiis*. Cambridge/London: Harvard University Press.

[27] Anna-Teresa Tymieniecka, "Transcendentalism Overturned: Life's Geo-Cosmic Positioning of Beingness", op.cit., pp. 9–10.

[28] Maija Kūle, "Logos and Life: Understanding of Rhythm", in *Analecta Husserliana*, Volume CX/Part II, op.cit., p. 682.

[29] Marcus Tullius Cicero, *De Officiis*, Harvard University Press, Cambridge/London, 1913, I, 100.

Henry, Michel. 1987. *La barbarie*. Paris: Editions Grasset.
Kūle, Maija. 2011. Logos and life: Understanding of rhythm. In *Analecta Husserliana*, vol. CX/Part II. Dordrecht/Heidelberg/London/New York: Springer.
Plato. 2000. *Timaeus*. Indianapolis: Hackett Publishing.
Scheler, Max. 1921. *Vom Ewigen im Menschen*. English translation: 1960. *On the Eternal in Man*. New York: Harper & Brothers.
Scheler, Max. 1927. *Die Stellung des Menschen im Kosmos*. Romanian translation: 2001. *Poziţia omului în cosmos*. Bucharest: Paralela 45 Publishing House.
Tymieniecka, Anna-Teresa. 1974. Imaginatio Creatrix, the creative versus the constitutive function of man and the possible worlds. In *Analecta Husserliana*, vol. III. Dordrecht: D. Reidel.
Tymieniecka, Anna-Teresa. 1977. The creative self and the other in man's self-interpretation. In *Analecta Husserliana*, vol. VI. Dordrecht: D. Reidel.
Tymieniecka, Anna-Teresa. 1983. The moral sense. A discourse on the phenomenological foundation of the social world and the ethics. In *Analecta Husserliana*, vol. XV. Dordrecht: D. Reidel.
Tymieniecka, Anna-Teresa. 1990. Phenomenology of life and the new critique of reason: From Husserl's philosophy to the phenomenology of life and the human condition. In *Analecta Husserliana*, vol. XXIX. Dordrecht: Kluwer.
Tymieniecka, Anna-Teresa. 1995. Measure and the ontopoietic self-individualization of life. In *Phenomenological inquiry*, vol. 19. Belmont: The World Institute for Advanced Phenomenological Research and Learning.
Tymieniecka, Anna-Teresa. 2000. *Logos and life, book 4: Impetus and equipoise in the life-strategies of reason*. Dordrecht: Kluwer.
Tymieniecka, Anna-Teresa. 2004. The triumph of imagination in the critique of reason. In *Analecta Husserliana*, vol. LXXXIII. Dordrecht/Boston/London: Kluwer.
Tymieniecka, Anna-Teresa. 2009. *The fullness of the logos in the key of life, book I: The case of God in the new enlightenment*. Dordrecht: Springer.
Tymieniecka, Anna-Teresa. 2011a. Inspirations of Heraclitus from Ephesus fulfilled in our new enlightenment. In *Analecta Husserliana*, vol. CX/Part I. Dordrecht/Heidelberg/London/New York: Springer.
Tymieniecka, Anna-Teresa. 2011b. Transcendentalism overturned: Life's geo-cosmic positioning of beingness. In *Analecta Husserliana*, vol. CVIII. Dordrecht/Heidelberg/London/New York: Springer.

The Inseparable Link Between "Cosmology" and the "World of Life" in the Philosophy of Anna-Teresa Tymieniecka: The Originality of a New Perspective on the "Real Individual and Autonomous Being" and a Comparison with the "Phenomenological Realism" of Hedwig Conrad-Martius

Francesco Alfieri

Abstract This study aims to examine Anna-Teresa Tymieniecka's original speculation on the "phenomenology of life" with reference to the anthropological question of the human being as a "real individual and autonomous being", considering its ontological centrality in the Cosmos. The investigations skilfully conducted in recent years by Tymieniecka show that the cosmological question is "directed to reality" and should not be associated exclusively with philosophical speculation. The strong link with the "world of life" will help us to work through her investigations and reach what she describes as "the trans-subjective universe".

Anna-Teresa Tymieniecka's plan is to rehabilitate phenomenology in accordance with a new approach determined by its reorientation as a philosophy of life in search of its *logos*: "Logos" and "Life" are conceived by Tymieniecka as inseparable. This plan is always present in the background to my own research.

In the analysis I will attempt a comparison between the philosophy of Anna-Teresa Tymieniecka and the "phenomenological realism" of Hedwig Conrad-Martius with reference to three constitutive elements: (1) the "phenomenology of life" and the "trans-physical world"; (2) the anthropological question of "individuation"; (3) an analysis of how the order and the forces of the cosmos are bound up with the creative forces of human beings.

F. Alfieri (✉)
Collegio S. Isidoro, Via Degli Artisti, 41, Roma, Italy
e-mail: frafrancescoalfieriofm@yahoo.it

A.-T. Tymieniecka (ed.), *Phenomenology of Space and Time: The Forces of the Cosmos and the Ontopoietic Genesis of Life: Book One*, Analecta Husserliana 116,
DOI 10.1007/978-3-319-02015-0_3,

It All Starts with Tymieniecka's "Novel Intuition"

Approaching the investigations of Anna-Teresa Tymieniecka and her philosophical project is certainly no easy task. However, the work is founded on her novel intuition that the intentional quality of consciousness is not the last word, but consciousness is *rooted* in the creativity of life.

The first achievement in her long research journey was her *magnum opus* entitled *Logos and Life*, published in four volumes.[1] This publication, like the others that followed, cannot be considered the last word of Tymieniecka's journey since, on reaching one goal, her thought always sets off again with renewed vigour towards new and unexplored horizons. As we shall see, she never declares the "end" – and quite rightly – of what I believe can be described as a philosophy of Logos and Life – life that is generated from a creative act and is destined to continue regenerating itself *ad infinitum*. Future generations of scholars will have to take account of Tymieniecka's philosophy; for now it is sufficient to note the recent arrival of two monographs, by Daniela Verducci[2] and Salahaddin Khalilov,[3] which provide a rigorous analysis of Tymieniecka's phenomenology of life. It is inevitable that new publications will continue to appear, since the phenomenology of life lends itself to extensive analysis precisely because it highlights the most artfully hidden corners of reality, prompting us not just to change our way of looking at the world, but also to reconsider our position in the cosmos. All these elements are closely connected to each other; none of them lend themselves to sterile speculation, and any investigation must aim to throw light on the dense relationships between them.

Over the years, Tymieniecka's sensitivity towards the phenomenology of life has led her to make continuous efforts to create a terminology and conceptual vocabulary which have made *Logos and Life* famous. Thanks to this terminology, the specific originality of her thought is well enough consolidated for it to be argued that it played a central role in the cultural panorama of the twentieth century. The originality of her thought and the full autonomy of her investigations are immediately clear from the difficulty experienced by every researcher every time they seek to fit Tymieniecka's thought into established conceptual schemes inherited from philosophy. Such attempts are unjustified given that all her work in recent years has spontaneously and naturally transcended the conceptual structures to which we

---

[1] Tymieniecka, A.-T. 1988. *Creative experience and the critique of reason. Logos and Life*, Book I, Dordrecht-Boston-London: Kluwer Academic Publishers (see also in *Analecta Husserliana* XXIV, 1988); Idem 1988. *The three movements of the soul. Logos and Life*, Book II, Dordrecht-Boston-London: Kluwer Academic Publishers (see also in *Analecta Husserliana* XXV, 1988); Idem 1990. *The passions of the soul and the elements in the ontopoiesis of culture. Logos and Life*, Book III, Dordrecht-Boston-London: Kluwer Academic Publishers (see also in *Analecta Husserliana* XXVIII, 1990); Idem 2000. *Impetus and equipoise in the life-strategies of reason. Logos and Life*, Book IV, Dordrecht-Boston-London: Kluwer Academic Publishers (see also in *Analecta Husserliana* LXX, 2000).

[2] Verducci, D. 2012. *La fenomenologia della vita di Anna-Teresa Tymieniecka* (*Percorsi di etica – Saggi*, 8), Rome: Aracne.

[3] Khalilov, S. 2012. *Phenomenology of Life or Life of Idea*, Baku: Azerbaijan University Press.

were previously accustomed. This only occurred because she was prepared to allow herself to work "interiorly", based on the experience that came from "life" in such a way that everything started from it and everything led back to it. Here lies the genius of a system of thought that has made its own way, while requiring its author to look at the things of the world in a completely new way.

In this context, the position of human beings in the cosmos acquires new meaning in that it is life itself that reveals its true collocation. Thus cosmology essentially addresses itself to the reality in which the human being is inserted. Cosmology becomes crucial to Tymieniecka's investigations and represents a problem for those who would approach her thought because she analyses cosmology without seeking to make any philosophical speculation; rather, all of her investigations in this direction "must" aim at achieving a greater understanding of the reality in which we find ourselves in a unique and unrepeatable way. It can without doubt be affirmed that Tymieniecka's great intellectual genius lies precisely in having helped divert all enquiry and intuition towards the human being, who carries within herself or himself an imprint of the original Logos.

Tymieniecka thus traces life back to its origins, or at least its "possible" origins. This is why every time we approach her analyses we must first and foremost remain faithful to her method. This entails "re-orienting" our gaze towards the philosophy of life without presuming to select what we desire to see or believing that we must investigate in detail only what we hold to be worth focusing on. This point requires a more effective explanation. Seeing questions of philosophical anthropology as a line of enquiry that simply cannot be neglected, Tymieniecka has always pointed out the inadequacy of certain contemporary tendencies which, blinded by a rigid rationalism, follow only investigations of phenomena that are considered to be worthy of scientific interest, precluding any "enlargement" of their investigations. It may be argued that such ruthless rationalism reduces life and all its possible connections to simple products to be quantified, thereby losing sight of the "qualitative fullness" inherent in every twist and turn of reality. The qualitative aspect of reality is a topic to which we will return, since it constitutes one of the pillars of Tymieniecka's doctrine regarding the original Logos of life.

Reality itself, inserted in a broader, cosmological vision, is bound up with a series of dense cascades within which the investigations of the human being must acquire a unified meaning. The individual and the construction of individuality become an integral part of the cosmic process: the dynamic relations that are created between the individual and the cosmos are possible in the light of the creative faculties, which constitute one of the pillars of Tymieniecka's thought. However, in order for this creative process to develop, it is necessary to go beyond the usual routines, which over time have become obsolete, of those who look at reality in a conventional way. On the contrary, it is necessary to take account of the "architectural" aspect of creativity, in which the rational model of the universe assumes a pre-ordained plan.[4] The architectural laws of the universe converge in a general law, which has its roots in the law of creativity.

---

[4]Tymieniecka, A.-T. 1965. *Leibniz' Cosmological Synthesis*. New York: Humanities Press, pp. 122–127.

For Tymieniecka, the creative faculties allow space for a vision of the world that is no longer based on a simple reading of how the world is, but a reading able to encompass the world "as it could be". This is why Tymieniecka's investigations cannot be pigeonholed in a conventional system; the creative act is characterised by a continuous movement that is able to "continuously intuit new processes" without ever fossilising in a sterile reading. When Tymieniecka looks at the world and all that lives in it, she "allows herself to be guided" by the creative act, which is the only act able to make each thing fully transcendent. Given that the creative act acts from within and permeates every reality, the things of the world are seen in a way that goes beyond what may appear to an inattentive observer. The things of the world are in their essence much more than what our consciousness attempts to grasp. This is why, in order to understand Tymieniecka's thought a little more closely, it is necessary to adopt her way of "seeing" the world, which I argue is characterised by "perspective". Indeed, perspective takes in not only the dimensions of the things of the world, but also sheds new light on the things themselves within the great cosmos, guided by the original logos. This is how everything can be traced back to the power of the creative act: only this is able to create a harmonious synthesis between life itself and the individual who lives it. Reality thus becomes a new dynamic process that acts in the individual, who has a well-ordered collocation in the already constituted world and in the world as it has yet to become. This is Tymieniecka's possibility of a double vision of the world in action: the already constituted world is linked to the possibility of a new reading of the world as it has yet to be. This is the perspective view of which we spoke above, i.e. what I see must be completed with what I will be able to see in subsequent developments. In this way the pre-established structures for seeing the world dissolve and new research horizons open up. The creative act is in a state of continuous activation, always opening up new and infinite possibilities of generating life.

To know the world, simple perception is no longer sufficient; for Tymieniecka, what is needed is "creative perception", which is able to reorganise every constituted perception precisely because it involves the entire cosmos.

The framework holding together what we have described so far is the awareness for Tymieniecka that true philosophy is only what considers "Logos" and "Life" as *inseparable*. In its movement – a continuous advance – the Logos makes its way without creating static structures along its route. Its flow is continuous and unstoppable. From a careful reading of this continuous flow we can understand more clearly how Tymieniecka overcomes each "banal reading of the world", which is what results from seeing it only in its external movement and thus in a superficial way, frequently degenerating into a relativism that does not help us to see the world in those particular movements that arise from within itself. Inherent in the world is a specific reality that is always active and lends itself to being investigated by the individual, considering that both the world and the individual share the same interior movement. This movement is common to both of them precisely because the forces of the cosmos intertwine with the creative forces of the individual. This arises from the reciprocal relationship between Logos and Life.

The Connection Between "Cosmology" and "World of Life" in A.-T. Tymieniecka

Seeking to set out the guidelines, or inspirational principles, that underpin Tymieniecka's reflections, we shall now analyse one of her works that captured my attention and to which the bulk of my contribution to the Paris Convention will be dedicated. I refer to *La plénitude du logos dans le registre de la vie. La métaphysique dans les nouvelles lumières*.[5] The work in question opened up a new line of research for me precisely because it showed how metaphysics acquires a new dimension in the light of the Logos. Metaphysics has always been a highly important theme for me, ever since, many years ago, I became acquainted with the work of the phenomenologist Hedwig Conrad-Martius,[6] who was a product of the Munich school and subsequently a student of Edmund Husserl in Göttingen. However, only at the end of this paper will I juxtapose the two authors, who are linked by similar research interests including cosmology, the philosophy of life, and metaphysics.

In the Preface to the work *La plénitude du logos*, Tymieniecka informs the reader that the investigations contained therein are the result of wide-ranging research that has "lasted her whole life" and has branched in multiple directions.[7] In the tradition of Plato, St Augustine, Husserl, Hedwig Conrad-Martius and even Heidegger, she stresses that the great philosophical questions that she deals with should be seen as part of "la perspective globale de la rythmicité et de la temporalité de l'être".[8] Time here is understood as being a constitutive feature that pervades human existence. It does not lend itself to simple chronometric conception, which frequently proves to be inadequate, but within it, the temporality of a time that acquires a fullness of

[5] Tymieniecka, A.-T. 2011. *La plénitude du logos dans le registre de la vie. La métaphysique dans les nouvelles lumières* (English translation by Hill, C. M. and Weber, L. M.), Paris: L'Harmattan.

[6] For a complete bibliography of Conrad-Martius' manuscripts, conserved in the *Bayerische Staatsbibliothek* archive in Munich, the reader is referred to the *Catalogus codicum manu scriptorum Bibliothecae Monacensis. Die Nachlässe der Münchener Phänomenologen in der Bayerische Staatsbibliothek*, E. Avé-Lallemant (ed.), Tomus X, Pars I, Wiesbaden: Otto Harrassowitz, 1975, pp. 193–256. In addition, for a systematic study of her impact and a complete overview of her books, articles, essays and translations, the reader is referred to: Alfieri, F. 2008. *Hedwig Conrad-Martius: A Philosophical Heredity Illustrated by Eberhard Avé-Lallemant*, in *Axiomathes* 18, pp. 515–531 (http://www.sprin-gerlink.com/openurl.asp?genre=article&id=doi:10.1007/s10516-008-9044-1); Idem 2010. *Nota Bio-bibliografica di Hedwig Conrad-Martius*, in Ales Bello, A. – Alfieri, F. – Shahid, M. (eds.). *Edith Stein – Hedwig Conrad-Martius. Fenomenologia Metafisica Scienze*, Bari: Edizioni Giuseppe Laterza, pp. 463–483; Idem 2011. *L'ancoraggio ontico tra "Natura" e "Spirito" nel Das Sein di H. Conrad-Martius. Una questione aperta*, in Baccarini, E. – D'Ambra, M. – Manganaro, P. – Pezzella, A. M. (eds.). *Persona, Logos, Relazione. Una fenomenologia plurale. Scritti in onore di Angela Ales Bello*, Rome: Città Nuova, pp. 346–362.

[7] Tymieniecka, A.-T. 2011. *La plénitude du logos dans le registre de la vie. La métaphysique dans les nouvelles lumières*, op. cit., p. 7.

[8] Ibid.

meaning – because it is crossed by the logos of life – is revealed.[9] This is why time escapes our grasp only when we reduce it to a static meter that merely marks the moments of our life. For Tymieniecka the starting point is clear: it is the Logos in its temporal manifestation, which reveals itself in a continuous becoming that gives meaning and fullness to time precisely by means of its movement. Its propulsive force marks time in a way that is never static or repetitive because its becoming does not include chronological repetitions. "Neither intelligence nor an observer are necessary to record the interval of time. Life proceeds and is temporalised without all that".[10]

This then is the point from which it all begins: "La vie dans son logos est le tremplin qui nous permet d'accéder aux origines des origines: au logos primordial".[11] As this is the starting point for all that follows, Tymieniecka travels a long way in order to descry, via the evolutionary meanders of the self-individuation of life, the "paths" that lead to the light of the original logos. The logos is the initiator and the promoter of life, and it generates life by creating space around itself and finding all suitable occasions for doing its work. Its movement cannot be stopped and in addition to the function of "creating" life, it can, once it has arrived at the peak of its action, re-orient its forces to the point of destroying all that it has created. "A-t-il des rênes avec lesquelles on peut le tenir, mais que le guide ne trouve plus?".[12] The movement then is aimed at both creation and destruction, but in all this there is a harmonious process able to contain this series of opposites by means of the fullness of meaning that resides in the logos. In the act of generating life the logos is the bearer of light, but to speak of "light" means saying that precisely in generation the logos must take account of the "darkness". This means that "light" and "darkness" enter equally into the generative act and in order for each to manifest itself fully, they need to enter a relationship that we may call "functional". Normally we speak of "light" and "darkness" as two opposites that repel each other; in contrast, in the movement of the logos they have equal value because they are each bearers of a specific "quality" that is essential to the other. "Light" would not be what it is without "darkness", and the function of the latter is to bring out the light in all its glory. Thus with respect to each other, "light" and "darkness" are fully complementary. Hence we can already see the fullness of meaning of their "relationship": "C'est le sens même qui émerge de ce jeu: c'est-a-dire le sens de *relation* émerge également. Chaque élément a besoin d'être apprécié dans tous ses degrés allant vers son opposé pour que ses proportions soient correctement devinées. Chacun a besoin d'être joint à son opposé qualitatif. Cela revient à donner un sens à la relation".[13]

---

[9] For a more detailed treatment of this theme the reader is referred to: Tymieniecka, A.-T. 1996. *Kronos e Kairos*, in *Tempo e Storia. Atti del VI Colloquio della facoltà di Filosofia* [1994] (*Dialogo di Filosofia*, 12), (Italian translation by Ales Bello, A.), M. S. Sorondo (ed.), Roma: Herder – Università Lateranense, pp. 287–307.

[10] Ibid., p. 294. (my translation).

[11] Tymieniecka, A.-T. 2011. *La plénitude du logos dans le registre de la vie. La métaphysique dans les nouvelles lumières*, op. cit., p. 23.

[12] Ibid., p. 24.

[13] Ibid., p. 25.

What is gained by harmonising opposites is the reorientation of our conception of the vital movement that we see as life flourishes and then perishes. This is the game of opposites that inevitably marks our life and cannot be stopped; herein lies the strategy of the Logos. When the action of the Logos of life reaches its peak with the formation of human life, a new phase begins that is no longer tied to determinism, but to a creativity that unfolds in complete liberty; our consciousness is rooted in it.

This strategy can be seen in the cosmos, but although the Logos has generated life, as far as can be perceived by us, it continues its action by planning new tissues that despite being subject to death are always recreated anew. Following this tortuous movement, we can fuel the hope of seeing, albeit imperfectly, the "primordial logos", which is defined by Tymieniecka as follows: "*raison de toutes les raison*, ne se rapporte à aucune forme de devenir, à aucune modalité de notre esprit, à aucune forme conceptuelle de notre raison intellectuelle spéculative, car il *ne génère ni ne possède d'origines*".[14] The way in which it can be seen is precisely by following the action of the logos of life, which in its movement is the "bearer" of the "fullness" of the primordial Logos. Life is affected by this dynamic flow and shares in the fullness of the primordial Logos, but as Tymieniecka has stressed many times in her writings, this flow has nothing to do with the wild flow of Heraclitus. The advancement of the flow of the logos of life is well-ordered and becomes stabilised in the cosmos in accordance with a temporal rhythm. This continuous flow is investigated by Tymieniecka with reference to the way in which Husserl conducted his own investigations, i.e. by means of the zig-zag procedure typical of an archaeologist who is seeking to discover the deepest layers that are not manifested for our direct observation. It should be pointed out here that Tymieniecka reviews all the pillars of Husserlian phenomenology including the concept of intentionality, as well as the results achieved by Husserl, in the light of the question of the logos of life, and finds the Husserlian reductions in the new research path chosen by her to be insufficient.[15] Above all the concept of intentionality is questioned because it is not able to grasp the roots of human life. It is not sufficient therefore to observe the results that the flow of the logos of life produces by its activity, but it is necessary to follow the flow backwards in order to arrive at its source, from where it draws its origins and thus its becoming.

A.-T. Tymieniecka and "The Third Phase of Phenomenology": Eco-Phenomenology

In recent years Tymieniecka has undertaken a necessary philosophical reflection that has helped her to overcome all the difficulties present in the Husserlian conception of intentionality: as we stressed earlier, for Tymieniecka the intentional quality of consciousness is not the last word, but consciousness is rooted in the creativity of life. Access to reality "est ouvert par l'expérience créative humaine et sa trajectoire

[14] Ibid., p. 30. (my italics).

[15] Ibid. pp. 31–44 (Chapter 2: *Le mode interrogateur du logos de la vie et sa révélation du logos universel*).

dans sa fonction d'établir le monde-de-la vie et l'être humain vivant en son sein".[16] The intention is clear: we must seek another level of investigation of reality that marks the birth of the "third phase of phenomenology". This is a new philosophy that is able to develop from the starting point of the human condition, understood in a cosmological sense. This new philosophy is "new" because it opens up a vast array of relationships: those between human beings and nature are now collocated on a "metaphysical level" of reality. Naturally, these two vital categories, "human beings" and "nature", no longer lend themselves to being investigated in isolation. Human beings are fully inserted in this vast panorama of reality – indeed, they are closely tied to everything around them – and it is their "relationship" with everything else that must take priority in the investigation. In the light of what has been said so far, despite being a mere summary, Tymieniecka concludes that "the human being unfolds and generates in a mutual contributive relation to all the other living beings".[17] The human condition must therefore be grasped in terms of the unity of everything that is alive, and in close connection with the laws that regulate the cosmos.

The starting point for Tymieniecka is "Life", and it is from this alone that the notion of "individual" must emerge. "L'individu est simplement le principe de singularisation qui crée la diversité dans le progrès de la vie [...]".[18] When Tymieniecka argues, as she has always done, that the notion of "individual" must be subjected to a new critical assessment, she means to direct the investigation of the individual towards its "ultimate foundation", which is life itself. This path marks a new era for "phenomenological realism" and is able to go beyond the ontological abstractions inherited from traditional metaphysics, which investigated the foundation of individuation in the individual as if this was cut off from reality. For Tymieniecka, this "reality" is a vital aspect that can no longer be left to the discretion of the observer. It is sufficient to follow Tymieniecka's rigour to realise that "le noyau essentiel de l'individualité [...] se structure, s'enrichit et s'étend dans des caractères producteurs de types le long de la voie évolutive de la vie".[19] The "consciousness" of each individual is firmly bound to "Life", and although in our investigations they may seem two separate elements, we must always consider them as different sides of the same coin.

The "Meta-Phenomenological Realism" of Hedwig Conrad-Martius

The discovery of Tymieniecka's philosophy, and in this I consider myself to be still a beginner, has broadened the horizons of my research and at the same time has helped me to re-read the writings of the phenomenologist Hedwig Conrad-Martius,

---

[16] Ibid. p. 45.

[17] Tymieniecka, A.-T. 2001. *The passion of the earth*, in *Analecta Husserliana* LXXI, p. 8.

[18] Tymieniecka, A.-T. 2011. *La plénitude du logos dans le registre de la vie. La métaphysique dans les nouvelles lumières*, op. cit., p. 94.

[19] Ibid. p. 100.

who I began studying in 2007 at the Munich state library. Specifically, there were two elements that prompted me to assess the similarities between the two authors: on the one hand there is Tymieniecka's philosophy of life and Conrad-Martius' phenomenological realism; on the other there is their common attempt to "recover" metaphysics in order to gain a greater understanding of the reality in which the individual is rooted in a dense network of "vital relationships".

I have come to realise that it is not always easy to fully understand the true nature of the links established between the founder of phenomenology and his disciples. In our specific case we are concerned with Hedwig Conrad-Martius in relation to the master Edmund Husserl.[20]

Without a shadow of a doubt, the "eidetic reduction" represented a real opportunity to go beyond the Kantian criticism that had argued for so long that it was impossible to explore "what is given" in such detail as to grasp its "essence" (*Wesen*). The Husserlian phenomenology of the *Logische Untersuchungen* opened up a season of studies and at the same time opposed those tendencies that sought to "reduce" *givenness* to parameters based on pure subjectivity. It is precisely this point that Conrad-Martius' aware and "autonomous" realistic phenomenology picks up on. Realistic phenomenology proceeds via research into the essence (*Wesenforschung*), in the sense of Husserl's eidetics, of reality itself and its various contexts. This is because true philosophical speculation, founded on real data, requires not only precise registration of empirical data but above all "observation of the essence" of the reality to be interpreted.[21] In this way it is easy to see how Conrad-Martius' first "expectation" that prompted her to follow Husserl was the Master's success in going beyond the naturalistic psychologism that prioritised psycho-physical knowledge processes at the expense of the real givenness of what manifests itself to me "here and now in flesh and blood". What this nascent phenomenology offered was an opportunity to "rebalance" the relationship – no longer antithetical – between "knowing subject" and "reality in its concrete manifestation", thus ensuring that

[20] From the autumn of 1909 until the summer of 1912 she studied Philosophy, Psychology and History of Art in Munich (2 semesters). She joined the *Akademischer Verein für Psychologie* founded by Theodor Lipps, whose disciples were part of the nascent phenomenological movement that was inspired mainly by Edmund Husserl's *Logische Untersuchung*. On the advice of Moritz Geiger she transferred to the University of Göttingen and attended four semesters with Edmund Husserl and Adolf Reinach. She was the first of three women – after her came Edith Stein and, in Freiburg, Gerda Walther – to be admitted to Husserl's group of students. From 1911 to 1912 she chaired the Göttingen Philosophical Society (founded in 1907 by Theodor Conrad, who was later to become Hedwig Martius' husband), whose members included Adolf Reinach, Max Scheler, Alexander Pfänder, Hans Lipps, Alexandre Koyré, Fritz Kaufmann, Dietrich von Hildebrand, Jean Hering, Winthrop Bell, and subsequently Edith Stein. She went on to win the Göttingen Faculty of Philosophy Prize for her essay entitled: *Die erkenntnistheoretischen Grundlagen des Positivismus*. The recognition she obtained did not help her to gain a doctorate from Göttingen, but Alexander Pfänder accepted her work and on the 20th of July she was awarded a Doctorate from the University of Munich, presenting a revised version of the afore-mentioned essay.

[21] Conrad-Martius, H. 1957. *Phänomenologie und Spekulation*, in *Rencontre-Encounter-Begegnung. Festschrift für F. J. J. Buytendijk*, Utrecht-Antwerpen, pp. 116–128; English translation, *Phenomenology and Speculation*, in *Philosophy Today* 3 (1959), pp. 43–51.

"reality" is no longer considered a mere "domain" of consciousness but is seen in its proper light.

Thus, for the first time, this new approach directed its gaze towards the *things themselves*, on which the observer did not "impose" his or her laws in accordance with some sort of constructivist tendency; rather the observer "received" the right way of proceeding from the laws inherent in the reality in question.

This "expectation" was soon to be called into play once more, following the publication of a short work by Husserl entitled: *Die Idee der Phänomenologie*.[22] It was with this work, containing the five lessons given by Husserl in Göttingen in 1907, that the first signs of "tension" between the master and his disciples began to appear. In these lessons Husserl highlights the "flow of consciousness", applying to it a sort of reduction that was comparable to eidetic reduction, generating in his disciples the suspicion that he sought to return to transcendental idealism. It is clearly inappropriate to cling rigidly to the Husserl of the *Logische Untersuchungen* in the face of the realistic position of phenomenology that was gradually forming. Nor can *Logische Untersuchungen* be seen as providing the only key in which to read the whole of Husserlian phenomenology.

Of the Husserlian construct as a whole, Hedwig Conrad-Martius focuses on purely eidetic phenomenology. This explains the difficulty of those who seek to study the investigations of Conrad-Martius: a difficulty attributable not only to the language, undoubtedly difficult, but also to her different approach to Husserlian thought, which makes no use of transcendental reduction. Conrad-Martius claims to accept only the first (i.e. the eidetic) moment of the reduction: for example the orientation of her ontology is static-descriptive, we may even say "geometrical"; in it, the concepts of "places" and "spheres" of the being can be "followed" only in the sequence of the levels of the complex real-ontological architecture. The difficulty is evident because in Conrad-Martius' investigations we can find no link leading from the more static aspects of the eidetic method to the "genetic" stage of the intentional analyses. The assumption of this final stage did not convince Conrad-Martius because it suggested a "return" of reality to a "subservient" role with respect to the sphere of consciousness.

Hedwig Conrad-Martius introduces a new way of conceiving of Husserlian phenomenology – that of "meta-phenomenological realism". Conrad-Martius' initial enthusiasm for the master's position was subsequently to be transformed into growing disagreement. Her disappointment was due to the fact that after the publication of the *Logische Untersuchungen*, the Master developed his thought in a direction that his pupils, in order to be consistent with their own ideas, could no longer follow.

Conrad-Martius' separation from the Master happened after the publication in 1913 of the first volume of the *Ideen*. The basic problem that widened the breach with the Master is the priority that he gives to consciousness with respect to

[22] Husserl, E. 1950 *Die Idee der Phänomenologie. Fünf Vorlesungen*, in *Husserliana* 2, W. Biemel (ed.), Dordrecht/Boston/Lancaster: M. Nijhoff, Den Haag.

phenomena that are manifested in the spatio-temporal world. For Hedwig Conrad-Martius and other disciples of the Göttingen Circle, this priority represented a change of course with respect to what Husserl had claimed in the *Logische Untersuchungen*. The fundamental question underlying this disagreement with the Master is why he had given such priority to consciousness as to make him affirm that reality, in terms of both things considered individually and the world as a whole, by its essence lacks selfsufficiency (*Ideen* I, § 50).

How are we to interpret Husserl's statement that "reality by its essence lacks selfsufficiency"? The response to this question can only lie in the position adopted by Conrad-Martius that characterises all her investigations. Indeed, Conrad-Martius detaches all analysis of reality from the narrow dependence on consciousness as conceived by Husserl. In this way the whole of Husserl's approach to phenomenology is heavily modified: for Conrad-Martius, "transcendental reduction" cannot play any role in her investigations of the real, because phenomena cannot simply be reduced to correlates of constitutive acts of consciousness. Thus phenomena are not such in their being because they are correlated with consciousness, but because they have ontic priority regardless of any correlation. In this way, for Conrad-Martius the "real" assumes an ontic autonomy in itself. In continuity with what has been set out so far, for Conrad-Martius the transcendental philosophy of Husserl not only "abstains" from any judgement on the real existence of the world (i.e. its existence outside consciousness) but even reduces it to a mere "world-*appearance*", thereby running the risk of losing sight of its real character altogether.

It is clear that for Conrad-Martius the starting point or rather the priority lies precisely in the "real", and that she intends to "go beyond" any idealistic position – the implicit reference in this passage is the Husserlian position – that contemplates the real only in close connection with the 'I'. This is where her efforts are leading: "My philosophy is always aimed at a problem that has interested me ever since the beginning of my autonomous philosophical activities: research into the essence of the real being as such, especially in nature".[23]

In this way Conrad-Martius seeks to give "concreteness" to the real, which cannot remain exclusively within consciousness, but may be assumed to have an autonomous structure of its own outside consciousness. Consciousness is therefore no longer the absolute, understood in the Husserlian sense, that "posits" the world and confers on it a structure of being. Conrad-Martius' grievance with Husserl is that although he has arrived at an authentic science of essences he has limited the confines of that science to "mere consciousness". However, Conrad-Martius does at least conserve eidetic reduction and *epoché* – "the starting points" – and the fundamental features of Husserlian procedure.

I would like now to move on to an analysis of Conrad-Martius' new perspective, starting with a little-known text that contains the main points of her investigations. In this context I will analyse only certain passages that are useful for our

[23] Conrad-Martius, H. 1959. *Die transzendendale und die ontologische Phänomenologie*, in *Edmund Husserl 1859–1959* (*Phaenomenologica*, 4), Den Haag, p. 177 (my translation).

purposes here, considering that a thorough appraisal of this work would require a far more detailed study.

"The human being passes through the world as a living being": this is the opening statement of the radio interviews conducted from 1949 to 1951 between the phenomenologist Hedwig Conrad-Martius and the doctor Curt Emmrich – known by his pseudonym Peter Bamm – which were subsequently published in a book entitled *Das Lebendige, Die Endlichkeit der Welt, Der Mensch*.[24]

The problem of the living can be read on a temporal level in accordance with a "before" and an "after": while the phenomenology of the essence conducts its investigations in the full "awareness" that the complexity of the "living" being must be investigated in all its parts (the before), on the other hand the experimental sciences aim only at the mathematisation of the phenomenon (the after), reducing it to a collection of "categories" and thereby losing sight of the deep connections binding the living to the world of nature. There is an "essential sphere" of meaning, which, as such, is an assumption that is always taken for granted, but for this very reason frequently neglected by the experimental sciences, which very often naively present their experimental world as the "real world". Conrad-Martius is clearly correct to argue that "the phrase "modern natural sciences" contains in itself the assumption that only that which can be reached by modern experimental science is worthy of being considered science of nature, and that all that cannot be demonstrated by means of experiments is not scientific, as if there were first class scientific truths, based on experimental results, and second class scientific truths, based on thought alone".[25]

Such reasoning begs the question: can the totality of the phenomenon in all its complexity be "described" solely by what can be experimentally determined and thus measured? Or is there something that escapes pure and "simple" mathematisation, such as the deep connections that are present and we "feel" but are not easily quantifiable? In addition, experimentation risks giving an incomplete account of the phenomenon, given that after having explained all the causal processes present within it, we still haven't begun to explain the intimate nature that constitutes the *primum movens* that gives meaning to every measurable manifestation. Hence the need for an objective assessment of the limits of what is commonly defined as "measurable", given that every phenomenon, in order to be "fully" known, requires the observer to descend through all of its different constitutive levels, intimately intertwined with each other, to reach the ultimate level in which the origin of its "meaning" resides, excluding the causal explanations typical of the "exact" sciences.

A complete dichotomy opens up at this point between the approach based on the method of the sciences, which use inductive reasoning starting from the mere "factuality" of the phenomenon, following a process linked to causality, and the phenomenological approach, which tends in contrast towards an investigation

---

[24] Conrad-Martius, H. 1951 *Das Lebendige, Die Endlichkeit der Welt, Der Mensch. Drei Dispute von Hedwig Conrad-Martius und Curt Emmrich*, München: Kösel Verlag.

[25] Ibid. pp. 15–16 (my translation).

that is "respectful" of the absolute objectivity of the "structure of the phenomenon's essence". In the latter approach, each "description" of the phenomenon is justified only when it adheres to the *givenness* of what is consigned to us, via our intuition, in its pure manifestation. Indeed, for Conrad-Martius it is clear that the "structure of the phenomenon's essence", i.e. its eidetic reduction, is the only way in which we can provide a foundation for what is given to us at a progressively deeper level: "[...] experiments must be *interpreted*, and even for this reason alone, it is necessary to define our terms precisely, and the definition of the terms is no longer a purely scientific process, but rather a philosophical one. In addition, if scientists come to an agreement on the definition of something, this in itself does not mean that with such a definition they have grasped the essence of the thing".[26]

I would like to state here the crucial point that was to lead Conrad-Martius to distance herself from her master Edmund Husserl, when the latter, after having argued for the absolute objectivity of the structure of the essence of things, switched in his work to giving exclusive importance to their constitution in the subjective realm. In fact, it would be more accurate to speak of an "apparent" shift from the eidetic analysis of regional ontologies towards the constitutive analysis of the structure of consciousness.

The eidetic analysis of regional ontologies, the first stage of phenomenological analysis, i.e. the eidetic investigation of reality, opened up the field to Conrad-Martius on the question of the "world" and "life". What is "life"? In itself, life can never be explained with reference to something else, which would be basically impossible. "Life" is never substantially real, only the living. Life is a way of existing, not something that can exist independently. Mere "life" does not exist. Only what is living is autonomous! Plants, animals, human beings live; God lives. However, in the living, the determinant element is precisely life. When Conrad-Martius says that "the essence of the living consists precisely in its life",[27] she is basically naming the determinant element of the *novum*. Though it may seem banal at first sight, it contains unexplored mysteries.

The basic problem is the nature of this *novum*, which is key to the question of the living but hard to "measure" because it belongs to a microcosm in which the synolon to be quantified has no space. Having ruled out any possibility of measuring the *novum*, the difficulty lies in explaining it, since despite its "simplicity", the *novum* cannot be grasped by means of external elements such as the "causality" that justifies and links phenomena to each other. *We have entered a microcosm that is so disarmingly simple in its pure manifestation that we can describe it as being the "indomitable foundation" in contrast to the "foundations" that are measurable and mathematised by the sciences*. This requires not only different foundations, but also a new language in order to explain what implicitly resides in the essence of the living thing: the closer the analysis comes to the essence of the foundation of the phenomenon, the simpler the language will need to be.

---

[26] Ibid. p. 16 (my translation).

[27] Ibid. p. 45 (my translation).

We come then to the question of the way in which the "essence of the living" can be grasped. In this regard Conrad-Martius suggested to her interlocutor, Curt Emmrich, that they follow the approach set out by the biologist Hans Driesch. The exchange went as follows:

> CONRAD-MARTIUS. – [...] Driesch was the first to demonstrate that the internal and external factors that biologists discover and analyse in their experimental research are not sufficient to truly explain all the phenomena of organisms.
>
> EMMRICH. – To explain his experiment with sea urchins, Driesch introduced the term "Entelechy".
>
> CONRAD-MARTIUS. – But speaking as a biologist, what do you understand by Driesch's term "entelechy"?
>
> EMMRICH. – It's not easy to explain! [...] By entelechy Driesch definitely meant a plan for what the organism becomes. Driesch's entelechy should be seen as something independent, *something positive (etwas Positives) that is added to matter* and makes use of the laws of matter. [...] However, if you now ask me to say something more precise, I realise that this entelechy remains something ungraspable [...].
>
> CONRAD-MARTIUS. – You're right. These are terms for which scientifically the problem lies in the fact that they are not physically or chemically definable. And this is where the ontologist comes in.
>
> EMMRICH. – I realise I am on shaky ground. But I accept the challenge! You have the courage to say here that we are making scientific use of terms that nobody is able to define precisely?
>
> CONRAD-MARTIUS. – And that's not all! The indiscriminate use of these terms obscures the true question of life.[28]

All this opens up a *radical metaphysical question*, since it is not only about understanding the various levels of the living, but the relationship between them and the newly introduced entelechial factor, a theme of clearly Aristotelian inspiration. Indeed, the path set out by the biologist Hans Driesch offers the possibility of reaching the essence of the living; by means of biological experimentation, he seeks to broaden his investigation by introducing the trans-physical factor as the constitutive foundation of the living. For Conrad-Martius, the entelechial factor – which can also be identified as an "acting power" – is not something formal but "real", except that its reality does not lie within the physical or psychic order, but is rather of the "metaphysical" type. The whole issue becomes clearer as soon as we return to the question of the living: where does the human being phenomenon originate from? Are the reading and measurement of all its developments and conditionings able to add anything to the question of the "living"? Indeed, the external determination alone of what manifests itself as living must assume "life" as the ultimate cause (entelechy) or the original phenomenon behind its birth and its vital manifestations.

It is thus the real-ontological aspects that escape the analyses of the "exact sciences", and with them there is also the risk of losing the full knowledge of the human being. The sciences will never be able to provide a univocal answer to the question of whether the method used by them to explain phenomena fully corresponds to the reality of the phenomenon itself. The basic problem, which justifies the

[28] Ibid. p. 36–38 (my translation).

validity or otherwise of any investigation, is the capacity to grasp the phenomenon, which means providing a reading that fully adheres to its real structure without the observer introducing any distortion that could disturb the delicate equilibrium between its "full manifestation" and the way it is grasped; when the description of the phenomenon that I investigate adheres to its full constitutive reality, I will have achieved a perfect and true reading of the phenomenon as it is in itself (what appears and the reading of it coincide, without conceptualistic distortions). This is why the full control over phenomena that mathematisation of them purports to achieve does not mean that we have acquired "full" understanding of the intimate nature of the thing in itself, since the simplification that the scientist arrives at in his or her reading of the world might no longer be seen in relation to the world "in itself". Between the "complexity" of the phenomenon in itself and every simplification of that phenomenon, such a gulf is created that we are obliged to wonder whether we are still dealing with the same phenomenon or with a "real world" and another phenomenon whose constitutive essence has been lost due to the process of simplification to which it was subjected. Thus we come to see in the same phenomenon the dichotomy between the real-ontological concrete and the "conceptual" arrived at by the scientist. Could it be that the "gap" that is created between the "concrete" and the "conceptual" corresponds to that part of the "fully real" that escapes the full control of those who seek only to mathematise the phenomena because they are not able to grasp it fully?

The results we have obtained here are sufficient to justify the conclusion that at the base of the human being there is an ultimate constitutive-entelechial structure – the entelechial factor, despite being a trans-physical element, belongs to natural reality – that designates it as a person in the full possession of himself or herself. At this point each broadening of the investigation to include metaphysical factors helps us to abandon the common prejudice that metaphysics does not have a solid foundation. On the contrary it is necessary because it is intimately linked to the real nature of the things, a constitutive element of reality itself. "A being which, like human beings, is autonomous and free from the world in its ultimate existential constitutive foundation must rest on itself, i.e. beyond nature. The human being is located on a metaphysical foundation".[29]

Some Conclusions from Which to Initiate Further Research

If one has the patience to follow the investigations of Anna-Teresa Tymieniecka in detail, one notices that her phenomenology of life tackles great philosophical questions, but also that it is necessary to "reorient" our traditional way of analysing them. Life, the cosmos and the individual take on new coordinates, and it is above all the relationship between these vital elements that interests her. This relationship

[29] Ibid. pp. 166–167 (my translation).

is able to demonstrate that only the vital lymph of the human condition reveals the original meaning of "life". The human condition must therefore be grasped in terms of the unity of everything that is alive, and in close connection with the laws that regulate the cosmos. In this network of relationships, each element occupies a very precise place.

With Tymieniecka the focus on the dimension of real existence means that her philosophy regenerates itself continuously, to the point that she frequently has to retrace her steps and question what she has acquired because the "adherence" and "sacred respect" for life in its continuous flow always require new corrections. Without a shadow of a doubt Tymieniecka's investigations belong to a form of thought that allows itself to be guided by the internal laws of life itself, thus going beyond all sterile speculation, which over time would be abandoned anyway. Her thought is alive, and this is demonstrated by her interest in the laws of cosmology, to the point that the current stage of her thought may be described as "eco-phenomenology".

This approach can help scholars interested in Hedwig Conrad-Martius to understand her "meta-phenomenological realism". Conrad-Martius' works are highly complex, and are mainly metaphysical in nature: what we may here call *metaphysical phenomenology* is however accompanied in her writings by an always-open dialogue with natural sciences. Today therefore, in an epoch in which phenomenology is in constant dialogue with the natural sciences, the question of a careful analysis of reality and its laws can no longer be separated from a serious investigation of the individual. There is another urgent matter: that of helping Conrad-Martius to emerge from the underserved ring of silence that surrounds her. In this, Tymieniecka's philosophy constitutes a solid basis for "rehabilitating" this thinker, thanks to whose efforts the cultural achievements of Husserl were harnessed to the investigation of the complexity of the real in a metaphysical key.

I would like to thank Professor Anna-Teresa Tymieniecka, who also kindly allowed me to be a guest in her private home in Vermont so that I might begin to acquire the necessary rudiments for comprehending her thought. Working "on location" I have been able to appreciate how eco-phenomenology is a lively current of thought whose flow is strong enough to regenerate our relationship with the entire cosmos. The experience has enabled me to feel an "integral part" of this vital flow.

The Forces of the Cosmos Before Genesis and Before Life: Some Remarks on Eugen Fink's Philosophy of the World

Simona Bertolini

Abstract "The forces of the cosmos" could be considered the title of the world-ontology Eugen Fink develops after World War II, after having been Husserl's research assistant during the 1928–1938 decade. The purpose of this paper is to clarify the specific meaning such a title has in the context of Fink's thought, analyzing its presuppositions and its limits. On the one hand, the development of a philosophy of the world alludes to a particular elaboration of phenomenology made in the light of the world-pregivenness-problem and its ontological reinterpretation. On the other hand, the priority of ontology seems to make Fink forget the wealth of phenomenological inquiries, insofar as his post-war considerations emphasize the analyzing of ontological structures without taking an interest in the genesis of those concrete distinctions which coincide with the self-display of these structures themselves. Fink also offers an interesting clue for a new kind of phenomenological description (with regard to the openness of reality to the "forces of the cosmos"), without really developing all its potentialities.

Fink's Meontic Phenomenology: World and Absolute Constitution

If we talk about "the forces of the cosmos" in contemporary philosophy, we cannot avoid thinking of the twentieth century's philosopher who held the concept of cosmos and world as the basis for his philosophical proposal: Eugen Fink. "The forces of the cosmos", in fact, could be considered the title of the world-ontology Fink developed after World War II. The aim of this paper is to clarify the specific meaning this title has in the context of his thought, analyzing its presuppositions and its limits.

S. Bertolini (✉)
University of Parma, Parma, Italy
e-mail: simona.bertolini@fastwebnet.it

A.-T. Tymieniecka (ed.), *Phenomenology of Space and Time: The Forces of the Cosmos and the Ontopoietic Genesis of Life: Book One*, Analecta Husserliana 116,
DOI 10.1007/978-3-319-02015-0_4,

First of all, it is not possible to understand the importance of the idea of cosmos in Fink's ontology without understanding its pre-war years' roots. As is well known, Fink was Husserl's research assistant during the last decade of Husserl's life (1928–1938), a period in which the young collaborator re-elaborated the phenomenological method by reconsidering the problem of world-pregivenness and its constitution. Fink's reasoning can be summarized like this: if the world itself is not an object, and therefore is not to be thought of as an aggregate of objects, then the phenomenological philosopher cannot seek its constitution within the same correlational intentionality in which the single things figure into experience. Thus, the particular way man understands the world becomes the indicator for an alternative direction of phenomenological inquiry. Ronald Bruzina, editor of the notes Fink wrote in those years[1] and translator of the English edition of the *Sixth Cartesian Meditation*,[2] on this subject states that "if the world is the all-comprehensive horizonality of the life of consciousness, then the modality of awareness regarding that horizonality would also have to be something all-comprehensive in the life of consciousness".[3] In other words, since the world is always a pregiven, every time we perceive a thing or a group of things, the modality of awareness regarding it cannot be considered as a modification of the consciousness of the objects (in this case it would not be pregiven anymore), but should rather be identified with an "openness-to-the-world" (*Weltoffenheit*) which lies *before* this consciousness, an "openness-to-the-world" which characterizes the whole of consciousness as such.[4] Consequently, when the philosopher questions the constitution of world-horizonality, this constitution cannot be conceived as the way in which the consciousness of horizons evolves, but it must be a pregiven just like its modality of experience. As Fink writes in some notes of the early 1930s,

> Die Konstitution der Weltganzheit ist nicht 'unterwegs', sondern grundsätzlich vorbei, vor der Gegenstands-Konstitution.[5]
>
> Nur weil die Welt konstitutiv 'fertig' ist, können fertige Objekte d.i. Seiendes zur Konstitution gelangen.[6]

---

[1] See Eugen Fink, *Die Doktorarbeit und erste Assistenzjahre bei Husserl*, ed. Ronald Bruzina, in *Gesamtausgabe*, Abt. 1 (*Phänomenologie und Philosophie*), Bd. 3 (*Phänomenologische Werkstatt*), Teilbd. 1 (Freiburg i. Br.-München: Alber, 2006); *Die Bernauer Zeitmanuskripte, Cartesianische Meditationen und System der phänomenologischen Philosophie*, ed. Ronald Bruzina, in *Gesamtausgabe*, Abt. 1, Bd. 3, Teilbd. 2 (Freiburg i. Br.-München: Alber, 2008), hereafter cited in text as EFM 2.

[2] See Eugen Fink, *Sixth Cartesian Meditation. The Idea of a Transcendental Theory of Method; with Textual Notations by Edmund Husserl*, Studies in Continental Thought, trans. Ronald Bruzina (Bloomington and Indianapolis: Indiana University Press 1995), hereafter cited as VI.CM.

[3] Ronald Bruzina, *Edmund Husserl and Eugen Fink. Beginnings and Ends in Phenomenology, 1928–1938* (New Haven-London: Yale University Press, 2004), p. 195, hereafter cited as RB.

[4] It is here evident the influence of Heidegger's lectures, from which Fink took notes for six semesters from 1928 to 1931.

[5] EFM 2, p. 16.

[6] Ibid., p. 68.

So the world is not only pregiven, but its pregivenness is always "done and ready", "fertig und vorbei". That is to say, from a constitutive point of view, the phenomenological philosopher, who examines the development of the intentional correlation between consciousness and its domain of sense, is always destined to find world-pregivenness behind his shoulder, without being able to explain its origin. In order to get over such a limit and to extend the phenomenological explanation to this further level of problems, it becomes necessary for phenomenology to overcome its initial stage and penetrate into a deeper transcendental field, situated above and beyond the subject-object correlation.

Fink describes such a depth with the following words:

> Welt ist, als Entnichtung des Absoluten, *ist* das Absolute *nicht*. Die Welt ist, als Bedingung des Nichtseins des Absoluten, die erste Stufe der Selbstverwirklichung. Erst die *philosophische*/Entnichtung des Nichts des Absoluten ist die eigentliche Ontifikation des Absoluten. So wird die Welt zur blossen Episode in der Selbstverwirklichung des Absoluten.[7]
>
> Solange man die absolute konstituierende Subjektivität am Leitfaden des mundanen Seinsbegriffs (oder des Seinsbegriffs überhaupt; denn die Welt ist das Universum des Seienden!) auszulegen versucht, ist die Gefahr unabwendbar, dass diese Auslegung misslingt. So z. B. ist der Begriff der 'Konstitution' nicht verständlich zu machen, wenn die transzendentale Subjektivität als seiend gedacht und somit auch die Konstitution als Verhältnis eines Seienden zu einem anderen Schon-Seienden oder erst gemachten Seienden verstanden ist. – Erst die meontische Fassung der absoluten Subjektivität zeigt die Konstitution als ein me-ontisches Verhältnis, als nicht-ontisches; nicht als Verhältnis zwischen Seiendem, sondern 'zwischen' Welt (Sein) und 'Nichts'; als Verhältnis von Ursprung und Entsprungenheit.[8]

In these quotations we may find two interesting claims. On the one hand we discover that the openness-to-the-world is based on an *absolute* constitution, the inquiry of which starts by following the "movement" of experience and finishes in a dimension which precedes not only this movement but also the first stage of phenomenology (Husserl's level of analysis according to the intentional model). In this second phenomenological stage, the central word is not consciousness any longer, but Absolute, for the self-origination of the totality of the world, in conformity with its modality of experience, must necessarily happen *before* the consciousness-wholeness and consequently be regarded as an origin which is ab-solute, free from the consciousness itself. On the other hand, since the world coincides with the space of Being, its constitution is also described as a "meontic" (*meontisch*, not-ontic) one, i.e. as an originative movement constituting beyond the whole field of beings, beyond the ontic field as such. The relationship of absolute origination which is being discussed is, therefore, between the Being and the Nothing, or more precisely, between the Nothing and the way the Being is structured in the "world's style" before the objects. According to Fink, this is the speculative conclusion towards which a coherent performance of the phenomenological reduction leads, in line with Husserl's adherence to the "things themselves".

---

[7] Ibid., p. 288.

[8] Ibid., p. 277.

So, to conclude this presentation of Fink's first period, phenomenology, at its ultimate level, becomes a *meontic phenomenology*, *eine Meontik*. This definition is surely one of the things in which Fink differs more radically from Husserl, who still explores the horizon-constitution from the point of view of an intentional modification of its consciousness,[9] without presuming the speculative constitutive ground Fink assumes in order to justify the horizonality-experience.[10]

"The Forces of the Cosmos" in Fink's Post-War Ontology

As we already mentioned, the meontic philosophy of the first period is at the basis of Fink's Post War world-philosophy, when the philosopher began to give lectures at the University of Freiburg. What is being debated here is a transformation of the *Meontik* into a form of ontology in which the traditional metaphysical questions can be answered thanks to phenomenological instruments (to be exact, through the instruments of meontic phenomenology), and can give solutions which differ from those proposed by the old metaphysics. The same structure of the previous period reappears as the framework of a theoretical proposal, the aim of which is to examine the idea of Being in all its components, considering both the ontological configuration of beings as well as their ontological ground. Thus the term "world" becomes the generic title of a new philosophical system which hinges on the ontological origin of world-horizonality. This title now includes *both* constitutive stages of Fink's previous phenomenology, the field of beings and the meontic Absolute, although treated within a philosophical context in which the phenomenological question about the origin of things-experience has evolved into a question regarding their ontological conformation, while the absolute constitution, even if it still concerns the relationship between Being and Nothing, has now the ontological importance of the metaphysical concept of ground. Furthermore, as far as the world is concerned (in the literal meaning of beings-wholeness), not only is it now considered as the pregiven whole of what is real and possible, but also, and in relationship to its position in the new ontological hierarchy Fink proposes, as alternative to traditional metaphysics. This hierarchy, schematically, is so composed: things (first ontological level), that are still understood as substances, cannot be thought of as closed ontological poles, but their substantiality is rooted in the world-horizon (second ontological level) and, more deeply, in the originative movement that makes it emerge (as an a priori structure of individuation and, at the same time, of human knowledge).

---

[9] See for instance the 28th paragraph of Edmund Husserl, *Cartesian Meditations: An Introduction to Phenomenology*, trans. Dorion Cairns (The Hague, Martinus Nijhoff, 1960).

[10] On the difference between Fink and Husserl see also Sebastian Luft, "*Phänomenologie der Phänomenologie*". *Systematik und Methodologie der Phänomenologie in der Auseinandersetzung zwischen Husserl und Fink* (Dordrecht-Boston-London: Kluwer, 2002); Guy van Kerckhoven, *Mondanizzazione e individuazione. La posta in gioco nella Sesta Meditazione cartesiana di Husserl e Fink*, trans. Massimo Mezzanzanica (Genova: Il Melangolo, 1998). An interesting proof is also Dorion Cairns, *Conversations with Husserl and Fink* (The Hague, Martinus Nijhoff, 1976).

The rigid separation between two things, or between subject and object, just like in the phenomenology, is so brought in flow (*in Fluss gebracht*) – but not denied – in its origin, while this origin, just like in the *meontic* phenomenology, is coincident with the continuous passage Being-Nothing through the mediation of horizonality.

We can say that there are two main passages which separate Husserl's formulation and the ontology of his last collaborator.[11] First of all, as we already saw in the first period, the difference between things and world, which Husserl emphasizes in his descriptions of the natural attitude[12], is transferred by Fink from the transcendental flow of subjectivity to a deeper constitution based on the world itself and on the switching between Being and Nothing. Secondarily, this phenomenological proposal is, through the mediation of Heidegger's ontological difference, treated in such a way that the distinction thing-world, more precisely the constitutive antecedence of the second term, becomes the way Fink re-elaborates the ontological distinction *Sein-Seiend*.

It's in this context that we can use the expression "forces of the cosmos". In describing the world-origin movement, Fink mentions two directions (we could say "forces") supporting his cosmological vision. The main work on this subject is the text of a lecture given at Freiburg in the winter semester 1950–1951, published posthumously under the title *Sein und Mensch. Vom Wesen der ontologischen Erfahrung*.[13] Fink asserts here:

> Im Kampf von 'Licht' und 'Nacht', in dieser Ur-Entzweiung ereignet sich die Einigung der Welt. Das besagt nicht, dass etwa zuerst zwei Seinsprinzipien anzusetzen sind, die alsdann in ihrem Streit nachträglich ein Ganzes bilden. Erde ist nur als streitend gegen Licht, und Himmel nur als streitend gegen Verbergung. Der Streit ist nichts Nachkommendes und Nachträgliches. Er ist ebenso ursprünglich wie die Streitenden selbst. Das Gegeneinander von Himmel und Erde aber hat nicht das Gehässige, das dem Menschenstreit anhaftet. Eher gilt hier das Hölderlinwort: "Wie der Zwist der Liebenden sind die Dissonanzen der Welt...", und wenn Empedokles die Bewegtheit des Seienden regiert sein lässt durch PHILIA und NEIKOS, so am Ende deswegen, weil das Endliche auf endliche Weise teilnimmt am NEIKOS, am Streit von Himmel und Erde, welcher Streit gerade als solcher PHILIA, die Liebe ist. Die Zwietracht der Seinsmächte ist die Eintracht der Welt.[14]

The two extremes of Fink's ontology are neither the world and its transcendent ground (like in the old metaphysics), nor the subject and the space of its experience (like in the modern tradition), but "light" and "night", "sky" and "earth", i.e. the vectors of the ontological/cosmological movement which founds and unifies the

[11] The specific way these passages develop is treated in Simona Bertolini, *Eugen Fink e il problema del mondo: tra ontologia, idealismo e fenomenologia* (Milano: Mimesis, 2012). A complete view of Fink's whole thought is in Anselm Böhmer (ed.), *Eugen Fink. Sozialphilosophie, Anthropologie, Kosmologie, Pädagogik, Methodik* (Würzburg: Königshausen & Neumann, 2006).

[12] We remember the 27th paragraph of Edmund Husserl, *Ideas Pertaining to a Pure Phenomenology and to a Phenomenological Philosophy, First Book*, transl. Fred Kersten (The Hague-Boston-Lancaster: Martinus Nijhoff, 1983).

[13] See Eugen Fink, *Sein und Mensch. Vom Wesen der ontologischen Erfahrung*, ed. Egon Schütz and Franz-Anton Schwarz (Freiburg i. Br.-München: Alber, 2004), hereafter cited as SuM.

[14] Ibid., p. 217–218.

cosmos. "Sky" is the world-openness where the individuation and the difference between things and men lie; "Earth", instead, is the dark bottom supporting the Being-movement and, at the same time, the destructive force bringing back everything to its unitary origin. "Sky", using a phenomenological terminology, is the dimension of the constituted natural attitude, the dimension of the world we live in, while "earth" indicates the most radical depth of its meontic constitution. Sky and earth, in other words, represent the structural moments of what Fink also symbolically calls "world-play":[15] a new conception of ground, aimless and structured at the same time, developed beyond metaphysics thanks to the ontological transposition of the priorities and the categories of a meontic phenomenology.

Cosmology and Genesis: Some Critical Remarks

Though differing from Husserl's criteria, to all intents and purposes, Fink's philosophy is still a phenomenology. Fink's aim is not to reject Husserl's analyses. His aim is rather to denounce the insufficiency of their methodological presuppositions, in order to discover the very ultimate ground that makes them possible. Fink starts by accepting the phenomenological life of consciousness, but, not satisfied, examines it *against the light*, and unearths the ontological/cosmological structure, based on the sky-earth dialectic, which lies ramified under its givenness and penetrates all real-fields.

We can say that the purpose of Fink's *Vorlesungen* given at the University of Freiburg is to cover step by step the whole of this ramification, seeking to construct a sort of "cosmological encyclopaedia" and a new descriptive method able to grasp the way different beings, in reference to their essence, reflect the "world-play".

All of the specific subjects Fink treats in these years (Thing-substantiality,[16] Man,[17] Society,[18] Relationship between family and state,[19] etc.) are consequently described considering their openness to such a "play", i.e. considering only their ontological structure.

I would like to quote an example from the work *Existenz und Coexistenz* (text of a *Vorlesung* given in the winter semesters of 1952/1953 and 1968/1969), the aim of

[15] Indeed the notion of play becomes the principal symbol able to indicate the dynamics of the world in relation to the ontological structure of man. On this subject see the following recent volume: Eugen Fink, *Spiel als Weltsymbol*, ed. Cathrin Nielsen and Hans Rainer Sepp, in *Gesamtausgabe*, Abt. 2 (*Ontologie – Kosmologie – Anthropologie*), Bd. 7 (Freiburg i. Br.-München, Alber, 2010).

[16] See SuM.

[17] See Eugen Fink, *Grundphänomene des menschlichen Daseins*, ed. Egon Schütz and Franz-Anton Schwarz (Freiburg i. Br.-München: Alber, 1995).

[18] See Eugen Fink, *Existenz und Coexistenz: Grundprobleme der menschlichen Gemeinschaft*, ed. Franz-Anton Schwarz (Würzburg: Königshausen & Neumann, 1987), hereafter cited as EuC.

[19] See Eugen Fink, *Grundfragen der systematischen Pädagogik*, ed. Egon Schütz and Franz-Anton Schwarz (Freiburg i. Br.: Rombach,1978).

which is to draw attention to the definition of human community and sociality. Fink summarizes the task of the work in this way:

> Philosophisch nach der Gemeinschaft fragen besagt nicht, ein jedem von uns bekanntes Phänomen 'phänomenologisch zu beschreiben', ein Geflecht sozialer Beziehungen und wechselseitiger Vorstellungen aufzuzeigen, die komplexen Intentionalitäten von Ich und Du und Wir zu analysieren; es gilt vielmehr die *Seinsweise* des menschlichen Miteinanderseins zu bestimmen, den ontologischen Charakter der Sozialität zu fassen.[20]
>
> Die Nachbarschaft von Mensch und Welt ist der ursprünglichste ermöglichende Grund aller Formen und Gestalten zwischenmenschlicher 'Gemeinschaft'. Weil alle zwischenmenschlichen Gemeinschaften bestimmte Weisen des Inderweltseins sind, bestimmte Arten, wie wir uns im Brauch, im Umgang mit den notwendigen Dingen, [...], in eine Welt teilen [...], deshalb gründen alle faktischen Gemeinschaften von Menschen in der Ur-Gesellung von Mensch und Welt.[21]

What is here at issue is an interpretation of reality based on the guiding function of the "forces of the cosmos". The human community, that we have chosen as paradigmatic instance, is explicitly examined by Fink in its *Seinsweise* and in reference to the "Ur-Gesellung von Menschen und Welt", that is to say, in its hidden ontological configuration, understandable in connection with the dynamics of the world-origin. To question the community philosophically does not mean "phänomenologisch zu beschreiben". With regard to this, however, the question arises as to which are the consequences for the phenomenological description if considered in its original meaning. If, according to Fink, "cosmology" supports the domain of Husserl's inquiries, then how can these be interpreted so as to be in accordance with such support? Given the fact that Fink's post-war considerations emphasize the analyzing of *structures*, what happens to the concreteness of the genesis and of the phenomenological distinctions which coincide with the actual self-display of the structures themselves?

It is with regard to this aspect that we come across the greatest limit in Fink's ontology. Every phenomenological description, every phenomenal shade, is here replaced with another kind of description, more universal, founding the "ontic" observations, while this adjective, "ontic", becomes a synonym of "philosophically insufficient and superficial". Fink does not doubt the validity and the importance of Husserl's analysis of natural attitude, as well as the results of static and genetic phenomenology. Nevertheless, according to him, philosophy, in the true sense of the word, is just ontology, and not phenomenology in the sense Husserl gives to it. It is for this reason that he devoted all of his research to finding the ontological structure lying at the base of the phenomenal world, moving exclusively on this "underground" level. The surface, on the contrary, is of no explicit interest. Thus, for example, in reference to *Existenz und Coexistenz*, Fink develops his remarks on sociality arguing that the human community consists of two components, a "panic" and an individualist one,[22] which are reverberation, respectively, of man's openness

---

[20] EuC, p. 191.

[21] Ibid., p. 193.

[22] Ibid., p. 173: "'Gemeinschaft' ist nicht einfach ein simpler Gegenbegriff zur Einzelexistenz. Vielmehr gibt es Gemeinschaften, die *Verbindungen*, Zusammenschlüsse von Einzelnen sind, und

to earth and sky. Every concrete community is the reflection of both ontological directions. But what about these concrete communities? What can be said about a social philosophy able to appropriate the results of a cosmological ontology without being itself an ontology (it is still the case in *Existenz und Coexistenz*)? How should a community be *concretely* with regards to the awareness of its openness-to-the world? Fink does not provide an answer to any of these questions.

The same is true for the concept of life as such. On the whole, we find two ways in which Fink understands this notion. On the one hand, above all in the writings on Hegel, the word "*Leben*" does not allude to a specific region of reality, but, in keeping with Hegel's terminology, is only intended as a finite symbol of absolute movement.[23] On the other hand, Fink's work indirectly shows the ontological configuration implied by the specific domain of life, inasmuch as it repeats the thesis Heidegger upheld in his lecture course of winter semester 1929/1930 (heard by Fink): "the animal is poor in world" and "man is world-forming".[24] The hierarchy inside the realm of living beings, according to Heidegger, as well as according to Fink, corresponds to a hierarchy which concerns the openness-to-the-world itself, having its culmination in the ontological role particular to man, in conformity with Heidegger's conception of *In-der-Welt-sein*. But we question again: what about the graduation of life in its concreteness? Also in this case, no mention is made of it.

What Fink's post-war ontology is lacking in, is the attention for genesis, for the way in which real concretely originates in displaying its ontological/cosmological presuppositions. It can be said that Fink offers and explores an interesting clue for a new kind of phenomenological description, without really using it. It is all the more surprising if we compare these considerations with paragraph 7 of the *Sixth Cartesian Meditation*, written in 1932, where the young philosopher focuses on the importance of completing Husserl's method through a new direction of investigation, called "constructive phenomenology", characterized by the union of the *Meontik* and the phenomenal exploration:

> The "objects" – or better, *the objects* – of constructive phenomenology are *not* "*given*". The theorizing directed to them is not an "intuitive having given", is not "intuitive"; but as referral to something that precisely by its transcendental mode of being is *in principle* deprived of "givenness", is "non-given", this theorizing is *constructive*.[25]

---

andererseits Gemeinschaften, die *vor* jeder isolierten Lebensweise liegen, – es gibt Gemeinschaften des Willens und Gemeinschaften des Blutes, – aber auch das nicht in einem unbezüglichen Nebeneinander. Das Verwirrende ist gerade, dass beide Gemeinschaftsformen sich durchsetzten, durchdringen, die eine in der anderen erst das Element ihres Daseins hat".

[23] See for instance Eugen Fink, *Hegel. Phänomenologische Interpretationen der "Phänomenologie des Geistes"* (Frankfurt a. M.: Klostermann, 2007), p. 154–155.

[24] See the second section of Martin Heidegger, *The Fundamental Concepts of Metaphysics. World, Finitude, Solitude*, transl. William McNeill and Nicholas Walker (Bloomington: Indiana University Press, 1995).

[25] VI.CM, p. 56. On this work see for instance the second section of Natalie Depraz and Marc Richir (ed.), *Eugen Fink. Actes du colloque de Cerisy-la-Salle 23–30 juillet 1994* (Amsterdam: Rodopi, 1997); Alfredo Marini (ed.), *La VI^a Meditazione Cartesiana di E. Fink e E. Husserl*, Atti del seminario di Gargnano 18–21 Aprile 1999, in "Magazzino di filosofia", 5, 2001; Martina Scherbel, *Phänomenologie als absolute Wissenschaft. Die Systembildende Funktion des Zuschauers in Eugen Finks VI. Cartesianischer Meditation* (Amsterdam-Atlanta: Rodopi, 1999).

Going back to the first paragraph of this paper, we can say that constructive phenomenology is the method that allows the phenomenological philosopher to explore the originating process of the meontic Absolute through the pre-given wholeness-structures of the world-experience. Its task, in fact, is the specific explanation of objects (of horizons) that "are not given" and are always presupposed in the "world-constitution in progress"[26] to which we already refer "when we phenomenologize" (in Husserl's sense within "egological" and "intersubjective" phenomenology). The adjective "constructive" literally alludes to the "construction" of a constitution we can never reach by following the reductively given transcendental life. A constitution we can only "construct"[27] since it identifies with the non-given[28] (and therefore intuitively unreachable) "beginning" and "end" of this life itself.[29]

But how does this mentioned construction develop? The answer to this question is particularly relevant to our argumentation. Indeed constructive recognition, although grounded in a speculative methodological dimension, does not display itself as a simple *leap* in a speculative domain, for in such a case it would not be a phenomenological method anymore. Rather, as Bruzina suggests, a speculative – constructive and meontic – phenomenology has to concretely be thought of as a "meontic interpretive integration",[30] as "the concrete way the results of exacting concrete phenomenological investigative analysis get redetermined".[31] Its field is consequently a *concrete* one, it is still the genesis of the realm of being-in-the-world, with the difference that such a realm, in accordance with its "wholeness-structures", is constituted by following an absolute and non-intuitive (given) trajectory. Fink's task in the *Sixth Cartesian Meditation*, in other words, is to explore the concreteness from a cosmological/absolute/meontic point of view...exactly the part that will disappear in his post-war philosophy.

Opposed to the first period and in keeping with the following acknowledgment of the priority of ontology, Fink's cosmology does not express the need of a constructive phenomenology any longer. Any reference to constructive phenomenology disappears as well as to phenomenology as such. The ontological presuppositions, by virtue of their antecedence, seem to forbid any attention for the domain *for which* they act as presuppositions. Considered that for Fink the world is the key issue in both stages of his philosophical inquiry, we can also say that such an issue, when it

---

[26] VI.CM, p. 62.

[27] Fink does not provide any clear definition of "construction", but just claims: "The concept of 'construction' must not, however, be understood here in an ordinary sense (such as hypothesis-making). It has no affinity to any kind of 'constructive' procedure as practiced in the worldly sciences, e.g., in mathematics, in paleontology, etc. The *transcendental* title 'construction' is only an allusion to the *modes of referral* on the part of phenomenological cognizing with respect to the theme that is here in question, modes of referral which are still completely obscure in their own special character" (ibid., p. 56).

[28] Ibid, p. 57: "In this context 'givenness' thus does not signify being-at-hand and lying before one, for instance, in the way things are given, are there, as objects of natural worldly experience; but it means *possible accessibility through the unfolding of the phenomenological reduction*".

[29] Ibid., p. 60.

[30] RB, p. 398.

[31] Ibid.

is "filled" with ontological consistency, does not require a proper phenomenological method (a method based on its preliminary constitution) anymore.

In conclusion what we would like to argue is that what is lacking from Fink's second ontology is actually the projection of a constructive phenomenology able to demonstrate the relation that the "forces of the cosmos" can establish with the "ontic" development of the genesis and – among the most important themes of a genetic phenomenology – with the concreteness and the variety of the whole realm of life.

Part II

Ontopoiesis in Ben Okri's Poetic Oeuvre and *A Time for New Dreams* (2011)

Rosemary Gray

Abstract Drawing on Ben Okri's *A Time for New Dreams* (2011), this paper adopts a literary aesthetics approach. Motifs of cosmic conjunctions inform Okri's *An African Elegy* ([1992]1997) and, more especially, *Mental Fight* (1999), subtitled "An anti-spell for the 21st century". The paper will suggest that inherent in this subtitle is the concept of ontopoiesis or the self-induced development of consciousness as propounded by Anna-Teresa Tymieniecka's phenomenology of life (2004). Ontopoiesis thus embraces the self-creative activity of consciousness, which Okri himself has emphasized: "All our creativity, our innovations, our discovery," says Okri in *A Time for New Dreams* (27), "come from being able first to see what is there, and not there; to hear what is said, and not said … the art of intuition". The article draws on a world view of a cosmic totality, perceiving of ontopoiesis as akin to Wole Soyinka's notion of self-apprehension: to "his gravity-bound apprehension of self," as "inseparable from the entire cosmic phenomenon" ([1976]1995: 3). The argument attempts to show that mythic conjunctions are inherent in this cosmogony. The paper uses selected poems from Okri's two anthologies in order to explore his ontopoietic aesthetics. "We ought to conjoin faith in evidence with a need for self-discovery," Okri avers in *A Time for New Dreams* (28). In his poetry ([1992]1997, 1999), a higher state of consciousness or 'illumination' is the basis for life's transitions, wrought largely through spirit awakenings via a retrieval of traditional geo-cosmic horizons. The argument shows that, in the poems in the tellingly entitled *Mental Fight* (1999), such transitions likewise accrue from a conscious reconstruction of the human self, affected by materialism pitted against the forces of the cosmos, and concludes by illustrating Okri's belief in the creative artist's civilizing role through a revitalization of psycho-spiritual life.

R. Gray (✉)
Department of English, University of Pretoria, Pretoria 0081, South Africa
e-mail: prof.r.gray@gmail.com

A.-T. Tymieniecka (ed.), *Phenomenology of Space and Time: The Forces of the Cosmos and the Ontopoietic Genesis of Life: Book One*, Analecta Husserliana 116,
DOI 10.1007/978-3-319-02015-0_5,

Towards an Understanding of the Symbiosis Between Poetry and Ontopoiesis in Ben Okri's Aesthetics and Poetry

Drawing on Ben Okri's most recent non-fiction publication, *A Time for New Dreams* (2011), this paper adopts a literary aesthetics approach. Arguing that poetry is an intrinsic physical human attribute, this Nigerian born writer asserts that "We are, at birth, born into a condition of poetry and breathing" (2011: 3). "Birth is a poetic condition," he continues in the same phenomenological vein: "it is spirit becoming flesh" (ibid.). Then, pointing to the mythic conjunctions inherent in such a non-dualistic cosmogony and the indissoluble cycle of life and death, he avows: "Death is also a poetic condition: it is flesh becoming spirit again. It is the miracle of a circle completed, *the unheard melody* of a life returning to unmeasured silence" (ibid. emphasis added). Although at first seeming radical, such pronouncements recall the holistic philosophies of ancient times: "In Asian and European antiquity … man did, like the African, exist within a cosmic totality, did possess a consciousness in which his own earth being, his gravity-bound apprehension of self, was inseparable from the entire cosmic phenomenon," asserts fellow Nigerian and Africa's first Nobel laureate, Wole Soyinka (*Myth, Literature and the African World* [1976]1995: 3).

Apart from the centrality of holism, two important tropes underpin Okri's aesthetics: the contiguity of life and death *and* music. For Okri, happiness ('the unheard melody') resides in identifying or coming to *understand* some sort of meaning in life. It is predicated on what Heidegger calls being-towards-death (Gadamer 1977). According to Heidegger, being-towards-death is pivotal to achieving *Dasein* or authentic human existence. This suggests that in order to lead a truly meaning life – and thus an authentically happy life – one must be acutely aware of the inescapable fact that one is going to die. In my February 2011 interview with the author, he prefigured "the unheard melody" of the cycle of life becoming death becoming life. He highlighted the synergy of all forms of life and music by sharing *his* consciousness of reality: "A piano with only five keys is a reality. But, if we include all the keys, the white keys and the black keys, this is a different reality." Not only does Okri perceive reality as being equated to a melody, it also "depends on our cultural perception of the keyboard of life".[1] Insisting that "poetry is not just what poets write" (2011: 4), Okri metaphorizes: "Poetry is also the great river of soul-murmurings that runs within humanity." The poet is an agent or channel, merely bringing "this underground river to the surface for a moment, here and there, in cascades of sound and suggested meaning, through significant form" (ibid.). Pertinently, the discussion of Okri's poetic *oeuvre* – which begins with his first anthology, *An African Elegy* ([1992]1997) – attests to a harmonious melding of poetic theory and literary product, to the interdependent web of life.

[1] See R Gray, "When chaos is the god of an era: rediscovering an *axis mundi* in Ben Okri's *Starbook* (2007)." (*Research in African Literatures* 44(1) 2013:128–145).

An African Elegy

A stanza, taken from *An African Elegy* ([1992]1997), *invokes* forces of the cosmos to *evoke* an ontopoietic promise of new life. Alan Singer and Allen Dunn (*Literary Aesthetics* 2000: 134) define phenomenology or the ontopoiesis of life as "an attempt to give a philosophically rigorous description of the essential features of different types of experience". Edmund Husserl (op. cit. 2000: 122) refines this thought, explaining that it is futile to attempt "to understand either the mind or the world in isolation from one another". Husserl's phenomenological method describes the interrelationship between the mind and experience reflected in Okri's "An Undeserved Sweetness", the third stanza of which reads: "And now the world is assaulted/With a sweetness it doesn't deserve/Flowers sing with the voices of absent bees/The air swells with the vibrant/Solitude of trees who *nightly*/Whisper of re-invading the world." ([1992]1997: 14; emphasis added).

This allegorical poem is a synechdochic force; at dusk, the trees surrounding the "malarial slums/In the midst of potent shrines/At the edge of great seas" form a heuristic [enabling discovery by oneself], inspirational catalyst for cultural revitalization. This is not a promise of a sudden revolutionary change, but an almost imperceptible, "nightly" evolutionary change wrought by the natural forces of life. As in so many of Okri's poems, this one invokes the idea of a metaphysical force, a "wind" that not only "... lifts the beggar/From his bed of trash", in the opening lines, but also "... bends the trees" into the poet's "dreams", in its closing lines. Here, a vibrant, vital Nature creates anticipation and opens up multiple and multivalent (many meanings) possibilities of renewal in both the poet's and the listener's heightened imagination. This is poetry that goes beyond the body towards the expanses of the soul in an implicit injunction for an expansion of consciousness. "Poets," avers Okri in an evocation of ontopoiesis, "want nothing from you, only that you listen to your deepest selves" (2011: 6). The poem is transformed into an Aeolian harp, enabling the wind to play "Handel on the set mysteriously", resuscitating a long forgotten memory of "... the long/Hot nights of childhood" so that "Dreams of the past sing/With voices of the future" (14). "Poetry," argues Okri (2011: 5) "is the descendant of the original word which mystics believe gave the impulse for all creation".

That poetry is synonymous with *logos* is open to debate, but Okri does intimate his belief in the ontology of leading a poetic life. "Poetry hints at the godlike in us, and causes us to resonate with high places of being" (ibid.), he declares. And, traditional oral African poetry was, like its Western counterpart, born as a song; it developed as praise chants to be heard, not read. "The voice of poetry," says Afro-Egyptian writer, Adonis (1992: 13), in corroboration, "was the breath of life – body music. It was both speech and that which went beyond speech. It conveyed speech and also that which written speech in particular is incapable of conveying". Poetic images are thus eidetic.

Informed by this tradition, a second poem from Okri's *An African Elegy*, recollecting a Spring fair on a wet Easter Sunday 1988, serves to illustrate the way in which the signifier is not just an isolated word or phrase, but a word or phrase bound to a voice, becoming "a music word, song word" as Adonis puts it. The opening gambit of Okri's "The Cross is Gone", reads: "It was a day of fairs/Yellow music on the wind, feathers/Of dead birds whirling beyond/The green trees." (15) These lines embody an energy replete with cosmic signs. As with those from "An Undeserved Sweetness", they are not merely an indication of a single meaning. Anticipating the climactic lines of the poem taken from The Book of Common Prayer: "Christ has died/Christ is risen/Christ will come again – " (18), the sentient trees "... bore/The features of dying men". The trees look back – recalling for the reader the paradox in the Anglo-Saxon poem, "The Dream of the Rood", in which the cross speaks of its honour of being, at once, the faithful retainer and the bejewelled instrument of death. They too have witnessed "... the sordidness/And the miracles ...". They "were heaving/Their comrades had fallen/The great spirits trapped in their monstrous/Trunks sang .../Songs of white mermaids/Corrupted beyond their time". Looking forward, these trees are depicted as "Breathing lamentations on the unforgiving earth/Into which they will not be reborn. Foretelling of an apocalypse, "The trees sang to us of a darkening age/With mysterious dying/And yellow spirits in the wind" (15).

These lines recall the poem's setting on "the graveyard of the Heath", with its sodden "deceptive grass" and "karmic hurricanes" (ibid.). The transferred epithets in these two metaphors are potent portents of inexorable change. The persona and his partner[2] – described as "the innocent journeyers/Into forbidden zones of dying gods" – reminisce about the "tyranny" of the past, "Noting the character and psychology/Of each surviving tree" (16). They gain the summit of Parliament Hill above the Thames, and gaze upon the cross, dreaming the city of London "... better/Than it dreams itself" (17). Then, in a *volte-face* and in an evocation of the awe-ful prophecies of the book of Revelations, the personae "... saw all the world laid out/Before us in the air", a city perceived "in a moment's enchantment" but "Whose history/weighed down with guilt and machines/Laughed all around us like ghosts/ Who do not believe in the existence/Of men" (ibid.). The dichotomies in the portrayal of the city of London, as full of both possibilities and tyranny, foreshadow the fateful closing lines of the poem, which sombrely declare: "The cross – that cross – is gone" (20). These lines point to or evoke a sense of a loss of faith, showing the organic link between poetry and lament, and indicate that Okri's poetry is at once hymnal and dirge. The symbols of London and the Cross, both rich in ritual and meaning, reflect a transcendent reality. They imply something which, though now lost, remains inscrutable. As the philosopher, Heinrich Zimmer suggested as long ago as 1942, "[symbols] hold the mind to the truth but are not themselves the truth" (625). This is because truth, or what Zimmer terms "the radiance of reality" is universally one and the same, but is mirrored variously according to the mediums in which it is reflected: truth thus "appears differently in different lands and ages" in accordance with "the living materials out of which its symbols are hewn" (ibid.).

[2] The poem is dedicated to R.C. – Okri's long-term partner, Rosemary Cluney.

The emotive and affective content of "The Cross is Gone" and "An Undeserved Sweetness" typify the profound congruence between the poetic voice and the acoustic values of speech in both *An African Elegy* and *Mental Fight*. This congruity is clearly articulated in *A Time for New Dreams* (ibid.): "Words, lighter than air, are as mysteriously enduring as lived time."

Mental Fight

Motifs of cosmic conjunctions thus inform Okri's *An African Elegy* ([1992]1997) but, more especially, his Blakean *Mental Fight* (1999). Subtitled 'An anti-spell for the 21st century', this second collection suggests that the notion of cosmic conjunction is inherent in the concept of ontopoiesis or in the self-induced development of consciousness as propounded by Anna-Teresa Tymieniecka's (2011a "Logos and Life" CX: s.p.) phenomenology of life. Ontopoiesis is seen to embrace the self-creative activity of consciousness: "All our creativity, our innovations, our discovery," says Okri in *A Time for New Dreams* (2011: 27), "come from being able first to see what is there, and not there; to hear what is said, and not said. Above all to think clearly; to be nourished by silence. And – beyond that – the art of intuition."

Section Two of Okri's epic poem for the new millennium, tellingly entitled "Time to be real", opens with an ontopoietic injunction: "Allow uncontemplated regions/Of time to project themselves/Into your sleeping consciousness,/Inducing terror, or mental liberation" (4). Drawing on varying responses to death-confrontation – despair, emptiness, enlightenment – the poet enjoins us to "explore our potential to the fullest"; to dispense with "our fear of death"…"[h]aving gained a greater love/And reverence for life/And its incommensurable golden brevity" (ibid.). Okri's idea of love and being-towards-death can here be seen as an a priori to embracing his anti-spell. He does not aim at realizing anything, at solving life's difficulties[3] but rather at awakening a realization of the marvellous: "… so long as poetry sends our minds into realms of gold and questions, and touches our deep and tender humanity," Okri (2011: 5) states, "then it will always be a force for beauty, for good, in the world, neutralising slowly the noise of guns and hatred." In this sense, the wind and the sentient trees are metonymic of poetry itself in the two poems from *An African Elegy* discussed at the beginning of this paper.

Prose too can have this same transformative quality. In conversation with Okri, I raised the import of an aphorism from his *Songs of Enchantment* (1993: 287). From the silence of momentary "unblindedness", Azaro's blind father, in conversation with his son, a 'spirit-child', an *abiku*, is moved to muse that "The light comes out of the darkness". Okri elucidated: "We either return to blindness or make a leap

[3] See Aristotle's *Metaphyics* (A,2,982). In Tymieniecka's words: "Aristotle emphasizes that that philosophy does not aim at realizing anything. It proceeds not from a desire to solve any life difficulties or arguments but from marvelling about the simplest of things, whose reasons escape us at first." ("The Pragmatic Test of the Ontopoiesis of Life," XXVIII: 6, 2004).

to a now. There are only two options: retreat to the cave or leap. This is primal. We have inherited from the ancestors. This must take us forward not back!" In *A Time for New Dreams* (2011: 27) Okri muses: "How to make those intuitive leaps that can transform humanity, how to make this mysterious faculty available to all – this will be the turning point in the future history of civilization." The implied call is for more evolved generations to come to live more responsibly, with greater sensibility.

"We ought to conjoin faith with a need for self-discovery," Okri (op. cit.: 28) exclaims, expressing true apprehension of self (to deploy Soyinka's terminology ([1976]1995: viii), mentioned earlier). True self-apprehension or self-individuation – also referred to as ego-development – correlates not only with the Zen notion of mindfulness, but also with ontopoiesis, as seen in lines from Section Two of Part Seven in *Mental Fight*, which read: "The world is not made of labels./The world, from now on,/Will be made through the mind" (55). "Practising mindfulness in Buddhism means to perform consciously all activities, including every day, automatic activities such as breathing, walking". It is to assume the attitude of "pure observation …" (Shambhala 1991: 145). This mindful infiltration is, at once, performative and transformative in that language brings man and his world into conscious existence. This emerges in the next lines which explain what Okri conceives of as the optimum operation of mindfulness. He admonishes us to "Accept no limitations to our human potential" as, in his opinion, transformation is wrought "Through great dreaming, great loving/And masterly application" (1999: 55) of spiritual values to our lives. Okri's language is charged with imaginative creativity. It is neither mimetic nor expressive, but is "pure poetry", characterized as a product of a synthesis of thought and imagination. Such language shapes consciousness and perception. "Poetry," says Okri (2011: 5), "incarnates that which shapes, changes, transforms."

In *A Way of Being Free* (1997: 2), Okri declares: "Poets seem to be set against the world because we need them to show us the falseness of our limitations, the true extent of our kingdom." In the poem preceding the one just discussed and in ontopoietic language that is simple, yet profound, calling a phenomenology of life into being, the poet emphatically asserts: "The mind of humanity is such a force./New worlds wait to be created/By free minds that can dream unfettered,/Without fear, turning obstacles/Into milestones towards luminous glories"(1999: 54). Comparably, Okri maintains in *A Time for New Dreams* that "We ought to step out of our old, hard casing. We think we are one kind of people, when in fact we are always creating ourselves. We are not fixed. We are *constantly becoming*. Constantly coming into being. Writers hold out a mirror to the bright visions of what can be" (2011: 18; emphasis added). In a similar vein in regard to what Tymieniecka terms, even more precisely, "beingness-in-becoming" (2004: 17), she has elaborated:

> Taking up the Kantian aspiration to a definitive critique of reason – albeit with a Husserlian twist – as we descend the phenomenological ladder of objectivities to the very end, but without postulating "things in themselves," the ontopoiesis of life reaches the incipient point of intentionality and unveils the prepredicative level of the becoming of life.

Throughout *Mental Fight*, Okri is searching for true conjunctions between the individual and the world, between national hermeticism and absorption into a cosmic unity beyond time. The above excerpts reiterate Okri's affirmations noted in the earlier discussion of *Mental Fight*. Part V opens with an implied *carpe diem* assertion: "Now is a material event/It is also a spiritual moment/And the blinding light of the real/Can pierce through and tear/Asunder the unreal./Every moment thus carries/The ordinary and the monumental" (1999: 7). Unblinded, the "celluloid" stripped from our eyes, "… behind it all we see things/As they could be" (ibid.). *Mental Fight* is, like Okri's first collection, clearly predicated upon ontopoieisis or the self-induced development of consciousness. Quoting from two other pieces in this collection, the new millennium is "richly potent" (18), because humanity is on the threshold, on the "cusp" of new beginnings, "transcending the political/Hinting at the evolutionary" (16). Clinching his argument, Okri affirms that "In time's ovulation/We are now at a rare intersection/That magic favours" (19).

The seventh or final poem in a section subtitled "Signs from the old times" points to the conjunction between the temporal and the divine, or between his/her story and poetry: "How often have great minds/In the past prayed, and wished/ For better favoured moments/In time to unleash their best/Gifts of humanity?" (20) The poet responds to his own rhetorical question, unequivocally declaring the ripeness of the time for spiritual awakening: "This is one such conjunction:/It fills the heart with too much humility/And amazement to behold." (20) And the poem closes with a plea: "… we must behold it, with minds calm,/With aspirations clear,/And with a smile in the soul/That only those fortunate people have/Who find themselves at the right time,/At the perfect mythic conjunction/That is also a living moment,/A moment lived through." (ibid.)[4] The injunction encapsulates Soyinka's notion of self-apprehension as the key to cultural and personal liberation. It prefigures the thrust of Okri's *A Time for New Dreams* (2011: 95) in which the entreaty is to "Let life inspire you, and teach you always how to be free, and to encourage freedom in others, if they so desire. All of humanity is really one person. What happens to others, affects us. There's no way out, but up. Let's all rise to the beautiful challenges of our age, and rise to our true mysterious luminosity". It is this quest for individual and communal freeing of the self that runs passionately through the writings of both Okri's anthologies of poetry.

Described as either "An anthem to mark the end of an age, or a hymn for the future" (dust cover), *Mental Fight* is shot through with a synthesis of celebration and heedfulness. It is described on the dust cover, which quotes *The Times*, as "An angry, hopeful, weary, wary, epic reveille to the human spirit".

---

[4] See R Gray. "Mythic conjunctions in transit: Ontopoiesis in Ben Okri's *An African Elegy* (1992) and *Mental Fight* (1999) and Wole Soyinka's *A Shuttle in the Crypt* (1972)". (*Journal of Literary Criticism* 28(4) 2012: 25–37).

Conclusion

The argument has attempted to show that ontopoiesis is inherent in Okri's non-dualistic cosmogony. It employs Okri's ontopoietic aesthetics as seen in *A Time for New Dreams* to interpret selected poems from his two twentieth-century anthologies of poetry. (His latest anthology was published in March 2012.) "We ought to conjoin faith in evidence with a need for self-discovery," Okri (2011: 28) asserts in his latest non-fictional work. In his poetry ([1992]1997, 1999), a higher state of consciousness or "illumination" is proffered as the basis for life's transitions, wrought largely through spirit awakenings via a retrieval of traditional geo-cosmic horizons,[5] that is, via the crucial link between the soul and the cosmos. In the poems in *An African Elegy* ([1992]1997) and the tellingly entitled *Mental Fight* (1999), borrowed from William Blake's visionary 'Jerusalem', such transitions accrue from a conscious reconstruction of the human self, affected by materialism, pitted against the forces of the cosmos. Intuiting Soyinka's affirmation: "(For let it always be recalled that myths arise from man's attempt to externalise and communicate his inner intuitions)" ([1976]1995: 3), Okri implies that the onus is on each one of us to "enter the new period" in our own way and to solve its questions for ourselves. In Blakean ontopoietic vision, he writes: "We need to become more adaptive mariners" (1999: 48) "... to make/This time a waking up event/A moment of world empowerment/To pledge, in private, to be more aware/More playful, more tolerant, and more fair/More responsible, more wild, more loving" (1999: 15) to be "Awake to our unsuspected powers, more amazing" (ibid.).

Aligning himself with Mario Vargas Llosa's Nobel acceptance speech (7 December 2010: 11), in which he credits writers with the ability to assist in preventing us from retreating "into the savagery of isolation", insisting that "... fiction [and by extension, poetry] is more than entertainment, more than an intellectual exercise that sharpens one's sensibility and awakens a critical spirit", Okri also attributes the transition from barbarism to civilization to creative artists. "Literature and civilization are different but not different," he stated in my early 2011 interview. He elucidated: "The society that gave rise to the Divine Comedy flowered as a renaissance; the society that gave rise to Shakespeare led to a greater civilization. Literature does not create it, but there is an inspirational link, like mercury and the alchemist's stone. Literature shows the presence and reveals and enhances a psychic strength, pulls it out to reflect a greater civilization." In *A Way of Being Free* (1997: 2), Okri credits creative artists with this same civilizing capability through a revitalization of one's psycho-spiritual life:

---

[5]A-T. Tymieniecka explains in "Logos and Life," (*Analecta Husserliana*, 110: 1) "Thus logos hitherto hidden in our commerce with earth is revealed in its intertwinings with the cosmos through the trajectories of the phenomenological/ontopoiesis of life. The crucial link between the soul and the cosmos, in the new geo-cosmic horizon, is thus being retrieved."http://www.springer.com/philosophy/philosophical+traditions/book/978-94-007-1690-2/ (accessed 6 November 2011).

> The poet turns the earth into mother, the sky becomes a shelter, the sun an inscrutable god, and the pragmatists are irritated.... The problem is with those who are frightened of the rather limitless validity of the imagination, frightened of people who continually extend the boundaries of the possible, people who ceaselessly reinvent existence; frontiers people of the unknown and the uncharted.

Poetry thus enables one to think beyond the canopy; it affords the opportunity to experience the transcendence of truth. At the risk of overstating the case, Okri's aesthetics and poetry coalesce in a reflective intelligence that should, as Tymieniecka (Low 2011: 32) reportedly avows, "have an important role in human life and cognition". It is, as the Peruvian laureate (2010: 11) asserts, "an absolute necessity so the civilization continues to exist, renewing and preserving in us the best of what is human".

References

Adonis. 1992. *An introduction to Arab Poetics*. Cairo: The American University Press.

Gadamer, H.-G. 1977. *Philosophical hermeneutics*. Berkeley: University of California Press.

Low, D. 2011. Merleau-Ponty's enchanted nature. *Phenomenological Inquiry* 35: 49–90.

Okri, Ben. [1992]1997. *African elegy*. London: Vintage.

Okri, Ben. 1993. *Songs of enchantment*. London: Jonathan Cape.

Okri, Ben. 1997. *A way of being free*. London: Phoenix.

Okri, Ben. 1999. *Mental fight*. London: Phoenix.

Okri, Ben. 2011. *A time for new dream*. Chatham: Random House.

Shambhala. 1991. *The Shambhala dictionary of Buddhism and Zen*. Boston: Shambhala.

Singer, A., and A. Dunn. 2000. *Literary aesthetics: A reader*. Oxford: Blackwell.

Soyinka, Wole. [1976]1995. *Myth, literature and the African world*. New York: CUP.

Tymieniecka, Anna-Teresa. 2004. The pragmatic test of the ontopoiesis of life. *Phenomenological Inquiry* XXVIII: 5–35.

Tymieniecka, Anna-Teresa (ed.). 2011a. *Astronomy and civilization in the New Enlightenment*, Analecta Husserliana, vol. CVII (8). Dordrecht: Springer.

Tymieniecka, Anna-Teresa (ed.). 2011b. *Phenomenology/ontopoiesis retrieving geo-cosmic horizons of antiquity*, Analecta Husserliana, 110. http://www.springer.com/philosophy/philosophical +traditions/book/978-94-007-1690-2/. Accessed 6 Nov 2011.

Vargas Llosa, Mario. 2010. *Nobel Lecture: In Praise of Reading and Fiction*, Trans. Edith Grossman. http://www.nobelprize.org/nobel_prizes/literature/laureates/2010/vargas_llosa-lecture_en.html

Cosmic Order and Exoneration of the Beautiful: Visions of the Problem in Contemporary Philosophy

Ella Buceniece

The fact that Cosmos is comprehendible, that it complies with certain regularities, is an awesome one. This is the most salient feature of God, revealing Himself in the harmonious order of being. When I inquire into some theory or other, I keep asking myself: if I were God, would I have made the same arrangement of the world.

Albert Einstein

Abstract Cosmos for the Greeks stands both for the orderly arrangement and for the manifestation of the beautiful, allowing to capture and to remember the real world: physics, metaphysics, anthropology and aesthetics have been closely bound together. However, with the passage of time these spheres have fallen asunder and have undergone re-grouping: 'cosmos' and 'the beautiful' are separated by a cosmic distance nowadays. The beautiful (pulchrē) (St. Thomas Aquinas) has been marginalized, placing the 'decorum' as an artificial ornamentation not related to the truth, to occupy the central place. Just like 'cosmos' has turned from an orderly beauty of the Universe into 'cosmetics', designed to overcome the natural order of things.

The attempts to recapture the orderly arrangement and the unity of both notions – that of the cosmos (micro- and macro-) and of the beautiful (in its ontological sense) by the present-day philosophy, and to exonerate their inherent unity, forms the central core of the present investigation.

E. Buceniece (✉)
Institute of Philosophy and Sociology, Latvian University,
Akademijas laukums, 1, Riga LV-1940, Latvia
e-mail: e.buceniece@gmail.com

A.-T. Tymieniecka (ed.), *Phenomenology of Space and Time: The Forces of the Cosmos and the Ontopoietic Genesis of Life: Book One*, Analecta Husserliana 116,
DOI 10.1007/978-3-319-02015-0_6,

"Philosophy views the world as a cosmos of being, and thereby the world becomes an objective world, as against our conceptions of it. Thus, philosophy begins as a cosmology, applying its theoretical interest to the corporal nature" – so writes Edmund Husserl in his work "Crisis of European Humanity and Philosophy".[1] Rightly understood – he goes on – the initial meaning of it (philosophy) means nothing more than a universal science – a science about cosmos, about the unity or singularity of everything-that-exists (Allenheit alles Seienden). However, it is only with the Greeks that we are confronted with a universal ("cosmological") life-interest in the manner of a novel "theoretical" form, which shows forth – due to its incipient causes – by way of essentially new philosophical and philosophy-related aggregate of sciences (mathematics, astronomy).[2]

Thus, the first understanding of the cosmos coming from the Greeks, according to Husserl, apprehends the cosmos as the wholeness of the world, as a universal order and orderliness, revealed by the theoretical placement, and simultaneously it also comprises the universal (field) "cosmos" of scientific cooperation.

Another feature of the understanding of the cosmos coming from the Greeks is concerned with the idea of ornamentation, of harmony as the visible manifestation of the "beautiful". This aspect is stressed by H. G. Gadamer in his work about the topicality of the beautiful: "The regular orderliness of the sky presents us with the most salient visualization of the order, that nature may offer. The passage of the years, the periodical changes of the monthly and daily rhythms provide experiences of continuity, of safe orderliness for our lives".[3] Philosopher and authoress Iris Murdoch also looks for the understanding of the beautiful in the direction of the ancient world, accentuating, as she does, the more anthropological facets. In one of her novels she makes a protagonist say:

> Plato says, that out of all the spiritual values that are concealed, but which become apparent by looking deeply into the human soul, the one most fully visible here on earth is beauty. It is only with difficulty that we can apprehend reason, wisdom. But each one of us is capable of simply viewing the external beauty, and there is no need for any special systems to apprehend and to love this beauty.[4]

Two main sets of problems become apparent in connection with the initial understanding of the cosmos as a universal orderliness. The first one is concerned with the macrocosmic domain, and is represented by the classical cosmological teachings. Another one belongs to the microcosmic domain and is anthropologically orientated.

---

[1] Edmunds Huserls, Eiropeiskās cilvēces krīze un filozofija. – Buceniece, E. Saprāts nav ilūzija (Rietumu filozofija modernisma situācijā). (*Edmund Husserl. Crisis of European Humanity and Philosophy. in: Ella Buceniece, Reason is not an Illusion. Western Philosophy in the Situation of Modernism*) Rīga: Pētergailis, 1999, pp. 236. (*in Latvian*).

[2] Ibidem, p. 239.

[3] Hanss-Georgs Gadamers, H.-G. Skaistā aktualitāte. Māksla kā spēle, simbols un svētki (*Hans-Georg Gadamer. H.-G. The topicality of the Beautiful* (*Art as Play, Symbol and Festival*)), R. 2002., p. 48. (*in Latvian*).

[4] Airisa Mērdoka, Vienradzis. (*Iris Murdoch, The Unicorn*) Rīga, Zvaigzne ABC, 2007, p. 106. (in Latvian).

Originally both domains – those of the macrocosmic and the microcosmic order – were viewed as structurally identical, and the distinctions between them were not sufficiently differentiated.

The essential link between macro- and microcosmic domains in a specific manner of articulation, is first undertaken in Kant's philosophy. Cosmic discourses in Kant's philosophy are introduced in a variety of ways. First of all, it is his cosmogonic hypothesis of the pre-critical period; the next stage is connected with the cosmological ideas of the critical period. In the works of the critical period Kant views the cosmos through the capacity of reason to perform a synthesis of the world in its universality. The transcendental cosmological ideas, according to Kant, are related to the "absolute fullness within the synthesis of appearances" that he designates as the concept of the world. But "the concept of the world itself is just an idea related to the appearance i.e. – the empirical synthesis, and is thus an unconditional wholeness. In the capacity of the absolute wholeness within the synthesis of all possible things, it leads to the ideal of pure reason, which – in turn – is completely different from the idea of the world, though being related to the latter".[5] And the antinomies of pure reason, the clashes taking place within the reason itself, reveal the transcendental basic regularities of seemingly pure (rational) cosmology. The presence of cosmological ideas, in distinction from transcendental paralogicality, produces one-sided appearances, and is concerned with our idea of the subject of thought. It places reason within its objective synthesis and offers possibility – according to Kant – for a completely new quality of reason – i.e. – the natural antithetics, which – naturally and inescapably – becomes the abode of reason, and is thus preserved from naked one-sided appearances, either succumbing to skeptical despondency, or sticking to dogmatic obstinacy, and refusing to listen to the arguments of the other side, and to evaluate its justifications. "Either of these positions – Kant considers – bespeak of the death of a wholesome philosophy, though the first one (i.e. – the skeptical despondency) at least could be called euthanasia of pure reason".[6] Thus, natural antithetics is a precondition of the existence of a wholesome philosophy. Thirdly, the "cosmic order" (A. T. Tymieniecka's term) appears in Kant's philosophy by way of providing a nexus between all three of his Critiques, while the last one – the Critique of Judgement – is positioned as a kind of synthesis, obtained with the help of the notions of "the beautiful" and "the noble": "Not only Kant's understanding of the subject, but also his Ideas of Reason belong to symptomatic matters, and are concerned with the "hysteric" aesthesis and its aesthetics". A similar subjectivising split is taking place within aesthetics itself – by the phenomenon of the lofty (sublime) and the experience of the split of the supra-sensitive Idea. On the other hand –a "softening" of this split takes place with the help of the beautiful and of teleology within the capacity of judgement – this is the view of Latvian philosopher Ansis

---

[5] Immanuels Kants, I. Tīrā prāta kritika. (*Immanuel Kant, Critique of Pure Reason*), R. 1931., p. 292. (*in Latvian*).

[6] Ibidem, p. 291.

Zunde.[7] This is where the connotation of the cosmos as ornamentation and the visual experiencing of the beautiful becomes apparent in Kant's philosophy. Thus, Kant's notion of the wholeness of the world preserves both the objective and the subjective dimensions and the non-univocal tenuous connection between macro- and microcosmos.

The fourth type of the cosmos discourse enters Kant's philosophy through his anthropology and the cosmopolitical dimension; this accentuates the essential belongingness of humans to the two worlds, being two forms of expression of human being that has acquired (by now chrestomathic) designation, about two kinds of fullnesses and two universalities – the starry heaven above and the moral rule within.

The division between two cosmoses – the macro- and the micro- ones – finds its way into philosophy and makes its abode there in two ways. First of all – when the fullness of the world breaks up into the world of things and into the field of cogitation (Descartes); or alternatively – as in modern phenomenological conceptions – when the world enters into the human being through his/her mind, or when the world gets concentrated in the microcosmos, thus losing any differentiation between micro- and macrocosm. Lately, it is exactly the human body that has become the starting point of human existence, or – to put it in another way – when the world manifests itself in human being via corporal existence (M. Merleau-Ponty) as the temporal centre of the world.

The second of the above divisions is discernable as a point at which the theme of macrocosm as world's universality tends to drift away from the concern of philosophy altogether.

One of the few modern authors who keeps cherishing the macrocosmic theme as a significant naturally/metaphysical idea of the world is Alfred Whitehead, whose work "Adventures of Ideas" contains a whole chapter dedicated to "Cosmologies". Here the author follows through the development of the cosmological conceptions as from the Greek antiquity, and draws some very telling conclusions. The first one: "This unshakable belief in the order of the world with its variegated history – Plato, Epicurus, the Gnostics, Alexandrian theologians, the rationalists of Antiochy and Mopsutia Maniheans, Augustine, Calvin – at last passes into the first phase of the modern age, beginning in the seventeenth century by way of an unassailable premise about the existence of a certain natural order, being open in each of its parts to human understanding.[8] The author concludes that the end-seventeenth century cosmological conceptions of I. Newton, W. Leibnitz and J. Locke have left a profound influence on the contemporaneous thought and have kept influencing the subsequent two centuries. Leibnitz's particular position is specific in that his monadology started to approach the cosmological problematics from the point of

---

[7] A. Zunde, "Gaisma no tumsas dzīlēm – filozofija "psihotiskā kosmosā". (Light from the abyss of Darkness – Philosophy within "Psychotic Cosmos": Philosopher between tradition and experience. In memoriam Pēteris Laizāns. Rīga, Zinātne p. 95 (in Latvian).

[8] Alfred Vaithed, Izbrannije raboti po filosofii. (*Alfred North Whitehead, Selected Philosophical Works*), Moskva, 1990. p. 530. (*in Russian*).

view of the subject, not just by way of obtaining objective knowledge. "Leibnitz was the first among the most important philosophers, who accepted the modern conception, and at the same time was aware of the difficulties it raises. He courageously eliminated God from the framework of his teaching (by inquiring not what the world of atoms is like for the contemplating intellect, but by attempting to understand what does it mean to be an atom. – E. B.) God and each individual monad are bound together in perpetual interaction. Thus the indirect connection between monads is augmented by direct connection of each one of them with God."[9] Whitehead's review of the development of the cosmological ideas is crowned by a conventional interpretation of the Laws of Nature.

The cognitive principle concerning the natural cosmic order, for a prolonged time – up to the Copernican turn – has been based on the notion of the Earth being an unmovable centre, a preconceived point of departure; yet with the coming of the theory of relativity and the changing of the whole system of the departure-points, the whole comprehension of the cosmic order was following suit. However, in spite of the relativistic changes permeating the spatio-temporal understanding, the Earth-orientated intentionality of mind remained intact. This new state of affairs – the changed cosmic, cultural and human situation – has become a subject-matter of phenomenological reflection looking for a new Archimedian point of support. The human body has become such a leverage point.

Husserl's approach to the human body as to the absolute zero and unmovable centre of the kinestetic space, is characterized by Paul Virilio as "an astonishingly abrupt severance of the philosophical and physical threads. The ancient geocentric view turns into Husserlian egocentricism. The centrality of the Earth – the chief point of coordinates of our ancestors – is substituted with a new central point: the living, here-and-now present being."[10] I have discussed phenomenological comprehension of space in connection with body in an article "Thinking with the skin: the Problem of Space in modern Philosophy: E. Husserl and A-T. Tymieniecka." – *Analecta Husserliana* LXXIX, Kluwer academic publishers."

A turn is being performed in order to restore the ties between philosophy and physics: it is no longer a human being in the world, or a microcosm as a kind of reflection of the macrocosm; instead – the world enters the human being, penetrates his/her mind and body. Paul Virilio in his work "Open Sky" quotes Merleau-Ponty: "Our own body is in the world as the heart is in the organism: it keeps the visible spectacle constantly alive, it breathes life into it and sustains it inwardly, and with it forms a system".[11]

One of the most fundamental theories incorporating the cosmic dimension into the field of philosophical cogitation has been worked out by A-T. Tymieniecka. The author tackles this theme within the context of a broad panorama of present-day philosophical, scientific and humanitarian views, because the development of philosophy is

---

[9] Ibidem, p. 533.

[10] Paul Virilio,"Un monde s'expose: fin de l'historie, ou fin de la geographie?" – Le monde diplomatique, August 1997, p. 17.

[11] Paul Virilio, Open sky. London – N.Y.: Verso, 1997., p. 28.

always a response to the challenges of the age. Tymieniecka acknowledges that the crisis of present-day civilization is concerned with the scientific creativity being realized, as it is, by way of technological improvements, and this requires that philosophy provides for the implementation of previously non-existent solutions. This is the concern of the people of all geographical regions, regardless of the geo-cultural fragmentation. The fundamental relationships of humans and nature have become dislodged in the result of achievements of nuclear technologies, genetics, environmental crises, species extinction and other factors. Human being and nature have taken up a belligerent stance, and the very life as such is endangered. Philosophy is called upon to renew the balance, therefore the author proposes to re-formulate the very notion of "nature", providing it with a new essence within a new contextual framework – within the framework of the philosophy of life and life-situation. In her work "Impetus and Equipoise in the Life-Strategies of Reason. Logos and Life", Book 4 (First part, third sub-part) she proposes a new formulation of the concept of nature-life which is open to the cosmos and culture.[12] By revealing the individualization of the logos in life, she reveals the rationale of the cosmos, where the "cosmic order" occurs in all phases of individualizing life, as it is transcribed in different keys. Cosmic order is manifested not only within cosmic spacing and scanning, but also in the order of the circuits of the psyche with its "lower levels", and the "higher" circuits of conscious life and intellect.[13] At the end of part 6 of the book Tymieniecka advances her "great vision of the all", and enters into a search for the "golden measure" in the form of "a new enlightenment" concept, based as it is on the three interconnected lines of logos and life development. These are the following: the vital, the Dionysian, and the Apollonian logos. The last two are of Greek extraction, yet their semantics is not limited only with the ancient meaning. The author marks the difference: "But in contrast to those Ancients, our explanatory insights, sparks, hints, formulations do not come out of speculative thinking about principles and final causes. They are working out in a concrete investigation, in the scrutiny of the real-first, in its subterranean workings, and later on, in the further manifest expansion of its circuits.[14] A concept appears in this citation, which is evolved by Tymieniecka in the next investigation, i.e. – she offers the concept of "the real" by way of elaborating "visions of the real" as securing great advances in contemporary philosophy and science. The author maintains that the foundation for the disentanglement of the very nature of reality is to be found by way of considering the flux and stasis. Such a stance is to be found in the three significant differentiations made by the Greeks: "in consideration of the media of becoming, of the first generative elements, and of composition amid everlasting transformation".[15] These roots – the author maintains – have nourished a whole lot of fascinating enigmas throughout the

[12] Anna-Teresa Tymieniecka, A-T. Impetus and equipoise in the life-strategies of reason. Logos and Life., Book 4, Analecta Husserliana, Vol. LXX., Kluwer academic publishers, 2000., p. 97–104.

[13] Ibidem, p. 64.

[14] Ibidem.

[15] Anna-Teresa Tymieniecka, Inspirations of Heraclitus from Ephesus Fulfilled in Our New Enlightenment. – Analecta Husserliana, Vol. CX/Part I, Phenomenology/Ontopoiesis Retrieving Geo-cosmic Horizons of Antiquity., Springer, 2011., p. 3.

history of Western philosophy up to the present day. One of the most salient examples of this kind of jigsaw-puzzle assembling is concerned with the manifold interpretations of the teaching of Heraclitus from Ephesus. Tymieniecka offers a very original interpretation of Heraclitus' thought, by formulating her conception of the New Enlightenment, designed to reach "the deepest level of the all-underlying unity of life, man and the cosmos".[16] The author accentuates a Heraclitean original approach – so different from his contemporaries – in that he singled out fire as the most fundamental element of cosmos, symbolizing – in one single wholeness – both the physical reality and the logos, for fire as an element is self-illuminative. Thus, logos is to be viewed as a "true account of the nature of things", but it also simultaneously reveals things, because "nature likes to hide itself". Only when in a state of enlightenment these two things come together, do we reach the complete sense of the logos."[17] The true nature is thus revealed as the relationship between the flux and the stasis, producing harmony within the disharmony of All. Flux remains an everlasting state of All, and the harmony perdures in its transformations. This is the deepest level of the recognition of things and nature. Heraclitus conceives of the logos, and of the illumination that it yields for the recognition of the deepest level of things and nature, as the underlying unity of the life of the cosmos and human life".[18] The author, by way of commenting on the Heraclitean idea about the correlation of the individual psyche's connection with the wider realm of the entire cosmos, concludes, that the human soul, growing without limits in its logos, is a microcosm interchangeable with the all-engulfing macrocosm.

Tymieniecka acknowledges, that it is exactly Heraclitus' teaching, that has provided the universal blue-print for her *mathesis universalis* of the phenomenology of life in continuing the search for the innermost depths of reality in logos.

I hereby offer a short insight into Tymieniecka's approach. Comprehension of reality requires understanding and recognition of the very phenomenon of reason, of reasoning, of cognizing. And, according to Tymieniecka, it envisages a new critique of reason in the new post-Kantian and post-Husserlian period, as a new Enlightenment of Reason – an Enlightenment allowing reason to emerge as all-illuminating logos: "In fact, the expression of scientific rationalities, and, in particular, their corroboration has imminently extended into a sphere of wonder and troublesomely dispersed queries carried on throughout centuries, a sphere that in our age has been recognized as existential counterpart of human reality, namely, the skies, that is, the heavens".[19]

However, the critique of reason as undertaken by Tymieniecka, in distinction from that of Descartes, Kant and Husserl, does not start with the initial approach to reason itself. Her point of departure is – "I live, therefore I am", because it is exactly the unique experience of life, inward/onward orientated as it is, that permits to encompass life in its ontopoietic process. Life is a flux within its various levels and forms of individuation ("unity of life, man and the cosmos"), that, correspondending to

---

[16] Ibidem, p. 4.

[17] Ibidem, p. 5.

[18] Ibidem.

[19] Ibidem, p. 7.

facts of subjective experience and observable to others, reaches the objective truth, the truth of things, the "naked truth", or the logos of each and all.[20] The author rejects Husserlian intentional structure of the mind, for this prevents approaching the "things themselves", while our cognition starts with the apprehension of the "vital order" of our existence and continues as recognition of the fullness of human experience in creative insights illuminated by reflection. Human life, being one of the realities alongside others – such as the Earth and the cosmic sphere, – the human condition as such, is the factor that produces the unity of everything-there-is-alive and its corollary – the cosmos. The stage for the acts of life is the human soul, encompassing the mind or consciousness as a kind of instrument for the projection onto and negotiation with life's horizons: "We can say that in this way the self-prompted and self-oriented human soul reflects the universal ordering of the All: from her originary ties to the earth's soil, to the congenital influences exerted on life on earth by the firmament".[21]

By providing an original interpretation and accentuating of the Husserlian teaching about cosmos as logos, and also by encompassing it within present-day modes of thinking with the help of the conception of the New Enlightenment, A-T. Tymieniecka shows forth the possibility of renewal of the tenuous ties between philosophy and physics. She demonstrates a new comprehension of reality in opposition to the ever-growing chaos, now turning even into chaosm (U. Eco) and asserting, that unity (after all) is possible – both inwardly as the content of our souls, and outwardly as things visible and invisible.

Latvian philosopher Konstantins Raudive, working in the philosophy-of-life manner, has also touched upon the theme of the Universe. K. Raudive (born in 1909 in Latvia, died – 1974 in Kronzingen, Germany) is a Latvian philosopher, essay-writer, translator (from Spanish) who has won international recognition on account of his novels, collections of essays, and especially due to his research on parapsychologigal phenomena. He has published 33 books in Latvian, German, English and Italian languages, among them a novel *Helligkeit und Zwielicht* (1967), collection of essays *Der Chaosmensch und seine Überwindung* (1951), etc. His parapsychological investigations have been especially concerned with the audial sphere – the paranormal voices being interpreted as coming from the "beyond" and providing contact with the souls of the deceased. The term "Raudive Voices" has become stock-in-trade in the present-day esoteric literature. In 1969 K. Raudive was awarded the highest prize of the Parapsychological Society of Switzerland, he was also elected a member of Tiberian Academy of Rome and offered a seat of professorship in parapsychology. His name is cited in *Lexicon der Weltliteratur* and also in *Who's Who* reference volumes in the USA. Raudive has discussed parapsychological phenomena in several books in German and in English (*Unhörbares wird hörbar*, 1968; *Überleben wir den Tod?*, 1971, *Breakthrough: An Amazing Experiment in Electronic Communication with the Dead*, 1971). These books have stimulated vigorous discussions and polemics up to the present time.

[20] Ibidem, p. 8.

[21] Ibidem, p. 11.

Thematically Raudive's essays and investigations may be grouped around two concepts: the personal and the ultra-personal, (corresponding to the heading of one of his essays.) The personal is characterized by the author with the help of several notions – it is the "centre of life" and at the same time it "radiates onto the whole existential circuit". Existence is a hermeneutically sealed circle containing thousands of cul-de-sacs, and cognition moves along by establishing the "I", the personal by way of simultaneously apprehending and forming ties with the beyond".[22] (Raudive K. *Divējādi dzīves ceļi*. Vesterosa: Dzintars, 1952: 26.). The personal for Raudive is the human personality of unmolested wholeness or of the total humanity; the number of such people is diminishing nowadays. Their place is taken by a type of human, whom Raudive designates as "people of despair", or by using M. Unamuno's concept – *desperados*. We are used by now to such code-terms for the characterization of modern culture as resentment (F. Nietzsche) and spleen (Baudelaire); Raudive proposes to augment this list of designations with the soul-searching Spanish word of *desperados*. He discusses "the personal" both in individual mode, and also thematicizes it in an inter-subjective modality as friendship, love, loneliness.

The concept of the "ultra-personal" constitutes the next stage in K. Raudive's thought. It is marked by such notions as ultra-being, spirit, culture, humanity, God, death (as a metaphyysical not a physical fact), life, universe. We notice that K. Raudive associates ultra-personal not only with the spiritual or with natural phenomena, but considers these spheres in a unified manner, recognizing both the existing distinctions, and also their amalgamation, confluence, in other words – he sees life as a way. Thus, for example, he does not consider life as a purely metaphysical category, instead – he amalgamates it with the mystery of life, and this is neither spiritism nor evalutionism. Rather – it is the general personal and universal, individual and cosmic "existence-together". This may be inferred from Raudive's use of such a term as "culture of life", or from his thesis that the origin of a personality does not mean severance from the wholeness, from the universal, from the cosmos, because "the human being by elevating into spiritual purity tends to get united with his original state – the ocean of humanity".[23] Raudive does not consider God and belief in God in categories of absolute spirituality, but speaks about the "mystery of faith", or "instinct of faith": "After all, it is not so very important how this cosmic order is designated – as a universal law, or instinct, or Spirit or even God. It does exist within its unfathomable state of ultra-being, it is the foundation of the first movement and development".[24] Life and death are connected in similar fashion, for there are two ways in cosmos – one of them leading upwards (life), another one – downwards (death). God is an invisible harmony well above the visible, perceptible harmony. Therefore ultra-personal, ultra-being is not being-in-itself; it is correlated with the personal, existential, with our capacity to reflect (or not to reflect) about these matters: "It is very often said that we do not understand things because our inner

[22] Konstantins Raudive, Divējādi dzīves veidi: filozofiska apcere. (*Konstantin Raudive, Twofold Life Styles: Philosophical Reflections*). Upsala: Dzintars, 1952, p. 26. (in Latvian).

[23] Ibidem, p. 33.

[24] Ibidem, p. 12.

senses with regard to the immaterial being are coarse, numb, as hard as glass. One should not resist the train of thought that exceeds the limits of human understanding; a person who dares to think more than our ordinary reason and logical cognition allows, is justified in seeking an answer from the One, who is the Creator of man and of all things".[25] Thus, we notice that Raudive's world-view contains elements of mysticism, esoterics and also certain facets of scientific thinking. Such an approach led him subsequently to parapsychological investigations that ensured his reputation within a certain sphere of scholarly culture.[26]

Tymieniecka has carried on Heraclitus' "torch" by elaborating the problem of the illumination of the invisible – the cosmos as "things that tend to hide", and their "logos". At the same time – consideration of the harmonious as the visible, permits us to postulate the beautiful – the concept spoken of earlier on in the present essay – as a manifestation of a distinctive feature of the cosmos. Up to the end of the Renaissance, and especially at the end of the eighteenth century, when special prominence was given to the fine arts, the beautiful was considered not so much as an aesthetical category, but as an ontological and a metaphysical one, apprehending the world as beautiful. Because the beautiful is not only man-made, but is given, and the discourse of the beautiful reveals nature and the natural.

Contrary to the aforementioned I. Murdoch's assertion to the effect that no teaching is necessary to learn to comprehend and to love the beautiful, thinkers and artists have exerted themselves in attempts to formulate systematic principles allowing to explain the phenomenon of the beautiful. It was in the ancient Greece that artists and thinkers evolved strict canons of art including also the subjective assessment. In particular, Socrates and Plato address the theme of the beautiful. Plato's sophisticated understanding of the beautiful gives birth to "two most important conceptions of the beautiful, undergoing subsequent development throughout the centuries: the beautiful as a harmony and proportionality of parts (borrowed from Pythagoras and developed in *Timaeus*) and the beautiful as brightness (understood as blessedness – an idea evolved in *Phaedrus*. The beautiful, for Plato, obtains of autonomous existence, separated from the physical vehicle, thus being independent of the concrete sense-perceptible object, but radiating its brightness all-around.[27]

Understanding of the beautiful as cosmos, and the idea of the beauty of cosmos is precisely delineated by St. Thomas Aquinas. He characterizes the beautiful with the help of three traits. First – with integrity (wholeness) or totality, for anything that lacks something is formless. Second – appropriate proportionality, or agreement (*consonantia*); and thirdly – clarity; this is why bright colour is called beautiful....[28] St. Thomas uses two concepts for the understanding of the beautiful – the

---

[25] Ibidem, p. 298.

[26] Gills N. "Raudives balsis". Konstantins Raudive un parapsiholoģijas pasaule. ("Voices of Raudive". Raudive and the World of Parapsychology, Filosofija. Year-book. Rīga, FSI, 2002, p. 98.

[27] Umberto Eko, Skaistuma vēsture. (*Umberto Eco, History of the Beautiful*) R.: Jāņa Rozes apgāds, 2009., p. 48. (in Latvian).

[28] Akvīnas Toms. Teoloģijas summa, (*Thomas Aquinas. Summa Theologiae*) I, qu. 39, a. 8. – Istorija estetiki. Pamjatņiki mirovoi estetičeskoi misļi, tom I, Antičnost, srednije veka, vorroždenije. (*History of Aesthetics*) Moskva, 1962, p. 290. (in Russian).

beautiful as the excellent (pulchrum), and the beautiful as ornamentation (decorum). God as absolute harmony and absolute clarity is characterized by a collation of both of these terms. It is a tragical story of the present-day situation that the beautiful has turned into mere decorum, thus losing other traditionally characteristic facets, including the giveness of the natural order. The narrowing down of the concept of the beautiful and even complete relinquishing of it, started by the aesthetics and philosophy of the modern era, that was initially justified (by way opening new alleys for unbridled contemplation of the "sublime", but actually robbing the world of self-creativity and ontopoietic power) has turned the world into an unpalatable construction. Art has been denied referential relations with the natural, and, consequently, it has lost the power of the beautiful and has turned into feeble beauty with completely marginal promise of beatitude (see: Adorno's interpretation of the Odyssean myth).

However, there is lately a noticeable, though a rather timid movement towards rehabilitation of the beautiful. Aesthetics is leaving behind its sojourn along the abysmal elevations of the sublime, and is slowly returning – not without resistance – to the "beautiful". Here is what Nicolas Bourriaud has to say in this connection: "Among the reactionary enticements in the present-day sphere of culture, first and foremost one notices the project of the rehabilitation of the beautiful. This notion may pass under different designations – art critic Dave Hickey proposes to return to the normativity, he speaks about an arrangement that produces visual pleasure in the viewer, and maintains that any theory of picture that is not based on the pleasure of the viewer, poses a question about its tenability, and is likely to remain unsignificant.[29]

Even so, Bourriaud calls the return to normativity and to arrangement a reactionary, insufficient enticement, for, according to him, such a tactic is being justified by the pleasure of the viewer. In my view, the feeling of the beautiful based on pleasurable emotions enhances the quality of one's appreciation of the natural, of the perfection (totality, completeness). It was Kant who spoke of "non-meaningful beauty", basing such a thesis on the beautiful to be found in nature – that is: in the experience of the beautiful, and not in the pure concept of it, or in art alone. Kant did not equate the experience of the beautiful with the general regularities of nature, neither did he consider this kind of experience to be of subjectively significant import only. H. G. Gadamer has this to say about Kant's position: "If I consider something to be beautiful, I think that this is beautiful. To use Kant's expression – "I demand everybody's acceptance of this"; which means, that anybody's senses have to be cultivated to recognize the beautiful, so as to develop a sense of discretion between the beautiful and the less beautiful".[30] This is what makes me take issue with Bourriaud's assertion about the beautiful as a reactionary enticement. Because anchoring of the beautiful in the pleasure of the viewer (observer), and returning to normativity should not be

---

[29] Nikolā Burjo, Attiecību estētika. *(Nicolas. Bourriaud, Aesthetics of the Relationships)* R.: Laikmetīgās mākslas centrs, 2009., pp. 61–62. (in Latvian).

[30] Hanss-Georgs Gadamers, Skaistā aktualitāte. Māksla kā spēle, simbols un svētki., R., 2002, p. 53. (in Latvian).

taken to mean the annihilation of the differences, or the extinguishing of discernment. Indeed, the evaluation of *the art of the present time* by using the standards of beauty could really be a reactionary temptation. However, Gadamer reminds us that "the concept of the beautiful even today obtains of various connotations, all of which to a greater, or lesser degree draw on the *kalon* notion of the ancient Greeks".[31] And the beautiful is not to be associated exclusively with art; to-day it regains its original meaning as the "harmony of celestial spheres" and as the visible manifestation of the invisible, and as a sense of wholeness of being, obtained through experience of subjective involvement.

Summing up, the changes produced by the human activities and the technological innovations in the environment, ecology, values and thinking demand a re-assessment of the "human placement in the cosmos" and our fundamental engagement in it. It requires a specific re-turning to the natural order of things, to the cosmic justice, based not only on moral principles, but anchored also in reality (from *res* – thing), in the manner undertaken by the ancient philosophers. We have to acknowledge that by attempting to be ourselves, we exceed ourselves. And regardless of all the fascinating visions about other possible worlds, created by scientific and artistic endeavours, we after all live in this one – the only possible world for us, for the upkeep of which we bear joint responsibility. One can only fully agree with the dictum of M. Merleau-Ponty: "There is no other world possible in the sense in which mine is, not because mine is necessary as Spinoza thought, but because any other world, that I might conceive would set limits to this one, would be found on its boundaries, and would consequently merely fuse with it".[32]

This is why it is impossible to-day not to think about the cosmic order in its various manifestations, for it marks the horizon of my vision solidified with the help of the visible things, that existed before my seeing them and will last much longer.

[31] Ibidem, p. 47.

[32] Maurice Merleau-Ponty, *Phenomenology of Perception*. London and New York, Routledge, 1996., p. 505.

The Law of Opposites in the Ontopoiesis of Life and in Language

Zaiga Ikere

Abstract The article considers the law of opposites as present in the philosophical ideas, cognition and language. In the Western philosophical tradition it has become accepted to draw a sharp distinction between the man as the subject and the world as the object. It is the philosophy of the ontopoiesis of life advanced by Anna-Teresa Tymieniecka where the Heraclitean intrinsic law of opposites is shown as the dynamic unfolding of forces in the self-individualizing process of life. Disparate elements are differentiated bringing about extreme points that oppose each other, namely cosmos and human world and human condition. For Tymieniecka, these two ultimate opposite ends of the developmental process are united in the unity-of-everything-there-is-alive. Language is the medium through which the human mind categorizes and conceptualizes the entirety of the world. Following Ferdinand de Saussure, structuralists hold binary oppositions to be one of the most important principles governing the language. The cognition of the world through binary oppositions seems to be characteristic of human psyche (e.g. subject – object, self – other). The principles of contrasting concepts: abstract – concrete and literal – metaphorical lie at the basis of shaping and understanding of the cognitive metaphor.

Philosophers throughout centuries have considered such opposites as cause and effect, essence and appearance, necessity and contingency, the particular and the general, differentiation and unity, being and non-being, subject and object, sameness and otherness, etc.

It is characteristic of the European mode of thinking to grasp the world through oppositions. First of all a notion of these oppositions can be learned from the phenomena of our everyday life experience. From our experience we get to know such oppositions as: day – night, dark – light, dead – alive, here – there, right – left.

Z. Ikere (✉)
Daugavpils University, 5 Ciolkovska street, Apt. 55, Daugavpils, LV-5410, Latvia
e-mail: zaiga.ikere@du.lv

A.-T. Tymieniecka (ed.), *Phenomenology of Space and Time: The Forces of the Cosmos and the Ontopoietic Genesis of Life: Book One*, Analecta Husserliana 116,
DOI 10.1007/978-3-319-02015-0_7,

We find the same opposing notions in mutual relationships, for instance, *love – hate*, *pleasant – disgusting*, *light – shadow*, etc.

People have believed in the fundamental character of binary oppositions since at least classical times. For instance, in his *Metaphysics*, Aristotle advanced the following binary oppositions: form – matter, natural – unnatural, active – passive, whole – part, before – after, being – non-being. The semiotician Daniel Chandler admits that in Aristotle's *Physics* the four elements of earth, air, fire and water were said to be opposed in pairs. For more than 2,000 years oppositional patterns based on these four elements were widely accepted as the fundamental structure underlying surface reality (Chandler 2003: 102).

The dualism of mental and material was given its first definitive expression in the seventeenth century by the French philosopher René Descartes. He divided reality into two distinct ontological substances – mind and body. These substances represented for him internal or 'mental' world and external or 'real' world. The theory of Descartes originated a number of associated dichotomies, such as reason – emotion, male – female, true – false, public – private, self – other and human – animal. The philosopher Nancy Mardas states that "Dichotomies that have plagued Western philosophy at least since Descartes are, for instance, the split between reason and passion, between the mind and the body, between a priori and a posteriori knowledge, between subject and object, between freedom and necessity, between noumenal and the phenomenal" (Mardas 2004a: XXi).

The ideas of Descartes contributed to the beginning of the elevation of humankind above nature, resulting in the split between the human realm and nature. In the Western philosophical tradition it has become accepted to draw a sharp distinction between man as the subject and the world as the object.

The philosophy of ontopoiesis of life as being propagated by Professor Anna-Teresa Tymieniecka and The World Institute for Advanced Phenomenological Research and Learning for more than 40 years has been trying to bridge the gap between the human realm and the kingdom of nature. When considering phenomenology of life and its inner workings, Tymieniecka discusses the role of the intrinsic law of opposites as held by Heraclitus. Reflecting on forces in nature and the being in general, Heraclitus has stated that all is in a perpetual flux; all things are in a process of a perpetual change. According to Heracletian vision, the differentiation of things occurs through the play of opposites. Tymieniecka sees the action of the law of opposites as the dynamic unfolding and construction of forces in "the self-individualizing progress that projects itself in a process" (Tymieniecka 1998: 13). Tymieniecka demonstrates how inner forces of life are governed by the law of opposites. Tymieniecka discusses such opposites as the flux versus statis, differentiation versus unity. Tymieniecka shows the principle of unity and the principle of differentiation that brings forth the self-individualizing progress. This progress is a continuous process, this process being: "the constructive vehicle of order within the flux".

She explains that "a process, indeed, remains in flux while its phases differentiate from each other. Each phase performs a distinctive segment of operations … Each actual phase, just now in performance, already anticipates the next, the one into which it passes …" (op. cit.: 13).

Tymieniecka concludes that the essence of the self-individualizing life is being- in-process. It is the human mind with its logics that cuts into this incessant flux and tries to insert points of stability and order. She asserts:

> The human creative mind in its creative manifestation of life cuts into this incessant flux, and by establishing correspondences with its "logic of contradiction" it establishes objectified reality according to a logic of its own, the logic of structures and essences… In the rhythm of taking, processing, absorbing, and radiating and rejecting [within life's poiesis] the rational principles of "sameness" and "otherness", of "inwardness" and "outwardness"… are projected and installed. (*op. cit.*: 14)

As stated by Tymieniecka, the advent of life occurs in the bringing forth of opposites in forces, qualities, tensions, etc. She holds that such opposites as "useful and noxious, hot and cold, light and dark, moist and dry, strength and weakness, etc., differentiated in the operations of the life-process … are, in fact, opposites in transition: they acquire their gradation of "opposition" in the play of vital forces that transforms substances" (*op. cit.*: 15).

Tymieniecka shows how the web of life is actually woven through opposing forces that ensure self-individualization of entities. Tymieniecka has widely discussed such opposites as differentiation and unity, namely, differentiation as self individualizing process within the unity-of-everything-there-is-alive. Considering the entirety of life's expansion (the world in its universality, according to Kant), Tymieniecka argues that within a common line of successive stages of development disparate elements are differentiated into an infinite gradation bringing about extreme points that oppose each other at the "opposite" ends of a common line. She explains:

> The game of life consisting in the play of trial and error, in the confluence and transformation of otherwise disparate elements unfolds the entire gamut of opposed tensions, and … they are differentiated into an infinite gradation of qualitative or operational intensities, forcefulness, etc. bringing about extreme points that oppose each other at the "opposite" ends of a common line … This differentiation flows, indeed, from the heart of the logos of life initiated with the cosmos on the one end and culminating in the Human Condition at the other. (*op.cit.*: 15)

To conceive Tymieniecka's notion of the coexistence of these two ultimate opposite ends of a line, i.e. the bios at the one end and human condition at the other, there might serve the presentation of the Chinese traditional concepts of yin and yang, when they are presented in colours as white and black and forming a unity within a circle. Likewise the opposing elements described by Tymieniecka could be visualized as not posited on a line, however, but situated within a circle and thus representing the entire flux of being around us in its differentiation and its unity.

The law of opposites governs different spheres of human life and one of those spheres is language.

Language as a means of expressing one's ideas was the object of philosophical interest since Aristotelian times in European philosophy. Language has been considered as a theoretical discipline beginning with the Swiss linguist Ferdinand de Saussure (1857–1913). Saussure is considered to be the predecessor of structuralism in linguistics.

Language for Saussure was a system of signs. For him, it is a system of functional differences and oppositions. In order to recognize a sign one has to differentiate it from the others to which it is related. According to Saussure, two signs are in opposition to each other. The entire mechanism of language is based on oppositions of this kind and upon the phonic and conceptual differences they involve. Saussure particularly emphasized negative, oppositional differences between signs. He argued that "concepts … are defined not positively, in terms of their content, but negatively by contrast with other items of the same system. What characterizes each most exactly is being whatever the others are not" (Saussure 1992: 115).

For Saussure, the sign involves two different notions, i.e. the sound pattern and the concept it signifies. He presents them as two facets of one system. Graphically it may be represented as:

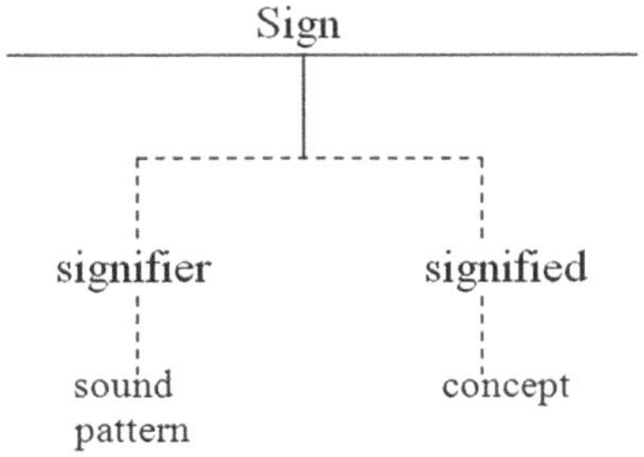

Likewise Saussure has contrasted two domains of language: the actual phenomenon of language or data of linguistics (as *langue*), and the actual use of language (as *parole*). He has also contrasted two planes of investigation of language, i.e. synchronic and diachronic aspects of research. Analyzing language as a system and as a structure, Saussure introduced the notion of syntagmatic and paradigmatic relations of linguistic units. The notion of paradigmatic relations is grounded upon the existing associative relations between linguistic units in a language system, whereas syntagmatic relations are simultaneously present in a structure and manifest the various ways in which linguistic units within the same text are structurally related to each other. Investigating human communication, he outlined the following distinctions: synchronic approach to investigations as opposed to that of diachronic and paradigmatic relationships among linguistic units as opposed to syntagmatic.

Thus, according to Saussure, in the analysis of language the following oppositions can be discerned which graphically may be shown as:

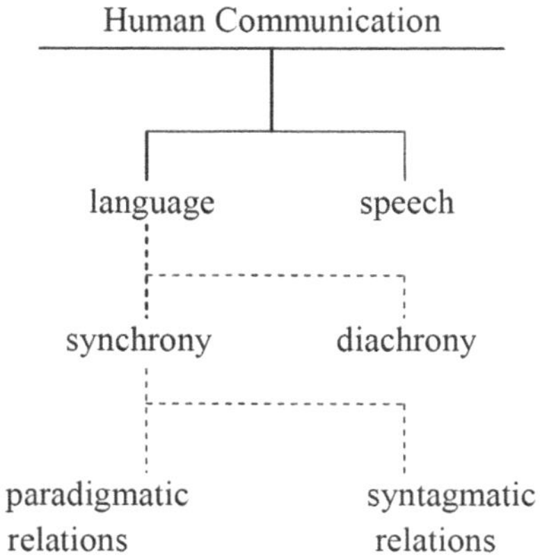

Poststructural theorist, literary philosopher Jacques Derrida followed in the footsteps of structuralists in regard to binary oppositions. For Derrida, the world is constructed in sets of binary opposites. Such an understanding echoes through Derrida's writings on the meaning as the play of differences. The philosopher Nancy Mardas points out:

> As well known, in Derrida's theory, the world is constructed in sets of binary opposites, each struggling to achieve the dominance of a central position, and against marginalization, striving for actualization and identity. In each case, what is present is privileged over what is absent. In the realm of language, in Derrida's classic formulation of the binary opposition of signifier to signified, the signified is internal, and the signifier external. The only way that the signifier gains identity is in its difference from other signifiers. (Mardas 2004b: 20–21)
>
> Saussure's views helped to shape structuralism and make it the dominant approach in European linguistics. Following Saussure, structuralists emphasized the importance of relations of binary oppositions. Daniel Chandler notes that Roman Jakobson proposed that linguistic units are bound together by a system of binary oppositions. As to Jakobson, such oppositions are essential to the generating of meaning, e.g. the meaning of "dark" is relative to the meaning of "light"; we consider "form" in relation to "content". Largely through the influence of Jakobson, the primary analytical method employed by many structuralist semioticians involves the identification of binary or polar semantic oppositions (e.g. *us – them, public – private*) in texts or signifying practices. (Chandler 2003: 101)
>
> As marked by Chandler, "binary oppositions for structuralists are considered to be pairs of mutually exclusive signifiers in a paradigm set representing categories which are logically opposed, e.g. alive – not-alive" (op. cit.: 224).

In respect to oppositions it should be noted that both in linguistics and semiotics they are grouped into mutually exclusive oppositions (e.g. alive – dead), which are termed "binary oppositions", and the ones representing categories with comparative grading on the same implicit dimension, e.g. good – bad where "not good" is not necessarily "bad" and vice versa. The latter are termed "analogue oppositions" (see Chandler, op. cit.: 223). A similar observation is made by the linguist John Lyons. Considering binary oppositions to be one of the most important principles governing the structure of languages (Lyons 1977: 271), he claims that certain distinctions can be made between the types of oppositions. He classifies them accordingly into *logical contradictories* and *logical contraries*:

- Oppositions (logical contradictories: mutually exclusive terms (e.g. alive – dead, where 'not alive' can only be 'dead');
- Antonyms (logical 'contraries'): terms which are comparatively graded on the same implicit dimension (e.g. good – bad, where "not good" is not necessarily "bad") (Lyons 1977: 270ff.).

Contrasting and dualism seem to be deeply rooted in the development of the human categorization of the world and mode of thinking. As noted by Chandler, Jakobson and Halle observe that "the binary opposition is a child's first logical operation" (Jakobson and Halle 1956: 60, cited in Chandler 2003: 101).

Contemporary psychologists hold that the cognition of the world through binary oppositions is characteristic of human psyche. As pointed out by Chandler "the opposition *subject – object* in human psyche is manifested as apprehension of *self – other*. The opposition of *self – other* (or *subject – object*) is psychologically fundamental. The mind imposes some degree of constancy on the dynamic flux of experience by defining "the self" in relation to "the other"" (Chandler 2003: 105). Chandler notes that the neo-Freudian psychoanalyst Jacques Lacan has argued that initially in the realm of "the Real" ... the infant has no centre of identity and experiences, no clear boundaries between itself and the external world. Lacan describes a defining moment in the imaginary which he calls "the mirror phase", when seeing one's mirror image (and being told by one's mother, "That's you!") it induces a strongly defined illusion of a coherent and self-governing personal identity. Chandler stresses that this marks the child's emergence from a matriarchal state of "nature" into the patriarchal order of "culture". As the child gains mastery within the pre-existing "symbolic order" (the public domain of verbal language), language (which can be mentally manipulated) helps to foster the individual's sense of conscious "self" residing in an "internal world" which is distinct from "the world outside" (Chandler 2003: 105). Chandler marks that "self – individualization process is realized with the help of language. It is our life experience manifested in our language that makes the individual, differentiating it from others. Subjectivity is dynamically constructed through discourse" (op. cit.: 105).

The opposition self – other can be expressed as the opposition us – them. To illustrate this common feature of the psyche to perceive the world in dichotomies and demonstrate how common for us it is to oppose "us" to "them", the latter being alien or even dangerous, there is an excerpt from a contemporary thriller:

> The tale of Eden itself had probably begun not far from here, somewhere in the parallel valleys of the Tigris and Euphrates that emptied into the Persian Gulf. Yes, if humanity were all one cast tree, then the oldest roots were right here, virtually in the center of the country he had just created.
>
> The ancients would have the same sense of centrality, he was sure. Here we are, they would have thought, and out there were ... *they* [author's emphasis], the universal appellation for those who were not part of one's own community. They were dangerous. At first they would have been nomadic travellers for whom the idea of a city was incomprehensible. How could one stay in one place and live? Didn't the grass for the goats and sheep run out? On the other hand, what a fine place to raid, they would have thought. That was why the city has sprouted defensive walls, further emphasizing the primacy of place and the dichotomy of *we* and *they* [author's emphasis], the civilized and the uncivilized.
>
> And so it was today Daryaei knew, Faithful and Infidel. (Clancy 1997: 721)

The intrinsic law of opposition manifests itself also in the cognition process, which then finds its realization in language structures. There are two different ways of deciphering reality, i.e. literal as contrasted to figurative or metaphorical.

Carl Gustav Jung, when investigating archetypes as reflecting universal human thought found in all cultures, turned his attention to symbols thus stressing the significance of figurative or metaphorical thinking. Ernst Cassirer in the investigation of symbolic forms in language and culture distinguished two forms of mental action, i.e. metaphorical (lingual and mythical) and discursive logical (Арутюнова 1990: 13).

Cassirer speaks of two ways of forming a concept as two different tendencies or modes of thinking. These are logico-discursive, and lingual and mythological. He explains that in the first case when forming a concept one can speak of widening of the range of notions and concepts referring to it. In the second case, however, we meet a different process, namely, the range of notions is not widened, but vice versa, they are squeezed together to focus in one point (Кассирер 1990: 37).

As concerns linguistics, nowadays it is cognitive linguistics where a number of linguists have turned their attention to elucidation of the ways in which linguistic structures reflect the manner in which human beings perceive, categorize and conceptualize the world. One of the topics of their investigation is the theory of cognitive metaphor.

Cognitive science views metaphor as a principle of thinking, as a key to understanding the basis of thinking. The theory of cognitive metaphor discloses the pattern how human beings arrange and structure their experience and knowledge. This process is determined by the ability to perceive abstract concepts metaphorically, i.e. by being able to compare them to something real. It is characteristic of human mode of thinking to compare and make contrasts. In order to comprehend and grasp the meaning of a new or abstract notion the mind tends to compare it with something familiar. The thinking process is organized in the way that the mind moves from the known to the unknown, from the concrete to the more abstract, for instance, some abstract notion is compared metaphorically to something well known from human everyday life. One notion is compared to another resulting in creation of an image schema. This is the way cognitive metaphors appear. The understanding about an abstract concept may exist in a form of a cognitive metaphor, which then is transformed into linguistic metaphor in communicative situations. The shaping of cognitive metaphor is being realized not in the verbal, but in the cognitive domain (Richards 1932). Metaphor expresses an abstraction (target domain) making use of more familiar concepts (source domain) pertaining to everyday life, for instance, in the well-known example LOVE IS A JOURNEY by George Lakoff and Mark Johnson (2003) which they employ to illustrate the conception of cognitive metaphors, where the concept of love is target domain and the concept of journey is source domain.

In conclusion, opposing ideas can be found in the history of the development of the philosophical thought, modes of thinking and everyday life experience. Language is the media through which the human mind categorizes and conceptualizes the entirety of the world around us. Language reflects the law of opposites in its linguistic structures.

References

Chandler, David. 2003. *Semiotics: The basics*. London/New York: Routledge.

Clancy, Tom. 1997. *Executive orders*. New York: Penguin.

Jakobson, Roman, and Morris Halle. 1956. *Fundamentals of language*. The Hague: Monton.

Lakoff, George, and Mark Johnson. [1980], 2003. *Metaphors we live by*. Chicago and London: The University of Chicago Press.

Lyons, John. 1977. *Semantics*, vol. 1. Cambridge, UK: Cambridge University Press.

Mardas, Nancy. 2004a. Creative imagination—the primogenital force of human life; Following Anna Teresa Tymieniecka's thread from the elemental stirrings to the human fulfilment. In *Analecta Husserliana*. The yearbook of phenomenological research Vol. LXXXIII, ed. A.-T. Tymieniecka, xxi–xiii. Dordrecht/Boston/London: Kluwer.

Mardas, Nancy. 2004b. The cipher as the unity of signifier and signified. In *Analecta Husserliana*, The yearbook of phenomenological research, vol. LXXXIII, ed. A.-T. Tymieniecka, 13–24. Dordrecht/Boston/London: Kluwer.

Richards, Ivor A. 1932. *The philosophy of rhetoric*. London: Oxford University Press.

Saussure, Ferdinand de. [1916], 1992. *Course in general linguistics*, eds Charles Bally, Albert Sechehaye, Abert Riedlinger, Translated and annotated by Roy Harris. La Salle: Open Court Publishing Company.

Tymieniecka, Anna-Teresa. 1998. The great plan of life: The phenomenology of life's return to the sources of western philosophy. In *Analecta Husserliana. The yearbook of phenomenological research*, vol. LII, ed. A.-T. Tymieniecka, 3–29. Dordrecht/Boston/London: Kluwer.

Арутюнова Н. Д. 1990. "Метафора и дискурс," в кн. Теория метафоры: Сборник, Пер. с англ., фр., нем., исп., польск. яз./Вступ. ст. и сос. Н.Д. Арутюновой и М.А. Журинской с. 5–32. Москва: Прогресс.

Кассирер Э. 1990. "Сила метафоры," в кн. Теория метафоры: Сборник. Пер. с англ., фр., нем., исп., польск. яз./Вступ. ст. и сос. Н.Д. Арутюновой и М.А. Журинской с. 33–43. Москва: Прогресс.

The Forces of Darkness and the Forces of Goodness: Jerzy Nowosielski's Concept

Katarzyna Stark

Abstract The paper aims at answering the question to what extent, according to Jerzy Nowosielski (the Polish artist, Orthodox theologian and philosopher [1923–2011]), the forces of darkness affect goodness. Despite the Manichean views that evil dominates the empirical reality, Nowosielski claims that goodness is revealed within this world and imposes a divinized status upon the cosmos. In this process a crucial role is played by art, which belongs to the sacred sphere. The Polish artist, following the Orthodox theology, states that it is only beauty and goodness that can save the world. Nowosielski focuses upon the Orthodox icon as the most important means of participation in the divinized cosmos. Thus art constitutes a true home for the artist within this evil empirical reality; it brings eschatological hope not only to human existence but also to the cosmic life.

Who Was Jerzy Nowosielski?

During the ceremony awarding the title of doctor *honoris causa* of the Jagiellonian University to Jerzy Nowosielski in Kraków on January 13, 2003, Mieczysław Porębski said in his laudation: "Not only does Nowosielski paint, he also writes. He writes, and converses. In his writings and conversations, we find everything that can be found in his art – existential and philosophical reflection, truly inspiring knowledge of the mysteries of faith, cults, liturgical rites of the European and non-European East as well as the West; of the glory of the *Mediterraneum* as well as the damp depths of the North."[1]

[1] M. Porębski, *Nowosielski* (Kraków: Wydawnictwo Literackie, 2003), p. 250.

K. Stark, Ph.D. (✉)
AGH University of Science and Technology,
Faculty of Humanities, Department of Culture Studies and Philosophy,
ul. Dominikanów 20, Kraków, 31-409, Poland
e-mail: stark@agh.edu.pl

A.-T. Tymieniecka (ed.), *Phenomenology of Space and Time: The Forces of the Cosmos and the Ontopoietic Genesis of Life: Book One*, Analecta Husserliana 116,
DOI 10.1007/978-3-319-02015-0_8, © Springer International Publishing Switzerland 2014

The reconstruction of the philosophical and theological views of the Orthodox icon painter is not easy. Nowosielski's knowledge relies on an understanding of many fields – most importantly: philosophy, religion and art. A synthetic formulation of his views becomes possible mostly through an insightful analysis of numerous interviews conducted with the artist. The gist of his thought can be expressed as a combination of the Eastern Orthodox tradition and elements of the gnostic knowledge. The theologian does not explicitly specify any gnostic sources of his vision of the world apart from Manichaeism.

He was an artist, an unorthodox Orthodox theologian, who expressed his views straightforwardly, unyieldingly, with astounding frankness and without academic support. Presenting his thoughts, he preferred to be heretic rather than orthodox. In his view, heresy is a different opinion, a right to freedom of speech, a right to oppose the official teachings of the Church.

Regarding himself as a heretic and a gnostic, Nowosielski stood in opposition to the official and binding "school" of theology of the Church. The views of the artist as a declared heretic place him in the movement of opposition to rigid orthodoxy. His numerous questions addressed to the Church are an attempt to "renew and resurrect" the original Good News.

Nowosielski, declaring himself as a gnostic, continued his studies in gnosticism throughout his entire life. Therefore, he is considered a heretic by his contemporaries. In this context, the words of Jerzy Prokopiuk prove valid: "The gnostics in various forms were thrown into the underground of Christianity. They gave rise to the so-called Esoteric Christianity (…). I think that the tragedy of Christianity in its entire history was the fact that gnosis was suppressed and thrown into the underground."[2]

If we assume that the statement of Gilles Quispel, who says that the gnostic movement regards itself as an "extra-ecclesiastical" Christianity, "Christianity without Church" is true, then the views of Nowosielski comply with this approach.[3] Gnosis is an answer to the questions which cannot be answered by the official teachings of the Church because of a dogmatic and one-sided formulation of Christianity.

For Nowosielski, gnosis is a form of knowledge, an initiation experienced especially in the process of reading the Bible. In the Christian West, "the Bible is read"; in the East, it is learned, contemplated. As Jerzy Prokopiuk says: "Gnosis (Gr. *gnosis*) is identical with a direct inner (non-intellectual) experience of God, the essence of things, self, others, as well as the world – through "enlightenment" or "initiation"."[4]

Nowosielski's assertions about the unknown God are not the only elements of gnostic origin. So are the statements about Sophia – the Creator of the world, about the dualism of light and the dark, the good and evil, spirit and matter, soul (mind) and body, and the teaching about the God-man.

---

[2] J. Prokopiuk, *Labirynty herezji* (Warszawa: Warszawskie Wydawnictwo Literackie, MUZA SA, 1999), p. 58.

[3] Cf. G. Quispel, *Gnoza*, trans. Beata Kita (Warszawa: Instytut Wydawniczy PAX, 1988), p. 105.

[4] J. Prokopiuk, *Labirynty herezji*, op. cit., p. 8.

As a gnostic, the author of *Mój Chrystus* did not share his world of mysteries with others. He kept the innermost hidden knowledge only to himself, and he emphasized clearly that he did not give interviews about his personal experiences. The knowledge of the gnostic is in reality a hidden knowledge, an individual revelation of the bond between the human and the Divine being. It relies on intuition, not on discursive thought. The following words of Gilles Quispel are pertinent in this context: "(…) the gnostic possesses a receiving apparatus not accessible to anyone else. The gnostic is proud of his apparatus, and he refers to it by the philosophical term *nous*. This term can be translated as super-consciousness, higher consciousness, clairvoyance or, most accurately, intuition."[5]

Based on his acquaintance with the artist, Piotr Sarzyński writes: "Nowosielski shared his thoughts with others willingly, though he might have kept the most vital ones to himself".[6] If we assume that gnosis is a cosmologic, anthropologic and eschatological knowledge, then Nowosielski's views satisfy these classification criteria. Regardless of the Orthodox pneumatological image of Christianity, the experiences and intuitions of Nowosielski comply with the gnostic movement because he values metahistoric and eschatological insight into reality over the plain historical course of events. It should be emphasized, that as an "Orthodox gnostic" he devotes most of his attention to the eschatological issues.

A hostile attitude towards the world one lives in is also a manifestation of gnostic thought. Anxiety, the feeling of being thrown into the world, and hatred towards this world are typical of gnosticism.

Nowosielski, following gnostic thought, takes a stance of negation of the existence in this world, which is infested by evil and suffering. This attitude does not arise from intellectual speculations or acquired theoretical knowledge, but from exceptional sensitivity of the artist to evil which takes over the entire cosmos and affects every aspect of life. "The Devil has to be believed in because we feel him every day. The whole tragedy of human and animal existence, the tragedy of nature, result from the blatant reign of Satan and fallen angels. Where is God, then? Where should we search for him?",[7] asks Nowosielski.

All Nowosielski's efforts are directed towards the overcoming of the evil of this world and breaking free from Satan's rule by means of the sacral power of culture and particularly of art.

The solution to this problem will come with the second cosmic catastrophe anticipated by Nowosielski, which will annihilate evil and open the gates of paradise, regained in the form of the entire cosmos transfigured (divinized) by the power of the Holy Spirit.

As an Orthodox theologian and partially a gnostic, the artist accepts a conception of the unknown, hidden Father-God of Jesus Christ. This unknowable God

---

[5] G. Quispel, *Gnoza*, op. cit., s. 93.

[6] P. Sarzyński, "Byty subtelne", *Polityka* 2797: 10 (March 5, 2011), p. 80.

[7] J. Nowosielski, *Sztuka po końcu świata. Rozmowy*, ed. Krystyna Czerni, (Kraków: Wydawnictwo Znak, 2012), pp. 363–364.

transcends the world; He is not present in this world. But He is also an immanent God because, as the theologian tersely says: "He divided Himself into two"; he begot his Son, who participates in human life, and therefore also in nature's life. Thus, Nowosielski clearly distinguishes the transcendence of God in His own being from His immanence in the created world. Through Jesus Christ, the unknowable God becomes immanently present in the world.

As Jerzy Prokopiuk writes: "(...) new gnosis is 'towards the world', that is, it wants to live according not only to the transcendent God's will, but also to the immanent Divinity's will, that is, the Deity involved in the world, in nature. It wants to unite heaven and earth, not to separate them."[8]

Jesus Christ united heaven and earth in His Divine and human nature, which means His participation in restoring the man to the state before the Fall, when the "image and likeness of God" made an integral and inseparable structure.

Nowosielski strongly emphasizes the suffering of God and man embodied in Jesus Christ. In Him God suffers together with man. The great merit of Christ lies in His *kenosis*. He descends to the world permeated by evil. If it were not for the evil of this world and human sin, the incarnation would not be necessary. The incarnation of God is a consequence of the Divine catastrophe, which humans also take part in.

According to Nowosielski, besides God-man, who assumed the role of a willing sacrifice for the sake of the world salvation, angels also participated in the sacrifice, permanently living their own hell.

The artist explains the origin of the overwhelming cosmic evil by referring to, among other things, gnosis and contemporary neognosis. He writes: "In Christianity, there are conceptions – of course deeply-rooted in gnosis – stating that God rules the world with His two hands. His right hand would be the Logos – Christ; the left one would be the Accuser, who did not like God's idea of the world's organization, and who caused the cosmic catastrophe and released a virus into the work of creation, which basically spoiled everything".[9] Elsewhere, the theologian speaks about Satan as the left hand of God.

Originally, Satan was a part of God, His servant or even His son. It was Satan who caused the cosmic catastrophe, rising up against the plans of God, the Divine plans of the cosmic organization. He aspired to hinder the development of the spirituality of matter.

Subtle intelligences cause real evil by invalidating certain elements of the primary reality, which is beyond good and evil.

The fallen subtle beings spoiled God's work of creation and control this world. Speaking of Satan, Nowosielski refers to the Gospel: "Besides, even in the Gospel he is called the prince of this world."[10] Elsewhere he says: "The entire empirical

---

[8] Ibidem, p. 59.

[9] J. Nowosielski, *Sztuka po końcu świata*, op. cit., p. 387.

[10] "Czy Bóg się wycofał? Z Jerzym Nowosielskim rozmawia Dariusz Suska", *Gazeta Wyborcza* No. 86 (1998), p. 14.

reality, both cosmic and planetary, the one closest to us, is precisely one giant *infernum*. It is hell."[11]

The Polish artist repeatedly emphasizes the fact that the tragedy of animal and human existence, the tragedy of nature, constitutes the evidence of the reign of Satan and fallen angels. The law of nature is an infernal law.

From the moment of "expulsion from the Garden of Eden", man is a sinner, a criminal who is aware of good and evil, and commits mostly evil. Nowosielski rejects Pelagianism. After the fall, man changed his state from being immaculate to being sinful in every aspect and he has to live perpetually with the awareness of his sinfulness: "We have to be aware of the fact that we are the last and the worst sinners (…). We cannot be better."[12] The evil caused by humans cannot be eliminated; it can only be lessened by doing as little harm as possible to people and animals.

Evil is not a lack of goodness; it is active and possesses actual power. The Orthodox heretic finds Satan everywhere in life. His existence is obvious, personal, and experienced directly. Moreover, since Satan is everywhere: "He is inside the table I am sitting at, inside the telephone, inside myself, inside anyone."[13]

According to Nowosielski, the world described as the "empirical reality" came into being as a result of two cosmic catastrophes – the fall of angels and the original sin of man. The theologian claims that the battle which took place between God and subtle beings became the reason why the "fallen world" formed. Furthermore, man also participates in the frightful cosmic battle on the Divine level.

Nowosielski says man was doomed to commit the original sin because of Satan's rebellion. It means that man carries not only the burden of his own fall, but also the burden of the rebellious angels' fall. It follows that man's sin is lesser than Satan's sin. This thesis is not binding in the Christian orthodoxy, where it is proclaimed that man is to blame for evil. Here the artist's words are completely valid: "We really are innocent; it is only the enemy who accuses our brothers before God, day and night. Regaining the consciousness of the lack of guilt is a fruit we can pick from the 'tree of life'."[14]

Man was doomed to commit the original sin, which is a "fortunate sin". One can thus speak of man's "unguilty guilt" as he is not ontologically and morally responsible for the cosmic evil and the evil within this world. He rebels against God and disagrees with Him, because he does not know why he is to blame and why he is sinful. In addition, man suffers the consequences of the cosmic catastrophe, including death.

Nevertheless, we owe the awareness of the good to Satan, who is directly experienced by us alongside the evil he creates. Experiencing evil opens up the horizons of goodness, the possibility of spiritual improvement, "Evil is, in a sense, blessed,

---

[11] Z. Podgórzec, *Mój Chrystus*, op. cit., p. 20.

[12] Ibidem, s. 45.

[13] "Mówi Jerzy Nowosielski: Jestem grzesznikiem, ale się tym nie chwalę". Rozmawiał Kazimierz Targosz, *Przekrój* 11 (1998), p. 17.

[14] Z. Podgórzec, *Mój Chrystus*, op. cit., p. 57.

since it is a necessary condition for being able to perceive the good and the Divine (…). In this sense we should be grateful to the devils because they introduce us into the world of Divine mysteries."[15]

The Orthodox theologian appreciates Manichaeism as one of humanity's basic religious experiences which has allowed man to possess the awareness of good and evil coexisting as "cosmological co-partners".

It is difficult to deny Nowosielski's Manichean intuitions, as he claims that the presence and reality of evil is commonly perceived in the world. The whole cosmic order is in a state of permanent catastrophe, and is by its very nature evil. Human nature is not evil as a result of the original sin, but because it constitutes a part of the cosmic order which is subject to the forces of darkness. It follows from the conviction that, as Jerzy Prokopiuk states, in Manichaeism, "planets, including the Earth, the whole mineral, floral and animal realm are products of the forces of Darkness. The human is one as well (…)."[16]

Nowosielski asserted that Manichaeism is viable because the entire reality in the cosmic dimension is evil; it is "one giant *infernum*." As the Manicheans proclaim: "Only in Manichaeism was the fall initiated by the forces of darkness. Anyhow, the formation of our world is the consequence of this fall."[17]

As his life passed, Nowosielski became increasingly concerned with the problem of good and evil. Asked in one of his interviews where the good exists he answered: "The good is the greatest mystery. Even where it comes from in this inherently evil world which has no positive natural law, and which has infernal natural law – is the greatest mystery for me, infinitely greater than the infernal mystery or the mystery of Satan's existence."[18]

For the Polish artist, the evil integrated into the structure of this world is obvious and palpable, while goodness and beauty that are perceived in this reality constitute a great mystery. But they also testify that another sphere of reality exists, which is far more valuable and permanent.

The artist declares with the utmost firmness: "Manichaeism, with its proclamation of permanent catastrophe, states that everything comprising the manifested world is evil and we should turn our backs against it. On the other hand, I think that certain elements of manifested reality or maybe even the entire manifested reality contains some fundamental good. In the same way, this good was depreciated by the cosmic catastrophe; it has to return to its constant equilibrium by means of some future catastrophe."[19]

The Manichean fight between good and the evil has to be settled. Nowosielski emphasizes that the splitting of the elements of good and evil was not primordial,

---

[15] Ibidem, s. 152.

[16] J. Prokopiuk, *Labirynty herezji*, op. cit., p. 44.

[17] Ibidem, s. 22.

[18] J. Nowosielski, *Sztuka po końcu świata*, op. cit., p. 391.

[19] Z. Podgórzec, *Wokół ikony. Rozmowy z Jerzym Nowosielskim* (Warszawa: Instytut Wydawniczy PAX: 1985), p. 61.

but resulted from the cosmic catastrophe. The essence of this catastrophe can be seen in the fact that the subtle beings possessed great demiurgic power and "spoiled" the cosmos, leading to the battle between good and evil.[20]

Nowosielski as an Orthodox heretic finds salvation for the human in self-knowledge and spiritual growth. He thinks it is gnosis that reveals the innermost essence of man and constitutes a direct experience of the bond with God on the spiritual level.[21] The awareness of Divinity, inherent in the Orthodox idea of *theosis*, is also reflected in gnosis. As Jerzy Prokopiuk states: "The knowledge of oneself is also the knowledge of God, for the spiritual ego of man, *pneuma*, this "spark of light", comes from the Divine Kingdom of Light, so the one who comprehends himself, his ego, comprehends also that he is the divine *ousia*. Accordingly, through the gnosis, the spiritual man becomes God again. Such man has always been God, but he has not remembered that."[22]

Nowosielski claims that the profoundly pessimistic Orthodox Church reveals its true character and overcomes its pessimism only on the artistic level.

The liturgy constitutes the heart of the Orthodox Church because the experience of liturgical mysticism is the core of Eastern Christianity. It results from the fact that the Orthodox Church is predominantly a liturgical-artistic practice in the domain of a cult, and not so much a doctrine. In the Eastern Church liturgy is a work of art, and is performed on the Earth as an icon of the celestial liturgy, or in other words – the cosmic liturgy. Here the artist refers to Fyodor Dostoyevsky, who made him realize the role of beauty in the process of the world "salvation" with his famous quote that "beauty will save the world". This is why art has a principal significance. Liturgical practices, art, and especially painting induce specific states of super-discursive consciousness, akin or even identical to a mystic experience of the Church community. It is mainly through art that the Orthodox Church uncovers the mystery of history and resurrection.

In accordance with the idea of *theosis* and *apocatastasis* embedded in the Orthodox tradition, the whole of reality will be divinized, "Therefore every manifestation of the divinized empirical reality being possible is a kind of mystagogy – is an introduction to the mystery of God's Church"[23] – says the Polish theologian states.

From the anthropological and cosmological viewpoint, Nowosielski most impatiently awaits the transfiguration of the world, his own resurrection and the

[20] Nowosielski thinks the return to the state before the "original sin" will invalidate the awareness of the good and evil, which is the result of Satan's action. The conception of apocatastasis invalidates all human speculations about reward or punishment in the afterlife. Eastern Christianity leaves this issue to God to resolve.

[21] Referring to the gnostic knowledge, he frequently recalls the division of people into pneumatics, psychics and hylics, considering himself the first one. Striving for spiritual development was a priority for him. Aware of the misery of existence, he perceived the spiritual experience with amplified intensity.

[22] J. Prokopiuk, *Labirynty herezji*, op. cit., p. 34.

[23] Z. Podgórzec, *Mój Chrystus*, op. cit., p. 129.

resurrection of the whole world. This world submerged in the sea of evil does not concern him. He looks forward to its end.

The key to understanding this resurrected reality is the person of Jesus Christ. Nowosielski writes: "Man will not repair the hell that includes the empirical reality, the hell of nature, the hell of animals, the hell of plants, the hell he dwells in. He cannot repair all that. It can be done only by the mystery of redemption, the catastrophe of the Cross and Redemption. It restores the state in which good and evil cease to have opposite labels."[24]

The reality we live in will be transfigured into the celestial reality, which for the theologian "comprises the great mystery of Christ."

The artist equates the problem of resurrection to the triumph of good and of beauty. He solves this problem by referring to the self-knowledge and creative activity of man, who is open to the Holy Spirit's acts. The image of the transfigured reality is formed by the acts of the Holy Spirit, who continues the impulse initiated with the resurrection of Christ.

The painter rarely says anything about the Holy Spirit, who is difficult to identify, but the essence of whom can be brought into light by referring to its acts. According to Nowosielski, the Holy Spirit's acts are best reflected in art. In other words, the Holy Spirit, the last link of the revealed truth of the Trinity, enables us to see the metaphysical reality through art.

For Nowosielski, the very existence of painting is the result of the Holy Spirit's acts: "The act of artistic painting is the act of the Holy Spirit. They are the same. The act of the Holy Spirit and the act of art are not separate. The entire art exists as a result of the Holy Spirit's acts. Without the acts of the Holy Spirit, art would not exist at all."[25]

The artist is the one chosen by God, acting under the Holy Spirit's inspiration, and he is a prophet of the Church. He is free: "The freedom of the artist is absolute; it has to be absolute since the artist is a certain apparatus that conveys the will of gods, the will of Heaven, the will of deities and, as we know, God is an absolutely free being. The freedom of the artist is derived therefrom; it depends only on the orders coming from more mysterious spheres."[26]

Nowosielski recognizes the creational power of consciousness in the process of the creation of beauty. The artist does not extract beauty from nature, but he ennobles nature, endowing it with the qualities of beauty. "Nature is neutral. We are the ones who have to introduce beauty by the acts of God and the Holy Spirit originating from inside us, and somehow we ennoble this nature and make it beautiful."[27]

As previously mentioned, the issue of how the human consciousness converts the elements of evil to good and to beauty concerned Nowosielski throughout all his life. The artist writes: "The issue does not lie in the questions of how evil can exist

[24] J. Nowosielski, *Sztuka po końcu świata*, op. cit., p. 309.

[25] Ibidem, p. 389.

[26] J. Nowosielski, *Sztuka po końcu świata*, op. cit., p. 112.

[27] Ibidem, p. 69.

in art, ethics, theology, but in the question of how art, ethics, theology are capable of existing in the sea of evil."[28]

Art is able to extract the essence of good from a reality that is fundamentally evil. Through art hell is saved, and we will be saved by art as well. Despite his pessimism and critical evaluation of the empirical reality, Nowosielski is convinced that this reality is never entirely spoiled.

From the existential perspective, for Nowosielski art constitutes an oasis of good; it is the affirmation of the world and its physical existence. It is also a real home for the artist. It helps him find hope in this gloomy reality.

From the eschatological perspective, Nowosielski wishes to solve the problem of evil in the world, referring to the creative capabilities of man, who becomes God's partner in the process of the creation of good and of beauty. Art is a window on another reality – the metaphysical reality. In art, he perceives the anticipation of the celestial reality – the real fatherland of man. Thus, the infernal reality can be "transfigured into the celestial reality." Henryk Paprocki writes: "Art finds its full expression when it is in harmony with the reality of heaven and earth. Thus it is the *Parousia* of the Kingdom, the icon of the Kingdom of God. And all temporal creation is endowed of an intransient quality, reflecting the element of eternity. Thus every artist is continually wandering towards the other side, casting us his works, in which shineth the dawn of the Kingdom."[29]

The artist says: "We have to prepare for the Second Coming of Christ. That means taking out all the elements of current human experience that can pass through the fires of the end of the world. And in pointing out these elements, I see the role of culture, philosophy and art."[30]

In Nowosielski's theological and artistic analyses, a significant role is played by the subtle beings, that is, the angels assuming the function of messengers between human consciousness and God, who is beyond our access. The artist emphasizes that in the Eastern tradition everything beyond God is, in a certain way, material. The angels possess luminous subtle matter different from human matter, and are genderless.

For Nowosielski, a direct contact with the world of subtle beings is possible because he himself possesses a gift that allows him to unite with the mystery of God and subtle beings through his spiritual experience. When he writes that art is an esoteric domain of man, he means a specific bond between the world of experience and the world of subtle beings, i.e. angels.

The Polish religious thinker leaves art under the guidance of angels. In the artist's oeuvre, abstract painting and abstract art in general are means of expression that enable an interaction with the spiritual reality, subtle beings and celestial powers. Nowosielski writes: "For me, abstract painting is a form of our human consciousness

---

[28] Z. Podgórzec, *Mój Chrystus*, op. cit., p. 13.

[29] H. Paprocki, "On Observing the Art of Jerzy Nowosielski. An Essay on the Nature of Art", in *Jerzy Nowosielski. Villa dei Misteri*, op. cit., p. 147.

[30] Z. Podgórzec, *Mój Chrystus*, op. cit., p. 22.

reacting to extrasensory consciousness – the extrasensory consciousness, which permeates us."[31] As Mieczysław Porębski writes, commenting on the Polish painter's art: "The artist is convinced that an abstract painting is also an icon, an icon of the angel, a recording of our sensations emerging from our contacts with the world of subtle beings, unmediated by symbols or prostheses."[32]

In Nowosielski's theological analyses of the empirical reality, pessimism and a feeling of life's misery dominate. When he talks about his painting, however, he seems to be fulfilled and happy. In the domain of art, happiness and experience of the good is possible. It is in abstract painting that Nowosielski achieves *katharsis* through contact with the spiritual world, luminous and good. He emerges from the shadows of evil and contemplates the otherworldly good and beauty. "(...) in painting abstraction I found peace and stability of contact with the world of good spiritual values, bringing about happiness, a sense of power, and felicity."[33]

Owing to his philosophical and theological faith in the existence of Sophia as a fallen soul of the world or the soul of man betrothed to the Creator, Nowosielski rehabilitates the fallen infernal reality. The elements of good and beauty existing eternally in the nature of Heavenly Sophia can be extracted from this reality and saved. Of course, for the artist-theologian this can be achieved in art, which is a symbol of the ultimate, the transcending of the Apocalypse and the consecration of the world of nature.

Mieczysław Porębski introduced the term of 'eschatological realism' to describe the artist's attitude towards reality, from which he tries to extract the elements of beauty and preserve them in his works. As Porębski writes: "Nowosielski's eschatological realism is a particular kind of acceptance of reality in its entirety, a solidarity with it, with both its "diurnal" and "nocturnal" beauty, bowing down before it in the face of the inevitable."[34]

For the artist, art is an affirmation of the world and also an expression of faith in the reality greater than its manifestation. To confirm his thesis, Nowosielski points to the historical fact that the Byzantine culture did not know the distinction between sacred and profane art. "In fact, I think the *sacrum* domain extends over the whole art of painting. The whole art of painting is sacred, related to the eschatological hope."[35]

Henryk Paprocki comments: "Jerzy Nowosielski's work forces you to stop and think about the phenomenon of culture, since the artist himself says that he only paints icons, and that there is no dividing–line between what we conventionally call 'religious' or 'sacred' art and 'profane' art. Professor Nowosielski relates his entire

---

[31] Z. Podgórzec, *Wokół ikony*, op. cit., p. 184.

[32] M. Porębski, *Nowosielski*, op. cit., p. 117.

[33] J. Nowosielski, "O abstrakcjach", in *Notatk*i. Part 2; J. Nowosielski (Kraków: Starmach Gallery, April 2000), (paper presented on April 27, 1983, at the Academy of Fine Arts in Kraków, and on April 30, 1983 at the State Higher Theatre School in Kraków), p. 40.

[34] M. Porębski, *Nowosielski*, op. cit., s. 134.

[35] J. Nowosielski, *Sztuka po końcu świata*, op. cit., p. 65.

oeuvre to the 'religious' domain".[36] It follows that, as he further claims: "If there is no division between the *sacred* and the *profane*: if the whole of art belongs to the realm of the *sacred*, then it is a manifestation of eschatological fullness. In this sense art is 'not of this world'."[37]

The artist finds the sphere of *sacrum* not only in icons, sacred murals, but also in his own, the so-called profane paintings – nudes, landscapes, which he also treats as true sacred paintings. He even said that he lit a candle in front of a landscape.

Mieczysław Porębski, commenting on the opus of Nowosielski, summarizes it briefly: "The Christological baseline requires an extension by at least another two lines: the angelological, associated by Nowosielski with his abstract painting, and the sophiological, which encompasses his entire figurative artistic output presenting secular themes – imaginative portraits, nudes, interiors, landscapes, still lives."[38] Consequently, for Nowosielski it is not only the icons which he paints for liturgical use that have sacred meaning; secular paintings manifest it as well. "Everything that is well painted is an icon"[39] – even a landscape, he says. What is more, he thinks that Malevich's or Mondrian's paintings would find their proper places in the church.

Portraits – Nowosielski's icons, as Mieczysław Porębski claims, represent his contemporaries "looking at us with their Byzantine eyes." The wonder of his portrait painting stems from the way in which he "brings back to life and updates the old icon tradition in a new and astonishing fashion."[40] Jan Stalony-Dobrzański adds: "But his people, both in his icons and those seemingly beyond them in his secular portraits, often constitute the source of pain and unrest. The source of the eschatological restlessness is a volcanic, hot, and still crystallizing material. But Nowosielski's inanimate world, his object, his interior, and in particular his landscape and his architecture, have already reached a conclusion. They have touched the glow of transfiguration. They are the Lamps of Tabor – the purest light of the icon."[41]

Therefore, everything recorded by our consciousness and preserved by its cognitive capabilities is resurrected – here and now – since there is no ultimate border between time and eternity. That which is resurrected is indestructible and immortal – in the human faces – the icons, the world of nature and the entire cosmos.

Nowosielski, reflecting on the history of Christianity, just like Nicolai Berdyaev, distinguishes between the "history accompanying" Petrine Church connected with Rome, and the eschatological Johannine Church, awaiting the Second Coming of Christ, the Church eternal forever. In the artist's opinion, the Johnian Church is the birthplace of the icon.

---

[36] H. Paprocki," On Observing the Art of Jerzy Nowosielski. An Essay on the Nature of Art", op. cit., p. 137.

[37] Ibidem, p. 140.

[38] M. Porębski, *Nowosielski*, op. cit., p. 115.

[39] J. Nowosielski, *Sztuka po końcu świata*, op. cit., p. 74.

[40] M. Porębski, *Nowosielski*, p. 27.

[41] J. Stalony-Dobrzański, "Villa dei Misteri", in J. Nowosielski, *Villa dei Misteri*, (Białystok: Galeria Słowiańska, ORTHDRUK Sp. z o.o., 1998), pp. 94–95.

The icon comes from supernatural reality and is a gift of the immediate vision of a resurrected reality. Thanks to the icon, man interacts with the supernatural reality, that is happier and more real than our human reality, which is unsteady and constantly threatened with decay. Thus, in the Orthodox *theosis*, divinization finds its material prototype in the icon. As Henryk Paprocki writes: "An icon comes from Heaven. If it is not given 'from above', then it never comes into existence at all. The painter of icons has the 'heroic virtue' of creating an artistic vision. The monstrosity of empirical reality is overcome, and the world becomes a theophany."[42]

The theological meaning of the icon in Nowosielski's oeuvre is aptly described by Adriana Kunka: "The icon is a mirror turned to the light of heaven. It reflects the realities of another existence, in which the dark and infernal aspect of human existence is also given a place in salvation. The icon is the ultimate rung on the ladder to heaven, but at the same time it is man's first step in his climb to God."[43]

Nowosielski in his questions, addressed mainly to the Orthodox Church, argues about the existence of God, Satan, and man immersed in the world of evil. We remember that the power of Satan, embodying the cosmic forces of darkness and evil, brings man within the range of the real reign of darkness, in the situation of the Fall described in the terms of a mythos. Nevertheless, he persists in his efforts to extract light and goodness from the darkness of the empirical reality he is immersed in. This effort is best visible and effective in culture and especially in art. As Wacław Hryniewicz aptly points out when analyzing Nowosielski's views: "Although the blackness of historiosophic pessimism is prevalent in this thinking, a bright tint of gold should be noticed therein – a thread of eschatological optimism and hope."[44] Therefore, in this reality art is that which allows the eschatic experience by transfiguring evil into good and to beauty. It shows us how beauty saves the world by overcoming the destructive, satanic power of evil. Through his painting, Nowosielski ennobles the whole world. Of special importance to him is the icon, representing the saved and transfigured reality. It is in the icon that the rehabilitation of all corporality of this world takes place. It prepares the space of New Jerusalem, where man will live. New Jerusalem will eventually appear after the second cosmic catastrophe, preceded by the purging fire of the end of the world. Although good will prevail, the divinized cosmos will most probably assume the form of existence beyond the dual experience of good and evil.

---

[42] H. Paprocki, "On Observing the Art of Jerzy Nowosielski. An Essay on the Nature of Art", op. cit., p. 142.

[43] A. Kunka, "On Contemplating Woman as Represented in the Art of Jerzy Nowosielski" in *Jerzy Nowosielski, Villa dei Misteri*, op. cit., p. 179.

[44] W. Hryniewicz OMI, *Kościół jest jeden. Ekumeniczne nadzieje nowego stulecia*, (Kraków: Wydawnictwo Znak, 2004) p. 161.

Part III

Anthropological Regression in the Modern World Versus Anna-Teresa Tymieniecka's Metaphysics of Ontopoiesis of Life

Jan Szmyd

Abstract Information and technical civilization and the neo-liberalist social and economic system embedded in it causes a number of unfavourable changes in the psychical, mental, personal, somatic and health spheres of life of the modern man.

They have a particularly negative impact on man's subjectivity and spirituality, quality and style of his life, internal development and traits that constitute humanity.

Among characteristic manifestations of changes discussed here it is necessary to include: emotional and empathic impoverishment of many individuals, weakening and intentional unilaterality of inter-human relations (transformation of such relations into instrumental relations, based on interest, making them formal and functional), trivialization and simplification of lifestyles, depriving life of deeper sense, falling into irrational consumerism and commonplace hedonism, into a life characterized by the premise of "to have" and not "to be", to become enriched internally and to aim for self-fulfillment.

Further manifestations of the changes that are of interest to us are the increasing symptoms of nervous weakening of many people, especially young people, disintegration or confusion of their personality, uniformisation and dulling of their mentality, lowering the skill of understanding and comprehending the reality in which we live, frequent confusion of what is real with what is virtual, increasing anxieties and internal frustrations, worrying types of various new neuroses and social diseases.

The basic causes of such anthropological changes which take place in the context of technical and information civilization in the area of modern globalization process are, *inter alia*, the overwhelming impact on mentality, stances and behaviour of an average media person (i.e. inadequate, unilateral and frequently deformed), picture of reality, excessively accelerated and continuously speeding up technological

J. Szmyd (✉)
University of Krakow, Krakow, Poland
e-mail: janszmyd@interia.pl

A.-T. Tymieniecka (ed.), *Phenomenology of Space and Time: The Forces of the Cosmos and the Ontopoietic Genesis of Life: Book One*, Analecta Husserliana 116,
DOI 10.1007/978-3-319-02015-0_9, © Springer International Publishing Switzerland 2014

developments whose results – in the form of various constructs, devices, items, IT programmes, etc. – are used so commonly and so mindlessly with respect to their internal mechanisms and technical functionality, especially from the side of various and long-term negative effects of their too frequent usability.

Among further causes of anthropological changes discussed here is the all-powerful impact on the mentality and emotionality of the modern man of various types of mass culture, frequently of very mediocre and spiritually blunt genre, successfully supplanting the impact of homo-creative high culture.

The discussed anthropological, social and cultural changes not only deform and distort the sphere of spirituality, consciousness and behaviour of man, posing a serious threat for correct development and survival of man as a species, but they also have a probable impact on a more long-term homo-creative process – on further social and cultural evolution and, to a certain degree, also on the natural and biological evolution of man.

1

Characteristics of a process which is here described as an "anthropological regression" – as opposed to semantically similar, yet different with respect to content – terms such as "regression of humanity" (K. Lorenz), "homo-destruction" (Z. Bauman), "disintegration of man's coherence" (H. Skolimowski), "fall of a human being" (Priest Józef. Tischner), etc. requires, primarily, recalling the basic meaning of the term evolution; evolution in a broader, not only biological dimension.

Following the leading inventors and theorists of this process (Ch. Lyell, Ch. Darwin, H. Spencer, H. Bergson, O. Spengler, A. Huxley, Teilhard de Chardin and other authors), here, we adopt such understanding where this process means – in its general tendency – a creative and, at the same time, a progressive change; a change that gives birth to structures and systems more and more complex and diversified, functionally richer and more efficient, a change that leads to better and better "learning" of its environment, "learning" the manners of becoming adjusted to it and lasting in it; speaking generally and slightly metaphorically – we are dealing with "upwards development", with a change that multiplies values.

The creative nature of evolutionary change – some theorists identify it with extraordinary and, fundamentally, mysterious creation[1] – and scientifically not thoroughly examined natural coming into being of something new, e.g. various forms of

[1] *Cf.* Konrad Lorenz, *Regres człowieczeństwa*, translated by Anna Danuta Tauszyńska (Państwowy Instytut Wydawniczy, Warsaw 1986), p. 35 (title of the original: *Der Abbau Des Menschlichen* (R. Piper & Co. Verlag, München 1983); J. Levine, D. Suzuki, *Tajemnica życia*, translated by Gasparska, B. Skarżyńska (KiW, Warsaw 1996), Henryk Skolimowski, *Święte siedlisko człowieka. O magii i pięknie życia*, translated by Robert Palusinski (Centrum Uniwersalizmu przy Uniwersytecie Warszawskim, Warsaw 1999) (title of the original: *A Sacred Place to Dwell Living with Reference Upon the Earth* (1993), op. cit., p. 77; Zdzisława Piątek, *Ekofilozofia* (Wydawnictwo Uniwersytetu Jagiellońskiego, Cracow 2008), p. 94–115.

life, consciousness, spirituality, etc. excludes – contrary to standpoints of certain early evolutionists, any type of predestination, or the course of this change, understood mechanistically or strictly in terms of regularity; however, it assumes that evolutionary change possesses amazing creativity, impressive inventiveness, certain "resourcefulness" and, therefore, unpredictability and inscrutability.[2]

Nevertheless, the above-listed features of evolutionary change do not exclude – which is noticed and emphasized more rarely – the characteristic feature, which generally speaking, may be determined as "downwards" change, i.e. its specific turn in the regressive direction; a direction that is opposite to the direction usually associated with evolutionary progression, i.e. the direction which, instead of multiplying values – decreases them, instead of enriching the created structures, impoverishes them, instead of improving their functional utility and increasing the "learning" skills of the environment and its better adaptation, causes gradual weakening of this vital and adaptive feature of its entities, etc.

This, paradoxically speaking, "downwards development" is described in a variety of ways in literature on the subject; e.g. it is called "regressive evolution", "involution", "regression", "de-generalization", "defective development", etc.

Here, we adopt a specific term to determine the "backwards" evolutionary process that is of interest to us that is most compliant – it seems to us – but obviously only approximately, with the requirement of relative adequacy, i.e. the term "evolutionary regression"; regression which, in reference to man's evolution, may be semantically reformulated as "anthropological regression."

This "deviation" or this "rebellious" turn in the dominant trend of the evolutionary process is no longer only a hypothetical assumption in modern evolutionary theories (there are many such theories), but an empirical, scientific establishment of an indisputable fact/phenomenon[3]; a fact/phenomenon that is scientifically confirmed both in the biological evolution of "simple" living organisms, as well as on a higher level of evolution, i.e. evolution of the human species, i.e. the psychical, cultural and social evolution.[4]

However, this fact/phenomenon is very intricate and complex, yet its significance is fundamental, as it determines the quality and survival of evolutionary entities.

---

[2] *Cf.* Henri Bergson, *Ewolucja twórcza*, translated by Florian Żnaniecki, Warsaw 1913 (title of the original: L'évolution créatrice, Paris 1907); Teilhard de Chardin, *Człowiek*, translated by J. G. Fedorowscy (Warsaw 1962) (title of the original: Le Phénomène humain, vol. 1, Paris 1955); *L'Apparition de l'homme*, vol. 2, (Paris 1956); Henryk Skolimowski, *Święte siedlisko człowieka* (Biblioteka Dialogu, Warsaw 1999); Konrad Lorenz, *Regres człowieczeństwa*, Translated by Anna Danuta Tauszyńska (Państwowy Instytut Wydawniczy, Warsaw 1986); Anna-Teresa Tymieniecka: Życie w pełni logos. Księga I, Metafizyka Nowego Oświecenia (Wydawnictwo Poznańskie, Poznań 2011) (title of the original: The Fullness of the Logos in the Key of Life, Book I, The Case of God in the New Enlightenment, Springer 2009).

[3] *Cf.* Konrad Lorenz, *Regres człowieczeństwa*, op. cit.; H. Skolimowski, Górecki J. K., *Zielone oko Kosmosu. Wokół filozofii w rozmowie i esejach* (Wyd. Atla 2, Wrocław 2003); Zdzisława Piątek, *Pawi ogon czyli o biologicznych uwarunkowaniach kultury* (Wydawnictwo Uniwersytetu Jagiellońskiego, Wyd. I, Cracow 2007); A. T. Tymieniecka, *Logos and Life*: *Impetus and Equipoise in the Life – Strategies of Reason* (Book 4, Analecta Husserliana, Volume C, Springer 2009).

[4] *Cf.* Konrad Lorenz, *Regres człowieczeństwa*, op. cit., p. 35, 158.

Let us present it here briefly and let us do it in full correlation with the concept of contemporary education and modern anthropological and pedagogical thought.

Therefore, what is the "anthropological regression" and what are the challenges posed by it for social sciences, including social and philosophical anthropology?

This question can be divided into several further questions, more detailed, such as:

- What happens to man at the current multi-dimensional (multi-course) i.e. biological, psychical, social and cultural development; in which degree is man – in the above-listed spheres of life – subject to regressive changes and what are the main manifestations and effects of such changes?

Taking into account the most characteristic manifestations of such unfavourable changes, the following question also arises:

- What, in the era of scientific, technical and IT revolution, common globalization, neo-liberal social and economic system (modern capitalism) and technocratic civilization, happens to the human nature, the human being, mentality and human behaviour, human lifestyle and existence in the world and, in particular, what happens with man's spirituality, subjectivity, identity and humanity, i.e. with these spheres of internal life of man which seem to be most prone to the process of clear regression?

Generally speaking, a question arises: What is actually happening? And specifically: What bad is happening along the way of evolutionary humanization of the human being?

- These questions may be expressed differently, i.e. by using the Latin phrase: *quo vadis homine*? i.e. where are you going? – assuming that the road sign may still be under man's conscious and rational control.

Further questions (some of them have already been formulated in the introduction to this paper):

- What is the quality of lifestyle of an average modern man, what is his attitude to the values of higher culture and significant accomplishments of humanity and how does this attitude translate to personal culture: intellectual, moral, aesthetic and subjective/spiritual?
- Are we not in a state of understated self-awareness, of one's own self and sense of existence, in excessive obfuscation of significant values and purposes of life and a radically lowered understanding of the world in which we live?
- Does the modern man, fully and on a level of evolutionarily shaped potential, realize his genetic, psychical and physical potential, in particular the personal and emotional potential?
- What is the status and the level of his spirituality and internal life in comparison to the standards achieved in prior historical periods?

Obviously, a prominent place in the circle of such fundamental questions about main directions and internal and external conditions (related to consciousness, intentions, society and economy, technology, information and civilization) of the process that is of interest to us and its predictable – yet to a greater extent unpredictable – consequences is taken up by the issue of dimension and depth of anthropological regression.

It is obvious that these and similar questions cannot be provided with exhaustive answers here; anyway, to many of them, there are no fully satisfactory answers in social and natural sciences, as well as in modern philosophy. In many cases, we are relying solely on intuition, imagination or, due to the very nature of this issue, uncertain anthropological and social futurology.

In this place, we need to restrict ourselves to experimental preparation of a register of main and, at the same time, well recognized manifestations, conditions and effects of anthropological regression and initial characteristics of selected examples.

Man "evolved" from nature, from a wide circle of his ancestors (in this place, we are not going to describe this rich and complex evolutionary genealogy), whose primary task was murderous fight for survival, preservation of life, acquisition of food, protection of own offspring, etc. and was primarily equipped by evolutionary mechanisms and processes in "hard" adaptation and vital skills. Apart from the drive to fight, they include self-preservation, egoism, etc.; there is also the natural drive not only to secure and protect life, but also that life is improved, comfortable and pleasant. It was only in the second place that the man procured strengths and predispositions that are necessary in co-existence with others – the so-called "soft" strengths and predispositions, such as an inclination for disinterested help, interest in others, taking care of others, spontaneous solidarity, respect and kindness, emotional bond, etc.

Man has been realizing these two, significantly opposing and, to a certain degree, contradictory sides of his nature with varying successes. At some times and in certain situations, he released the "hard" drives in a greater extent (e.g. during wars, social revolutions, internal conflicts, etc.), whereas in others, he was able to inhibit them and gave vent to "soft" pro-human drives (as a matter of fact, the entire history of humanity could be described in a very interesting manner from this point of view).

However, in this place, it is necessary to emphasize that man, as a result of coincidence of historical aspects (trade, geographic and discovery, colonial and occupational, technical, scientific, philosophical – including philosophical and scientific rationalism and utilitarianism – [Descartes, Newton, Voltaire, French Encyclopedists, English ethics and economists – J.S. Mill, J. Bentham et al.]), but not as a result of inevitable necessities and historical regularities – came across – already in the early modern times, in the fifteenth, sixteenth and seventeenth century – a social and economic system, developed dynamically and expansively to date – a system known as capitalism (currently, it can be called global neo-capitalism or post-capitalism).

This system, as no other, maximally released the above-mentioned "hard" tendencies and drives of the human nature. It created huge and efficient stimulation for them and relatively optimum conditions for their development, along with convenient manners, tools and measures for satisfying them. At the same time, it efficiently inhibited, restrained and frequently even blocked the "soft" drives and impulses – also (as mentioned above) integrally inherent in the human nature. Here, we are talking not only about the natural predisposition for altruism, disinterestedness, solidarity, assistance or ethical and aesthetic sensitivity, but also such subtle traits of the human nature and mentality as the study of reality, metaphysical and religious contemplation, deeper internal spirituality, lively emotionality and empathy, etc.

The excess release of "hard" drives and tendencies, including exuberant egoism, excessive calculation, absence of care for others, predatory competition, disproportionate belligerence for "trophies", aggression and violence, immoderate desire for possession and profit, making them, as a rule, dominant in the sphere of human desires, plans and life aspirations, has, to a certain extent, "knocked out" evolution of man from its internal balance, harmony and directed its certain evolutionary tendencies to a side track, significantly disrupting the correct nature of human evolutionary process.

As a result of this fundamental disruption of the evolutionary process, it is gradually falling into – in the modern times, this is happening very quickly – an inharmonious and irregular course of development of its natural tendencies and predispositions; it is sinking into certain evolutionary "dilemma" and into a specific evolutionary antinomy and into evolutionary inhibitions and withdrawals which are dangerous for its sustainable development, which are called "anthropological regression" here. This condition denotes deep deformation of comprehensively and harmoniously understood human being. Precisely, it denotes the progressing disintegration of "cohesion of internal life" of man, i.e. the growing distance between the pace of intellectual, rational and conceptual development and emotional and spiritual development.[5]

Moreover, it denotes a very dangerous dilation between the relatively slow pace of philo-genetic development and the accelerated functional and civilizational development of man; between the standstill and even regression of deeper spirituality of man and very quick pace of development of his functional and technical skills; between life pragmatism and the sphere of experiences of deeper values and vitality of spiritual transcendence: metaphysical, religious and aesthetic which is called by some – with good reason – the "damaging" and the "falling down" of humanity.[6]

H. Skolimowski emphasizes: "(…) destruction of important values – in the nineteenth century – and the atrophy of beauty and the loss of sense – in the twentieth century – are aspects of the same process; they are the symptoms of the fall of man as a transcendent and spiritual being".[7]

One more statement from this excellent philosopher: "We are currently witnessing not only intellectual and aesthetic confusion, but also confusion inside our soul; not only the collapse of the ideals of beauty, but also the collapse of coherence in our

[5] *Cf.* Henryk Skolimowski, *Święte siedlisko człowieka*, op. cit., p. 12, 162.

[6] *Cf.* Ks. Józef Tischner, *Spór o istnienie człowieka*, op. cit., p. 67–78.

[7] *Cf.* Henryk Skolimowski, *Święte siedlisko człowieka*, op. cit., p. 164.

lives; not only the loss of traditional handicraft, superseded by tacky plastic products, but the loss of spirituality and God".[8]

2

The most explicit and probably the most meaningful manifestation and, at the same time, coefficient of "anthropological regression" is what Jose Luis Sampedro [9] – a well-known Spanish thinker and banker, as well as one of the spiritual leaders of "the indignant" movement – called a "dwarfish" – overwhelming the modern man – social outlook, i.e. such a general view on the human world according to which everything is limited to money; where money is the total reference for everything; it is the final criterion for the majority of values and assessments; where market and economic values are deemed most important; interest, profit, profitability, property, ownership of material items, desire for power and influence, dominance and ruling are the leading motivation for efforts and public actions; economic efficiency and the so-called "economic progress" is treated as the most important and completely indisputable determinant of social and civilizational development and, in this context, development of man.

The fact that this social outlook is really "dwarfish" and overly one-sided, not to say mentally and morally handicapped, has not been realized by anybody in a situation of almost common and surprisingly efficient indoctrination of a majority of human minds (mainly by the mass media) – excluding a definite minority of people who are not completely subjected to such super-indoctrination. This group includes a few intellectuals, excellent thinkers, philosophers and artists and the increasing number of movements of "the indignant", alter-globalists, anti-globalists and other radical modern social groups.

At the same time, people indoctrinated by this truly "dwarfish" and – as we said – intellectually handicapped outlook, are generally not aware of the obvious fact that this outlook is constantly losing not only its theoretical and moral reason, but also its "coverage" in natural and energy resources of Nature (*Gaia*) and intellectual, moral and humanistic acquiescence of people who are deeply reflective and truly concerned about the future of man and his best cultural and civilizational accomplishments.

Some critical analysts and shrewd observers of the contemporary world, i.e. its current social and inter-human relations, perceive as very dangerous – from the point of view of correct personal and spiritual development of people (and it is difficult to disagree with them) – usually unsuccessful attempts made by young people consisting in a transfer (transposition) of the "digital world" (computer, Internet) to the "non-digital world" ("outline") i.e. the real world. A majority of models of life,

---

[8] Ibidem, p. 162.

[9] *Cf.* Jose Luis Sampedro, Nadchodzi lepszy świat, "Gazeta Wyborcza", 21–22 April 2012, p. 15–16.

thinking and behaviour in the "digital world" that are easy to execute for obvious reasons do not correspond to their hard equivalents in the "non-digital", real world and, as a rule, require great effort and – obviously – do not create a convenient area for realizing non-realistic, virtual and, at the same time, simplified proposals for the "art of life" and "being-in-the world." In fact, they weaken self-identification, individual self-determination of personality, autonomy and internal identity, feeling of trust to others and personal credibility, responsibility and affiliation to community, solidarity and reliance, respect and dignity – indispensable conditions for shaping the appropriately strong and integrated personality, richer spirituality and, at the same time, main pillars of communal life.[10]

Therefore, inevitably, a question arises: What is the way out from this – generally speaking – regressive anthropological situation?

Everything indicates that there is only one way out – the only one – i.e. a change or thorough repair of the social and economic system which inevitably generates such situations and constantly aggravates them – with all the negative and destructive effects; which, on a massive scale, directly or indirectly, influences the progressing regression in the sphere of human mentality, spirituality, behaviour and existential creativity.

This change or radical reform is – contrary to pretences, intellectual habits and stereotypes shaped in the last centuries – indispensable, without alternative and necessary, if the regression of humanity is to be effectively stopped or at least significantly decreased.

However, this systemic, social and civilizational change is, obviously, a change that is very difficult and it is unknown whether it is still possible.

As of now, there are no significantly strong social, political and ideological forces that could perform this unprecedented historic revolution. What is more, there is no great determination and ideological inspiration to realize it, unless the issue is going to be decided by a chain of some great and unexpected social events or total economic crises or some sudden, today unimaginable, civilizational collapses, or possibly the fact that such movements as the movement of the "indignant" or anti-globalist groups or similar future movements will gather enough strength to perform this fundamental and – as it seems – necessary from the point of prospective human interests – revolutionary or evolutionary turn in the current and – as we tried to show before – significantly disrupted development trends of the human species.[11]

Without voicing any, due to obvious reasons, specific judgment or even an assumption on such unpredictable and even completely unimaginable civilizational

[10] *Cf.* So Is There Going to be a War? Tomasz Kwaśniewski asks and Professor Zygmunt Bauman answers, "Gazeta Wyborcza" daily, April 15, 2012, p. 14–17.

[11] *Cf.* Teodor Krzysztof Toeplitz, *Dokąd prowadzą nas media* (Wydawnictwo ISKRY, Warsaw 2006); Wiesław Sztumski, *Quo ruis, homo? Środowisko życia, czas, ludzie* (Wydawnictwo Naukowe "Śląsk", Katowice 2008); Zygmunt Bauman, *Socjalizm. Utopia w działaniu*, translated by M. Bogdan, Warsaw 2010; (title of the original: Socialism: The Active Utopia, George Allen & Unwia Ltd., 1976); Jan Szmyd, *Odczytywanie współczesności. Perspektywa antropologiczna, etyczna i edukacyjna* (Oficyna Wydawnicza AFM, Cracow 2011).

transformation (a transformation with basic consequences for the correct and unthreatened development of the human being) – it is now possible it seems to me – to formulate a justified thesis that the issue discussed here is gradually becoming a great challenge for open-minded and responsible intellectual and ideological elites of the contemporary world; for outstanding influential representatives of social and natural sciences, for creative and humanistically oriented philosophers, moralists, artists and clerics. One should not really count upon the current political, financial and media elites (elites of power, influences and ownership), in the majority of cases due to well-known reasons.

3

This transformation of the general condition of man which is taking place in the contemporary world, thorough and very characteristic, yet generally negative in its consequences, the transformation of life and spirituality of man which is here described as the "anthropological regression" requires – along with the entire civilizational process in which it is taking place – due to obvious reasons – also a philosophical approach. It is only from the philosophical perspective that it is going to obtain – or may obtain – a fuller and more in-depth explanation – both from the side of its nature and sense and basic significance for man, including its homo-destructive effect. Therefore, a question arises: What can philosophy, in its current not-so-good cognitive condition – condition that is greatly diverging from the possibilities and expectations that are traditionally ascribed to it – may really do in this situation? These possibilities, in spite of the common pessimistic convictions, are quite numerous:

- First of all, philosophy can (and should) simply start dealing seriously with this current process that is so important for people; a process that is laden with various consequences that are dangerous for the human species – both immediate and long-term ones – in particular threatening for man and his spirituality. Here, we are talking about philosophically specified explanations of the nature and the character of this process, its specific modalities and anthropological functions, effects and consequences for the correct development of the mental and spiritual condition of man. Philosophy should also become involved in an in-depth and versatile philosophical recognition and explanation of the main external – civilizational, economic, social and technological – conditions, stimulators and mechanisms of the process discussed here.

Moreover, it should formulate convincing and reliable evaluations and critical philosophical illuminations of the homo-destructive tendencies of this dangerous and ominous anthropological change and announce communicative, convincing and, therefore, socially effective warnings and cautions with respect to this unfortunate affliction of the modern civilization.

In this place, it would also be good to present a proposal for thought-out and properly arranged means and countermeasures which could efficiently stop or at least significantly restrain the pace of growth of this regressive change which is, in many ways, unfortunate for man.

It would also be helpful to have modern – walking hand in hand with the newest revolutions in natural sciences (in particular biology, molecular physics, cosmology, theory of life) – broadly and fully using a wide range of rich, often surprising cognitive results that disrupt the stereotypes of scientific and philosophical thinking, that are multi-disciplinary and that throw new light on the evolution concepts of the cosmos, inanimate matter, biosphere of the psyche, etc., and at the same time confirm two basic theses: first of all that "there is no sanctity in the world and no theory is inviolable"[12] and secondly that even though "the extent of our knowledge is increasing, yet, simultaneously, the extent of our ignorance is growing bigger too"; "we know the secret of the Cosmos in 4 %, yet in 96 % it remains unknown to us".[13]

These newest concepts of evolution of various manifestations of the Universe may provide a very important and instructive point of reference for proper understanding of this peculiar disruption of man's evolution, in particular his spiritual status, marked by us with the term "anthropological regression."

Let us add that in this cognitive radicalism, attention is focused not only on deeper description, explanation and interpretation of the process discussed here, but also on influence on the desired and possible correction of its course, in line with the still valid motto that the obligation of philosophy is not only to "interpret" the world, but also to contribute to its change.[14]

This double, i.e. theoretical and explanatory and, at the same time, constructive and practical function may be performed by philosophy – as it has proved many times during its rich history – with greater or smaller success, because it is (it may be) not only a more or less clear "reflection" ("mirror") of reality, but also the expected "tool" or "instrument" of such transformation.[15]

Obviously, not every philosophy, not every type of philosophy is able to fulfill these dual and difficult cognitive and practical functions efficiently with respect to this or an other area of reality and its manifestations that are important for man; in the specific case discussed here with respect to the so-called "anthropological regression."

---

[12] Mark Henderson, *Cząstka Boga i prędkość światła*, "Gazeta Wyborcza", February 1, 2012, p. 7.

[13] A Different Point of View, Interview of G. Miecugow with Professor M. Różyczka of PAN, astronomer (TVP 24, 1.01.2012)

[14] This postulate of K. Marx formulated in the nineteenth century regarding the role of philosophy (philosopher) as can be seen, has not lost its significance at the beginning of the twenty-first century.

[15] *Cf.* Jan Szmyd, *Odczytywanie współczesności. Perspektywa antropologiczna, etyczna i edukacyjna* (Oficyna Wydawnicza AFM, Kraków 2011); Chapter II: Wielokierunkowość filozoficznego i naukowego odczytywania świata ludzkiego; Chapter III: Czy filozofia współczesna "odzwierciedla" kulturę moralną; Chapter IV: Cywilizacja współczesna z perspektywy filozofii edukacji, p. 35–68.

This reservation with respect to philosophy gains special significance in the era of its clear crisis and cognitive sterility of traditionally understood philosophy, including traditionally understood philosophical anthropology, i.e. maximalist philosophies and speculative intellectual constructs; constructs that try to reconstruct the conceptually hypothetical beings and their certain features and elements, including speculatively understood human being, its nature, etc. This type of philosophical reflection, insensitive and cognitively closed to infinite complexity, variety and changeability of reality, which is encrusted in old intellectual paradigms and obsolete manners of practicing (only intuitive, ideological and speculative) – omitting or neglecting the wealth of life experiences, practical struggles with reality and, in particular, growing abundance of cognitive data of various scientific empirical studies – has almost completely lost its cognitive power. According to A.-T. Tymieniecka, "credibility of old, fossilized chains of terms and theories has been loosened and weakened, along with pre-determined concepts with respect to the human nature, the world, the environment, the moral standards, laws and principles of ethics; their power of convincing takes place under the pressure of new perspectives open to scientific progress".[16] She adds: "The multitude of empirical studies of reality that is spreading everywhere is not conducive to the great principles determined a long time ago by speculative imagination, as far as dealing with questions of philosophy is concerned and the conduct of most personal searches for wisdom; thence, these principles have no application today. In our post-modernist era, they are just an anachronism".[17]

One more alarming statement of the author of the *Life of Logos*: "the current state of affairs is calling for salvation deriving from the mind itself; it is calling for a philosophy that would release us from the impasse in which we are stuck and that would lead us further".[18]

In the "new world", the "old philosophy" seems to be completely lost, cognitively and practically helpless due to the fact that the new world is fundamentally different from the old one and, at the same time, so changeable and complex that it is clearly escaping the current human cognitive potential and, in consequence, leaving man in a state of growing uncertainty. It may be stated briefly that "living in unique times, we are submerged in a great variety of new ideas, experiences, customs and intuitions. We have to devote a lot of our time and effort to get to know them, understand them and apply them in practice. It seems that it is not only difficult for us to get our bearings in the tangle of constant changes, but that we are also not able to keep up with and embrace the constantly new and surprising form of reality. The growing knowledge about nature, the world, cosmos, as well as about people, keeps humanity in a state of constant insecurity".[19]

---

[16] Anna-Teresa Tymieniecka, Życie w pełni logos. Księga I, Metafizyka Nowego Oświecenia. Translated by M. Wiertlewska, op. cit., p. 14

[17] Ibidem, p. 15, 16.

[18] Ibidem, p. 17.

[19] Ibidem, p. 11.

The above-quoted author adds: "(...) quick telecommunication and abundance of equipment that facilitates and accelerates the pace of our every-day life, not only transformed our existence in many ways, but also put us on our guard in expectation of further miracles and shocks. The entire humanity is simply awaiting, but also fearing the constant, progressing transformation of life".[20]

4

However, there is hope that the expected new philosophy could be somehow helpful in strengthening this cognitively weakened embrace of the quickly changing reality (human reality) and in soothing the increasing fear and insecurity with respect to the future of man of modern times. Yet, this has to be a completely new philosophy! A philosophy that, *nota bene*, has been created in its basic premises, creatively and with great impetus, in several fundamental modern philosophical works.[21]

Its *novum* consists in several innovative, original and exceptionally inventive and, at the same time, fundamentally different from the majority of hitherto philosophical directions, including the main trends of classical phenomenology (M. Scheller, E. Husserl and R. Ingarden et al.) solutions.

These are:

1. Ability for cognitive embracing and reasonable "reading into" the deeper layers and mechanisms of the complex and differing reality; mainly the reality of the human world and the ability of "listening attentively" to the "spirit of our times".[22]
2. Appropriate to this skill, the ability to pose new philosophical questions, at the same time maintaining the validity of many traditional, to a certain degree, universal questions, such as, *inter alia*: "What is being?", "What is its nature?", "What is the cause of its changeability?", "What can we know?", "How can we know it?", "Who is man?", "What is man's nature?", "What is the place and the role of man in the world?", "What can we hope for?", etc.

   As far as new questions are concerned, they are formulated in the following manner: "What is the actual "source" of being?", "What is the nature of creative ontopoiesis of being?", "What is the place and the role of life in the evolutionary transformation of being?", "What does the process of self-individualization of life consist in?", "How can I be?", "What makes our being possible?", "What is the logos of life?", etc.

   To these and many other new questions of metaphysical, epistemological and anthropological nature, posed inquisitively in the new works of A.-T. Tymieniecka, it is necessary to add a few more questions, closer related to the issue undertaken in these divagations, i.e. anthropological regression, such as: "Is anthropological

[20] Ibidem.

[21] This mainly refers to previously quoted works of Teilhard de Chardin, K. Lorenz, H. Skolimowski, Z. Bauman, Priest J. Tischner and others.

[22] Cf. Anna-Teresa Tymieniecka: Życie w pełni logos. Księga I, Metafizyka Nowego Oświecenia, op. cit., p. 11–23.

regression an inevitable and irreversible process in the entirety of man's evolutionary transformations?", "What is the final human-anthropological result that it can lead to?", "Are there any efficient forces, manners and measures for counteracting this process?", "What can be done by sciences in this respect and what can be done by philosophy understood in modern terms?"

3. It can be easily perceived, even in the context of the above questions that the object of such philosophy and its fundamental cognitive and practical objectives has to be maximalist, containing almost all of the basic philosophical areas: metaphysics, theory of cognition, ethics, logic, philosophical anthropology and moreover, many new philosophical specializations, such as philosophy of life, bio-ethics, eco-philosophy, etc. This has to be a philosophy whose range of issues and questions is very broad and whose cognitive insight has to be very deep, reaching to the primeval source of everything and to the sense of sense itself, to the "first principles", to the primeval mobility of all-being, to the primordial sources of Life and Spirituality, to deepest mechanisms and conditions, relations and ties of all forms of being and life.

The demand for such "source" philosophy, integral and maximalist, has not been exhausted against all appearances; in contrast – in our times it has evidently grown stronger. And such philosophy – which is proved by the works of A.-T. Tymieniecka – is, contrary to the opinions of post-modernist or shallow pragmatic realist skeptics, fully possible; however, it requires a unique talent and huge, or even heroic, intellectual effort.

Let us recall what is written about it by the main creator of this type of philosophy: "And yet today, as in the old times, furtive secrets of reality and human cognition are still being clarified and the "first principles" have not completely disappeared from the field of vision".[23] This thought is expressed more broadly in the following manner: "(...) even though our perception of reality and human involvement in it has diametrically shifted from the uplands of speculative reason to primeval concreteness and its sources, the roads leading from such sources direct our searches towards the final, so ostentatiously cast questions. Even becoming acquainted with the history of development of philosophical reflection encourages us to consider the "eternal return" of *human affairs*, i.e. issues and ideas which awaken the response of our mind. The manner of formulating them – emphasizes the author – is constantly changing, transforming their sense and manner until complete undermining of their reliability, as not corresponding in their intended depiction with what is "real" and their replacement by other issues and ideals. Investigations are changed, but they last. And the issues return, expressed in this or another manner".[24]

Finally, let us emphasize that it is justified to expect that the "ontology of being and becoming" and the "metaphysics of life" of A.-T. Tymieniecka presented here will probably turn out to be very helpful in further development and justification of

---

[23] Ibidem, p. 13.

[24] Ibidem, p. 16.

the concept of anthropological regression that is of interest to us. In particular, they may contribute to throwing light onto many problems that have so far remained unsolved; problems which have been signalled in several basic questions of this paper. Therefore, the original and innovative philosophical structures of the author of "life in the fullness of the logos" deserve further in-depth studies and analyses.

Biologically Organized Quantum Vacuum and the Cosmic Origin of Cellular Life

Attila Grandpierre

Abstract We point out that the origin of life cannot be understood without a closer look at the nature of life. Therefore we present here, for the first time, ten fundamental biological facts opening new avenues to address the question of the cosmic origin of cellular life. We find that all living beings, including cells, have a genuine biological autonomy that acts with the help of spontaneous, physically indeterminate or arbitrary cellular decisions that initiate quantum effects in support of biological aims. We propose that these biologically initiated spontaneous processes are assisted by vacuum processes. A natural corollary arises, telling us that biological processes organize the quantum vacuum processes of living organisms from below the physical level, which in turn strongly suggests that biology is more fundamental than physics. We point out that just as the functions played by the forwards in a well-trained football team cannot be assigned externally by a series of physical forces acting on the bodies of the football players, biological aims and functions cannot be attached to physico-chemical structures of the first cell of the Cosmos by mere physico-chemical processes, but must be assigned by a more general cosmic life form pre-existing before the first cell and containing it like the mother its foetus. This would indicate that the Cosmos is not only the source of stars, galaxies, and cosmic clouds, but also of biologically initiated and organized cosmic "forces" pre-existing in the vacuum and, ultimately, the Cosmic Subject.

Introduction

How can we fulfil our own human nature? How can we understand ourselves as living beings? How can we understand life from the viewpoints of science and philosophy? We find our life in a cosmic context. Basic questions are towering

A. Grandpierre (✉)
Konkoly Observatory, P. O. Box 67, H-1525 Budapest, Hungary
e-mail: grandp@iif.hu

A.-T. Tymieniecka (ed.), *Phenomenology of Space and Time: The Forces of the Cosmos and the Ontopoietic Genesis of Life: Book One*, Analecta Husserliana 116,
DOI 10.1007/978-3-319-02015-0_10,

before us, posited by Nature, ultimate questions which, if left unanswered, then, as a result, our life can miss its genuine, original aim. These ultimate questions impress us human beings trembling from the same cosmic powers resonating within us throughout the life of the cosmic-minded species, the Homo sapiens. We cannot escape from the powers of the Cosmos disquieting and intriguing us with ultimate, cosmic questions, arising from the creative forces of the Universe that form and drive the life of our innermost identity as it is given by Nature. Facing the cosmic questions of life, our mind becomes driven by cosmic forces arising from the ultimate openness of the Universe towards inner and outer infinities. This ultimate openness of the whole of cosmic existence brings forth vast questions about our very human nature: What is the nature of life? What is the origin of life? In order to answer these ultimate, cosmic questions, we have to free ourselves from all prejudices and find the suitable, cosmic perspective in which we can contemplate these ultimate cosmic questions impartially, learning from Nature by our best, Nature-given abilities. A first observation that we can obtain and learn by contemplating these two questions is that we cannot find the answer to the question of the origin of life without being aware of the relevant aspects of the nature of life, without having a closer idea of what we are asking about.

Biological Teleology

Before presenting a tentative list of the most basic problems of life, we have to say a few words about one of life's perhaps most characteristic and nowadays hotly debated properties, teleology (Glasersfeld 1990; Wouters 2005). Recently, Kane (2002: 9) pointed out that due to the development of quantum physics, universal determinism has been in retreat in the physical sciences. At the same time, biology, neuroscience, psychology, psychiatry, social and behavior sciences have been moving in the opposite direction. In this conflicting situation the concept of teleology has played a central role.

It was not a scientist, but the socialist-journalist Marx, who exclaimed that Darwin dealt a "death blow to Teleology" (Dennett 1995: 126). There is a general belief that teleology is scientifically bankrupt, and that history shows it always has been. This belief is based on the widespread opinion that physics is incompatible with teleology (Russell 1946; Taylor 1964, 1967; Walsh 2000; Johnson 2006: 23–24, Illetterati and Michelini 2008, back cover); yet, as we will argue below, this claim can have only a limited validity, even in physics. It is important to keep in mind that there is also the widespread opinion that "it is now also popular, perhaps more so, to defend teleology" (Kreines 2009). There is absolutely no doubt that every effect in the universe can be explained as satisfying final causes, with the aid of the method of maxima and minima (the action principle), as it can from the effective causes (Euler 1744, cited in Lemons 1997: x). The action principle represents the contemporary descendant of final causes (ibid.). Actually, opponents of the least action principle have expressed a hostility toward introducing the concept of teleology into physics, for this notion has usually served as a wedge to infiltrate religious and metaphysical ideas into what should be a purely physical discussion (Barrow and Tipler 1986: 150). Nothwithstanding, Barrow and Tipler (ibid., 123–218) have shown that, on the contrary, teleology has on occasion led to significant scientific advances.

We point out that the rejection of genuine, teleological biological function has, in a certain context, its own but limited scientific basis. As the *Encyclopedia Britannica* tells (in its entry "action"): "Motion, in physics, may be described from at least two points of view: the close-up view and the panoramic view. The close-up view, describing motion by differential equations, involves a local, instant-by-instant charting of the behaviour of an object. The differential equations are statements about quantities localized to a single point in space or single, instantaneous moment of time. By contrast, the action principle is not localized to a point; rather, it involves integrals over an interval of time and (for fields) an extended region of space. The panoramic view, offered by the action principle, reveals not only a complete picture of the actual behaviour of an object but also all the possible routes of development connecting an initial situation with a final situation. Each route between the two situations is characterized by a specific numerical quantity called its action. The principle of least action states that for "small" variations of the paths, the end points being fixed, the action S is an extremum, in most cases a minimum" (Brown 2005: xiv). Teleology is defined in the *Encyclopedia Britannica* as "explanation by reference to some purpose or end". The fixing of the final state gives the action principle a kind of teleological character, since the motion of a physical system is determined in the action principle formulation by both the initial and the final states of the system (Barrow and Tipler 1986: 149).

Importantly, all the fundamental laws can be derived from the least action principle, including Newton's equations, the wave equation, the diffusion equation, Poisson distribution, and each of Maxwell's, Einstein's and Schrödinger's differential equations. General relativity and quantum mechanics both originated from variational principles (Lemons 1997: 111). Depending on the actual branch of physics, the content of the Lagrangian point characterizing the interactions of the system and the auxiliary conditions will be different. The basic textbook of physics serving for university students worldwide, written by Landau and Lifshitz (the famous ten-volume Course of Theoretical Physics series 2000), is based on the least action principle as the core idea. The action principle turns out to be universally applicable in physics. All physical theories established since Newton may be formulated in terms of an action. The action formulation is also elegantly concise. The reader should understand that the entire physical world is described by one single action (Zee 1986: 109). It is a widespread view that the least action principle is equivalent with the fundamental physical equations. We note that in an important sense, the equivalence is not complete, since teleology is not present at the level of fundamental differential equations. Therefore, in the narrow picture of physics containing only the physical conditions plus the differential equations of physics, teleology is, indeed, not present, and this gives an apparently irrefutable scientific basis for the opinion that all forms of teleology must be excluded from physics. Yet if we step out from this narrow picture, we can work in a more complete conceptual framework of physics, in which the least action principle is also available as an explanatory tool, then teleology is actually present in a specific, mechanical form that is fundamentally different from genuine biological and human teleology.

Even acknowledging the endpoint-oriented, teleological character of the least action principle, there is strong resistance to accept this teleology as real, and, consequently, many consider this teleology as being only 'apparent'. The popular attitude against teleology led scientists to regard the least action principle as 'puzzling' because of its 'seemingly' teleological quality: Given a set of initial and final conditions, one is able to find a unique path connecting them, as if the system somehow 'knows' where it's going to end up and how it's going to get there. We point out that this opinion arises from confusing physical, biological and human teleology, which, actually, have a fundamentally different nature. Certainly, elementary particles do not 'know' where to go in the same way as some of us human beings know. The path integral method, worked out by Feynman (1942, 1948, 1964, 1994) offers a kind of explanation in terms of quantum superposition as to why the least action principle works so similarly yet differently than the way we humans do. The system (for example, a photon in the two-slit experiment) explores every possible path to any possible endpoint with the help of virtual particles that are freely created from the quantum vacuum, and the path integral simply calculates the sum of the probability amplitudes for each of them. Interference effects guarantee that only the contributions from the stationary points of the action give histories with appreciable probabilities, and the most probable path corresponds, remarkably, just to the least action. The system does not have to know its endpoint in advance, as a human being has to know where to go, since it does not have to decide about its path, because the physical path arises without the active contribution of the system. Therefore, it may seem that the least action is the result of a simple and mechanical summation of the probabilities of all paths. Yet these probabilities themselves were calculated on the basis of the least action principle (with the help of equations derivable from it). Definitely, the process is similar to a human decision process in which the first phase corresponds to exploring all the possibilities, and the second phase to sum them up, weighed up by our own principles of evaluation. With this addition, Feynman's argument makes it clear for us why and exactly in what respect physical teleology is different from the human one. Physical teleology—although the reference to the end is explicit—is not apparent, but automatic and mechanical; the endpoint is determined not by the system (as in the case of a human being) but by the initial (and boundary) conditions on the basis of physical laws. We emphasize: The physical system is passive in this respect, it does not contribute to determine its endpoint.

We point out that in biology the case is already different. All living organisms, from cells to plants, animals and humans, actively contribute to maintain their lives, flourish, and determine their biological processes (most biologists accept teleology in biology, see e.g., Ruse 2012). As we will see below (Sect. F4), a kind of teleology different from the mechanical, physical one is a basic fact in biology. An example may be helpful to shed light on the difference between physical and biological teleology.

A stone falling from the Pisa tower cannot contribute anything to the selection of its endpoint. In comparison, a living bird dropped from the same height actively contributes to select the endpoint of its trajectory, unlike a similar but dead bird, which cannot. Definitely, living organisms could not maintain their life if they could not contribute to the determination of their structures, functions and processes. Teleology is ubiquitous already at the cell's life. Indeed, "little occurs in the cell on the basis of chance" (Agutter et al. 2000). For example, hormones, neurotransmitters, and other signals must be directed towards their receptors; and if so, then these processes are, by their very nature, actively teleological, endpoint-oriented, since the living cell actively determines their endpoint (Kawade 1992; Grandpierre 2012). Usually, a neurotransmitter must reach a suitable receptor. Otherwise, the experience of pain elicited at the end of our finger from a candle flame burning it would not be able to transmit the sign to our brain and back, so that we may withdraw our finger away from the fire. Extending Feynman's argument, we propose that in biology an extended version of the action principle is at work, namely, the greatest action principle (Grandpierre 2007). By our proposal, in a living organism, virtual particles also map the whole situation, exploring every possible trajectory, like in the case of the least action principle; but in the case of a living organism, the organism actively contributes to determine the endpoints of its biological processes on the basis of the generalized action principle, the greatest action principle. The organism selects the outcomes corresponding to the greatest action, since this is the path securing the most energy for the longest time period—that is, the greatest action. Yet there is potentially a large number of biologically equivalent solutions, each satisfying the greatest action principle and the biological functions in an equal rate. Therefore, living organisms must be acknowledged as actively contributing to the selection of the actual path of their biological processes. This biological selection process does not involve human-like self-consciousness, since in it the natural principle of greatest action plays the determinative role, and the contribution of the organism is, usually, only complementary, secondary, selecting from biologically quasi-equivalent versions.

Such a genuine biological teleology is obscured not only by the widespread, but, as we argue here, unsound opinion claiming that every type of teleology is excluded by physics; but also by confusing biological teleology with an even more familiar but fundamentally different kind of teleology, namely, the teleology characteristic to self-conscious beings, humans. Human beings have the ability to recognize conditions and foresee the consequences of their actions initiated by their free will (Kane 2002; Brembs 2011). (Regarding the free will debate, see Clarke 2008; Doyle 2011). It is important to note here that recently the existence of free will became experimentally demonstrated. In two novel experiments Cerf and Mackay (2011) had shown that subjects are capable of overriding external sensory input with internal imagery, and can directly control the firing rate of individual neurons in the medial temporal lobe. Human teleology manifests itself in self-conscious, representational control, planning and careful, responsible realization of human purposes—a process in which individual contributions become dominant.

A Set of Fundamental Biological Facts and Problems

We present here, for the first time, ten basic biological facts (facts 1–10, termed F1–F10) transcending the present conceptual framework of biology. We point out that in the last decades biology has reached a turning point and we need to re-evaluate the theoretical framework of the nature of biological autonomy, aims and functions.

F1. Biological aims/Functions. A cell cannot be alive if its proteins could not have their functions, like defense against germs, facilitation of biochemical reactions, coordination of activities, storage, synthesis and transport of biomolecules, bodily movement, or structural support (Grandpierre 2013). Let us define biological functions as coherent systems of biological processes serving biological aims, ultimately, the survival and flourish of the organism as a whole. Two types of such functions may exist: (i) biologically completely determined (assuming there really are any biologically completely pre-determined phenomena, such as, perhaps, digestion); and (ii) incompletely pre-determined (e.g. problem-solving) ones. Incompletely pre-determined functions correspond to biological autonomy. Again, we find ourselves in a conflicting situation. On the one hand, as the entry "biology" in the *Encyclopedia Britannica* (Green 2012) states, living organisms cannot exist without biological functions: "Living things are defined in terms of the activities or functions that are missing in nonliving things." Accordingly, it has been argued that "'Nothing in biology makes sense, except in the light of teleology'. This could be the first sentence in a textbook about the methodology of biology" (Toepfer 2012). In living organisms, very special structural changes are permanently produced that as boundary conditions harness the material forces (quantum mechanical-obeying forces) to the purposeful pursuits of organisms (Strohman 1997). On the other hand, there is strong resistance to the idea of a genuine biological teleology in principle, based on the objection that the very idea of biological function seems to be inconsistent with the conceptual scheme of physics: "The biological concept of function appears teleological, implying goal directedness or purpose…. Ever since the scientific revolution, however, teleology has become exiled from science" (Buller 2002: 393). We note that the reason beyond this argument is that teleology has appeared inconsistent with deterministic physical equations. Here is the problem to be solved: how to make teleological functions compatible with present-day physics? We point out that teleology cannot arise from physical conditions and laws, since it represents a fundamentally different type of causation. The resolution of this fundamental problem directs us not only to a generalized action principle, the endpoint of which can be determined biologically (Grandpierre 2007), but also to the indeterminacy of quantum physics. Actually, genuine biological determinations can act only on physically indetermined, that is, quantum processes. This means that *it is quantum indeterminism that may open the possibility for the physical realization of biological aims.*

F2. Functions/Quantum indeterminism. In a heap of radioactive material, parcels of matter are interchangeable. In a living organism, different organs or limbs are not interchangeable. Within quantum physics, the chances of physically undetermined processes average out. This is to be expected when all subsystems are independent. Yet in the case of living organisms, evidently subsystems are highly dependent on each other in a specific manner that makes the life of a highly complex, composite being possible. *Therefore we have an objective basis to consider the proposition that, with respect to biological functions, the chances of different, physically undetermined microprocesses can be systematically changed and utilized for realizing the biological aims of an organism in a physically arbitrary, that is, not a physically completely predetermined manner.*

F3. Functions/Machines. There are no natural machines. Complex machines cannot arise from spontaneous physical processes. The construction of machines cannot arise from an initial state as a result of random accidents and physical determinations, since the function of any machine requires a consequent series of physically arbitrary steps. On the basis of random physical processes, the chances of selecting any of the exact individual steps suitable for building up a machine are infinitesimal; their joint occurrence in the right order would be exponentially improbable. Similarly, if all you have to go on is what physics allows, the individual steps building up a machine by an engineer are extremely improbable, and their occurrence together in the right order would be exponentially more improbable if the engineer himself can work only with entropic processes. This means that the engineer's actions in constructing a machine cannot be completely determined by physics; therefore, the decisions reifying these actions must be physically arbitrary.

F4. Functions/Gratuity. The same 11 human beings can form in the same spatial arrangement a football team, an orchestra, a crew of sailors, or a family club, depending on their internal focus of attention. At variance with machines, the same living organism in the same situation can behave in many different ways. This means that the material structure of a living organism does not determine completely the biological behavior. The description of a behavior belongs to a logical type of a greater order than the description of a structure (Nobili 1997: 7). This means that until decisions about the biological behavior occur, structure does not determine function. Defining functions as aim-oriented organized systems of biophysical processes serving the survival and flourish of the organism as a whole, we can observe that such a type of logical relation between parts and wholes is completely missing in physics.

Physically arbitrary phenomena occur not only in quantum theory and at the construction of machines, but also in biology. Jacob and Monod (1961) discovered that there is no chemical necessity about the chemical composition of the molecules regulating the functions of allosteric enzymes; e.g., which inducers regulate which genes (see also Monod 1972: 78). "On such a basis…it becomes possible for us to grasp in what very real sense the organism does effectively transcend physical laws—even while obeying them" (ibid. 80). The structure and function of a molecule (of protein or any other substance) are associated with each other, with various degrees of arbitrariness, as are the content and expression of a sign in general. Namely, the activities or the sign functions of biological molecules are determined by the organized system they belong to, and not vice versa (Kawade 1992). The bridge between genes and proteins was provided by molecules called adaptors (transfer RNAs) that have two recognition sites: one for a group of three nucleotides (a codon) and another for an amino acid. The crucial point is that the two recognition sites are physically separated and chemically independent. There is no deterministic link between codons and amino acids, and a one-to-one correspondence between them could only be the result of conventional rules (Barbieri 2008). A sign is a sign only when it stands for something that is other than itself, and this otherness implies at least some degree of independence (Barbieri 2008). It can represent something else only if it is able to step out of its immediate physical determinations. Similarly, Maynard-Smith (2000: 193) observes that there is no physico-chemically necessary connection between the form (chemical composition) and regulatory function (genes switched on and off) of proteins. Moreover, there is no chemical necessity between biochemical structures and their biological functions (like those of hormones). The same chemical structure (of e.g., adrenaline) could, in principle, raise blood pressure or decrease it; the connection between its chemical and biological properties is, in this sense, arbitrary.

The physically arbitrary nature of biological functions is made more remarkable in light of models depicting self-organizing networks that lead to emerging global patterns without an apparent corresponding function (Keller 2007). Designing models with externally assigned functions is a procedure too arbitrary to explain the systemic properties of biological phenomena (Krohs and Callebaut 2007). "In biological systems self-organization is a process in which pattern at the global level of a system emerges solely from numerous interactions among the lower-level components of the system. Moreover, the rules specifying interactions among the system's components are executed using only local information, without reference to the global pattern" (Camazine et al. 2003: 8). Now, functions by their very nature refer to the global level of the living organism (Beckner 1969: "Functional ascriptions describe the role played by a part or process in the activities of a larger or more inclusive system"). Therefore it is clear that physical self-organization is not suitable to explain biological functions. We note that the following problem needs to be resolved: A biological aim as such is teleologically given by the organism, stepping out from its immediate physico-chemical determinations; thus it cannot arise from physical processes. Given that insight, we arrived at a basic problem: What kind of process assigns functions to biomolecules (Grandpierre 2013)? The next problem is: How can physics account for "things" so alien, unobservable and theoretically undescribable by physics such as biological aims? And how do such unobservable "things" become, however, perceived, understood, decided about and realized in the physical form of a function? Moreover, regarding the set of biologically useful functions as a tiny subset of all possible (useful and non-useful) functions, an additional question surfaces: How do all the many thousands of teleological biomolecular functions (which are necessary for an individual cell to exist and function) arise? To answer that question by reference to one or more "frozen accidents" in the history of inanimate matter on the Earth (Crick 1968) is not only insufficient but misses the point: Frozen physical accidents cannot assign biological functions to physico-chemical structures.

F5. Functions/Organized Complexity. All living organisms interact with a complex environment that is indefinitely rich in unexpected challenges. Therefore, all organisms must continuously solve newly encountered problems in their daily lives. Such an achievement requires the internal generation of new algorithmic information (Grandpierre 2008a), corresponding to new biochemical structures serving newly created aims. This means that already the smallest living organisms, the unicellular organisms, are not machines. They can't be machines, since they act as their own engineers, spontaneously generating their own aims and functions, continuously re-engineering their internal structures in a biologically suitable manner. Accordingly, numerical estimations indicate that a significant part of the thermodynamic potential of cells is utilized in order to generate new information (ibid.). The complexity of living organisms has not only static, but also a systematically varying component.

F6. Creativity/Lawful variability. Life is by its very nature a creating power. Despite this basic fact, it is usual to consider that living organisms are like physical machines, just much more complex (Vogel and Angermann 1984: 1) and their complexity is unfathomable (ibid.; Hempel 1966: 101). In the physicalist bottom-up approach such an unfathomable and time-variant complexity is considered contingent and thus cannot be described. In contrast, an approach searching for the universal law of biology has been successful: The fundamental principle of time-variable biological complexity has already been formulated by Ervin Bauer (1935/1967). He was able to derive all the fundamental life phenomena from the mathematically formulated version of his principle (ibid.; regarding biological laws, see also Beloussov 2008).

F7. Control by Information. Life is based on information processing (Hoffmeyer 1997; Maynard-Smith 2000; Mayr 2004; Ben-Jacob et al. 2006; Shapiro 2009; Binder and Danchin 2011). We point out that it is a fundamental problem how the static information content of e.g. the DNA can be transformed into the continuously changing information that directs the time-dependent *dynamic behavior* of material molecules (Grandpierre 2008a). At all levels of analysis living organisms from the global to the molecular level represent high-information-content (low thermodynamic probability) entities. So far as their internal dynamics is concerned, most biochemical processes are channeled or "directed" rather than random processes, which further suggests that little occurs in the cell on the basis of chance or as a simple consequence of the law of mass action (Agutter et al. 2000). There is nothing random about the assemblage of a bacterial cell (Harold 2001: 10). The great irony of molecular biology is that it has led us inexorably from the mechanistic view of life it was believed to confirm, to an informatic view that was completely unanticipated by Crick and his fellow scientific pioneers (Shapiro 2009). It is the cell as a whole that is an active agent utilizing and modifying the information stored in its genome (ibid.). How the cell as a whole can act on its parts is a fundamental problem, the solution to which leads necessarily to the extension of the conceptual framework of biology beyond that of physics.

F8. Qualia/Subjectivity. Cogito, ergo sum – I think, therefore I am, as the famous saying of Descartes formulates. All our knowledge arises from our personal experiences. All objective physical properties such as mass, form, color, temperature, pressure of physical bodies observed through our outer senses, originate from subjective perceptions of these phenomena. Despite this genetic connection between subjectivity and objectivity, physicalism *ab ovo* rejects the concept of "subjectivity." Qualia, the subjective experience of phenomenal qualities, are utterly expunged from the objective viewpoint of physics (Chalmers 2006: 4). As Thomas Nagel formulated in his famous essay "What is it like to be a bat?": Instead of grasping the concept of subjectivity, physicalism rather seeks to sidestep it. Thus the theoretical background of the "subjective" is missing (Nagel 1974). Every subjective phenomenon, however, is essentially connected to a single, first-person point of view. And it seems inevitable that an objective, physicalist theory cannot accommodate such a point of view. "The problems raised by subjective consciousness are perhaps the most baffling in all philosophy" (Dawkins 1998: 283). In biology, the subject is given in the form of the biologically autonomous living organism. It is biological autonomy that creates biological aims, functions and realizes them. Therefore biological autonomy can be regarded as the first scientifically exact formulation of the 'subjective', or consciousness itself. If genuine biological autonomy exists, as we argue in this paper (see also Grandpierre and Kafatos 2012), then this is a concept that can open a perspective towards developing the first comprehensive scientific theory regarding this subject. In quantum physics, the subject enters as the 'observer'. In this way, a plausible new idea arises for us, namely, to extend quantum physics to biology. Indeed, it is already indicated that the new biology will be a more general science than quantum physics (Wigner 1969, 1970; Grandpierre 2007; Josephson 2012).

F9. Cellular intelligence/decision making. It is well known that the rate of self-conscious information processing of the human brain as a whole, when measured in bits/s, is around 1–100 bits/s. Since our brain consists of ca. 10^{11}–10^{13} neurons, this yields an average rate for the self-conscious information processing as 10^{11} bits/s/neuron. In comparison, neurons of the visual system process an information rate of ca. 3 bits/s/neuron (Anderson et al. 2005). The intrinsic activity of the cells is estimated to process roughly 10^6 bits/s/cell (Grandpierre 2008a). Therefore, cells process information not only in a biologically appropriate rate (Ganesan and Zhang 2012), but at a rate that is ca. one hundred quadrillion (10^{17}) higher than that of human self-consciousness. Indeed, self-consciousness represents only an infinitesimal fraction of our ability to process information (Norretranders 1998). This means that our cells manifest a remarkable degree of intelligence.

The phenomenon of cellular intelligence seem to be not widely known. Notwithstanding, the topic of "cellular intelligence" has a rapidly growing literature (Quevli 1917; Albrecht-Buehler 1980, 1985, 1990, 2005, 2009; Mathieu and Sonea 1996; di Primio et al. 2000; Ben-Jacob et al. 2004; Ford 2004, 2006, 2010; Hellingwerf 2005; Ben-Jacob et al. 2006; Shapiro 2007). Cells can demonstrate both anticipatory and contemplative behavior (Tanaka and Nakagaki 2011). Bacteria are shown to be able to solve newly encountered problems, assessing the given problem via collective sensing and recallable stored information of past experience, as well as solving optimization problems that are beyond even what individual human beings can readily solve (Ben-Jacob 2009). They can generate new genes that allow the bacteria to cope with new environments (Ben-Jacob et al. 2006). The ability to assign contextual meaning to externally gathered information is a fundamental semantic function of natural intelligence that every organism must have (ibid.). Bacterial chemical communication also includes assignment of contextual meaning to "words" and "sentences" (semantic/syntax functions) and conduction of "dialogue" (ibid.). With regard to bacteria, semantics would imply that each bacterium has some freedom (plasticity) to assign its own interpretation to a chemical signal according to its own specific, intercellular state and external conditions (ibid.). Cells can perceive self and group identity and act accordingly to self and group aims (Ben-Jacob et al. 2004), sense their external and internal environment (Ben-Jacob et al. 2006: 514), and monitor their internal states (Shapiro 2009: 9). Cells demonstrate the capability of collecting and integrating a variety of physically different and unforeseeable signals as the basis of problem-solving decisions (Albrecht-Buehler 2009). They can respond and make biologically useful, efficient decisions (Linder and Gilman 1992; Strome and Lehmann 2007; Ngalim et al. 2010; Sanges and Cosma 2010; Hyduke and Palsson 2010; Ford 2004, 2006, 2010; Balazsi et al. 2011; Bandyopadhyay et al. 2011). Decision-making is a central feature of the cell (Shapiro 2009). Although Ben-Jacob (2003) and Ben-Jacob et al. (2006) seem to suggest that bacterial freedom is related to physical self-organization, we emphasize that it is the cell itself that makes decisions about cellular processes. It is widely agreed that the phenomenon of play requires freedom (Hughes 2003, 2010: 4–5). Therefore, it is important that play is manifest in animals (Brown and Vaughan 2009: 23), plants (Mancuso 2010), and it is suggested to extend from cells to the Universe (Brown and Vaughan 2009: 37–38). "*The chemical forms are utilized as symbols that allow the cell to form a virtual representation of its functional status and its surroundings*" (Shapiro 2009). Any successful twenty-first-century description of biological functions will include control models that incorporate cellular decisions based on symbolic representations (ibid.). We point out that since all organisms are either cells or build up from cells, the ability of the cells to act according to their aims and to make spontaneous decisions with the help of biologically meaningful symbols representing their functional states and their surroundings while transcending physical determinations means that all living organisms are autonomous.

F10. Quantum physics/Observer problem. Dirac (1927) remarked that the 'freewill' of the observer seems to play a crucial role at the preparation phase of measurement (Bacciagaluppi and Valentini 2008: 188, 493). Von Neumann (1955: 351) demonstrated that the action of the observer is physically undetermined, arbitrary (see also Bishop 2011). We also have Wheeler's (1978: 14) famous saying, which states: "No phenomenon is a phenomenon until it is an observed phenomenon". If so, the physically arbitrary observer–observed relation has a fundamental significance, leading to the idea of a participatory Universe in which the observer participates in creating the observable Universe (Wheeler 1981; Kafatos and Nadeau 2000).

Now let us summarize these fundamental findings. On the basis of the facts showing the existence of cell intelligence (F9), and keeping in mind the arguments presented above as well, we can realize that the cell as a biological entity has the ability to observe and represent its internal states in a symbolic form and to initiate biologically useful changes. On the basis of facts referred to in F1, F4, F8, F9 and F10, we propose that these biologically initiated cellular changes utilize quantum effects. Such spontaneous, physically indeterminate or arbitrary cellular decisions that initiate quantum effects assisting biological aims will be called here quantumbiological interventions. We can realize that F1 corresponds to functions, teleology, and biological aims; F2, F3, F5, F6 and F7 to complexity; F9 to autonomy, F3, F4, F8 and F10 to arbitrariness; F1, F4, F8, F9 and F10 to a quantumbiology that is more fundamental than physics. We note that the physically arbitrary character of biological processes assigning function to material carriers (F4); the ability of cells to make intelligent decisions on the basis of symbolic representations (F9); the necessity that the cell must assign a certain kind of teleology to its processes (F1); the physically arbitrary nature of the subject's actions (F8); and the relation between the subjective nature of the observer and the objective nature of the observed (the observer "intends" what is to be observed) can be traced back to a common basis. We propose that the common basis is that biologically initiated, physically spontaneous decisions exert their determinative power through quantum processes in the context of their extremely high and time-variable complexity (F2, F3, F5, F6, F7). Physicists' measurements are indirect observations of quantum processes, amplifying them into observable macroprocesses. In contrast, biological quantum interventions correspond to the cell's observation of its own microprocesses directly. These observations serve as the basis for cellular decisions that are then transformed, through quantum effects, into macroprocesses, changing the macroscopic behavior of the cell. Therefore, while physical measurements yield indirect output data about microprocesses, quantumbiological interventions produce input and boundary conditions to the cell's global control of microprocesses, modifying them into a form suitable to serve a time-variable output fulfilling biological aims.

It is a popular idea that quantum mechanics is a complete theory and so it must be capable of explaining biological phenomena. In contrast, we point out that, in the absence of guidance by measurements or observations, the time development of the wave functions is either determined by Schrödinger's equation, or changes randomly due to vacuum fluctuations that average out. Yet, as we can see from F1-F10, the fulfilment of basic biological aims requires biological determinations transcending physical determinations and randomness. This means that the assumption of the adequateness of quantum physics with respect to biological phenomena is in sharp contrast with the ten basic biological facts presented above. Similar to the insufficiency of classical physics to explain quantum phenomena, quantum physics is unable to explain genuine biological phenomena. We find that biology represents a deeper layer of Nature, beyond the level of the quantum vacuum: The development of science must proceed along the line exploring how quantum physical processes are organized in living organisms.

The Origin of Life in a New Light

At present, the dominant view of the origin of cellular life on Earth is abiogenesis (Ricardo and Szostak 2009; Panno 2010: 20). Yet the theory of abiogenesis completely leaves without account the genesis of first cellular *functions as such* (Grandpierre 2013); and, as Davies (2006: 300) has said, the origin of life in such approaches remains "a completely unexplained bonus".

Biological Meaning Is Assigned to Biochemical Structures

Recently, Barbieri (2008) has shown that three basic ideas of modern biology—namely, (i) the model of the cell as a biological computer made of genotype and phenotype; (ii) the physicalist doctrine that everything in life must ultimately be accounted for by physical quantities; and (iii) the idea that all biological novelties have been brought into existence by natural selection—are already ruled out by experimental facts. In short, the genesis of biological meaning—i.e., biosemiosis—is not only a fact of life but is 'the' fact of life that allows life to emerge from inanimate matter. Biological teleology, as we have shown in F4, cannot arise from physical conditions on the basis of physical laws. But can it arise spontaneously in a physical process, in an extremely rare random 'accident' at the early Earth, as generally assumed? *We argue here that biological teleology cannot arise from physically pre-determined and spontaneous physical processes since such processes cannot attach a biological function to biochemical structures.* We point out that the physical production of a protein means only the production of a molecule having the same chemical composition as that of a 'living protein' that already has one or more biological functions assigned to it. Since biological functions, such as the defense of the cell against germs, cannot be produced in a physical process, it stands to reason that a yet functionless protein-produced physico-chemical to obtain an ingredient for the first cell in the abiogenetic picture, cannot be a biologically useful protein. We conclude that the 'cell' consisting of such functionless ingredients cannot be viable.

The case is similar to the case of a football player who has a function, for example, to play the role of a goal-oriented forward. Forwards, also known as strikers, are the players on a team in association football who play nearest to the opposing team's goal, and are therefore principally responsible for scoring goals. In order that proteins can work in a biologically useful manner for their host cell, they cannot work on the basis of their chemical affinities. They must be driven by information. But in the linear sequence of base pairs of the DNA only static information is present. Yet at the same time, the protein is expected to act dynamically, driven by forces governing their position, conformational state, energetic and electronic states. And this dynamism, within the physicalist picture, must be driven by physical forces. Yet it is impossible to plan a machine that could exert a series of physical forces to the body of the forward in order to guide its behavior successfully for the football team. Similarly, it is impossible to realize any physico-chemical structure that could exert just the required series of physical forces on the structure of the protein in order to fulfil its biological function, such as defense against germs. This means that in order for the first cell to be viable, all the necessary biological functions must already have been assigned to its biomolecules. In the hypothesis of abiogenesis, such a process is missing. Even if we assume that the first cell in the cosmic genesis of life by good luck possesses all the necessary ingredients just in the right place and in the right shape formed spontaneously in a purely physical process, any such first cell would not be viable, since its biomolecules would not have acquired any biological functions, since all that happened is simply a series of purely physicochemical processes.

Our argument shows that the biological control of living cells cannot be realized by physically governed conditions acting on biomolecules. This means that a physically uninterpretable entity like a biological aim must act as a determining factor within the process of attaching a biological function to certain physicochemical structures. As Abraham Maslow once noted (1966: 15), "I suppose it is tempting, if the only tool you have is a hammer, to treat everything as if it were a nail." Now, in the light of our argument above, we think it is obligatory to allow that Nature may employ a tool yet to be acknowledged by mainstream science; and that this very tool, in conjunction with effective physical forces, is what gives rise to biological teleology, as represented by biological aims and the functions serving them. This natural tool will be identified below later on.

Our result tells us it is not possible to determine the function of a protein, for example, to defend the cell from bacteria, in a physical process. The assumption of abiogenesis rests on the ground that once the suitable physicochemical structures of the first cell are formed, their biological functions are automatically attached to them. Our argument presented here shows that such an assumption lacks a suitable physical basis.

Fundamental Role of Life in the Universe

There is an increasing number of scientists recognizing the fundamental role of life in the Universe. The anthropic principle (Barrow and Tipler 1986) tells us that "our location in the Universe is necessarily privileged to the extent of being compatible with our existence as observers". Moreover, there is the oft-repeated claim nowadays that life is 'written into' the laws of nature (Davies 2003). There is now broad agreement among physicists and cosmologists that the universe is in several respects 'fine-tuned' for life. This claim is made on the basis that existence of vital substances such as carbon, and the properties of objects such as stable long-lived stars, depend rather sensitively on the values of certain physical parameters, and on the cosmological initial conditions. There are many facts indicating that the universe is 'biophilic' (Rees 2001) or 'bio-friendly' (De Duve 1995). Shapiro (1986) refers to the idea of optimal biophilicity of the universe as 'biological determinism'. It is the assertion that life will be almost inevitable given earth-like conditions. The three main schools that interpret biological determinism are assuming that either the probability of the origin of life is extremely enhanced by autocatalytic cycles (Eigen 1992: vi), or by self-organization (Kauffman 1995: vii) in open, far from equilibrium systems (Grandy 2008). Yet we point out that all of these mechanisms represent physical determinations, and as such are incapable of giving an account of genuine *biological* phenomena. As we pointed out on the basis of Bauer's principle (Bauer 1967; Grandpierre 2008b), it is the most basic characteristic of biological processes that they deviate from the ones expected within the given physical conditions on the basis of physical laws from time step to time step, mobilizing all their internal energy resources against the equilibration processes that should occur on the basis of physical (physico-chemical, thermodynamic, quantum physical, etc.) laws.

Recently, Davies (2006: 300) wrote that the bio-friendliness of the universe may arise from an overarching principle that constrains the universe to evolve towards life and mind. It has the advantage of 'taking life seriously', treating it neither as a completely unexplained bonus, as in the popular physicalist picture of abiogenesis; nor as a passive selector, as in the theory of 'multiverse', in which infinitely many 'universes' exist and we just happen to live in the one suitable for life. The apparent disadvantage of the life principle, in the opinion of Davies (2006: 300) is that it introduces "teleology that represents a decisive break with traditional scientific thinking, in which goal-oriented or directional evolution is eschewed as anti-scientific." We note that explaining the biggest unsolved problems of science – among them the existence of biological teleology – might require, indeed, a thorough revision of some deeply ingrained traditional assumptions. Moreover, as we argued in Sect. 2 (Biological teleology) of this paper and elsewhere in more details (Grandpierre 2012), there are different types of teleology (physical, biological and psychological) which are not to be conflated or confused. Additionally, a crucial circumstance in favor of the life principle is that it has already been formulated in a mathematical form (Bauer 1967; Grandpierre 2007), and all the fundamental life phenomena can be derived from it (Bauer 1967). We think that these fundamental facts will be useful in establishing the life principle—as first formulated by Ervin Bauer, the Hungarian-born biologist—firmly in modern science.

It is important to observe that cosmic evolution—the idea that the universe and its constituent parts are constantly evolving—has become widely accepted in the last 50 years (Dick 2009). In the last decades, it has become increasingly clear that biological (and cultural) evolution has been an important part of cosmic evolution on Earth, and perhaps on many other planets (Dick and Lupisella 2009; Davies 2009).

Indeed, Davies (2009: 383) presents arguments showing that the long-held prevailing view claiming that living systems have no particular significance in the cosmic scheme of things is "profoundly wrong". Biological organisms are a product of a very basic organizational principle even in the Darwinian theory of evolution—replication with variation plus selection, a principle that applies anywhere in the cosmos. Moreover, the ability of living organisms to construct a computational representation of the universe makes them capable of manipulating their environment on a large scale. Therefore, "life (…) and mind is a key part of the evolution of the universe" (ibid.).

Biology Is More Fundamental Than Physics

As Tymieniecka (2011: 4) recently indicated, Heraclitus claimed that flux is more basic than stasis. This means that instead of 'stasis', which we can reformulate here as physical states, it is 'flux' that is the fundamental driving factor of change in physical states. Regarding that, the factors driving natural changes are the most fundamental laws of Nature, and, as we argued above, these are the first principles of physics and biology, this means that these first principles are more fundamental than observable phenomena and objects. We can obtain some further insights into the relation between physics and biology, physical matter and life.

In this paper, we found that initiating a biological process starts by a purely biological act that precedes physical processes. As Michael Polanyi (1968) argued, life harnesses the physical laws controlling the physical processes. We add that biological teleology can be regarded as an extension of the physical, mechanical teleology. While in physical teleology the endpoint is determined by the boundary and initial conditions on the basis of physical laws, ultimately, by the least action principle, in biology, the endpoint is determined by the greatest action principle (Grandpierre 2007) together with the autonomous decision of the living organism itself. With the help of an example, a bird dropped from the Pisa tower in the extended version of the Galileo experiment will not fall down freely as a stone or a dead bird. Instead, in the absence of any other biological aims, it will approximately regain its height, investing the minimal amount of energy, obeying the greatest action principle, securing its vitality. Yet there is an infinitely large number of biologically equivalent endpoints corresponding to the same height. Therefore it is the bird itself that decides which to select. Once the bird itself decides the direction of its flight, on the basis of the greatest action principle, the endpoint becomes fixed; and so the physical principle of the least action becomes relevant. Indeed, the greatest action principle tells the bird, if we express ourselves in a somewhat simplified form, to live as long as possible with as much energy as possible. Therefore, once the endpoint is determined in accordance to the greatest action principle, the bird must apply the least action principle to realize the already fixed endpoint. Thus we obtain this result: The biological principle logically precedes the physical principle.

This means, on the one hand, that one of the basic characteristics of biological teleology is that it corresponds to the biologically autonomous decisions of living organisms themselves. On the other hand, our result indicates that in a biological process it is the biological principle, together with the biologically autonomous decisions, that acts first, and the physical actions only follow the biological decisions as secondary events. In light of the fact that the physical actions are already determined by the physical conditions on the basis of physical laws in the realm where physical determinism prevails, the biological decisions must correspond to the physically indeterminate realm. That is, at or below the quantum level. Considering that in quantum physics the vacuum effects are random, while biological actions are not random, we obtain a result that biological actions must start from below the quantum level, from a layer of Nature where the requirement of randomness does not apply. This means that the level of biologically autonomous decisions is below the quantum level. Thus we discover, for the first time, a deeper level of Nature beyond the quantum level: the biological level. This achievement has fundamental consequences.

If biology is more fundamental than physics, then there is no need to 'naturalize' (i.e., 'physicalize') teleology and biological autonomy. This circumstance offers an inference of first importance for us: *If biology is more fundamental than physics, then it is not necessary to explain the origin of life within the physicalist framework.*

From Acausality to Free Will: A Natural Shift

It is important to be aware that there are three basic conceptual frameworks and corresponding mental toolkits by which to consider the problems of determinism, 'acausality', autonomy and 'free will'. In the first and narrowest conceptual framework, corresponding to strict physical (Laplacian) determinism, only physically determined processes are available as tools of explanation. In such a narrow context, the spontaneous quantum processes must arise acausally since there are no physically determined processes to explain phenomena such as spontaneous radioactive decay. In a somewhat wider context that includes vacuum processes, radioactive decay can be explained by spontaneous vacuum processes. In that second context, the apparent 'acausality' (indeterminacy) is shifted from radioactive decay to vacuum fluctuations. In this paper, we attempt to outline a novel, third, even wider context, in which vacuum processes can be initiated by biological autonomy, since, as we argued here, biologically initiated vacuum processes are also available as tools of explanation. In this widest, biological context the apparent 'acausality' is shifted from vacuum processes to biological autonomy. Indeed, 'acausality', or, more precisely, physical and biological indeterminacy is the characteristic property of biological autonomy, leading to an unexpectedly simple and natural explanation of 'free will'. Indeed, the 'acausal' decisions of living organisms are physically and biologically not completely pre-determined—that is to say free, to a certain extent. We point out that to understand biological autonomy and consciousness requires a mental shift from the narrowest Laplacian mental toolkit of classical physical determinism, through the conceptual framework of quantum physics to the widest, biological context of actual reality.

How Is Biological Meaning Attached?

We are considering now the cosmic genesis of cellular life, more precisely, the cosmic genesis of the first, protein-based cellular life form. Before the existence of such a cell, there were only atoms and molecules present, in their physico-chemical states, without biological functions. Biological functions always refer to the organism as a whole (Beckner 1969). Therefore, before the existence of the first cell as a whole, cellular biological functions could not exist. This means that the protein molecule, if it existed before the existence of the first cell, cannot have the biological function e.g. to defend the cell from germs.

Certainly, the function of e.g., proteins cannot arise by chance in the course of evolution, since evolution can select only the living organism as a whole, but cannot attach physically arbitrary rules assigning biological aims to biomolecules within cells. The number of biological functions of biomolecules is vast, and these functions by their very nature do not show a general, simple correspondence with their physical or biological properties; the relation between them is physically arbitrary at a certain degree. Fundamental laws of Nature contain a few variables only, and they, being laws, constitute the same relation between the corresponding properties in all cases. It follows that the biological functions cannot be attributed to each of these biomolecules on the basis of biological laws alone.

Our results show that biological functions cannot be assigned to their host biomolecules on the basis of physical or biological laws. Moreover, we have seen that the first cell cannot assign these functions either, because the existence of such functions is the precondition of the existence of the first cell. The problem, apparently, involves a kind of circularity: Biological functions of the components of cells can be assigned only by their host cell, but the first cell cannot arise without these functions being already assigned.

We propose a novel solution to solve this *problem of circularity*. The biological functions must be assigned indeed by an inclusive, host living organism; but this host organism cannot be the first living cell, but must be another, more inclusive, cosmic life form. On the basis of Bauer's principle, we have explored the idea that, alongside protein-based cellular life forms, other more general types of life forms may also exist in the vast Universe, such as plasma-based stellar life forms, intermittent life, microlife and the vacuum itself (Grandpierre 2008b). This last, the cosmic vacuum, is an especially favorable candidate for fulfilling the task of assigning biological functions to the first cell, because it is not only more inclusive, involving in itself all biomolecules, but also because its existence precedes the existence of biomolecules. If so, then the relation between the cosmic vacuum and the functioning first cells is similar to that of a mother and her foetus, since the mother involves the foetus in her organism, and the existence of the mother precedes the conception of the foetus. In a sense, we can even say that the cosmic life form is what drives the functioning of the living cell. Therefore, by our proposal, it is the cosmic vacuum that acts as the 'soul' of all cellular and multicellular life forms. If this is the case, then it is not so much of a reach to say our soul is naturally attracted to the Cosmos; and this can explain the origin of cosmic sympathy as well as the circumstance that "the soul is in the cosmos, and the cosmos is in the soul" (Tymieniecka 2011: 11).

Willpower Beyond the Quantum Vacuum

How can the cosmic life form act, if not with the help of biological laws or Bauer's principle? We find the solution in the concept of biological autonomy. Indeed, we argue that genuine biological autonomy is present in all living organisms (Grandpierre and Kafatos 2012). Biological autonomy is, in the cosmic context, the ultimate, cosmic subject, playing a fundamental role in the Cosmos that can be comparable to that of the laws of Nature. On the basis of our proposal, biologically autonomous decisions must be able to initiate vacuum processes that are, in the following, suitable to govern and realize biological aims, triggering and coordinating physical processes according to the decisions made. This means that biologically autonomous decisions must occur from beyond the vacuum level itself. If this is the case, then the biologically autonomous decisions of living organisms would be even more subtle than the virtual processes of the quantum vacuum which are already the manifestations of these decisions. Certainly, subjective experience tells us that our autonomous self has a certain degree of energy, a kind of mental energy, like willpower, by which we can freely decide our actions.

Subjective Tools Capable to Act on Matter

We mention that more and more evidences have been accumulating indicating that it is possible to act on the physical states of living organisms by subjectively accessible tools of biological autonomy (*aims*, *beliefs*, *expectations*, *emotions*, *thoughts*) that are not really effective in the external world; and which are not restricted to the production of slight deviations from the physically expected changes (Miller 2011). It is known that beliefs and expectations (e.g., *placebo effect*) can markedly modulate neurophysiological and neurochemical activity (Beauregard 2009; Pollo et al. 2011; Meissner et al. 2011). Neural correlates of *emotional states* such as sadness or depression have already been identified (Fortier et al. 2010), as well as measurable skin-conductance, heart rate and event-related potential changes (Balconi et al. 2012). It has been shown that emotions can induce secretion of hormones and thus influence behavior (Marin et al. 2010; Martin et al. 2010). Rossi and Pourtois (2012) demonstrated that converging electrophysiological and brain-imaging results show that sensory processing in V1 can be modulated by *attention*. We think these facts indicate that living organisms actually have suitable subjective tools that are effective—through the occurrence of biologically initiated spontaneous vacuum processes—in acting on physical matter and producing physically measurable outcomes. If such subjective tools are already demonstrated to be effective in acting upon matter, and the material effectivity of free will is experimentally proved (Cerf and MacKay 2011), then autonomous decisions of living organisms can also be effective in a similar manner.

Higher Dimensions and Subjective World

Acknowledging the reality of the subjective tools of mind in shaping vacuum processes that are suitable to modify and govern physical processes within living organisms, we are led to a new and unprecedented picture describing the forces of the Cosmos and the cosmic genesis of life. In this new world picture, the Universe (to distinguish it from the physical universe, we capitalize it as Universe when it includes biological aspects, too) extends to a level "beyond" the quantum vacuum; it involves mental energies of cosmic life forms organized under such factors as decision making, attention, and biological aims. Moreover, since decision, attention and aims are tools of a biologically autonomous self, the utilization of such tools depends on the existence of a biologically autonomous self capable of deciding by itself, in a free decision; i.e., one not predetermined by physical, biological and psychological conditions and laws. This self is, as we argued above, beyond the cosmic quantum vacuum, therefore it does not exist in physical space and time. Thus it can be regarded as a cosmic self transcending the already familiar 3+1 spatio-temporal dimensions. Research on the nature of time (Saniga 1996, 1998, 2000, 2005) strongly underpins, using detailed mathematical descriptions and their empirical substantiations, that our 3+1 dimensional physical world is but one part of the whole mathematically describable mental reality existing in a multitude of higher spaces that serve as the framework of experiences in the corresponding mental states, and our mind is capable of switching between these spaces.

Transforming Autonomous Decisions to Biological and Physical Forces

We are now able to formulate a fundamental enigma present in every biological action: How is biological action possible? How can something as 'immaterial' as a biological aim or purpose become transformed into physical forces capable of causing actual physical changes? How can a cell move itself? How can we move our arm? Without doubt, our arm can move at will. Apparently, our decision is not effective in moving physical objects external to our organism; yet, within our organism, it can. Since our arm is, at least in a certain sense, a physical object, we can say that a physical object can be moved at will. It sounds something like magic – at the same time, we know that our arm cannot be moved in the absence of physical forces. This means that our decision must be transformed somehow into a physical force capable of moving our arm.

Our solution offered here suggests an unprecented solution that solves the mind-body problem. There must be biological causes beyond physical causes. The physical world is not closed, it is not restricted to physical causes only, but, through the quantum door of indeterminism, it is open to biological causes as well. Our solution proposes that our will (more concretely, our genuine biological decisions) cannot move physical objects directly, but it can move physical objects indirectly—with the help of one and only one tool: by the mediation of quantum vacuum processes that are initiated by our decisions. This unique achievement is possible only because our living organism is unified by biological organization in such a way that every part of it is coupled to the whole organism. Therefore, if our decision acts at the global, organismal level, then, utilizing the whole dynamic network of biological couplings, it can exert effects to each of its coupled subsystems by initiating quantum vacuum processes that interact with the matter of our body in a systematic and organized manner. Indeed, it has been well known since the advent of quantum electrodynamics that all the physical forces can be described as exchange fluxes of virtual particles generated from the quantum vacuum by the physical laws. Our explanation of the mind-body problem makes it reasonable why such effects can occur regularly and systematically within living organisms.

Interaction Between Our Self and the Cosmic Life Form

Since the results presented here tell that our will can act on our body only through the quantum vacuum, which is organized as a cosmic life form (Grandpierre 2008b), then, inevitably, all our decisions have a cosmic context. This circumstance may have a central significance for shedding more light into the relation between life and the Cosmos. The mental energies utilized in our decisions may be regarded as creating and organizing new structures in the cosmic quantum vacuum. Therefore, there exist a next layer beyond the quantum vacuum, consisting of creative mental, subjectively accessible energies capable of organizing the quantum vacuum. Although this newly discovered layer of the Cosmos is found beyond the quantum vacuum, at the same time it is intimately connected with our innermost mentally accessible universe. The external and the internal universes show up as intimately related. Our personally accessible inner world has a fundamental relation to the cosmic quantum vacuum, to a mentally accessible, universal, cosmic dimension, which can act within us as a source of our inspirations, of our intuitive, creative energies, as the inexhaustible source of our personally accessible mental powers. Our result shows that cellular life did not originate from physical matter, or from physical laws, but in the interaction of the universal biological principle and cosmic biological autonomy with the organized cosmic life form, the biological quantum vacuum. Our naturally autonomous, physically undetermined, free self is itself shaping the cosmic dimensions, being rooted in the vast cosmic realm forming an interface with the quantum vacuum. We propose that this cosmic connection is what lies beyond the eternal call of the Cosmos, establishing a most personal contact between Homo Sapiens and the Universe.

Acknowledgements The author wishes to express his deep gratitude to his friend Jean Drew for the continuous inspirations, galvanizing discussions and lecturing the English.

References

Agutter, Paul S., P. Colm Malone, and Denys N. Wheatley. 2000. Diffusion theory in biology: A relic of mechanistic materialism. *Journal of the History of Biology* 33(1): 71–111.

Albrecht-Buehler, G. 1980. Autonomous movements of cytoplasmic fragments. *Proceedings of the National Academy of Sciences of the United States of America* 77: 6639–6643.

Albrecht-Buehler, G. 1985. Is cytoplasm intelligent, too? In *Muscle and cell motility*, vol. VI, ed. J. Shay, 1–21. New York: Plenum.

Albrecht-Buehler, G. 1990. In defense of non-molecular cell biology. *International Review of Cytology* 120: 191–241.

Albrecht-Buehler, G. 2005. A long-range attraction between aggregating 3T3 cells mediated by near-infrared light scattering. *Proceedings of the National Academy of Sciences of the United States of America* 102: 5050–5055.

Albrecht-Buehler, G. 2009. Cell intelligence. http://www.basic.northwestern.edu/g-buehler/FRAME.HTM. Accessed 4 Aug 2011.

Anderson, C.H., D.C. Van Essen, and B.A. Olshausen. 2005. Directed visual attention and the dynamic control of information flow. In *Neurobiology of attention*, ed. Itti Laurent, Rees Geraint, and Tsotsos John, 11–17. Burlington/San Diego/London: Elsevier Academic Press.

Bacciagaluppi, G., and A. Valentini. 2008. *Quantum theory at the crossroads*. Cambridge, UK: Cambridge University Press.

Balazsi, G.A., A. van Oudenaarden, et al. 2011. Cellular decision making and biological noise: From microbes to mammals. *Cell* 144: 910–925.

Balconi, M., L. Falbo, and V.A. Conte. 2012. BIS and BAS correlates with psychophysiological and cortical response systems during aversive and appetitive emotional stimuli processing. *Motivation and Emotion* 36(2): 218–231.

Bandyopadhyay, K., P.K. Parua, et al. 2011. Studies on Escherichia coli HflKC suggest the presence of an unidentified lambda factor that influences the lysis-lysogeny switch. *BMC Microbiology* 11: 34.

Barbieri, Marcello. 2008. Life is biosemiosis. *Cosmos and History: The Journal of Natural and Social Philosophy* 4: 29–52.

Barrow, John D., and Frank J. Tipler. 1986. *The anthropic cosmological principle*, 132. Oxford: Oxford University Press.

Bauer, Ervin. 1967. *Theoretical biology*. Budapest: Akadémiai Kiadó, in Hungarian; in Russian, 1935, 1982, 1993, 2002.

Beauregard, M. 2009. Effect of mind on brain activity: Evidence from neuroimaging studies of psychotherapy and placebo effect. *Nordic Journal of Psychiatry* 63(1): 5–16.

Beckner, M. 1969. Function and teleology. *Journal of History of Biology* 2: 151–164.

Beloussov, L.V. 2008. Mechanically based generative laws of morphogenesis. *Physical Biology* 5: 015009.

Ben-Jacob, E. 2003. Bacterial self-organization: Co-enhancement of complexification and adaptability in a dynamic environment. *Philosophical Transactions of the Royal Society A* 361: 1283–1312.

Ben-Jacob, E. 2009. Learning from bacteria about natural information processing. *Natural Genetic Engineering and Natural Genome Editing Book Series: Annals New York Academy Sciences* 1178: 78–90.

Ben-Jacob, E., I. Becker, Y. Shapira, and H. Levine. 2004. Bacterial linguistic communication and social intelligence. *Trends in Microbiology* 12: 366–372.

Ben-Jacob, E., Y. Shapira, and A.I. Tauber. 2006. Seeking the foundations of cognition in bacteria: From Schrödinger's negative entropy to latent information. *Physics A* 359: 495–524.

Binder, P.M., and A. Danchin. 2011. Life's demons: Information and order in biology. What subcellular machines gather and process the information necessary to sustain life? *EMBO Reports* 12: 495–499.

Bishop, Rober C. 2011. Free will and the causal closure of physics. In *Visions of discovery: New light on physics, cosmology, and consciousness*, ed. R.Y. Chiao, M.X. Cohen, A.J. Leggett, W.D. Phillips, and C.L. Harper Jr. Cambridge, UK: Cambridge University Press.

Brembs, Björn. 2011. Towards a scientific concept of free will as a biological trait: Spontaneous actions and decision-making in invertebrates. *Proceeding of Biological Sciences* 278(1707): 930–939.

Brown, Laurie M. (ed.). 2005. *Feynman's thesis. A new approach to quantum theory*. Singapore: World Scientific.

Brown, S., and C. Vaughan. 2009. *Play. How it shapes the brain, opens the imagination, and invigorates the soul*. New York: Avery Books.

Buller, D.J. 2002. Function and teleology. In *Encyclopedia of life science*, 393. London: Macmillan.

Camazine, S., J.-L. Deneubourg, N.R. Franks, J. Sneyd, G. Theraulaz, and E. Bonabeau. 2003. *Self-organization in biological systems*, 8. Princeton: Princeton University Press.

Cerf, M., and M. Mackay. 2011. Studying consciousness using direct recording from single neurons in the human brain. In *Characterizing consciousness: From cognition to the clinic?* ed. Dehaene Stanislas and Christen Yves, 133–146. Berlin: Springer.

Chalmers, D. 2006. *The conscious mind: In search of a fundamental theory*. Oxford: Oxford University Press.

Clarke, Randolh. 2008 Incompatibilist (Nondeterministic) theories of free will. In *The stanford encyclopedia of philosophy*, ed. Edward N. Zalta (Fall 2008 edition). http://plato.stanford.edu/archives/fall2008/entries/incompatibilism-theories/

Crick, F.H.C. 1968. The origin of the genetic code. *Journal of Molecular Biology* 38: 367–379.

Davies, P.C.W. 2003. How bio-friendly is the universe? *International Journal of Astrobiology* 2: 115.

Davies, P. 2006. *The goldilocks enigma. Why is the universe just right for life?* 300. London: Allen Lane.

Davies, Paul. 2009. Life, mind, and culture as fundamental properties of the universe. In *Cosmos & culture: Cultural evolution in a cosmic context*, ed. S.J. Dick and M.L. Lupisella, 383–398. Washington, DC: National Aeronautics and Space Administration.

Dawkins, R. 1998. *Unweaving the rainbow*, 283. Boston: Houghton Mifflin.

De Duve, Christian. 1995. *Vital dust*. New York: Basic.

Dennett, D.C. 1995. *Darwin's dangerous idea: Evolution and the meaning of life*, 126. New York: Simon & Schuster.

di Primio, F., B.S. Müller, and J.W. Lengeler. 2000. Minimal cognition in unicellular organism. In *SAB2000 proceedings supplement*, ed. J.-A. Meyer, 3–12. Honolulu: The International Society for Adaptive Behavior.

Dick, Steven J. 2009. Cosmic evolution. History, culture, and human destiny. In *Cosmos and culture: Cultural evolution in a cosmic context*, ed. S.J. Dick and Mark L. Lupisella, 25–59. Washington, DC: National Aeronautics and Space Administration.

Dick, Steven J., and Mark L. Lupisella (eds.). 2009. *Cosmos and culture: Cultural evolution in a cosmic context*. Washington, DC: National Aeronautics and Space Administration.

Dirac, P. 1927. In Bacciagaluppi G, Valentini A. 2008, ibid.189.
Doyle, R. 2011. *Free will: The scandal in philosophy*. Cambridge, MA: I-Phi Press.
Eigen, M. 1992. *Steps towards life*. Oxford: Oxford University Press.
Feynman, R.P. 1942. The principle of least action in quantum mechanics (thesis). Princeton University. Also in: Brown, L. M. (ed.) Feynman's Thesis: A New Approach to Quantum Theory. Singapore: World Scientific, 2005), pp. 1–69.
Feynman, R.P. 1948. Space-time approach to non-relativistic quantum mechanics. *Reviews of Modern Physics* 20: 367–387. Also in Brown (2005, pp. 71–112).
Feynman, R.P. 1964. The principle of least action. In *The Feynman lectures on physics*, vol. 2, ed. R.P. Feynman, R.B. Leighton, and M. Sands, 19–1–19–14. Reading: Addison-Wesley. Chap. 19.
Feynman, R.P. 1994. *The character of physical law*. New York: Modern Library.
Ford, B.J. 2004. Are cells ingenious? *Microscope* 52: 135–144.
Ford, B.J. 2006. Revealing the ingenuity of the living cell. *Biologist* 53: 221–224.
Ford, B.J. 2010. *Single cell intelligence*. Mensa Mag, February 2010, 6–7. Cambridge: Mensa At Cambridge.
Fortier, Emilie, Anne Noreau, Franco Lepore, Michel Boivin, Daniel Pérusse, Guy A. Rouleau, and Mario Beauregard. 2010. Early impact of 5-HTTLPR polymorphism on the neural correlates of sadness. *Neuroscience Letters* 485(3): 261–265.
Ganesan, Ambhighainath, and Jin Zhang. 2012. How cells process information: Quantification of spatiotemporal signaling dynamics. *In Protein science: A publication of the protein society* 21(7): 918–928.
Grandpierre, A. 2007. Biological extension of the action principle: Endpoint determination beyond the quantum level and the ultimate physical roots of consciousness. *Neuroquantology* 5: 346–362.
Grandpierre, A. 2008a. Fundamental complexity measures of life. In *Divine action and natural selection: Questions of science and faith in biological evolution*, ed. J. Seckbach and R. Gordon, 566–615. Singapore: World Scientific.
Grandpierre, A. 2008b. Cosmic life forms. In *From fossils to astrobiology*, ed. J. Seckbach and M. Walsh, 369–385. Berlin: Springer.
Grandpierre, A. 2012. On the biological origin of design in nature. In *Origin(s) of design in nature: A fresh, interdisciplinary look at how design emerges in complex systems, especially life*, ed. L.S. Swan, R. Gordon, and J. Seckbach. Dordrecht: Springer.
Grandpierre, A. 2013. On genuine biological autonomy. ATINER Conference, May 27–30, 2012. Athens: ATINER'S Conference Paper Series, No: PHI2012-0197. http://www.atiner.gr/papers/PHI2012-0197.pdf
Grandpierre, A., and M. Kafatos. 2012. Biological autonomy. *Philosophy Study* 2(9): 631–649.
Grandy Jr., Walter T. 2008. *Entropy and the time evolution of macroscopic systems*, 29. Oxford: Oxford University Press.
Green, Edna R. 2012. *"Biology". Encyclopædia Britannica. Encyclopædia Britannica ultimate reference suite*. Chicago: Encyclopædia Britannica.
Harold, F.M. 2001. *The way of the cell. Molecules, organisms and the order of life*, 10–11. Oxford: Oxford University Press.
Hellingwerf, K.J. 2005. Bacterial observations: A rudimentary form of intelligence? *Trends in Microbiology* 13: 152–158.
Hempel, C. 1966. *Philosophy of natural science*, 104. Englewood Cliffs: Prentice-Hall.
Hoffmeyer, J. 1997. Biosemiotics: Towards a new synthesis in biology. *European Journal of Semiotics Studies* 9: 355–376.
Hughes, F.P. 2003. Spontaneous play in the 21st century. In *Contemporary perspectives on play in early childhood education*, 21–39. Greenwich: Information Age Publishing.
Hughes, F.P. 2010. *Children, play, and development*, 4th ed, 4–5. Thousand Oaks: SAGE.
Hyduke, D.R., and B.O. Palsson. 2010. Towards genome-scale signalling-network reconstructions. *Nature Reviews Genetics* 11: 297–307.
Illetterati, Luca, and Francesca Michelini (eds.). 2008. *Purposiveness. Teleology between nature and mind*. Frankfurt: Ontos Verlag. back cover.

Jacob, F., and J. Monod. 1961. On the regulation of gene activity. *Cold Spring Harbor Symposia on Quantitative Biology* 26: 193–211.

Johnson, Ransom Monte. 2006. *Aristotle on teleology*, 23–24. Oxford: Oxford University Press.

Josephson, Brian D. 2012. Biological observer-participation and Wheeler's 'Law without Law'. In *Integral biomathics: Tracing the road to reality*, ed. Plamen L. Simeonov, Leslie S. Smith, and Andrée C. Ehresmann. Berlin: Springer.

Kafatos, M., and R. Nadeau. 2000. *The conscious universe: Parts and wholes in physical reality*. New York: Springer.

Kane, Robert. 2002. *The Oxford handbook of free will. Oxford handbooks in philosophy*, 118–119. Oxford: Oxford University Press.

Kauffman, Stuart. 1995. *At home in the universe*. Oxford: Oxford University Press.

Kawade, Yoshimi. 1992. A molecular semiotic view of biology – interferon and 'homeokine' as symbols. *Rivista di Biologia/Biology Forum* 85(1): 71–78.

Keller, E.F. 2007. The disappearance of function from 'self-organizing systems'. In *Systems biology. Philosophical foundations*, ed. F. Boogerd, F. Bruggeman, J.H. Hofmeyr, and H.V. Westerhoff, 303–317. Amsterdam: Elsevier.

Kreines, James. 2009. The logic of life: Hegel's philosophical defense of teleological explanation of living beings (Chap. 13). In *The Cambridge companion to Hegel and nineteenth-century philosophy*, ed. F. Beiser, 344–377. Cambridge: Cambridge University Press.

Krohs, U., and W. Callebaut. 2007. Data without models merging with models without data. In *Systems biology. Philosophical foundations*, ed. F. Boogerd, F. Bruggeman, J.H. Hofmeyr, and H.V. Westerhoff, 181–213. Amsterdam: Elsevier.

Landau, L.D., and E.M. Lifshitz. 2000. *Course on theoretical physics* Vols. 1–10 (trans: Sykes, J. B., and J. S. Bell). Oxford: Butterworth and Heinemann.

Lemons, Don S. 1997. *Perfect form. Variational principles, methods, and applications in elementary physics*. Princeton: Princeton University Press.

Linder, M.E., and A.G. Gilman. 1992. G proteins. *Scientific American* 267(1): 36–43.

Mancuso, S. 2010. The roots of plant intelligence. TEDGlobal Talk. Transcript. http://www.ted.com/talks/stefano_mancuso_the_roots_of_plant_intelligence.html

Marin, Marie-France, Kamala Pilgrim, and Sonia J. Lupien. 2010. Modulatory effects of stress on reactivated emotional memories. *Psychoneuroendocrology* 35(9): 1388–1396.

Martin, Martins Joao J., Sónia do.S. Vale, Florbela F. Ferreira, Maria Joao M.J. Fagundes, Isabel do.I. Carmo, Carlota C. Saldanha, and E. Silva J. Martins. 2010. Plasma corticotropin releasing hormone during the feeling of induced emotions. *Neuroendocronology Letters* 31(2): 250–255.

Maslow, Abraham H. 1966. *The psychology of science: A reconnaissance*, 15. Richmond: Maurice Bassett Publishing/Zorba Press.

Mathieu, L.G., and S. Sonea. 1996. Time to drastically change the century-old concept about bacteria, Science Tribune August 1996, http://www.tribunes.com/tribune/art96/math.htm

Maynard-Smith, J. 2000. The concept of information in biology. *Philosophy of Science* 67: 177–194.

Mayr, E. 2004. *What makes biology unique?* Cambridge, UK: Cambridge University Press.

Meissner, K., N. Kohls, and L. Colloca. 2011. Introduction to placebo effects in medicine: mechanisms and clinical implications. *Philosophical Transactions of the Royal Society B – Biological Sciences* 366(1572): 1783–1789.

Miller, Neil R. 2011. Functional neuro-ophthalmology. In *Handbook of clinical neurology* Vol. 102, ed. Vinken, P.J., and G.W. Bruyn, 493–513. New York: John Wiley & Sons.

Monod, Jacques. 1972. *Chance and Necessity: An Essay on the Natural Philosophy of Modern Biology,* 80. Translated from the French by A. Wainhouse. New York: Vintage Books.

Nagel, T. 1974. What is it like to be a bat? *Philosophical Reviews* LXXXIII(4): 435–450.

Ngalim, S.H., A. Magenau, et al. 2010. How do cells make decisions: Engineering micro- and nanoenvironments for cell migration. *Journal of Oncology* 2010: 363106.

Nobili, Renato. 1997. The conceptual basis of theoretical biology. Annales biotheoretici L. Galleni and M. Rizzotti Editors, SEU (Servizio Editoriale Universitario di Pisa) 1: 15–68.

Norretranders, Tor. 1998. *The user illusion: Cutting consciousness down to size*. New York: Viking.

Panno, J. 2010. The cell: Nature's first life-form (New biology). Revised edition. New York: Facts on File. An imprint of Infobase Publishing.
Polanyi, Michael. 1968. Life's irreducible structure. *Science* 160(3834): 1308–1312.
Pollo, A., E. Carlino, and F. Benedetti. 2011. Placebo mechanisms across different conditions: From the clinical setting to physical performance. *Philosophical Transactions of the Royal Society B – Biological Sciences* 366(1572): 1790–1798.
Quevli, N. 1917/2007. *Cell intelligence*. Whitefish: Kessinger Pub.
Rees, Martin. 2001. *Our cosmic habitat*, 175–179. Princeton: Princeton University Press.
Ricardo, Alonso, and Jack W. Szostak. 2009. Origin of life on earth. *Scientific American* 301(3): 54–61.
Rossi, V., and G. Pourtois. 2012. State-dependent attention modulation of human primary visual cortex: A high density ERP study. *NeuroImage* 60(4): 2365–2378.
Ruse, Michael. 2012. *"Biology, philosophy of." Encyclopædia Britannica. Encyclopædia Britannica ultimate reference suite*. Chicago: Encyclopædia Britannica.
Russell, B. 1946. *A history of Western philosophy*, 86–87. London: Allen & Unwin.
Sanges, D., and M.P. Cosma. 2010. Reprogramming cell fate to pluripotency: The decision-making signalling pathways. *International Journal of Developmental Biology* 54: 1575–1587.
Saniga, M. 1996. Arrow of time and spatial dimensions. In *Cosmological constant and the evolution of the universe*, ed. K. Sato, T. Suginohara, and N. Sugiyama, 283–284. Tokyo: Universal Academy Press.
Saniga, M. 1998. Time arrows over ground fields of an uneven characteristic. *Chaos, Solitons & Fractals* 9(7): 1087–1093.
Saniga, M. 2000. Algebraic geometry: A tool for resolving the enigma of time? Appendix. In *Studies on the structure of time: From physics to psycho(patho)logy*, ed. R. Buccheri, V. Di Gesu, and M. Saniga, 301–306. New York: Kluwer/Plenum Publishers.
Saniga, M. 2005. A geometrical chart of altered temporality (and spatiality). In *Endophysics, time, quantum and the subjective*, ed. R. Buccheri et al., 245–272. Singapore: World Scientific.
Shapiro, Robert. 1986. *Origins*. New York: Summit Books.
Shapiro, J.A. 2007. Bacteria are small but not stupid: Cognition, natural genetic engineering and socio-bacteriology. *Studies in History and Philosophy of Biological and Biomedical Sciences* 38: 807–809.
Shapiro, J.A. 2009. Revisiting the central dogma in the 21st century. *Natural Genetic Engineering and Natural Genome Editing Book Series: Annals New York Academy Sciences* 1178: 6–28.
Strohman, Richard C. 1997. The coming Kuhnian revolution in biology. *Nature Biotechnology* 15: 194–200.
Strome, S., and R. Lehmann. 2007. Germ versus soma decisions: Lessons from flies and worms. *Science* 316: 392–393.
Tanaka, Y., and T. Nakagaki. 2011. Cellular computation realizing intelligence of slime mold physarum polycephalum. *Journal of Computational and Theoretical Nanoscience* 8: 383–390.
Taylor, C. 1964. *The explanation of behaviour*. London: Routledge & Kegan Paul.
Taylor, C. 1967. Psychological behaviorism. In *The encyclopedia of philosophy*, ed. P. Edwards, 516–520. New York: MacMillan.
Toepfer, Georg. 2012. Teleology and its constitutive role for biology as the science of organized systems in nature. *Studies in History and Philosophy of Science Part C: Studies in History and Philosophy of Biological and Biomedical Sciences* 43(1): 113–119.
Tymieniecka, A.-T. 2011. Inspirations of heraclitus from ephesus fulfilled in our new enlightenment. Phenomenology/ontopoiesis retrieving geo-cosmic horizons of antiquity. Logos and life. *Analytical Husserl* 110: 3–12.
Vogel, G., and H. Angermann. 1984. *dtv-Atlas zur Biologie*, vol. 2, 1. Munich: dtv.
von Neumann, J. 1955. *Mathematical foundations of quantum mechanics*. New Jersey: Princeton University Press.
von Glasersfeld, E. 1990. Teleology and the concepts of causation. *Philosophical* 46: 17–43.

Walsh, D.M. 2000. Chasing shadows: Natural selection and adaptation. *Studies in History and Philosophy of Biological and Biomedical Sciences* 31: 135–153.
Wheeler, J.A. 1978. The 'past' and the 'delayed choice' double-slit experiment. In *Mathematical foundations of quantum theory*, ed. E.R. Marlow, 9–48. New York: Academic. p. 14.
Wheeler, J.A. 1981. Beyond the black hole. In *Some strangeness in the proportion*, ed. H. Woolf. Reading: Addison-Wesley.
Wigner, E.P. 1969. Are we machines? *Proceedings American Philosophical Society* 113: 95–101.
Wigner, E.P. 1970. Physics and the explanation of life. *Foundations of Physics* 1: 35–45.
Wouters, Arno. 2005. The function debate in philosophy. *Acta Biotheoretica* 53(2): 123–151.
Zee, A. 1986. *Fearful symmetry. The search for beauty in modern physics*, 107–109. New York: Macmillan. 143.

The Cosmos of Yolanthe: Knowing Without Seeing

Detlev Quintern

Abstract *Yolanthe*, Peter Ilyich Tchaikovsky's last opera (1891), is based on the lyrical drama *King Rene's Daughter*, written by the Danish writer and poet Henrik Hertz in 1845. It entails layers of philosophical dimensions, which invite to be unraveled while applying the "tools" of the phenomenology of life (Anna-Teresa Tymieniecka) in order to achieve not only a more detailed understanding of philosophical messages in the lyrical drama, but likewise to utilize the phenomenology of life as a starting point for a cross-cultural re-reading of *King Rene's Daughter*. Yolanthe, the central female protagonist, lost her eyesight at the age of 1 due to a traumatic experience. At 16 she owes the restoration of her vision to the creative interplay of self-knowledge and the communicative forces of life by experiencing the *unity-of-everything-there-is-alive* (A.-T. Tymieniecka), echoing divine love.

The opera *Yolanthe* belongs to the group of late works by the Russian composer Peter Ilyich Tchaikovsky (1840–1893) and became the last opera to be premiered in Tchaikovsky's lifetime. Its one-act libretto, written by Modest Tchaikovsky, the composer's younger brother, in the second half of 1891, almost a year prior to its first public appearance in St. Petersburg, is based on Vladimir Zotov's Russian adaptation (1864) of a lyrical drama by the Danish writer and poet Henrik Hertz (1797–1870).

Originally entitled in Danish *Kong Renes Datter*, meaning *King Rene's Daughter*, the one-act verse drama enjoyed such a tremendous popularity following its first publication in 1845 that it was subsequently translated into almost every European

I gratefully acknowledge Jana Nittel's indispensible help with the Enghlish version of this paper.

D. Quintern (✉)
Director for Development and Training
Prof. Dr. Fuat Sezgin Research Foundation for the History of Science in Islam,
Gülhane Parkı İçi, Eminönü, Fatih, İstanbul 34122, Turkey
e-mail: cdq@uni-bremen.de; http://detlevquintern.wordpress.com

A.-T. Tymieniecka (ed.), *Phenomenology of Space and Time: The Forces of the Cosmos and the Ontopoietic Genesis of Life: Book One*, Analecta Husserliana 116,
DOI 10.1007/978-3-319-02015-0_11,

language. A German translation was published in Bremen in 1871,[1] in addition to the several English translations, including those by Jane Frances Chapman (1845) and Theodore Martin (1850).[2] With Maude Fealy and Harry Benham in the lead roles, *King Rene's Daughter* was first transferred to the screen in 1913, as a silent film of the same title produced by the New York based Thanhouser Company, one of the first motion picture studios at the time. *Love's Light*, or *Das Licht der Liebe*, a DEFA film produced in German/Czech collaboration in the 1990s presents itself as a more recent rendering, beautifully but firmly turning the original drama into a fairy tale for children.

According to the German language edition of 1871 of *King Rene's Daughter*, Henrik Hertz briefly outlined in an introduction the historical context of his work, which is seemingly set in fifteenth-century Europe: King René of Anjou, the Count of Provence comes to an understanding with Count Antoine of Vaudémont regarding the succession to the Duchy of Lorraine by strategically engaging in political match making. The dynastic alliance between both houses was to be established through the union of King Rene's daughter Yolande and Count Antoine of Vaudémont's son.[3] Although it has become generally accepted that the story features historically documented figures of the close-knit fifteenth-century European nobility, such as of the Lorraine, the Provence and even beyond, the question of potentially earlier literary precursors to the drama by Hertz, either Provençal or likewise older fictional representations has yet to be settled. Several hints and traces seem however to encourage further research.

In this light, the story of the one-act libretto *Yolanthe* of Tchaikovsky's opera might presumably or probably be based on a scarcely documented narrative of Provencal-Arab origin/nature, its plot almost uniquely predestined to invite an intercultural or cross-cultural reading of the text. Besides its pronounced mystical strand, the libretto offers also a contextualization regarding the history of science, here specifically of ophthalmology, a medical branch dealing with the anatomy, physiology and diseases of the eye, which advanced greatly during the Golden Age of Arab-Islamic sciences. This holds particularly true for Al-Andalus (711–1492), located in close proximity to the South of France, a name that stood synonymously for a peaceful coexistence of cultures and belief systems over extended periods of time and not least therefore fostered significant developments and breakthroughs in the sciences, medicine and philosophy. Even so, while transcending intercultural and/or cross-cultural dimensions as well as representations of the history of science, the libretto *Yolanthe* also stipulates further philosophical and epistemological enquiries into the relationship of knowledge of God, universal knowledge and knowledge of reality in the context of love as motivating and instigating force.

[1] Hertz, Henrik, *König Rene's Tochter*, translated into German by Willatzen, P.J. (Bremen: Kühtmann's Buchhandlung, 1871).

[2] Martin, Theodore. *King René's daughter: a Danish lyrical drama*, (Boston: W. Crosby and H.P. Nichols, 1850).

[3] Hertz, Henrik, König Rene's Tochter, translated into German by Willatzen, P.J., (Bremen: Kühtmann's Buchhandlung, 1871).

Explored on a deeper, in other words, more fundamental and philosophical level, *Yolanthe* can be described as an exemplary didactic play that invites a cross-cultural re-reading in the context of the *phenomenology of life*, as developed by Anna-Teresa Tymieniecka in recent decades.

Yolanthe's Cosmos: Knowing Without Seeing

Emotional learning, intuition, cognition and applied sciences, in this case, ophthalmology, and finally metaphysics are entwined in the story of Yolanthe, forming an interwoven web of knowledge. Ultimately it is the conscious love, which enables Yolanthe to recover her previously lost eyesight. But how does she arrive at such a state or condition, in other words, this conscious love and what are the links between conscious or knowing love and the "science of eyes", the literal translation of ophthalmology, as practiced by Ibn Yahya, the trusted friend and teacher who Yolanthe relies on since the early days of her childhood? A brief summary of the plot will support any further attempt to approach these questions.

King René's Daughter: The Drama's Historical and Cultural Context

Yolanthe, daughter of King René, lost her eyesight during an accident in her parent's castle. While the castle was engulfed in flames, the infant child, wrapped in a pillow, was dropped from one of the castle's windows. Although this brave act spared the life of Yolanthe, it also left the 1-year-old blinded. At this point the reader is not privy to the specific reasons or causes of the impairment. Shielded from any knowledge of what seeing might be and discouraged to engage in associative contemplations on this sensory mode, the grown-up Yolanthe remains wholly unaware of her blindness. Since the traumatic events of her childhood she has been placed in the remote and secluded valley of Vaucluse, lovingly brought up and looked after by Bertrand and his wife Martha. King René, her father visits frequently whenever he happens to pass through the area.

Ibn Yahya, in the New York edition of 1867 referred to as a so-called Moorish physician, is a famous Arab scholar whom her father had sent for from Cordoba. As an intimate friend and teacher he not only treats Yolanthe medically and therapeutically, but provides in addition lessons on music and musical instruments, probably the Arab *Qanun*, a rather popular instrument in Al-Andalus that has found its way to the Provence and on to Europe, before it arrived in the form of the zither in Alpine folk music. Yolanthe was also trained in poetry, contemporary troubadour songs and lyrical composition; as well as in the natural sciences which is implied through her close knowledge of zoology, botany, and mineralogy. Her teacher emphasizes at one point, the superiority of the laws of nature and, indirectly, the superiority of their empirical observation over astrological speculation. "Yet the stars [...]. They influence [t]he fortunes of mankind, yet do they not [r]ule Nature's laws with absolute control."[4] "Sie (die Sterne, D.Q.) üben wohl auf das Schicksal Einfluss, doch bezwingen sie's nicht mit Macht und wider die Natur."[5]

This comprehension of science points also to the prevailing thinking and understanding of science in Al-Andalus, which was undoubtedly guided by ideas regarding the individual's responsibility for his or her own actions and the desire for acquiring knowledge and insight about the developments of nature. One day, however, the royal messenger, Sir Almerik, announces the visit of the king that is Yolanthe's father and that of the physician, Ibn Yahya, the famous Moor.[6] Ibn Yahya had initially forecast a complete cure of the girl before her 16th birthday, but was dissuaded from his undertaking since he believed that Yolanthe's awareness and recognition of her own blindness was an essential precondition for the restoration of her eyesight. Yolanthe "[m]ust comprehend what she till now has lack'd, Must learn this very day that she is blind."[7] With these very words he attempts to convince King René that Yolanthe's ignorance of her condition will fail to establish, if not prevent the happiness of his daughter. Just this day, for the lyrical drama's timeline lasts from midday to sunset,[8] in other words on the day which Ibn Yahya suggested to be the day of revelation, two strangers enter the valley. Count Tristan of Vaudemont and his friend, Sir Geoffrey of Orange, himself a famous troubadour, and each slung a zither around their shoulders, gain by chance access to the valley through a mysterious entrance, a moss-covered gate, concealed in a cliff. Once arrived in this heavenly place, Tristan discovers the sleeping Yolanthe in her chamber and enthralled by her gracefulness and sweetness, falls immediately in love with her. In vain Geoffrey tries to convince Tristan of the necessity to leave this mysterious and enchanted place. Whilst softly kissing her, Tristan removes the stone that always rests on Yolanthe's chest during her induced sleep. In the past it was only Ibn Yahya who placed the stone on Yolanthe in order to transport her into a deep sleep and it was only he that awakened Yolanthe by removing it. This task however is now unexpectedly performed by Tristan. King René's daughter, thus awoken, seems no less impressed by the sight of her two guests and offers wine and fruits as refreshment. At the end of the poetic dialogue that unfolds between the two central protagonists, Tristan overwhelmed by devotion to the beloved Yolanthe, decides to abandon his given promise of betrothal to King René's daughter, blissfully ignorant that she is the very woman he so longingly pursues. In due course he becomes aware of Yolanthe's visual impairment by asking her first for a red and then for a white

[4] Martin, Theodore. *King René's daughter: a Danish lyrical drama*, (Boston: W. Crosby and H.P. Nichols, 1850), p. 27.

[5] Hertz, Henrik, *König Rene's Tochter*, translated into German by Willatzen, P.J. (Bremen: Kühtmann's Buchhandlung, 1875), p. 23.

[6] Ibid., p. 4.

[7] Hertz, Henrik, *King René's Daughter: A Danish Lyrical Drama*. Translated by Theodore Martin. (New York: Leypoldt and Holt, 1867), p. 27.

[8] Ibid., xiii.

rose as a keepsake. The king's daughter who is yet to learn of the multitude of colors presents him on each occasion with a differently colored blossom. When finally departing the valley in order to rejoin his friend Geoffrey, Tristan vows to return and to win Yolanthe's affections.

In the fifth scene of the lyrical drama, Martha by returning to the palace makes the startling discovery that Yolanthe has awakened by herself, an observation the latter strongly affirms. She relates her experience with the two strangers to Martha, a conversation secretly overheard by her father and Ibn Yahya. Soon after King René makes his presence noticed and announces Ibn Yahya with the words: "Thy tutor, *Ibn Yahya*, comes with me".[9] They, too, now officially learn of the appearance of a stranger who has raised her from her slumber and subsequently acquainted her with the knowledge of her blindness. This is the moment that Ibn Yahya had anticipated for Yolanthe's cure. The ensuing dialogue between King René and his daughter regarding the nature and relationship of seeing and knowledge will be at the centre of a more detailed discussion in the later parts of this article. Tristan, still strongly in love with Yolanthe, returns. He kindly requests to be relieved from his first promise of marriage and finally discovers that his future wife and the object of his desire are identical. Now the moment of healing as anticipated by Ibn Yahya has finally arrived and Yolanthe regains her eyesight. Initially overcome by the intense brightness that surrounds her she first lays eyes on her father whose voice she immediately recognizes. Moments later she is frightened by the sudden and unfamiliar sight of the date palm trees, but is calmly consoled by Ibn Yahya. King René blesses the union of both lovers by handing over his daughter to Tristan, who in turn gladly embraces her. In Tchaikovsky's opera, unlike in the numerous literary precursors, two mezzo-sopranos join the stage as Yolanthe's female friends Laura and Brigitte in the final moments of the opera. Fully aware of Yolanthe's regained eyesight, Brigitte breaks into a song of gratitude and jubilation: "Heaven, hear mercifully our song of gratitude. You didst send her the light. You didst send her the light. You are omniscient in the smallest things; magnificent just as the water drops, the full sun shines bright…"[10]

Ophthalmology in *King René's Daughter*

At this point I would like to return to some of the ophthalmological references the text seems to offer. When King René refers to what he has recently read in a book kept in his possession, "[t]hat oftentimes an unsound eye is cured [b]y application

[9] Hertz, Henrik, *King René's Daughter: A Danish Lyrical Drama*. Translated by Theodore Martin. (New York: Leypoldt and Holt, 1867), p. 75.

[10] *Jolanthe*, Lyrische Oper in einem Aufzuge, nach Henrik Hertz'König Renés Tochter von Modest Tschaikowsky, deutsche Umdichtung von Hans Schmidt, Textbuch (Hamburg, Leipzig: D. Rather), p. 40.

of the surgeon's knife",[11] he underlines his knowledge and possession of works on cataract surgery. Ibn Yahya however rejects such a method at this point by responding: "'Twould aid us little, should I have recourse [t]o instruments".[12] This singular statement equally implies that he is quite familiar with the option of surgical intervention of the eye lens.

Ibn al-Haitham (965–1040/1041), the founder of modern optics, developed a theory of seeing or visual perception which explored the theories of light in relation to the structure of the eye and the transmission of visual perception to the brain, thus correcting classic misconceptions. By doing so he revolutionized the understanding of how human vision operates as well as of the movements of light. The awareness of seeing in Antiquity was characterized by a discussion whether the eye sends rays towards the object or not (e.g. *Aristotle*); whereas the Arabic-Islamic Sciences in the first quarter of the tenth century verified that it is the object which reflects the light. It was Ibn al-Haitham who emphasized that nothing comes out of the eye, but that the outer world is perceived through straight rays of light; their ends are focused (bundled) in the centre of the eye. Thus the objects' forms are "transferred by lines, which the mathematicians call seeing rays."[13] Ibn al-Haitham verified his hypothesis by experiments with different forms of mirrors (concave, convex etc.) in order to gain this groundbreaking new understanding which also required as a precondition, precise knowledge of the eye's anatomy with its spherical form of the cornea and the visual nerve that connects it to the brain. His integrated and deep understanding of the interplay of light and visual perception enhanced the ophthalmology and cataract surgery. Below, in the context of perception and cognition we will come back to Ibn al-Haitham. As early as at the end of the tenth Century *al-Mausili* described the radical operation of the soft cataract by suction with a metal syringe (cannula, a hypodermic needle) he had devised for that purpose.[14] Although eye surgery had been undertaken prior to al-Mausili's innovative invention, it remained merely cosmetically in scope and failed to improve the quality of the patient's eyesight.

But beyond ophthalmological questions the physiological absence of eyesight in the drama *Yolanthe* seems to function as an indicator for the yet to be obtained divine love. Blindness serves here as a metaphor for Yolanthe's lack of self-knowledge and knowledge and love for God. Nevertheless, it is Ibn Yahya, who initiates and supports this process of transformation. The interaction between two pathways to the knowledge of one and the same truth, namely the scientific

---

[11] Ibid., p. 24.

[12] Ibid., p. 25.

[13] Sabra, Abdelhamid I., *The Optics of Ibn al-Haytham, Book I-III, On direct Vision* (London 1989), S. 82, cit: Belting, Hans, *Florenz und Bagdad, Eine westöstliche Geschichte des Blicks*, (München: C.H. Beck, 2008), S. 113.

[14] Hirschberg, Julius, *Geschichte der Augenheilkunde im Mittelalter* (Leipzig 1908), cit.: Sezgin, Fuat, *Science and Technology in Islam*, Vol. I, Introduction, (Frankfurt a.M.: Institut für Geschichte der Arabisch-Islamischen Wissenschaften), p. 21.

and the spiritual one, is highly characteristic of the Islamic philosophy which transcends various denominations and its corresponding understanding of sciences in Al-Andalus.

Knowing Without Seeing

Perception and Cognition

In the ensuing dialogue between Tristan and Yolanthe, who greatly endeavors to pick the right-colored rose Tristan wishes to receive, appears a moment of uncertainty when Yolanthe having contemplated on what he might mean by asking for a red rose responds: "Take it thyself!"[15] With these words she invites Tristan to imitate her pathway to knowledge, a path, lying beyond the visualization of the color spectrum that includes the feeling objects' texture, their shape, as well as the perception of the fragrance, which in turn enable her to distinguishes between a rose and a carnation. Yolanthe also identifies and distinguishes between the different species of singing birds by way of differentiated hearing of the various sounds and chirping patterns. Color perception, however, seems in this context to be an almost negligible level of cognition, for Yolanthe's description of plants for example are so much more detailed, accurate, diverse and multidimensional than they would be if she were able to visually perceive these. If so, she would perhaps focus entirely on the color of things whilst neglecting to explore their internal structure, and returning to our example, the internal, microcosmic structure of plants. Yolanthe proceeds by asking Tristan to teach her the application of her eyes: "Yet with the help of sight? They told me not of that. An Instruments [f]ashion'd by art, or but a tool, perhaps? I do not know this sight. Canst teach me, then, [i]ts use and purpose?"[16]

At this point, the lyrical drama reveals various epistemological dimensions, which are referring to Arab-Islamic sources, if we are indeed to believe that Yolanthe has followed the instructions of her tutor Ibn Yahya. The Arab-Islamic philosopher, astronomer, mathematician and physician, Ibn Tufail (1110–1184), primarily based in Al-Andalus, and author of *Hay Ibn Yaqzan* (philosophus autoditactus), draws on an analogy: "Just imagine a human being born blind, who has an excellent natural disposition, a strong intuition, a good memory and excellent ideas".[17] This visually impaired man is capable of being perfectly oriented in the city by way of increased development of his remaining senses. The colors, however, he only knows as a name and by the definitions that refer to them. Ibn Tufail further clarifies and

---

[15] Hertz, Henrik, *King René's Daughter: A Danish Lyrical Drama*. Translated by Theodore Martin. (New York: Leypoldt and Holt, 1867), p. 61.

[16] Ibid., p. 64.

[17] Ibn Tufail, Abu Bakr, *Der Philosoph als Autodidakt, Hayy ibn Yaqzan, Ein philosophischer Inselroman*, transl. by Schaerer, P.O., (Hamburg: Meiner, 2004), p. 6.

allegorically compares the visually impaired individual with those thinkers and philosophers who have failed to attain a metaphysical level of familiarity to God. "If him, now that he has reached this stage, the eyes are opened and he gains vision or eye sight, then he will find, so he chooses to wander through the city and look around, everything just as he had imagined, and nothing will be unfamiliar to him, [...] new to him (are D.Q.): First, the increased clarity and brightness, and second, the overwhelming pleasure. The condition of the theorists who have not reached such familiarity, equals the condition of this blind man before gaining sight, and the colors that are known to him in that state, are according to *Ibn Bagga* likewise of the following nature: they are too sublime, as that they may be related to the everyday, physical world – God bestows these gifts unto the servants of his choice".[18] Familiarity or intimacy is according to Ibn Tufail, the highest metaphysical level of knowledge of God – a hierarchy of knowledge that is already reflected in the tenth-century curriculum of *Ihwan as-Safa*: "If sense perception is the starting point of the cognitive process, abstraction is then a higher stage of knowledge which is a common feature to all the sciences of the curriculum. Within abstraction there are further degrees of knowledge, which are organized in a hierarchical pattern: the more the seeker advances on the path of knowledge, the higher the knowledge he acquires. The highest of all sciences is metaphysics which, in a hierarchical manner, comprises of (1) the knowledge of God and His attributes, (2) the knowledge of the soul, (3) the knowledge of resurrection and of closeness (mugawara) to the Merciful."[19]

Following the Arab-Islamic theory of learning as established by *Ihwan as-Safa*, training of vision or eye sight marks the highest level of sensual perception. Visual images are attributed to reason in the system of understanding and comprehension. In the book (Kitab) al-Manazir (of Optics), an all-encompassing work on optics, which was first translated into Latin in the twelfth and thirteenth centuries, *Ibn al-Haytham's* findings, which he collected, using mathematical, geometric and experimental methods, while combining anatomic and optical observations, paved the way for a breakthrough in the history of optics. The "spherical form of the cornea [...] it alone guarantees the unbroken penetration of the rays which advance from all sides of to the centre of the eye and to the centre of vision."[20] Except that objects are only perceived effectively and only exist if they are mentally appropriated, whereby the knowledge, which we have of the objects, our visual perception classifies. Yolanthe's father outlines aspects of a theory of seeing/vision in the fifth scene that already vaguely resemble the existing studies at the time of *Ibn al-Haitahm*: "Like wind and storm, [light] doth descend unto us from above, [a]nd like to these, with

[18] Ibid.,

[19] Bafioni, Carmela, "From Sense Perception to the Vision of God: A Path towards Knowledge according to the Ihwan as-Safa", in *Arabic Sciences and Philosophy*, vol. 8 (1998), (Cambridge: University Press, 1998), p. 216.

[20] Schramm, Mathias, "Zur Entwicklung der physiologischen Optik in der arabischen Literatur", in Sezgin, Fuat, *Science and Technology in Islam*, Vol. IV, Medicine, Chemistry, Mineralogy, (Frankfurt a.M.: Institut für Geschichte der Arabisch-Islamischen Wissenschaften), p. 19.

swiftness uncontroll'd, [t]he objects which it touches gain a new [s]ignificance, and a peculiar stamp, [a]nd oftentimes with warmth 'tis closely blent. 'This through the eye it finds its way to us, [a]nd by the power of seeing it we gain [a] true perception of the universe, [a]s it went forth from the Creator's hand, [a]nd apprehend His wisdom and His goodness."[21] It is this knowing of the wisdom of God and his goodness, to which Ibn Yahya referred in Yolanthe's teachings, the recognition of the inner eye as the highest level of the path of knowledge: "Her inward eye must first be opened, ere [t]he light can pour upon the outward sense. [a] want must be developed in her soul; [a] feeling that anticipates the light – [a] craving sense; for know, my noble liege, [t]hat nothing is on mankind e'er bestowed, [u]nless for it he feels necessity."[22]

The love for God that transcends from Yolanthe's inner being ignites love's desires in Tristan. As such it is an encounter that already anticipates Yolanthe's transformation as her kindling love for Tristan seems to reflect universal love as a communicative interaction. Preluding each verse with a few notes of his zither, Tristan begins to reveal his affections with the following words: "For all things have life through thee alone, [f]or all things will only be thine own, and close their eyelids when thine do rest."[23] "Dein Herz schlug ja in jedem Sein, Alles beseeltest Du allein – mit Deinem Aug'war aller Aug' geschlossen."[24]

When considering the knowledge of the all-creative, here in this lyrical drama figuratively outlined by Yolanthe's regaining of her eyesight, as the highest metaphysical level in the hierarchy of knowledge, it becomes clear that Yolanthe has already surpassed all of the other subjacent levels that the play refers to. And although, according to the Arab-Islamic development and learning theory, taste, touch, smell and hearing predate seeing, for it is referenced as the final stage of sensual perception in the early childhood development, Yolanthe's accurate perception of the environment that is so familiar to her, emphasizes the interaction of levels of perception and cognition in the process of learning as well as how the *unity-of-everything-there-is-alive* (Anna-Teresa Tymieniecka) in the universe, progressively communicates engaged in a continuous process of sensing, learning, knowing and recognizing.

> The real existence of every being is connected by countless circuits of accompanying efforts, desires and powers at every turn of transformation, at every moment of its development. They formulate both being and its world.[25]

---

[21] Hertz, Henrik, *King René's Daughter: A Danish Lyrical Drama*. Translated by Theodore Martin (New York: Leypoldt and Holt, 1867). p. 77.

[22] Ibid., p. 28.

[23] Ibid., p. 50.

[24] Hertz, Henrik, *König Rene's Tochter*, translated into German by Willatzen, P.J. (Bremen: Kühtmann's Buchhandlung, 1875), p. 25.

[25] Tymieniecka, Anna-Teresa, cited: Szmyd, Jan, "The role of Anna-Teresa Tymieniecka's Philosophy in den "Post-Modern World", Cognitive Optimism, Innovativeness and Creativity" in *Phenomenological Inquiry*, Editor in Chief: Anna-Teresa Tymieniecka, Editorial Board: Angela Ales Belo et al., A review of Philosophical Ideas and Trends, Towards the Skies, Vol. 35, October 2011, (Hanover, New Hampshire, USA: The World Institute for Advanced Phenomenological Research and Learning, 2006), p. 41.

Yolanthe's Cosmos

It is in the context of this cognitive theoretical dimension of the story of Yolanthe that the phenomenology of life (Anna-Teresa Tymieniecka) is applied in this paper in order to achieve not only a more detailed understanding of philosophical statements in the lyrical drama, but likewise to utilize the phenomenology of life as a starting point for a re-reading of *King René's Daughter. Anna-Teresa Tymieniecka* emphasizes in various chapters of her study *The Fullness of the Logos in the Key of Life*, the sentient and at the same time communicative part of the Logos as "[t]he communicative and Dionysian Logos of sharing-in-life"[26] or "[t]he communicative rays of the Logos present at the origin of individuation"[27] It is in particular this communicative force of life in the unity-of-everything-there-is-alive, communicative because it is sentient, of the all-pervasive Logos of Life, which allows Yolanthe to recognize various plants so accurately without actually visually seeing them. Not unjustly Yolanthe asks: "Can I see [w]ith these my eyes the nightingale's thick note, [w]hereon I've mused so oft, and vainly striven [t]o follow it in thought, away, away?-Or is her song a flower, whose fragrant breath I know, but not its root, and stem, and leaves?"[28] "Ist's eine Blume, deren Duft ich kenne, doch Wuchs und Stengel nicht und Blatt?"[29]

Awareness regarding the precise nature of things can only be developed through the knowledge of the microcosmic structure of beingness. How would it be possible to capture the structure and ornamentation of a rose without the communicative encounter/interaction in the *Logos of Life*, individualized here in two different forms of life, namely the flower and the human being? This communicative force initiates "interactive exchange": "Symbiotic linkage and interactive exchange offer the web of life, which among its many functions serves as it were a nervous system, and make the unity-of-everything-there-is-alive, without which no living being would come to exist. In this sense 'Once is all'".[30]

However, the path of knowledge as announced for Yolanthe does not end at the cognitive stage of existence, therefore she confronts her father with the following words shortly before her departure and the imminent restoration of her eyesight: "Yet on one point [t]hou doust mistake. 'Tis not the eye that sees; [h]ere, close beside the heart, our vision lies; [h]ere is it seated in remembrance sweet, [a] reflex

[26] Tymieniecka., Anna-Teresa, *The Fullness of the Logos in the Key of Life*, Book 1, The Case of God in the New Enlightenment, (Dordrecht, NL: Springer, 2009), p. 99.

[27] Ibid., p. 243.

[28] Hertz, Henrik, *King René's Daughter: A Danish Lyrical Drama*. Translated by Theodore Martin. (New York: Leypoldt and Holt, 1867). p. 82.

[29] Hertz, Henrik, *König Rene's Tochter*, translated into German by Willatzen, P.J., (Bremen: Kühtmann's Buchhandlung, 1875), p. 71.

[30] Tymieniecka., Anna-Teresa, *The Fullness of the Logos in the Key of Life*, Book 1, The Case of God in the New Enlightenment, (Dordrecht, NL: Springer, 2009), p. 189.

of the light that pierced my soul, [t]he light I go with bounding hope to meet."[31] "Doch in einem irrst Du: Gewisslich, mit dem Auge sieht man nicht. Hier, nach dem Herzen, hier ist das Gesicht. Tief im inneren ruht, wie freudige Erinnerung, ein Nachklang jenes Lichts, das mich getroffen, dem ich nun hoffnungsvoll entgegengehe."[32] The cosmic ontogenesis of vision and sight meets the light, which references the origin of creativity. Finally, moments before regaining her eyesight, Yolanthe repeats the following sentences as dictated by Ibn Yahya: "Mysterious being, who to me hast spoken [w]hen darkness veil'd mine eyes, teach me to seek Thee [i]n Thy light's beams, that do illume this world; Still, in the world, teach me to cling to Thee! […]".[33] The German translation differs here notably (a comparison with the original sources in Danish would be necessary to prove congruence), when identifying "mysterious being" with "Weltengeist" (universal reason or intellect, universal spirit): "Du Weltengeist, der zu mir redete, indes die Nacht mein Aug umhüllte, lehr mich dich suchen in den Strahlen dieser Welt! O, lehr mich, Dich zu lieben in der Welt![34]

The philosophy of life, unfolded by Anna-Teresa Tymieniecka in the last decades, proves to be a key for a new in-depth scientific research into the dimensions of reality, unraveling the "hidden networks of life."[35] Against the backdrop of a New Enlightenment the "*ontopoiesis of beingness in generative flux*" (Anna-Teresa Tymieniecka) cannot be traced back to a "mysterious being". By overcoming the enduring division between empirical and spiritual approaches towards reality, the philosophy of life deciphers the sentient and communicative-creative power of the *Logos of Life* towards the blossom of individualized beingness. In this re-reading of Tchaikovsky's last opera, in which the central figure's blindness might also signify the traumatic loss of her beloved mother, Yolanthe recovers her vision through the communicative-creative interplay of growing sentience into the real. In other words, she attains her self-knowledge while being inspired by the imagination of the Logos of Life flowing through the unity-of-everything-there-is-alive.

---

[31] Hertz, Henrik, *King René's Daughter: A Danish Lyrical Drama*. Translated by Theodore Martin. (New York: Leypoldt and Holt, 1867). p. 82.

[32] Hertz, Henrik, *König Rene's Tochter*, translated into German by Willatzen, P.J., (Bremen: Kühtmann's Buchhandlung, 1875), p. 73.

[33] Hertz, Henrik, *King René's Daughter: A Danish Lyrical Drama*. Translated by Theodore Martin. (New York: Leypoldt and Holt, 1867). p. 97.

[34] Hertz, Henrik, *König Rene's Tochter*, translated into German by Willatzen, P.J., (Bremen: Kühtmann's Buchhandlung, 1875), p. 85.

[35] Tymieniecka., Anna-Teresa, *Reason, Spirit and the Sacral in the New Enlightenment, Islamic Metaphysics and Recent Phenomenology of Life*, (Dordrecht et al., NL: Springer, 2011), p. 5.

Philosophical Hermeneutics Confronted by That Which Is Different

Aleksandra Pawliszyn

Abstract Our intention is to reveal that the human being is always surprised by the world's changeability, his existence being like footprints on the sand... And that this continuous changing reality is experienced by the human as a game – a kind of a movement organized by a specific formula. We refer to the motive of a game – the ability to express the essence of occurring being – which is taken from philosophical hermeneutics. Let us remind ourselves here that the first part of Gadamer's *Truth and Method* was devoted to explaining the special movement of existence: "here and there" as a game without any aim – like existential gambling.

How can the human being learn from such a changing world? We look at inspiration in phenomenological meditation, as a special kind of human contact with such a reality, giving understanding to surrounding occurrences.

The meditative attitude leads us into the world of linguistic expressions, which one can – after Heraclitus and Paul Ricoeur – name the world-text – a world the human being is still learning to read.

We also consider the hermeneutics grasp of language from the psychoanalytical perspective, in order to touch upon – somehow – the mystery of what is inexpressible. Next, we relate Lévinas' reflection on epiphany in the face of the other, to the authentic otherness of death. From the above it follows that contact with this mystery is a discreet contact, based on our resignation absorbing the other human being existence.

The human being is continually surprised by changing reality, which as it were expresses the human as a transitory creature, his existence being like footprints in the sand... On the one hand, common meaning as permanent meaning proves that it is a truth. And on the other hand, if we add, that the world has an unquestionable influence the human – one can suppose that the meaning of the human condition says something important about the condition of the world.

A. Pawliszyn (✉)
Institute of Philosophy, Sociology and Journalism, University of Gdańsk, Gdańsk, Poland
e-mail: wnspaw@univ.gda.pl

A.-T. Tymieniecka (ed.), *Phenomenology of Space and Time: The Forces of the Cosmos and the Ontopoietic Genesis of Life: Book One*, Analecta Husserliana 116,
DOI 10.1007/978-3-319-02015-0_12,

Continuous changing reality, is sometimes described as a movement organized in a cycle, characterized by a special formula – as in every game.

The Game as an Essential Symptom for the Hermeneutic Explanation of Existence

In the frame of philosophical hermeneutics the game motif is used to express the essence of an occurring reality. The first part of Gadamer's *Truth and Method* was devoted to explaining the special movement of existence: "here and there" as a game without any aim – like existential gambling. And in that kind of game nobody can win – one can only talk about a specific aim (to win) – the game's existence itself.

Gadamer notices that the game occurs when it is being played, and play cannot be ruled by the consciousness of the player. The game can only be based on various existing consciousnesses. So, the human being – as a kind of existence of expression – must take into consideration that he is immersed in the variety of entities. The pain of understanding every new constellation of events ought to connect with the grasping of a sort of *arche* of happenings experienced by the human being. It seems to us, that if the human being reaches the source (*arche*) of experienced events, he might feel settled in the face of a continually moving existence.

Reading the World-Logos

According to Martin Heidegger, understanding in its ontical meaning: "'to be able to manage a case', 'to grow up to it', 'to be able to do something'"[1] – is the way of being *Dasein*, as the possibility of being. Here it is important for us that "understanding can first set about opening the world".[2] Because of this the world belongs to the human being (*Dasein*) – as being-in-the-world,[3] so the human being finding himself among a kaleidoscope of events should connect with an attempt to "open" here his own world. According to Kant's reflection, one can say that the activity of human learning is realized as a kind of process, cutting a profile in the eternally changing world, which is the home of humanity (after Heidegger). The household of the human being – speech[4] – would be, generally, the human world, created by knowledge, following on from the human learning and being in the world.

---

[1] Martin Heidegger, *Sein und Zeit*, Max Niemeyer Verlag, Tübingen (17. Ed. 1993). [*Being and Time*, Oxford: Blackwell, 1964]. [Our translation is after the Polish translation by Bogdan Baran in the Polish issue *Bycie i czas*, Warszawa 1994, p. 203.]

[2] Ibid., p. 207.

[3] Ibidem.

[4] See Martin Heidegger, *Wozu Dichter*, in: *Holzwege*, Frankfurt am Main 1962. [Our translation is after the Polish translation by Krzysztof Wolicki, *Cóż po poecie*? in: *Budować, mieszkać, myśleć. Eseje wybrane*, Warszawa 1977, p. 212.]

Let us notice that the will to reach the source of existence's impulses must take into account that direct access to it is impossible. Therefore a way must be found of breaking open even if only relatively the influence of murderous, changing reality. In such a case it is useful to refer to Descartes' meditative way of philosophizing, which was specified by Husserl's phenomenology. It continuous the presumption that in the case of learning through meditation we are dealing with a special relation between the human being and the world – which is a conscious looking, penetrating the surface of phenomena, manifesting itself in front of the eye of consciousness, in order to grasp the essence of these phenomena.

Let us stress, that only because of meditative distance can one grasp the essence of that, which is changing – so that it is possible to imagine that changeability in general. So, the attempt – let us name here the archeology of events – must be based on the phenomenological model of the proper - meditative (excluding changeability) – contact between the human being and surrounding reality. This kind of contact should be treated as being on the border of change.

Thus, phenomenology introduces the philosophical hermeneutics and leads into the human household of speech – the world of linguistic expressions, which, according to Heraclitus' logos (Logos) one can – after Paul Ricoeur[5] – name the world-text – the world the human being is learning to read. The human being learns to read the world, which is the world of culture, so, among other things, the world of linguistic expressions is in an actual *here and now*, as well as consisting of tradition. But when tradition appeals to presence, it has a chance of occurring in the actual *here and now*.

Let us sum up: the archeology of events based on the meditative method is to give the experience of reading the cultural world's occurrences. That world should be grasped as if being lit up by the spiritual tradition, which appeals to the human being. To read that world is to make sense connections by attentive meditative looking, trying to cut through the changing winds, to read and tell the plot (logos)[6] of these mutable tracks of the human lot...

The Hermeneutic Grasp of Language from the Psychoanalytical Perspective

To explain the particular hermeneutic grasp of language, it is useful to compare it with Freud's psychoanalysis[7] – to be more specific – with the psychoanalytic grasp of the human *psyche*. It will reveal then, that except for the sphere of expressing

[5] See Paul Ricoeur, *La métaphore et le problem central de l'herméneutique*, in: "Revue philosophique de Louvain", 70 (1972).

[6] See Paul Ricoeur, *La fonction heméneutique de la distanciation*, in: *Exegesis*: *Problème de method et exercises de lecture*, Neuchâtel 1975.

[7] See Aleksandra Pawliszyn, Skryte podstawy rozumienia. Hermeneutyka a psychoanaliza, Gdańsk, 1993. "... es mag aber sein, daß es sowohl bei Freud wie In meinen eigenen Arbeiten verschiedene Aspekte sichtbar sind, die Ihr Interesse geweckt haben" – form Hans-Georg Gadamer's letter to me, Heidelberg, 13th of February 1989.

words (linguistic expressions), one can also consider it in the frame of language grasped in a metaphorical (non-linguistic) way, a pre-expressive sphere (a kind of disposal, a potential possibility to act, when we will it), and an inexpressible sphere, which reveals itself when we are confused and our speaking is not connected with understanding – as a symptom of our being lost in the present world.

We accept the above metaphorical meaning of language, connected with an assumption of its ontological (according to Heidegger) status, and we admit that the dimension of what is inexpressible is in the world; so, that it is as if a dim, inarticulate, dark mystery determines the frame of what is light and articulate – as a sign of something that leans out of the dark background…

An Inscrutable Murmur of Being and a Signal of Existence Revealing Itself in the Human Face: Plurality of Entity

In the phenomenological tradition there are philosophers, whose reading of the situation of the human being in the world is like participation in a kind of a game of existence. Gadamer writes about the existential game *here and there*, which seems to be an ontical moment of existence (Heidegger) and sometimes takes the shape of an artist's game. As for Lévinas, one can say that he describes such an ontical moment of existence – an existence without an existing human (*il y a*), which occurs like a murmur, soufflé, whispering, which still is not a voice, but as if charged with the power of being a voice.

The human being is a being in a gusty current of change – only nearby, on the rim of the source of existence. So, human existence could only be a sort of wiping away of the mystery of existence, which reveals itself to humanity as a murmuring entity…

First (in the presence) an unguessed murmur can later (in the future) be a signal of existence, manifested, for example, by the epiphany of the Other face (Lévinas) – the signal exposes individuality and sovereignty only when surprise is essential to an authentic future – the otherness of death. And this otherness of death can reveal the special relation face to face, that of revealing the Other. Let us notice, that Lévinas' reflection causes ontological pluralism,[8] because it emphasizes an ontical connection between an authentic future and the absolute otherness of death, which breaks down present life.

That pluralism could be read as an indication to the human being to change radically his attitude to the whole of the surrounding world (to *epoché* his own existence) in order to meet authentic otherness, which enriched and set free is the essential individuality of the human being.

Therefore, essential humanistics is an attitude of openness and tolerance in relation to that, which is not us, but shakes the foundations of our being in order to retrieve them from a sort of indefiniteness.

---

[8] See, Emmanuel Lévinas, *Le temps et l'autre*, PUF, 1998.

The Fusion of Horizons Guarantees Diversity

Philosophical hermeneutics sometimes as an artistic skill of understanding, as a skill of meditative (methodical) penetrating of the always unpredictable constellations of events, must often measure that which is incomprehensible, or even hermetic, like e.g. Paul Celan's poems.

Philosophers searching the language of people's communication (Gadamer, Ricoeur), teach us on the one hand, humility before otherness, on the other hand, responsibility for both dignity of that which is different, and of our own dignity. Through those attitudes manifest in sovereignty and freedom, it is worth stressing the essential connection between that which is ours, and that which is different – although everything is contained in the formal frame of one world. In philosophical hermeneutics that connection is named "the fusion" (Gadamer, Ricoeur), and can be recognized as the proper symptom of human freedom.

So, the fusion of different worlds does not rely on the absorbing of one another, but is rather a kind of guarantee, a mutual peculiarity and sovereignty, which can cause unique contact – the mutual epiphany of what is mutually different.

A proper, authentic relation with otherness occurs when it can be realized that that which is different will never be ours, for all that it will be; and my world is not able to take possession of that which is different.

Can We Learn That Which Is Different?

It is beyond question that the mystery of existence resists illumination before the light of learning, because it is based on an intentional, here possessive relation to the world, so, it must be revaluated. The point is to weaken the tension of intentionality as much as to release that which is in the darkness, which is not under learning's power, so that it has a chance to announce itself in some way. One could talk about a sort of learning activity, suspending claims to take possession of the object of interest, to reveal that which manifests itself from itself. This situation recalls a kind of modifying phenomenological proceeding, which reveals that which does not submit to the power of learning, what is dark, what is other…

Let us recall here, that both Gadamer's hermeneutics and Lévinas' philosophical reflections, lead on from phenomenological considerations.

Contact with Mystery Is Discreet

According to Gadamer's and Lévinas' analysis, the proper contact between the human being and the world relies on acceptance by the human and an open attitude to every otherness in the world. So, every contact with the world is as with authentic otherness, which sometimes unconceals itself (Heidegger) to the human being, to reveal the mystery of existence (being). However, only at that time can we really experience the entity which announces itself, as if stoking our existence.

So, contact with this mystery is discreet – when we share our world with the other, we do not manifest it noisily, we do not forget that it is a gift. Only that kind of contact can give us an opportunity to taste the game of continually searching for the satisfaction of existence. This situation can be illustrated by the metaphor of whispering existence, seducing us to "touch" someday the ravishing mystery of that which continually runs away – the mystery of continually happening existence…

The occurrence of that which is still not here – the end, which means: the final drawing out of the energy of life, the definite use of the potential we have to participate, the occurrence of death, of that, which is authentically other – the need to restore proper contact with another human being, different from than our world, and because of that we can in a circular way reach our own essence (the hermeneutics circle).

Let me stress here: it is impossible to posses, violate, or master authentic otherness. One can only experience a kind of syntony, consonance, harmony between dissimilar, sovereign "elements", which like difference can only coexist; so, when our desire to master is replaced by love – it becomes the unconditional protection of existence of what is different, because of its existence.

To Make Contact with Otherness Is to Resign from Absorbing Nearness

So, the essential symptom of love is to resign from annihilation of otherness, so, to resign from absorbing nearness, which reduces (totals) otherness to that which can be learnt (in the classical meaning). Resigning from learning nearness then is like a game behaviour – the game of that which is inexpressible, which emerges from the darkness, and continually separates, stoking discreetly our world of light. Like a game between light and darkness, emanating in a variety of the world's constellations, always only fragmentary, which the human being is still learning to play…

An important rule for the contemporary humanity derives from the experience of the holocaust – which calls human beings to measure swords with the riddle of darkness, and through suffering to learn the difficult joy of existence. In inter-human relations this rule appears as a new form of communication with an estimation of otherness – giving room to mysterious silence, sometimes to the mystery of silence…

Authentic Love Lets Otherness Retire

So, in some cases the contact estimated by otherness relies upon allowing that which is authentically different (actually unintelligible) to retire. We can here talk about a relation different from intentionality, or, perhaps, about something more fundamental than a learning relation between the human being and the world, which relies on belief in the power of the Logos – Transcendentalism. Also on sensitivity, on the vivifying joy following from freedom – connected with a responsibility for everything which is. It is worth stressing that responsibility is not connected with power, but rather with an agreement to do away with mysterious otherness – to express an authentic acceptance of existence – love.

So, otherness in human existence can be treated as a symptom of the end – the otherness of death, which can also be described (G. Bataille) as tempting humanity with "an infectious power" to life; which generates an inexpressible joy, following on from the experience of lack "the bottom and the limits of the universe".[9] However sometimes otherness appalls by its sable darkness, generating anxiety before that which is always out of limits...

To find yourself among the things of the world is here, to pull oneself out of the abyssal powers of nonexistence, to be in an oasis of reasonable penetration in existence – which through a game with the limits of darkness, seduces humanity, searching for sense, by promising to resolve the riddle of existence...

To Love the Difficult Freedom of Existence

Our human fantasy in a land of brightening rationality is only a fantasy. For the dark mystery is swaddled at the core, everything appearing on the lighted surface. So, the light can never be without a heavy dark background. As if the darkness of the abyss of nonexistence fights with glimmering light, which at that very moment becomes the light of existence. And, it is exactly in that world that we come to live, to feel, to love, that which simply is – to love the difficult freedom of existence...

[9] See, Georges Bataille, *Œuvres complètes*, t. VIII, Paris, Gallimard, 1976. [Our translation is after the Polish translation by Ireneusz Kania, *Historia erotyzmu*, Kraków 1992, s. 143.]

Anna-Teresa Tymieniecka's and Max Scheler's Phenomenology as the *Ontopoietic Genesis of a Manager's Life*

Bronisław Bombała

There are two kinds of intelligence: one acquired, as a child in school memorizes facts and concepts from books and from what the teacher says, collecting information from the traditional sciences as well as from the new sciences.

With such intelligence you rise in the world. You get ranked ahead or behind others in regard to your competence in retaining information. You stroll with this intelligence in and out of fields of knowledge, getting always more marks on your preserving tablets.

There is another kind of tablet, one already completed and preserved inside you. A spring overflowing its springbox. A freshness in the center of the chest. This other intelligence does not turn yellow or stagnate. It's fluid, and it doesn't move from outside to inside through conduits of plumbing-learning.

This second knowing is a fountainhead from within you, moving out.

(Jalaluddin Rumi, *Two kinds of intelligence*)

Abstract This article proposes to apply the phenomenological method in the analysis of management sciences, especially those that are used to diagnose human problems. A major barrier in this process is the strong technocratic paradigm (instrumental rationality). It seems that we should abandon in education and managerial action the technocratic model and introduce the personalistic-phenomenological model. Personalistic phenomenology allows the construction of stable foundations of management science and creates a coherent philosophy of managerial education. Crucial elements in phenomenological analysis are: Anna-Teresa Tymieniecka's phenomenology of life with her concept of "soul", Max Scheler's phenomenology

B. Bombała (✉)
Institute of Political Science, University of Warmia and Mazury,
Szrajbera 11, Olsztyn, 10-007, Poland
e-mail: 1dak@wp.pl

A.-T. Tymieniecka (ed.), *Phenomenology of Space and Time: The Forces of the Cosmos and the Ontopoietic Genesis of Life: Book One*, Analecta Husserliana 116,
DOI 10.1007/978-3-319-02015-0_13,

with "moral flight", Karol Adamiecki's concept of management with the law of "harmony of spirit" and Ken Blanchard's concept of servant leadership. It should be emphasized that the vision of good leadership requires for its realization – as stated by Max Scheler – *moral flight*, and this is done today in the sphere of praxis in servant leadership and in "economy of communion".

Anna-Teresa Tymieniecka's Phenomenology of Life and Management Science

Our era is characterized, on the one hand, by amazing technological development and, on the other, by many disturbing events in the world of turbo-capitalism. As noted by Anna Teresa Tymieniecka, the azimuth which one allows to determine how to make good use of advances in science and technology is missing. There are different views on human nature, which, instead of a coherent vision of man, have increased areas of relativism and permissiveness. Phenomenology proposes that the need is to understand the "object", which requires investigations into its genesis. The question of the origin of life (a derivation) is significantly associated with the search for understanding of what life itself is; life as a factor which is the attitude of science, culture and personal fate – the fate of the manager. Is life just an empty game? Unfortunately, the contemporary world is dominated by an empirical and naturalistic approach to reality. Science has descended from the heights of speculative reason to the original concreteness. However, the current state of affairs cries out for help, stemming from the reason itself, and calls for a philosophy with which to relieve us from the impasse in which we stand and which could conduct us further (Tymieniecka 2011: 17).

Tymieniecka rightly states that rationality is not a privileged way of knowing. There are instinctive and intuitive signposts in the course of a life. According to Bergson, she states that the various forms of life are focused on increasing the freedom of action. Tymieniecka has reached her intuition pertaining to life through a detailed analysis of the creative act of man. Understanding, or properly Creative Experience, is a necessary result of the human condition. This is reason, which dissents from the narrow, traditional frame of reference and is open to creativity and appreciation of the many new rationalities, in order to address the changing currents of existence, to form the criteria of validity, predictability and measurement (Tymieniecka 2011: 15).

Mankind tries to solve the various issues related to its survival and to deal with the diverse rationality which is offered by scientific discoveries. Then there is the spiritual, personal desire to find meaning in one's own life and its fulfillment. All of this calls for a network of "communication", which will satisfy the interdisciplinary activities and break the inter-theoretical incommensurability. Tymieniecka rightly states: "Man's search for wisdom, search for the meaning of what we accept on faith, is torn to pieces by an intellectual program of "deconstruction" on the one hand, and the resurgent religious distrust of reason, on the other. This is a situation that demands a thorough revision of the foundations of our reality, built by faith and reason, because the traditional standards of morality, customs, practices, goals and perspectives are blurred in the spirit of progress and stability of expectations disappears from our view" (Tymieniecka 2011: 18). We can make such revision, reaching – as Tymieniecka claims – the sense of meaning, i.e. to the ancient *logos*.

The science of management seeks philosophical, ontological and synthetical determination of its findings. The more economics, sociology, anthropology and other disciplines delve into the intricacies of the management of the world, the more lost they have become in a maze of details. They have lost sight of the unifying principle. And here comes to the aid Tymieniecka's phenomenology of life: "I claim that the living being identifies himself as "self", not by a cognitive act, but by "being alive" – experiencing their surroundings inside your being, guided by their instincts and desires, identifying elements of the surrounding world [...] and finally, most importantly, identifying themselves as the active center of the universe of existence as an independent entity causative, who directs the universe of existence from within, through experience, observation, reflection and reflection on the course which follows, and who ultimately gives this course of moral and aesthetic values, and on the wings of the spirit of seeking to understand the reasons for all this and rises high in the field of metaphysical and spiritual, bringing in a deeply felt conviction that to be is to be living" (Tymieniecka 2011: 26). In this way, we find the basic observation platform, not in knowledge but in the human creative act.[1] The *logos* of life is not a product of the noetic functions of the mind, but the efforts of the whole sentient human person in his logical expansion. Differential/compositional *logos* and operational/co-formational rationality activate the disclosure of the *ANIMUS* through *life-feeling-logos*. Everything achieves full meaning with the explosion of life, through which the logos illuminates the darkness, arranging the stage for the spectacle of life. Animated vital *logos*, sacred *logos* (logo-teic), noetic *logos* (triplex-noetic), communicative Dionysian *logos* (feeling/co-participating), Promethean *logos* (inventive) – they all allow you to complete the spectacle of life

[1]There is a similarity between the philosophies of Anna-Teresa Tymieniecka and Józef Maria Hoene-Wroński. Hoene-Wroński's starting point was the assumption of rationality of the world. If the world was not rational, theoretical study and technical work would be pointless. The rational world order can be explained only that the world exists and is growing by some law. This law, Hoene-Wroński claimed to have discovered and named it *the law of creation* (*loi de Création*). It not only involves the structure of reality– the auto–creation (*autogenia*), but also to establish processes and manage them – the auto–establishing (*autotezja*). It provides a method for error-free learning and action. Hoene-Wroński divided philosophy into the chrematic and achrematic. The first is derived from the world of created things and looking for its conditions until the condition is independent of anything, that is, up to the Absolute, which does not fit into the world of created things. This stream should be – according to Hoene-Wroński – all previous philosophy. The starting point of the second type of philosophy is that achrematic philosophy originates from outside the realm of things. It is independent of experience and constitutes the world as a set of conditions existing in the mind of the creator of the universe. This philosophy transcends created things and gives an analysis of basic principles. It reaches the most important principles of all existence and abolishes the gap between knowledge and being, forming their synthesis in immediate insight. This bears a striking similarity to Husserl's transcendental reduction (Murawski 2006: 143–150).

(Tymieniecka 2011: 37). The force of life, *logos*, emerges in this constructive *élan*. We find, in this way, the link between the constitutive intentionality of its noetic root on the one hand, and the emotional sensory-emotive subsoil sphere on the other. This approach considers both areas and their specific roles and functional effects.

Max Scheler's Phenomenology: An Inspiration for Management

Max Scheler in his anthropological investigations focused on the discovery of human beings. He sees man as a vital dilettante (unlike animals) and decadent that also transcends all, even his own life; this is the intention and gesture of transcendence. One factor that can exceed biopsychic life is spirit, manifesting itself in various intentional acts. The basis of the person is dynamic and acts, and especially the act of "moral flight" (Scheler 1987: 276). Access to the nature of the entity itself is achievable only through the act of "moral flight". In this act, the whole man is involved as a spiritual person. The result of cognition depends upon the purity and strength of "moral flight". As noted by M. M. Baranowska: "The man for Scheler is a psychophysical being [...]. The living body and psyche are two sides: inside and outside the same thing. That recognized as a man belongs to the animal kingdom and is part of nature, which is its environment. But the specificity of man is, according to Scheler, something quite different. Man is a place of emergence of a completely different order of nature, it is the essence of order, the order of the spirit – the spirit, "which is manifested only in man," and the "entity's constitution" which expresses "ideas and the right sense and nonsense, truth and falsehood, right and wrong"" (Baranowska 2009: 79).

The starting point in moral development is, for Scheler, a natural worldview, which is a correlate of reality. Scheler states that the "common point of departure for all types of higher spiritual activities (scientific, philosophical, aesthetic, artistic, religious, moral) – facing the group values, which in my ethics called the "spiritual values" – is a natural human worldview and the data in this worldview what is existing, and what valuable" (Scheler 1987: 282). It includes every possible existence, both external and internal space-time, and refers to the divine and into the realm of ideal objects. The world in the individual dimension appears as the "environment", which creates some community. This precinct includes the products of the surrounding culture, and various types of objects, ideas about space and time, tradition, natural to the culture system of thought, perception, language and "common sense".

Scheler also distinguishes three different worldviews: the worldviews relatively natural, which correspond to the term "environment", worldviews based on a scientific perspective of the world, which are surface-mounted on a natural worldview, as learning and thinking about being less relativized. The last is a philosophical worldview which goes beyond the previously listed and tries to get to **being-the-world-itself**. "Therefore" – Scheler states – "you need a flight with a set of moral acts: These acts are needed to make the spirit essentially abandoned of this being relativized vitally, being for life (and in it for the man as a creature), to make it begin to participate in being [that], which is in-itself" (Scheler 1987: 285–286). Flight contains three moral acts which are a condition of philosophical knowledge. The first act is to love all spiritual persons, to the absolute value and existence. This act frees man from relativism, which is associated with a natural worldview. Therefore, an act of love is here the core of the whole flight. The second act is "humiliation (*Verdemütigung*) natural self and ego (*Selbst*)", which removes pride inherent in man, the natural pride of a person entering into a state of humility. In this way, the path to philosophical knowledge is revealed and liberated from the things that exist in an individual and uncertain way, leading to a "pure content entity's world". The third act is self-control, which seeks to tame the natural instincts and desires, because they covered a relevant view of the objects of knowledge (that insight they cause is characterized by a lack of fullness and the object is only implicit). The application of this operation makes the transition from the inadequate presentation of content objects to a perfect and adequate presentation of the content of the world. Moral acts are a precondition for obtaining correct and full knowledge and each is responsible for another aspect. As well, love is responsible for the type and degree of ontic relativization of its subject, the rooting is responsible for granting knowledge of the essence and the self-control for the adequacy of knowledge (Scheler 1987: 286–287).

Scheler's act of moral flight fulfills a similar role to that of Husserl's reduction. However, as stated by Peter Orlik: "It should be noted that although Max Scheler speaks of phenomenological reduction, it is not the headline for the condition of the phenomenological immediacy. For Max Scheler, moral conditions are paramount. When moral conditions are fulfilled, the operation of the phenomenological reduction is made in passing, as it were ("automatically")" (Orlik 1995: 22). Fulfillment of moral conditions gives you access to essential knowledge, whose source is obvious insight into the nature of any possible existence – also the essence of being a manager.

The basis of the analyzed issue is Max Scheler's approach to love and hate as the original acts in relation to the knowledge of an object. A man reacts emotionally, bestowing an object of knowledge with the primary emotion such as love or hatred (Scheler 1986: 228–318). Only on this basis is true knowledge possible. Thus, Scheler's statement that the *aficionado* is always the forerunner of the researcher (Scheler 1987: 272), worthy of attention. This statement contains a deeper meaning, because it is not a purely sensual feeling, but rather the emotional spiritual experience that needs and creates value. Only then can one grasp the value clearly and directly. The acts of love and hate are a prerequisite and foundation of all the emotional acts, on the basis of which values are grasped: "Therefore" – Scheler states – "we can just add that specific types of "emotional" acts of our spirit – which first presents us with values and are a source of material for all secondary value assessments for all standards and duty statements (Sollseinssätze) – it constitutes a common link between both for the whole of our practical procedure, as for our whole theory of learning and thinking. However, because within a group of emotional acts love and hate are the most primary varieties of acts (Aktweisen) including all other types of acts [interest in something (Interessenehmen), feeling something, preference, etc.] and which is their foundation, they are, therefore, also a common source our practical and theoretical procedure, they are the only basic acts, where our theoretical and practical life finds, and retains, its ultimate unity" (Scheler 1987: 275).

For Scheler, love is a basic axiological experience, which is focused on values, and the object of love can only be that which is the bearer of values. There is a clear tendency for him to settle on love not only feelings but also the intellect and will. In this way, love becomes a great act.

Moral Flight as an Ontopoietic Genesis of a Manager's Life

The basis of an *Ontopoietic Genesis of a Manager's Life* can take Scheler's act of love as the cause and condition of all knowledge and action, since love always raises cognition and volition – the mother of the spirit and the intellect. Management-by-values may also be based on Scheler's hierarchy of values: the utilitarian (to make life easier), then the hedonic (fun and use), higher vitality (health), even higher spiritual (culture, knowledge, morality), and last – the absolute values (religious). Scheler's phenomenology seems to be a solid basis of what we refer to today as Emotional Intelligence, as a basis to more ethical behavior and integral personal development, similar to the ancient Greek concern for promoting virtuous character. Emotional life ought not be viewed as a chaotic impediment to reason, but rather should be understood as a sort of "sixth sense"; what Scheler termed our "Ordo Amoris" or "Logic of the Heart" (Scheler 1998: 25). As a value being and bearer of values, every person is as unique as a snowflake. This is why Scheler's ethics are commonly referred to as Material Value-Ethics as opposed to Kantian formal ethics. Values are emotively intuited, intuited by the consciousness before the whole of "something" can fully be rationally known or assimilated.

There are three types of intentional emotional experiences: feeling something, preferences, and love. Love opens us to the values, but not an act of cognition in the strict sense. Cognitive act is a feeling of something (value) and its preference. At the same time, preference should not be identified with the desire, will or decision. The preference is more primary. Also, if you want to know the value we refer to it directly, not through the perception made by someone – that may be illusory. Feeling the values and preferring are rooted in love, but love, even though it is a creative experience, does not create value. Values are the specific, irreducible properties of things and, as ideal entities, determine the importance and the validity of people and their acts. Values are timeless, unchanging and transcendent to all experiences, including the act of love. Love precedes the knowledge of values. Everywhere "the *aficionado* is always the forerunner of the researcher". Love designates an area in which value can become known. It is the aim of the knowledge of yet-unknown values. It discovers and reveals the emotional cognitive authority, without which it would be completely covered. Feelings allow us to penetrate certain areas of existence, where reason is powerless.

When we apply Scheler's statement "the *aficionado* is always the forerunner of the researcher" on the basis of management science, we discover what Scheler called resentment: the cult of modern work and modern humanitarianism. Humanitarianism, according to Scheler, is illusory love, in fact, based on hatred and escape-from-oneself. Resentment is also manifested in subjectivization and relativization of values in general. Not knowing how to read the objective order of values and live according to it, modern man takes revenge on an idea of values by denying its objectivity and the proclamation that all values are subjective. Illusory perceptions of value:

- Concerning the quality axiological – the illusion of cognition of something as being a value, which is not a value,
- An erroneous reading of the axiological order as "misstated table values",
- An extreme form of interference knowledge is blindness to value itself, the inability to experience axiological quality (Scheler 1997: 27–44).

Let's look at the issue, which is called "human resource management" (workforce management). Its core is the relationship: supervisor – subordinate (slave or free man?). Phenomenological reflection shows that the management of the organization shall be primarily by other people. This means that the "subject" of the work, i.e. the manager, acts on the 'object' work, which is not the thing, but also a man – a person. Each member of the organization, each of its stakeholders should therefore be treated as an entity and not as an object of labor. In this way, leadership becomes the core of management, namely, personalistic leadership and servant leadership (Fig. 1).

The phenomenology of leadership establishes the basic principle of "moral flight", i.e. it is assumed that the improvement of the organization starts with yourself. Access to the entity-in-itself is achieved by means of an act of "moral flight". This act involves the whole person, not merely psychophysically, but at the person's spiritual core. Phenomenological reflection leads to the conclusion that leadership is not the exercise of power in the traditional sense. It is important that leadership refers to people, not things. A particular type of leadership – personalistic leadership – occurs when leaders and their followers fall in each other's relations, which amounts to a higher level of morality (Bombała 2011: 11–33).

The leadership that has led to the moral flight is based on moral values. For a full understanding of the significance of this fact, it is necessary to move beyond an economic interpretation of management. The concentration of activity in the transformation of the outside world leads to the oblivion of the "inner world". The prospect of his own spirituality eludes managers from view, since Western civilization is dominated by utilitarian and practical values. The changes in an entity may be in the negative direction. Clearly, this refers to the manager, whose profession makes it difficult to distance oneself against the outside world with an attitude of reflective, self-centered interior. Sometimes he/she is often confronted with a dramatic choice between economic and moral values. An excessive preference for economic value

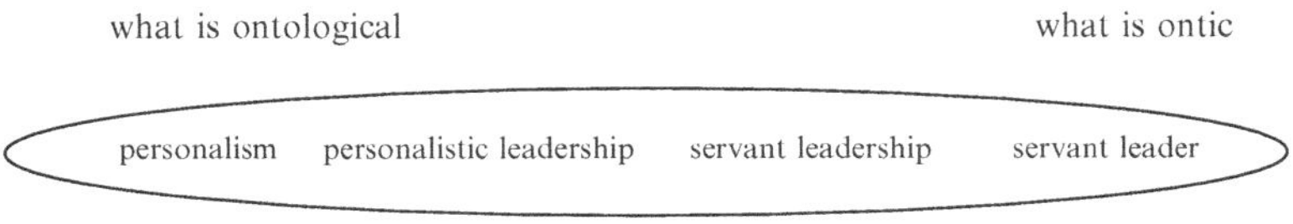

Fig. 1 Leadership in the phenomenological lens (In my phenomenology of management (phenomenological praxeology), the main instrument is the "phenomenological lens". A phenomenological lens focuses on what is ontological and what is ontic, existential and existentic – in Heidegger's sense. It allows for more accurate analysis of the object – both from the philosophical (ontological) and scientific (ontic) perspective. Phenomenological lens is the key concept of the phenomenology of management and, at the same time, a crucial instrument in the diagnosis and development of an organization. As a meta-method, it gives a view of the object from different perspectives and acts as a "binder", linking diverse factors affecting this object)

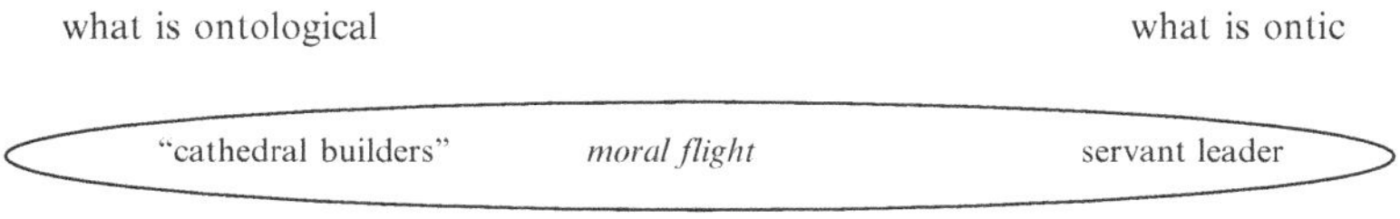

Fig. 2 Semler's autobiography in the phenomenological lens

causes a negative transformation in the entity and leads to immoral management (Bombała 2002: 335–346). It should be emphasized that a morally ascending manager is provided not only with his personal development, but also with the conditions for development of other participants in the organization. A good example is the **internal transformation** of the Brazilian entrepreneur Ricardo Semler, which began an entirely new management philosophy.

Phenomenological analysis leads to the conclusion that a Semlerian leadership style is the exemplification of the idea of personalistic leadership and servant leadership (Bombała 2010: 121). By placing his autobiography (Semler 1995) in a phenomenological lens, we see how he managed to overcome the technocratic attitude and make the moral flight. The idea-value, which determined the direction of the path was the metaphor – "a cathedral builder". As a result of an internal transformation, Semler became the "cathedral builder", i.e. a leader who serves – it was a condition for conversion of others into "cathedral builders". As a result of the moral impetus, Semler's company was transformed into a kind of "cathedral builders' guild" – community of work (Fig. 2).

Community of work is created through dialogue. And this is where phenomenology as a philosophy of dialogue, which, rightly, Jan Galarowicz considered as one of the greatest spiritual and intellectual events of the twentieth century (Galarowicz 2006: 185) helps with an in-depth analysis. The philosophy of dialogue has raised concerns, the existence of the close presence of another, the need to feel responsible for him and – to recognize that this responsibility is a meaningful ethical imperative. There are three basic conditions for the meeting: co-existence, reciprocity and openness (Wieczorek 1989: 26). The meeting should always have the character of being "face-to-face", without masks or veils. The phenomenological postulate, "back to things," in the philosophy of dialogue takes on the form of "back to another human being." (Tischner 1978: 73–75).

Closure: The Question of manager's Creative Condition: When *Poiesis* Becomes *Techne* ('Bringing-forth')

The main cause of the crises that plague the contemporary world is the modern fascination with mathematical formula, i.e. – instrumental rationality. Absolutization of the significance of mathematics in the design of economic strategies has become one of the main determinants of economic crises (Husserl 1987: 48–49). A similar pattern of thinking dominates in managerial education. It seems that we should give up on the "one dimensional man" model and introduce the personalistic-phenomenological model (Harciarek 2011: 120–127). It has high educational and ethical values, and most importantly – allows you to create a coherent philosophy of managerial education (Bombała 2012: 50–60). It should be emphasized that the key issue in management is ethical leadership. Phenomenological analysis shows that the foundation of ethical leadership is – as stated by Max Scheler – the moral flight, and this is done most fully in servant leadership.

There are high expectations of business ethics and CSR. Serious moral crises in the business world, however, indicate that the reduction of ethics to ethical codes and ethical programs has proved to be a far-reaching simplification. The pressure on business makes firms adopt various proposals of business ethics, but often only declaratively. Researchers have drawn attention to the weakness and inadequacy of academic business ethics. It turns out that the issue of ethics in business is less dependent on the number of codes of ethics, and more on the perception of their roles of entrepreneurs and managers. Often, business ethics is treated instrumentally and used as a kind of a screen for fraudulent practices. Whereas a personalist phenomenology allows to build stable foundations of business ethics, in both theory and practice.

Of great importance is an introduction to management analysis, an analysis of the manager's life, to what constitutes a man – a "soul". For Tymieniecka, the soul is the totality of sensory-psychic-creative-noetic complex living individual, who shows them his roots in the unity of life, and in an interactive network of the community of existence (Tymieniecka 2011: 279). This definition allows us to understand the imaginatio creatrix that breaks strands of constructive systems of nature, paving the way for limitless spiritual possibilities. The *imaginatio creatrix* is derived from the nature-life and is perfectly adapted to different life situations, while autonomy is manifested in the proposed ontopoietic wealth of opportunities for development. Tymieniecka introduces three new factors: the noetic sense, the aesthetic sense and the moral sense, which together inspire the human mind (Tymieniecka 2011: 297). The noetic sense is responsible for the world order and human communication. The aesthetic sense is responsible for broadening the

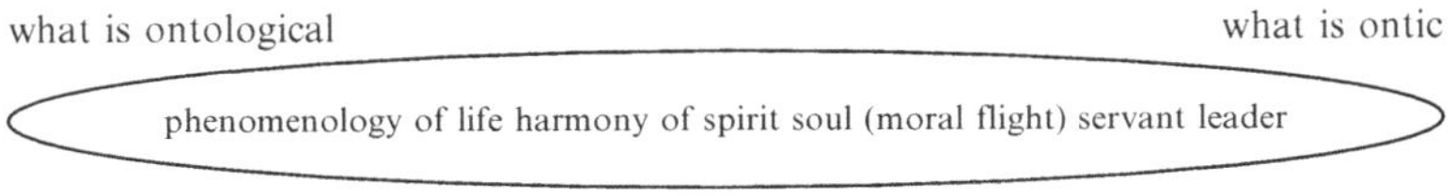

Fig. 3 Phenomenology as the ontopoietic genesis of managers life

experience beyond the strictly pragmatic understanding of what serves the vital interests of being, for the opening of a specifically human sphere of beauty, ugliness and sublimity. The moral sense, which is central to the metamorphosis of vital existence in the human condition. The world-as-a-human-community explains the moral sense. It is an engine of human design and carries the seeds of religious exploration efforts (Tymieniecka 2011: 298).

The soul is also the basis for Ken Blanchard's concept of servant leadership. In his theory of leadership the terms "soul", "heart" and "servant" are primary concepts. Blanchard's concept of leadership to a large extent is based on Jesus' teachings. His words were an inspiration and served as a basis of Blanchard's theory of servant leadership. Studying the essence of leadership should be sought to answer the question: "Am I a leader who serves (the other), or serving a leader?" (Blanchard and Hodges 2003: 17). Blanchard is not the first proponent of spirituality in management. Polish engineer Karol Adamiecki at the beginning of the twentieth century, spoke of the need for "harmony of spirit" in an organization. In 1903, he published a public article on how to organize collective work, in which he says: *one of the fundamental principles of action is fair and honest conduct*. Harmony of spirit – in his conception – is built on the basis of solidarity between workers and employers (Bombała 2006: 125–132).

In a phenomenological analysis (phenomenological lens), Blanchard's concept of leadership is an ontic dimension, and Tymieniecka's *Phenomenology of Life* has an ontological dimension, with the "soul" being the link between them (Fig. 3). The question of the creative condition of a manager is a question about the strength of his moral flight.

In *An Ontopoietic Genesis of Manager's Life* as a starting point, we take the study of creative/inventive human virtuality, so as to enter the core ontopoiesis, where the subject of managerial work is: nature and culture. The manager-as-inventor/creator emerges inside the system vitality, while at the same time modifying it. The arrangement of these forces: (1) carries self-individualization of life, (2) is responsible for the unity-of-all-that lives and (3) shows the layout of the existential relationship with the cosmic laws and the logos. Nature brings upon itself the burden of the development of life and meets with human interference in the universe of life. The vitality of survival-oriented, requires noetic (morally, aesthetically, socially) significance, the interaction between human beings (Tymieniecka 2011: 284).

In this way, somewhere in the depths of the technopoly, a Heideggerian rescue – a *poiesis* is sprouting. In spite of the technocracy, a "community of work" is growing, i.e. affective commitment to the action and the creation of human benefits, aspirations and values. The company embodies the purpose of the whole community and the common good, which is the development of its members and help for the excluded – for instance in "economy of communion" (Bruni 2002). It means to realize itself (exist), through the manager and the community of work. This phenomenon is concerned with freedom and personal development and is most fully expressed by Norwid's idea of "art-work", which symbolizes the spiritual relationship of people forming a "church of work" (Bombała 2007: 123). This analysis leads to the conclusion that the personalist phenomenology can be an important source in discovering the meaning of a manager's activity.

References

Baranowska, M.M. 2009. Man as a spiritual person in terms of Max Scheler. *Fenomenologia* 7: 73–82.

Blanchard, K., and P. Hodges. 2003. *The servant leader. Transforming your heart, head, hands & habits*. Nashville: Countryman.

Bombała, B. 2002. The autocreation of a manager in the process of transformational leadership. In *Life – truth in its various perspectives ... (Analecta Husserliana LXXVI)*, ed. Anna-Teresa Tymieniecka, 335–346. Dordrecht: Kluwer.

Bombała, B. 2006. Karol Adamiecki's harmony of spirit as a basic ethical organisational culture. *Annales. Etyka w Życiu gospodarczym* 9: 125–132.

Bombała, B. 2007. In search of the sources of humanity. In *Phenomenology of life from the animal soul to the human mind. Book I: In search of experience (Analecta Husserliana XCIII)*, ed. Anna-Teresa Tymieniecka, 385–396. Netherlands: Springer.

Bombała, B. 2010. *Phenomenology of management. Leadership*. Warsaw: Difin. in Polish.

Bombała, B. 2011. Phenomenology of the leadership: To be somebody – to make something. *Prakseologia* 151: 11–33.

Bombała, B. 2012. Phenomenology of management – didactic aspects. *Management and Business Administration. Central Europe (MBA.CE)* 3: 51–60.

Bruni, L. (ed.). 2002. *The economy of communion*. New York: New City Press.

Galarowicz, J. 2006. E. Lévinas: Inspirations and confrontations. *Kwartalnik Filozoficzny* 3: 185–196.

Harciarek, M. 2011. Phenomenological method in study of human and social capital. *Zarządzanie* 3: 118–130.

Husserl, E. 1987. *The crisis of European sciences and transcendental philosophy* (trans: Polish). Kraków: PAT.

Murawski, R. 2006. The philosophy of Hoene–Wroński. *ORGANON* 35: 143–150.

Orlik, P. 1995. *The phenomenology of axiological consciousness. Max Scheler – Dietrich von Hildebrand*. Poznań: W drodze. in Polish.

Scheler, M. 1986. *The nature of sympathy* (trans: Polish.). Warsaw: PWN.

Scheler, M. 1987. *Letters of philosophical anthropology and the theory of knowledge* (trans: Polish.). Warsaw: PWN.

Scheler, M. 1997. *Ressentiment and morality* (trans: Polish.). Warszawa: Czytelnik.

Scheler, M. 1998. Ordo amoris. In *About love*, ed. M. Grabowski, 13–47. Toruń: Wyd. UMK.

Semler, R. 1995. *Maverick! The success story behind the world's most unusual workplace*. New York: Warner Books.

Tischner, J. 1978. The phenomenology of the meeting. *Analecta Cracoviensia* 10: 73–98.

Tymieniecka, A-T. 2011. *The fullness of the logos in the key of life*. Book I: *The case of god in the new enlightenment. Analecta Husserliana*, Vol. C (trans: Polish.). Poznań: Wydawnictwo Poznańskie.

Wieczorek, K. 1989. Towards a definition of a meeting. *Studia Filozoficzne* 10: 25–38.

Part IV

Comparative Phenomenology of Singing and Dance as Involving Artistic "Instruments" Incorporated Into the Body of Their Performer

Alessia Rita Vitale

Abstract This essay provides a basis for reflecting upon the sensible body when it becomes a "body-case" (Vitale 2009c), as in the specific instance of its incorporation into the body of the performer (interpreter), understood as transformed thereby and so making of him both a creator of art and, at the same time, an artistic product. This is how the body can become both the physical and the psychic "*locus*" (Winnicott 1971) of a training procedure (carried out across a more and more meticulous regulation of gestures aimed at the accomplishment of an aesthetic composition). But our investigation also deals with the "body-case" when it has the double value of engendering the artistic work, even while interpreting it. This is why the body is subject to the double constraint of being both the body of the transmission and the transmission of the body, in such incorporated arts as dance and singing (Vitale 2007c, 2012b). This incorporation entails a whole series of dynamics such as the need for an "other than self", indispensable to enabling the self to get a distance from itself with a view to adopting a proper perspective vis-à-vis the instrument that it both incarnates and seeks to improve.

We will start out with our research on the art of singing – on the voice as instrument (see the *Instrument-voix* in Vitale 2007b, c, 2008a), the only musical instrument incorporated in the "body-case" of its performer – to then go on to extend the investigation to the art of dance across parallels concerned with what is specific to incorporated instruments, and this by analysing their phenomenological dynamics. We will also take into consideration other arts that are not located within the body of their interpreter, with a view to analysing how the relation between perception, gesture and memory impacts back on the artist's sensible experience.

A.R. Vitale (✉)
Université de la Sorbonne – Paris IV, France
e-mail: dr.alessia.vitale@gmail.com

A.-T. Tymieniecka (ed.), *Phenomenology of Space and Time: The Forces of the Cosmos and the Ontopoietic Genesis of Life: Book One*, Analecta Husserliana 116,
DOI 10.1007/978-3-319-02015-0_14,

Introduction

> The passage from prose to verse, from speech to song, from walking to dancing is a moment both enacted and dreamt. Dance is not meant to get me from here to there; and the same applies to verse and to pure singing. They exist to make me more present to myself, more handed over to myself, uselessly expended, so following upon me; and all other things and sensations cease to make any sense. One particular movement liberates them; and, infinitely mobile, infinitely present – they then rush forward, feeding the flame. This is why there are such things as metaphors, these stationary movements!
>
> Valéry[1]

This article should be seen in the light of a larger investigation whose objective is to reconstitute – across an epistemological research – the mosaic that makes up the phenomenological dynamics proper to the learning of such artistic instruments as are incorporated into the body of their interpreter.

The goal of this research is to understand "how" a human being organizes its *savoir*, its *savoir-faire* and its *savoir-faire-faire*, with regard to such incorporated instruments as are involved in singing and dancing. Obviously we are talking of a knowing that requires a type of appropriation "other" than that with which the individual makes its body own, even though it is its own body which is in question, that body with which it has been living since birth. In this sense, the study of singing and dance share certain points in common; not only do they interconnect, they are also interdependent, the body acting at the crossroads of Time and Space.

This article does not lay claim to being exhaustive on so large and complex a subject, but I do hope to draw attention to certain typically phenomenological approaches, which bring to light certain epistemological modalities belonging to the corresponding forms of knowledge.

Method

This research has been developed at the crossroads of the human sciences, and in particular at the point of intersection of philosophy, of a clinical psychology based on phenomenology, as well as of semiotics and pedagogy. It owes its inspiration to detailed research on the singing lesson and to an understanding of the dynamics of learning how to deploy the only fully incorporated musical instrument, research I conducted over a period of several years in the field of singing training in France, and this in many different musical conservatories, dividing equally my time (10 years) between periods of observation and periods of analysis.

My method of clinical observation relies on video supports and adopts a procedure of investigation based on interviews, questionnaires and also on my own experience

---

[1]Valéry, Paul (1912/1974). *Art et Esthétique*, in *Cahiers*, vol. II, Judith Robinson (Ed.), Paris, Gallimard, Bibliothèque de la Pléiade, p. 932. *Own Translation*.

Fig. 1 Paradigm (Vitale 2007c, 2008a)

BODY
↕
PRE-VERBAL ↔ VOICE ↔ VERBAL
↕
UNCONSCIOUS

as a singer. On the basis of a longitudinal-transversal method (this longitudino-transversal method consists in continually observing a sample of persons with the help of film material over an extended period), I have taken a sample of persons and observed them over time thanks to filmed support – in this case, students and their professors. My rhythm was essentially weekly, which made it possible for me to develop a phenomenology of the improvements over time lasting several years.

My analysis of languages, of their spatial and temporal dynamics, of the phases I call "structuring structures" (Vitale 2007c), lies at the root of this research, and offers thematic approaches leading from singing and its gestures to the body and its figures. The study of the *instrument-voix* bears on all these rituals, thereby conferring an intrinsic anthropological value on this research.

Since this field had not been explored in this fashion in recent research, it was necessary to create a terminology making it possible for me to deal with the subject in a precise and pertinent way, so applying an appropriate language to the analysis of multifunctional and polyvalent instruments. It was for this reason that, in 2004, I created the notion of the *instrument-voix*.

By *instrument-voix* I mean the voice when it is studied as having the value of an instrument of music, therefore examined in all its polyvalence and all its multi-functionality. The notion of the *instrument-voix* makes it possible to distinguish the learning process of the singing voice as such (the only musical instrument incorporated into the body of its performer, with all the implications that this incorporation carries from a phenomenological point of view but also from the point of view of perception and of the psyche taken in all senses of that word), from the processes that intervene in the learning of other musical instruments.

In view of the fact that this approach had never been taken into consideration up until now, it proved necessary to create a new paradigm, where Time and Space constitute a nodal interface (Fig. 1).

As is shown by this paradigm, whose value is both methodological and epistemological, the voice emerges at the cross-roads of different elements, such as the *body* (bone, muscle …), the *pre-verbal* (babbling, gurgling, meaningless words, lover's language …), the *verbal* (linguistic faculties) and the *unconscious*, with the result that the voice has been *the* privileged "instrument" of psycho-analysis from its very beginnings (Vitale 2007c, 2008a). The research I have been carrying out takes this complexity into consideration, this "multi-vocity" of the voice.

On what can the mysterious and complex process of learning to sing be founded? How can the learning of the vocal instrument be structured given that the teacher has to ensure that his pupil knows it, recognizes it, integrates it and maintains a distance from it to be able to use it? Does the singer play the vocal instrument or is he played by it? What emerges out of the long course of study of the singing voice?

It should be noted that these problems arise again in the dynamics of the study of dance.

The Body

The "Body-Case"

The study of singing and the studying of dancing share in common the phenomenon of the in-corporation of the artistic instrument: the singer, just like the dancer, is both the instrument and the performer of the same instrument. The singer and the dancer are both subject to the difficult task of carrying the instrument "*in*" *themselves* and making use of it "*for*" *themselves* that is, of being a "body-case" (Vitale 2009c). Hence the difficult enterprise of "taking up a distance" vis-à-vis oneself and of one's own instrument; for the latter is lived "in the flesh", "in one's tendons", in one's "most intimate self-perceptions", which tend to inscribe an ever more visceral relation with what remains one's own body, with its invisible parts "hidden" even to its own possessor. Singing and dancing make it possible for their performers (interpreters) to be both *solo* "instruments" and *ensemble* "instruments". I play with my body, play with my internal space as a "place" of resonance, my body touches sounds and is in turn touched by them. It emits, it frees and it receives … to the point of becoming what it plays. The instrument and its different parts, both visible and invisible, parts of the person involved in the process of learning how to sing and dance, tend to intensify progressively their link with the body, thereby improving the quality of this link; like husband and wife, *these incorporated instruments strive towards a conjunction of opposites, which explains their unifying character.*

I worked out the concept of the "body-case" in French ["human-body-case" in English (Vitale 2009c)], employing a neologism underlining the importance of these "casing" functions, whether physical or psychic. This "case" is indispensable, since it protects the artistic instrument one makes use of and on/with/in which one has learnt to develop a specific discipline, peculiar to oneself. This process of protection is peculiar to singers, dancers and actors and concerns, for example, the style of life, the hygienic aspects of life (for instance, culinary methods, physical exercises and so on).

The "body-case" is not given in advance (in this respect it differs from the concept of the "Ego-skin" [*Moi-peau*] developed by Didier Anzieu 1985). In fact, according to my hypothesis, the "body-case" is constructed by the apprentice, while the incorporated instrument is being learned, in proportion as the learning process progresses and the instrument is built up in oneself (Vitale 2012b). The concern with protection also has to be pursued as a function of age and the danger represented by certain physical exercises.

> Know that the dancer is not a woman who dances and that, for other reasons, she is not even a woman but a metaphor summing up elementary aspects of our form - blade, cutting edge, flower etc.; and that she does not even dance, merely suggesting by prodigious short-cuts or brilliant leaps, and with a corporeal writing, what it would take entire paragraphs in carefully structured prose to describe, to express, in words: a poem disjoined from any writing apparatus.
>
> Mallarmé[2]

The degree to which one develops and makes progress in the study of any artistic instrument is determined by one's ever growing intimacy with the instrument, so much so that one is in a position to calibrate each micro-movement in such a way as to anticipate gestures and to effect more and more minute adjustments, at the limit even imperceptible improvements whose results are however clearly detected and appreciated by the public.

> With great artists, the true technique simply speaks for itself: how easy it all seems! Sensible, sensitive, it never relies on producing mechanically the scales that make life so difficult for the neighbours. With the help of a system of contradictory tensions, it confers depth upon every gesture.
>
> David[3]

The Passage from the "Body-Case" to the Instrument: The Plural Body

The singer, just like the dancer, must, from within the vast multi-functionality of his or her own body, activate gestures accompanied by a certain kind of concentration allowing for all those transitional processes, running from the "ordinary" employment of the body to one that is quite "specific", and which are realized in daily practice in that temporal phase-structure called "warming up". My experiences based on phenomenological observations have led me to formulate the hypothesis according to which *warming up is both physical and psychic* (Vitale 2007c, 2009c). For we are talking of a set of processes that can be seen to take place both in the body and the psyche. It is not just a matter of warming up the muscles but also of what, in my view, takes on the value of a "disposition" or of a "transition" towards the acquisition of a physical and psychic posture appropriate to the task to be accomplished. This task requires quite specific acts which are both cognitive and sensori-motor.

> The number of physical movements that most people employ throughout the entire course of their life is extremely limited. They fall back on a series of habits, which hardly ever vary, having repressed and disciplined their movements from the first stages of childhood. In the same way, their mental activities also respond to a series of often repeated formulae. Through this repetition of physical and mental movements, they limit their expression right

---

[2] Mallarmé, Stéphane (1887/2003). *Crayonné au théâtre*, in *Œuvres complètes*, vol. II, Paris, Gallimard, Bibliothèque de la Pléiade, Bertrand Marchal (Ed.), p. 171. *Own translation.*

[3] David, Catherine (2006). *La beauté du geste,* Arles, Actes Sud, p. 21. *Own translation.*

> down to the point of resembling actors who, every evening, play the same role. Their life runs on with the help of these few stereotypical gestures, without their ever suspecting for a moment that they are losing sight of an entire world of dance.
>
> Duncan[4]

Like the dancer, the singer proceeds each day into the spatio-temporal studying routine across a gradual detachment from its surrounding world, with a view to concentrating upon its own body and its proprioceptions. The "*locus*" of study is a physical *and* a psychic space, where both external and internal processes take place, the external being readily visible, while the internal are so only to the performer and, across transitional dynamics, to his or her teacher also, but only if the latter is capable of a high degree of empathy-sympathy.

> I look at external objects with my body. I manipulate them, inspect them, move around them. As for my own body, I do not observe it in itself. For that I would need a second body, which would not itself be observable ....
>
> Merleau-Ponty[5]

My concept of the space of apprenticeship, understood as a didactic space, rejoins the concept of psychic space developed by Winnicott.

Psychic Space in the Learning of Singing and of Dance

In light of my clinical observations, the singing lesson appeared to me both a physical and a psychic "space" (Vitale 2007c, 2009c, d, 2012b). This approach, which seems to me fundamental, explains the way I analyzed the research conducted on the basis of the observations mentioned above. The concept of psychic space – or psychic "*locus*" – was introduced by Winnicott. By that is meant the initial "possession" of an object as located between the "inside" and the "outside", and which makes it possible to localize the experience of self. A space without which experience can never be lived out as self-generated, as discovered, as created.

In pedagogy, psychic space is, as far as I am concerned (Vitale 2007c, 2009c, d, 2011b), the "*locus*" where both teacher and pupil are reunited, for example, the "*locus*" where the teacher prepares his lecture (Vitale in press-b), this in accordance with a hypothesis I have myself developed (Vitale 2007c, 2009c). In the same way, it is also the "*locus*" where the lesson is continued *in* the pupil, so that the latter can deepen his or her reflections, uncover deeper semantic layers, or prolong the "presence" of the teacher *in* his or her self. A similar kind of phenomena can be found on the side of the professor. It is the space of "reverberation" and of a meeting, where the "other than I" continues to live "*in* me", in a "mineness" that accompanies me everywhere (Vitale 2007c, 2009c, d, 2013a). Psychic space therefore appears primordial in the processes of co-construction, inasmuch as it builds a pathway between

---

[4] Duncan, Isadora (2003). *La danse de l'avenir*, Paris, Ed. Complexe, pp. 73–74. *Own translation.*

[5] Merleau-Ponty, Maurice (1945). *Phénoménologie de la perception*, Gallimard, Paris, p. 107. *Own translation.*

past experiences and this potential space in which future experiences are prepared. So, within the singing lesson as well as the dancing lesson, the presence of an "other than I" is more significant than ever, since it assumes the role of a physical and psychical "mirror" thanks to the quality of its interactions and its competencies. In this way, through its attentive awareness, it enables me to redefine the way in which I "see" and "listen" to myself. The quality of its presence leads me to sound the depths of my possibilities and to concentrate all my energies so that – made aware of myself – the result exceeds my expectations. In so doing I can observe myself outside of myself (Vitale 2007c). This observation is accomplished by "watching", "listening" to oneself as to an other (Ricoeur 1990), and reciprocally, another as oneself. One is there in the attention paid to the creative act. So psychic space implies a quite personal relation to time, which can "expand" just as much as it can "contract" the experience. We are talking of a Space-Time created by the person, a sort of "psychic envelope" in which both our representations of the world and our relations with it are co-constructed.

Transitionality

The concept of psychic space refers right away to that of transitionality, also worked out by Winnicott (1951/1971). In what concerns us here, it is a matter of phenomena, linked to a transitional dynamics, my research "in the field" with singing classes largely confirms.

> Inasmuch as a violinist hands his violin over to his pupil, the singing professor likewise hands over his body, his intimate being.
>
> Vitale[6]

A genuine procedure of transfer and transitionality has to be put in place if the pupil and the teacher are going to be able to feel each other's physical sensations. The professor has to experience those of the pupil in order to understand what type of correction, what kind of improvement in gesture and attitude, he should propose. For his part, the pupil is all the better able to understand the meaning of the corrections to be adopted, the more he succeeds in investing his body with a work of projection into the gestures of his teacher.

> The context of the dance class allows one to see gestures as "fixed" physically, as learnt corporeally across a relation to space, to time and to others in this very body to body encounter which is itself linked to linguistic interactions, sometimes even appealing to reflectivity and to knowingly organized conscious attitudes. Little by little, this bodily knowledge gets set up, at first clumsily, then ever better and better until eventually it gets fixed in the sensori-motor and kinaesthetic memory.
>
> Faure[7]

---

[6] Vitale, Alessia R. (2007b). "Gestes et mémoires dans l'apprentissage du chant", in *Temps, geste et musicalité*, M. Imberty & M. Gratier (Eds.), Paris, L'Harmattan, p. 115. *Own translation.*

[7] Faure, Sylvia (2000). *Apprendre par corps*, Paris, La Dispute, p. 9. *Own translation.*

This transitionality is initially activated by very small gestures. Within warming up sequences, the singer, like the dancer, painstakingly carries out each day the "first steps" (the *abc*) on its body, and so on its own self. As if he were impelled by an obscure necessity to rediscover gradually the vowels, the consonants then the syllables, the little words, with a view to constructing the "phrases" in the artistic narrative process. Like the baby who, with its gurglings, its broken words ... progresses towards a coherence appropriate for his or her stage of development (Vitale 2007c, 2009a, 2013a).

> True movements are not invented, they are discovered, as are harmonies in music.
>
> Duncan[8]

Structuring Sequences of Gestures in Time and Space

These fundamentally primary gestures have to be confirmed and developed each day if one is to make any progress in the construction of one's own instrument at the highest levels, with a view to attaining an ever increasing virtuosity in more complex passages. A paradox? Perhaps! However, if paradox there is, it's an unavoidable, and irreplaceable, paradox. This shows the need to begin again each day from the *abc* of fundamental gestures, whose function it is to facilitate the re-assembling, the re-construction of the instrument, piece by piece, until a unity is recovered.

This sequential phenomenon (structuring at both the spatial and the temporal level) has led me to propose a hypothesis according to which warming up consists of an ordered "series" of physical and physiological gestures, all of which are reflected in corresponding mental stages, activated in the present moment of the physical phase in which one finds oneself (Vitale 2003b, 2007c, in press-b). The present moment is both a *summa* of past experience and an anticipation of future experiences. In this way the global "posture" – physical and psychic – becomes the container, the envelope, the macro-structure capable of accommodating the sequential events ordered according to a series appropriate to the level of the pupil, to his or her physical and psychic state, to the difficulties of the moment, to technical imperfections.

The construction *ad personam* of this "sequence" of gestures is proposed by the teacher following upon an anamnesis bearing upon the global state of the pupil, and on peculiarities disclosed in the moment, the *hic and nunc*. Collective "warming ups" consist of a sequence of gestures whose efficacity is largely proven. The basic warming up is followed by exercises whose aim it is to work at some specific technical aspect or on a combination of such technical aspects. But it is in the one-to-one relation with the student that the teacher devises the series of exercises designed to reinforce the weak points. Thus, in singing, the creation of singing exercises makes

---

[8] Duncan, Isadora (2003). *La danse de l'avenir, op. cit.*, p. 44. *Own translation.*

it possible to assume this difficult task. For the singing exercise gets transformed into a sort of "puzzle", to be combined with the musical and linguistic elements that the teacher derives from his or her own tool box, with a view to creating an appropriate vocal exercise – according to the quite specific requirements of each separate occasion (Vitale 2003a, 2007c). After having examined the rationale behind the technical-gestural-attitudinal default, the teacher proposes didactic strategies apt at getting the pupil to become conscious of the faulty gestures and, in what follows (which can closely succeed the preceding) he proposes the right gesture designed to replace the former. At this point the teacher proposes exercises to be carried out under his direct control to verify the correct execution.

This didactic sequence is generally carried through in two steps:

1. The teacher executes the figure or the passage with a view to provoking an imitation, an introjection (imitative modality).
2. The teacher adopts an even more maieutic procedure, by seeking to arouse the right gestures by means of willfully metaphorical verbal indications, or accompanying gestures.

The gesture and the metaphor have this in common, that they are both openly polysemic and multi-functional "transitional objects".

For the purposes of this study, and in order to bring to light the dynamics common to the study of both dance and singing, the notions of *habitus* and of gesture are essential. These two topics are presented here in a systematic ordering.

Habitus

The notion of *habitus* precedes that of posture and gesture.

In Latin *habitus* (from the verb *habere*: to have) is a word defining a way of being, a general atmosphere, an attitude, a mental disposition.

The *habitus* is a primary state of being, of being in the world, of one's being there. The *habitus* not only guarantees our physical presence in the world, it also reveals us to the world and reveals the world to us. It features as a sort of envelope, containing our being and contained by being. The *habitus* is a physical and psychic skin delimiting the limits and therefore the existence of the person. These limits can only interact with the world, according to a sort of porousness that receives, accommodates and gives back events that have been metabolized, metaphorized (for a detailed analysis of the *habitus* and the development of the singing voice, see Vitale 2012a, b).

With Aristotle, *habitus* (*hexis* in Greek) designates that authentic knowledge which engages the soul of a being in its entirety and not its habits, its "routines", for however powerful the latter might be, they still do not get written into our own very being. St. Thomas Aquinas understands by *habitus* the subject's ability to interiorize the perfection he aspires to, and which is revealed in his practical activities. For Pierre Bourdieu, the *habitus* is more than just a simple conditioning, one that leads us to reproduce mechanically what one has acquired. The *habitus* is not a habit. In fact, the dispositions it implies are more like the grammar of one's native language. Thanks to this grammar (acquired through socialization), the individual is able, in fact, to construct an infinity of sentences, enabling him to face up to all kinds of situations. He does not simply tirelessly repeat the same phrase like a parrot. The dispositions engendered by the *habitus* are of the same kind: they are schemas of perception and action making it possible for the individual to produce a whole range of new practices adapted to the social world in which he finds himself.

The Phenomenology of Gesture

Gestures in the Art of Dancing and in the Art of Singing

The notion of gesture is basic in this type of research. But what is the nature of gesture?

Imberty has this to say: "*Gesture is then characterized by the temporal profile of the movement that supports it. But gesture cannot be reduced to this movement and, moreover*, all movement is not gesture. *Gesture has to be defined as a more or less complex intentional movement*, oriented towards a determinate goal *that gives it its sense, whether individual, social or historical*".[9] Imberty tells us here that the gesture is a bearer of meaning. The gesture goes beyond the domain of the art of singing and the art of dancing since it is a form of language (Vitale 2007a). But I do not want to undertake a philosophical research into language here, since this would take us too far out of our way.

My proposal is to study gesture insofar as it belongs to singing and to dance and so from an interdisciplinary point of view. When do we resort to gestures? Is the gesture only a bodily imitation? Is it not rather an imitation of the voice, of an intentionality, of a physical or psychic posture? What is the place of gesture in those situations in which singing or dancing is learnt?

Does the gesture have a didactic or maieutic role? Can the gesture assume a didactic (pedagogical) function even with students having perfect hearing? At what moment does the gesture intervene from a phenomenological point of view? Why just at that moment? What do we express across our gestures? Do gestures have a voice? And can the gesture be conceived and perceived as such?

In what concerns the voice, gestural activity is inherent in the learning of the *instrument-voix*, especially when the latter is compared with the learning of other musical instruments. The vocal gesture is indeed an essential element in the deployment of the voice. All the same, to avoid making too approximate a use of an often debased expression, I propose to talk of vocal gestures in the plural, and this by distinguishing two types: internal vocal gestures and external vocal gestures.

[9] Imberty, Michel (2005). *La musique creuse le temps*, Paris, L'Harmattan, p. 90. *Own translation*.

(For a detailed presentation of vocal gestures, see: Vitale 2007b, c, 2008a). This suggestion is connected with an earlier work where I advance the hypothesis that gesture stems from a double origin, from an external vocal gesture *in the text* and an internal vocal gesture *in the score* (Vitale 2004, 2007b, c, 2008a, 2009b).

No vocal production is possible in the absence of internal and external vocal gestures. I consider the immobility of a singer and even the aptitude for immobility as a veritable (vocal) gesture since, being a "*habitus habité*", this very immobility contributes to the construction of the vocal sound (Vitale 2007b). The immobility I am concerned with here relies on this type of analysis and consideration. A state of immobility on stage is always the fruit of a present, direct and living attention. Every external gesture, whether mobile or immobile, is full of intensity, responds to an intentionality and is oriented in a certain direction: It is projected in a condition of fixity, and then worked by an intrinsic intention that brings it to life. This is not the case with the rigidity of a lifeless body, which, for this reason, is devoid of intentionality. In this sense posture should be considered as a veritable gesture (physiological and theatrical) capable of facilitating (or not facilitating) the production of gestures; for the rest, the study of posture has a significant impact upon both the communicative contact with the public and on the production of the vocal sound (Vitale 2007c, 2012a). By communicative posture singing teachers mean a disposition towards communication given by the orientation and the living and magnetic presence of the look. The look exerts both a power of concentration and of abstraction. In communicative posture the body disposes of an antenna that both receives and emits.

> Nietzsche said he could not believe in a God who didn't know how to dance. He also said that we should consider the day we did not dance as a lost day. By dance he did not mean the execution of pirouettes, for he was talking of the exaltation of life in movement.
>
> Duncan[10]

Gesture in the Dynamics of the Transmission of the Art of Singing and the Art of Dancing

Often, the gesture is also used as a didactic tool, either by the teacher, who offers it to the student as an example to imitate or, more directly, by the pupil. The conscious use of gestures is directly linked to the autonomy the pupil has to acquire progressively, as also to his personal ability to develop strategies helping him through the period of his apprenticeship.

Does the gesture explain the spoken word or is it not rather the spoken word that explains the gesture? What is the influence of one on the other and what is the relation of the one to the other? It all comes down to a painstaking operation designed to reinforce the relation to the body across its own self-perception,

[10] Duncan, Isadora (2003). *La danse de l'avenir*, *op.cit.*, p. 68. *Own translation.*

thanks in particular to the continuous adjustment of the *Self Ideal* to the *Ideal Self*, and this in a perpetual confrontation between what one would ideally want to realize on the aesthetic plane and the real possibilities of developing such possibilities in the present moment, *hic et nunc*.

The gesture turns out to be an excellent didactic tool for different reasons. On the one hand, it offers the advantage of completing the spoken word by superimposing itself upon it and so lending it a semantic complementarity linked to the "inexpressible". On the other hand, the gesture becomes a silent instrument making it possible to intervene – with corrective finalities – without interrupting the action.

Amongst other things, the gesture can be very eloquent, for it can revive and express a bodily lived experience or a sensible experience so difficult to express as a whole (relation to time and to the quality of the pedagogical intervention "of the moment"), even while retaining its "openness", its polysemy. The gesture is a trace, a trace of the in-corporated experience. The gesture involves the body (bone, tendons, skin) and metaphor.

The gesture is a metaphorical inter-semiotics intervening between the abstraction from the body and the corporality of thinking, this chiasm that confers an interdependent structural unity upon the whole. And what other discipline could pull together so effectively the intimate relations that subsist between thought, imagination and its expression in the instrument, the instrument that coincides with the body of its performer? Dance, singing and the theatre arts, each in their respective ways and according to the variations introduced by different schools, incorporate truths in interdependent semantic forms, each retaining its separate identity. The word "is made" flesh. In turn the body spiritualizes the text thanks to a relation to time, time as both immanent and transcendent, the time of an obsolete body and also the overcoming of obsolescence through gestures of immortality (Vitale 2012b).

Truth becomes incarnate – The body looks for the truth.

Gesture extends reception by developing a sort of kinaesthetic reception: one sees and one feels a gesture (Vitale 2011c, 2012a, b), as is confirmed by the theory of mirror neurons (Rizzolatti et alii). It is not limited to encoding and decoding. Through this relation a narrativity gets created and in this way communication relies upon gesture as the primary element in the human exchange (Vitale 2011c, 2012a, b).

Life plays with gesture. Gestures influence the memories of those who carry them out *and* those who are their recipients. Life is transmitted thanks to gestures. Learning gestures, gestures designed to perpetuate and to transmit knowledge and, more generally speaking, also values. "Conferring value" automatically means not wanting to lose anything. So the gesture intervenes to "say", to "give", and so to "perpetuate" its apparently ephemeral condition. To leave a trace of one's self and of other selves who, in turn, have insinuated themselves deeply into our being to the point of marking it with "expressive" similitudes retained, unconsciously absorbed, through emulation. The perpetuation of values, making up a shared heritage. The body acquires the lineaments, the expressions of persons who have "marked" our life, who left upon us a mark whose gestures one wants to transmit (even if sometime unconsciously). The body reveals the incisive side of encounters.

Between the infinite possibilities of improvement and the *hic et nunc*, between the present and the immanent, the real and the ideal, a polysemy gets set up, the polyvalences of an unconscious that unites the body and the soul.

Merleau-Ponty writes: "The thickness of the body, far from competing with that of the world is, on the contrary, the only way I have of going to the heart of things, by making me be world and making them be flesh."[11]

The relation between what is concrete, the weight of the body, the beating of the heart, the control of breathing, emotions that alter the rhythms of the heart, salivation, the faculty of abstraction making it possible for me "to see" my own voice propagating itself spatially in the air, to giving colour to my own voice, the entire fabric of these links makes it possible to actualize a synaesthetic modality of thinking, in which an inter-relation between the physical anchorage (of the senses) and the faculty of abstraction gets confirmed.

> In order to understand how the senses communicate amongst themselves it is enough to appeal to the imagination, which thereby acquires the status and function conferred upon it by Hume: it is the place where the real gets constituted through associative links.
>
> Dufrenne[12]

A Grammar of Gestures

If gestures can replace verbal language, is it legitimate to talk of a genuine *Grammar of gestures*? I refer here to texts in which I present the discovery and analysis of a *Grammar of Gestures* (Vitale 2010, 2011c, 2013a).

Can gesture completely replace the human voice in all its aspects, as *non-verbal gestural discourse*?

For example:

1. *Inflexions* of the voice → *Inflexions* of discourse by gestures
2. *Rhythm* of the discourse by the voice → *Rhythm* of the discourse by gestures
3. *Elements of punctuation* by the voice → *Elements of punctuation* of the discourse by gestures (Vitale 2010, 2012b, 2013a)

The expression of the gesture complements the expression of the voice in a vital way. The gesture offers an unsaid enrichment to the expression of the voice: it *amplifies*, *contradicts*, *underlines*, *affirms* and *differentiates* the latter (Vitale 2011b). Thanks to gesture, the voice becomes polyphonic, engendering languages whose meaningful figures remain in an internal correspondence with the voice.

---

[11] Merleau-Ponty, Maurice (1964b). *Le visible et l'invisible*, text established by Claude Lefort, Paris, Gallimard, p. 118. *Own translation*.

[12] Dufrenne, Mikel (1991). *L'œil et l'oreille,* Paris, Ed. J.-M. Place, p. 122. *Own translation.*

Using the voice is a gesture – and the gesture can be perceived as a voice. In what concerns the voice, the use of a gestural apparatus belongs to the learning of the *instrument-voix*, especially when it is compared with the learning of other musical instruments. The vocal gesture is in fact an essential component of the use of the voice.

Gesture and Metaphor

The gesture intervenes *silently*, which makes it possible for it to act alone or in conjunction with the verbal expression. It is polysemic, and "open" to several interpretations. Hence the need for a "shared code" in order that this language should first of all be *understood* then *retained* and finally *applied*, this by putting into practice the message it carries, especially in a didactic situation (Vitale 2010). By its semantic openness, the gesture appears as a metaphor and can in consequence be adapted to a large public.

If the gesture is a metaphor, can metaphor – in turn – replace a gesture?

The gesture always refers to something else; it is a passage essential for symbolic communication and as such it is situated on the threshold between oral and written language (Vitale 2009b, 2010, 2013a). It is "written" in the air, in time and space, in our memories. The gesture creates a "choreography" (Vitale 2011c). According to my hypothesis, the gesture is remembered since it is able to create not only visual but also sensori-motor impressions (Vitale 2007b, 2008a); in fact, the gestural modality awakens an empathic relation (already present at the pre-verbal stage of the baby). The gesture stems from the body and is situated at the interface between ideas and their instinctual representations in a sort of instantaneous impact (Vitale 2007c). The gesture is evidently a communication bordering on transitionality: which, for Winnicott means that the experience, the knowledge, the affectivity it expresses gets transmitted independently of all physical contact (Winnicott 1957/1971). One learns through the body. *This knowledge inhabits the body, traverses and modifies it* (Vitale 2011c). Moreover, as child psychologists tell us, before learning to "speak", the baby naturally grafts a "gestural counterpoint" upon these pre-verbal stages. Thus the sonic discourse is accompanied by an other discourse where the looks, facial expressions, the movements of the body all get written into a globally polyphonic communication (Vitale 2011c).

The Temporal Phenomenon of Gestures

The gesture is born and enacted at the crossroads of several temporalities, which is why it makes an excellent didactic tool. In fact the gesture can:

1. Act *by anticipating* a mistake made by the pupil and well known to the teacher, so helping him to avoid the error.

2. Intervene *during* the action by correcting it in the present moment.
3. Intervene *immediately after* an error has been noticed to correct it (Vitale 2007b).

In this way gestures are able to insinuate themselves into time, as in the movement of a pendulum: this is what makes them such an effective didactic tool. But the gesture can also be employed by the student himself with several goals in mind: maieutic, recollective, explicative ...

> This person who dances shuts herself up, so to speak, in a duration she engenders herself.
>
> Valéry[13]

The Gestures of the Body: The Body of the Gestures

In addition, the gesture facilitates the dynamics of "affective attunement/*accordage affectif*" (Stern), relying on an intuition of the felt-lived sensori-motor activity of the "other than self". This need to "be with", to feel as the other feels and perceives, is necessary and belongs to *anamnesis* – in its etymological meaning of remembering – as formulated by Plato. In our case, to be able to grasp the state of tension-distension exhibited by the other across rather little supportive evidence (micro-indications) facilitates the understanding and the anticipation of the creative dynamics linked to what is tactile in our gestures.

Thanks to a transitional dynamics, I can try to feel what the "other than I" experiences with a view to better understanding what generates this behavioural pattern, the "musical terrain" in which we interact with each other. Transitional dynamics are, therefore, absolutely indispensable in cooperative practices of artistic interpretation.

Gestures are fundamental tools in music just as well as in dance, both in the learning phases (solitary or accompanied) as also – quite evidently – in the phases of execution. They turn out to form an extremely dense and precise language, as much for the production of movements destined also to enhance the imagination-creation-generation of sounds as for their contribution to the artistic interaction. Just to take our ability, developed throughout the first period of our lives, to decode the gesture of a partner (see Trevarthen); what this means is that we are able *to anticipate* his or her musical-gestural interactions. This makes it possible for us to install a musical discourse into the micro-context of the *predictable unpredictability* of events that are both corporeal and sonorous.

Amongst singers as well as dancers, a gesture (however microscopic, that is, hardly perceptible it might be) is employed as a truly parallel language both for its precision, its variety and its composition, just as long as a sufficient period of time has elapsed to make it possible to harmonize these languages – verbal and non-verbal – whose public and private codes thereby become shared.

[13] Valéry, Paul (1936/1957). *Œuvres*, vol. I, Jean Hytier (Ed.). Paris, Gallimard, Bibliothèque de la Pléiade, p. 1396. *Own translation*.

Fig. 2 Figure representing the interactive functions between the body, the mind and the realization of the artistic act, whether sung or danced

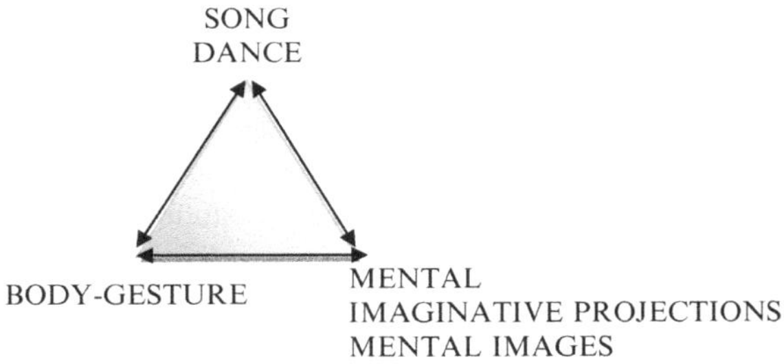

The success of a concert or of a ballet depends largely on the feeling engendered by, and between, artists, that is, on their complicity, synonymous with a massive togetherness accomplished with just a few gestures. A look can have the value of a gesture, just as can a more visible inspiration … We are talking of micro-indices making it possible for us to "be with" in order to co-create the event. The gesture does of course also have the value of an *anticipation*. Across this or that other gesture it becomes possible to predict the following gesture, the *emerging gesture*, the gesture that my colleague has not yet realized but which he is in the process of realizing. Phenomena such as these are particularly interesting and contribute much to the practical structure of improvisation, in music just as in dance. In this sense the gesture, as the sensori-motor transposition of an internal pattern, whether a musical or a danced figure, is linked to the global value of the mental image.

> The image is a psychic event serving to restore the figurative appearance of objects or events outside the material conditions of their realization in the perceptual field.[14]

The mental image, in a larger sense, can constitute the idea of a musical "pattern" or representation of a musical phrase. To be more exact, for us, this both represents and implies an intrinsic struggle between, on the one hand, the need to uphold physiological needs (linked for example to the limits imposed by breathing) and, on the other, to overcome these same limits with a view to improving one's own performance (Vitale 2007c) (Fig. 2).

The specificity of gestures relies on certain quite precise characteristics:

1. Gestures are rapid signals. The relation they have with time makes their presence in musical art essential since the latter is, *in primis*, an art of events and of micro events unfolding over time. Gestures can therefore be written into the temporal event without interrupting it, thanks to their silent character which, for all that, does not reduce their eloquence.
2. Even though gestures are born in the body, they tend to overcome the limits imposed by the threshold established by this same body, understood as the physical and psychic envelope on the basis of which the gestures are engendered.

---

[14] Denis, Michel (2000). "Image mentale" in *Grand Dictionnaire de la Psychologie*, (Ed.) Bloch Henriette et al. *ii*, Paris, Larousse, p. 451. *Own translation.*

In the singing lesson the gesture assumes a quite specific and all the more irreplaceable role, given that it makes it possible not to have to switch continually from the singing to the spoken voice. This rapid transition to a completely different dimension and employment of the voice (to offer rational explanations) is of little benefit to the vocal chords since they only get more tired, nor does it help the concentration, on account of the interruption. And as for the sonorous event, it simply gets more fragmented as a result (Vitale 2007b, 2008a, 2012b, 2013a).

Gestures of Creation: The Creation of Gestures

> The word is to the idea what the gesture is to the sensation: its sensible manifestation. Ideas do not exist outside words, nor sensations outside the body. The word is at one and the same time the incarnation of the idea and the artisan of its conception; the gesture is, at one and the same time, the expression of the sensation and its source.
>
> David[15]

What kind of order is to be found inherent in the knowledge that develops as an artistic experience takes shape? In other words, for example: does the idea of a sound come first, with the result that the appropriate type of functional gesture tends to realize the idea?

For musicians, including singers, the tuning adjustments have a dual function:

1. Adjustment of the gesture after the emission of the sound
2. Adjustment of the sound after the gesture.

And this makes it possible for the aesthetic-sonorous realization to fully encompass the original idea that gave birth to it. No doubt there is a prior preparation of the body, its tissues, its tonality, the quality of its (emotional) presence, with a view to being able to produce the movement that responds to the aesthetic creation one wanted to give birth to.

Before emitting sounds with its *instrument-voix*, the singer (even more so the debutant) commits himself to a process of anticipatory creation of the sounds. Thanks to a synthesis he owes to his past experiences, he "previews" the sounds, feels them in advance and so, to some extent, selects them, even if this is done unconsciously. He creates the sounds in the space of his imagination, in a process of abstraction whereby an ideal of the sound and an aesthetic ideal are brought together, intersect and, at the same time, intermingle with each other. We now find ourselves in a psychic space where the artist, especially during the artistic performance, lives his *Self-Ideal*. This *Self-Ideal*, lifted up above his real possibilities, makes it possible for him to enact a collection of dynamic procedures of abstraction, which optimize the in-corporated knowledge to attain the highest artistic beauty he is capable of realizing at that moment in his life. As the acme of his past experience, a *Self-Ideal*

---

[15] David, Catherine (2006). *La beauté du geste, op.cit.*, p. 18. *Own translation.*

emerges, one which he hopes to surpass with the new challenge he takes upon himself (Vitale 2007c).

This space of creation is that of any form of execution, whether it concerns a single note or an entire *cantata*. The singer "seeks" the sound and, before producing it physically, he imagines it (by means of synaesthetic processes) in his space of abstraction, to the point of being able to feel it physically even before he emits it. He "creates" it in his imagination (the imaginary possessing a highly concrete value for musical processes, and especially the singing voice). Following thereupon, the body, with its gestures, prepares to actualize, to execute, the input made available by the imagination.

The gestures realized before the vocal execution (only a fraction of a second before) are the gestures which, even while anticipating the difficulty, tend to find a solution for it, a counterweight, an antidote to the difficulty itself: we are talking of "paradoxical gestures", that is, gestures going in the "opposite" direction to that taken by the voice. For example, in singing, if a phrase has to be attacked with a sharp sound, the hand anticipates the attack (a hand gesture is made even though silence still prevails) before the sound occurs and in an opposite direction: if the attack is to be high pitched, the gesture moves downward, in a *moto contrario* to that of the voice. Only when the singer has anchored the gesture, "foreseen" the personal difficulties and proposed a kind of "compensatory remedy" will he then proceed to the execution of the attack, which for him presents a certain difficulty (Vitale 2007b, 2008a).

Phenomenology of Perception and Learning Processes

In the first place, studying one's own voice as a singing voice requires from the start that one should establish a new relation with oneself, a relation of attention to one's self and to one's body. The study of the *instrument-voix* requires the human musical body to perceive and to become aware of very subtle sensations, which have to be remembered in such a way as to be identifiable. This equilibrium between incorporating the *instrument-voix* and being the principal spectator of oneself by means of it requires quite specific perceptive faculties.

On the basis of my preceding research on the body and the *instrument-voix*, parallels can be drawn with the study of the bodily instrument in dance, a piece of research never before attempted.

The development of the details linked to the movement in question and implied by the gestures and their combinations seems to require an appeal to a network of physical and psychological knowledge that needs to be extended. In this respect, a specific relation to time and to space constitutes one of the learning modalities in which the dynamic structure of the danced body is anchored. And so we bring to light a body that experiments with those dimensions which intersect through its activity, to the point of opening up fields where the dynamics of the evolution of the art emerge.

Memories get structured in this way in a temporal space according to points of reference in a spatial time in which experience is fixed. Out of this chiasm is born a learning dimension which can be reiterated in the incorporated instrument and placed at the disposal of the artist who recuperates it, in its sensible structure. This synergic coordination of the senses cannot be taken for granted. How often the pupil attempts to reproduce at home the experience lived through in the presence of the teacher, who remains the true guide, and even an *alter ego*, enabling the pupil to make progress. In the study of singing in particular, this attempt to reconstruct the action in solitude encounters great difficulties, especially with beginners.

An alternation between a phase that is felt and a phase that is thought about later, in a time one might call *otium* – to make use of the Latin with its opposition to *negotium* – is constitutive of the singing lesson: there is no *negotium* without an intelligent *otium*. A cyclical alternation has to be envisaged between, on the one hand, the experimental activity and, on the other, an "inhabited silence" that makes it possible for synthetic processes to function even while the perceptual sensations settle down over time and get sedimented in memories. So a time organized cyclically into an alternation of periods of activity followed by periods of reflection. A time of silence during which the sound of the musical experience is lived in a reflective way, allied with the impact of the sound made by the voice of the teacher (Vitale 2007b, c, 2009d).

The genuine apprenticeship consists in taking an experience up a posteriori with a view to nourishing it after the event with reflections having a quite precise aim: discovering intellectual keys making it possible to resolve tasks left unresolved up until then. This is how the pupil will work out personal solutions, develop hypotheses he can work with the next time around.

$$\textbf{SENSIBLE EXPERIENCE} \rightarrow \textbf{REFLECTION} + \textbf{ELABORATION} + \textbf{ABSTRACTION} \rightarrow \textbf{SENSIBLE EXPERIENCE} \; (\textit{Vitale}\,\text{2007b})$$

This period of reflection is also one during which precise questions can be formulated, questions that can be addressed to the teacher at the next meeting to clarify doubts and get confirmation of personal intuitions.

$$\textbf{SEDIMENTATION} \rightarrow \textbf{FERMENTATION} \rightarrow \textbf{CONNECTIONS} \; (\textit{Vitale}\,\text{2007b})$$

The process of learning to use incorporated instruments constantly solicits the "body's sensible structure". The "musical body instrument" generates aesthetico-sonorous ideals and cooperates in their realization, this over a period of gestation and artistic birthing.

> Que me fait un art dont l'exercice ne me transforme pas?
> What's the use of an art that does not transform me?
>
> Valéry[16]

[16] Valéry, Paul (1939/1973). *Ego scriptor*, in *Cahiers*, vol. I, Judith Robinson (Ed.). Paris, Gallimard, Bibliothèque de la Pléiade, p. 293. *Own translation.*

The Phenomenology of the Transmission of the Art of Singing and the Art of Dancing

Transmitting Through the Body: Body of the Transmission or Transmission of the Body?

The singing teacher, like the dancing teacher "passes" his gestures on to his pupil, those that accompany him and which belong to his body, which were transmitted to him, at least in part, by his own teachers and which ever since, have been integrated in his being inasmuch as he has incorporated them (Vitale 2007b). Every pupil works out and then personalizes his gestures, adding his own contribution to the chain of oral bodily memory, which remains the most important means of transmission, even though writings on dance and singing are more and more frequent.

The body, a human instrument, remains the privileged vector, just as the quality of the living presence of the "other than I" also remains fundamental, that is to say, irreplaceable in the learning process. Interaction in singing and in dancing remains indissolubly linked to the intercorporality, so dear to Mearleau-Ponty.

Here one finds situations that are well known in the teaching of all musical instruments, but the incorporation of the voice considerably amplifies their primacy. The study of the *instrument-voix* and the *danced-body* as incorporated instruments requires an extreme generosity on the part of the teacher who "lends" his body to the pupil, right up to looking into his own mouth to demonstrate the movements of the uvula, thereby revealing the interior of his mouth, the limiting point of the visible. Ever more necessary for beginners, this aptitude contributes powerfully to explaining the dependence of pupils with regard to their teacher. The quality of the presence of an expert is, moreover, essential *to improving the quality of the presence to self across the presence of the other*, the other as the mirror of the self.

Merleau-Ponty states: "I do not hear myself as I hear others, the sound of my own voice for myself is, as it were, badly folded; it resounds as an echo of its own articulation, it vibrates across my head rather than outside myself."[17]

Merleau-Ponty brings into relief the problem of distance, a problem common to singers, dancers, actors, *a* distance which has to be understood in the *plural*. We are talking about a distance that is perceptual (auditory, visual) as much as it is psychological and physical. This brings us to the interesting phenomenological subject of the irreplaceable "other than self", to which a human being has to resort in such circumstances. It is interesting to consider at what moment in one's life one has the feeling of having reached the limit of self mastery, of self sufficiency, and to check, in connection with other events, to what an extent this feeling is urgent, thereby implying decisions that cannot be put off "until later". In other words, one has to ask how far it is possible to explore alone that instrument one not only did not acquire but possessed from birth. The "other than self" develops "in me" a number of

[17] Merleau-Ponty, Maurice (1964a). *Le visible et l'invisible*, *op.cit.*, p. 194. *Own translation.*

(cognitive, perceptual, abstractive, etc.) faculties, but *how* does it manage to do so? What didactic strategies can be of help here? (Vitale 2007b, 2008a, 2009c, d, 2012b).

An important heuristic aspect needs to be borne in mind, an aspect linked to the knowledge and the modalities across which knowledge (the degree of knowledge and the degree to which it is capable of mastering the incorporated instrument) gets developed. In addition, this knowledge contains an epistemological aspect. It also establishes a definite relation with experience, the personal experience that influences each act, each choice, each important gesture.

Personal experience is anchored in psychic life, where not merely time, but also the relation to time is lived out according to certain modalities, subjective rather than objective. The time in question here is not clock time; rather, it is a time that is dilated or restrained, depending upon the feeling associated with the personal time of the subject. In this sense, the scholarly program calls for a considerable effort, that of abstracting from the personal necessity of living in a certain relation to time, living at one's own rhythm, the rhythm of one's body (which is a transformed body). An important compromise is needed. A compromise that can be carried out in the living and inter-corporal presence of the teacher-midwife, who serves as a mediator enabling the pupil to slide from a subjective temporality over to another temporal mediation. Just as there are many bodies in one body, there are many temporalities in one day. Acquiring knowledge means being able to make the passage from the "I-subject" to the "I-object". Such a passage can moreover always be effected in both directions. It lends itself to comings and goings, that is, if it stays opens and does not "close its bridges". Verbal or non-verbal communication between what is said and not said opens up horizons structuring the personal space and time of development.

Mirror

As an object in the singing and dancing class, the mirror is always there – proving how much it is needed –, though it is much more rarely present in classes devoted to other musical instruments. *What makes the mirror so indispensable to the study of singing and dance*? What function does this didactic instrument assume? What activities does it allow for? What makes it so indispensable? Citing Merleau-Ponty:

> I am always on the same side as my body; it is always there for me in a perspective that never changes.[18]

I am not in a position to see my body and a great part of it eludes my eyes. I am in a position to use rather than to see it. The body is made in such a way as to be available for use by its possessor rather than to be observed by him. Across my body I know the world, I "am", I exist in the world and the world exists for me.

[18] Merleau-Ponty, Maurice (1964a). *Le visible et l'invisible*, *op.cit.*, p. 194. *Own translation.*

In the study of singing and dancing, the look has to have at least two functions:

1. The internal regard has to be developed. We are not talking of the mental eye but rather of an "introspective-introjective" capacity to look at what I call internal gestures; as for external gestures, they cannot be observed in a global way except across a mediation or a mediator, a physical or human mirror.
2. A play of reflections, of global (macro) and detailed (micro) references, makes it possible for me to become conscious of gestures or postures (physical or mental) I was not conscious of before. The mirror is "other than I", that alterity which makes it concretely possible for me to adopt (even if only partially) the right distance vis-à-vis myself, the distance needed for me to perceive myself from without, to correct myself.

> In his teaching, R. Noureev paid particular attention to the direction given to the look (*regard*), since it makes it possible for the dancer to compose the mental space he wants to develop on the basis of the real space.
>
> Pietragalla[19]

A poietics and a poetic of the gestures adapted to realizing the artistic action is studied and perfected by taking note of the aesthetic ideal lying at the root of motor impulse.

Filming oneself (new technologies now make this possible) is certainly enough to establish a movement of return, a feedback. But that is not the end of the matter. For although the camera can reproduce our action, only the teacher can interact with us. The teacher is the only "mirror" capable not just of reflecting but also of interacting, by contributing strategic suggestions, solutions, proposing variations, etc. So it is important that the mirror should not be "distorting" but "limpid", faithful, so that one can find oneself in it exactly and, on the basis of this true reflection, work at improving oneself.

Intercoporality

The notion of Intercorporality, so dear to Merleau-Ponty, is absolutely fundamental in the phenomenological dynamics of dance and singing.

The body plays with time, with space, with itself and with other bodies. Making it possible to study the body with a view to optimizing its function as the instrument of dance and/or of singing presupposes a conscious or unconscious process of perceptual transformation, therefore a transformation of the vision bearing on oneself, that is, it presupposes a dynamic metamorphosis of one's own self-representation. An in-depth solitary study, at different levels, is periodically accompanied and progressively perfected by the presence of a master, who has to present and represent a totality of capacities and competence, linked to the discipline and the modalities of their transmission.

---

[19] Pietragalla, Marie-Claude (1999). *La légende de la danse*, Flammarion, Paris, p. 90. *Own translation.*

The teacher of singing and dancing manages, willy-nilly, and across the dynamics of Intercorporality, to place himself between the instrument and its interpreter, between the art and the body of the other. Transferential learning dynamics necessarily make their appearance in certain phases of the learning process in these *in*corporated arts. A space of ambiguous transmission gets set up, one which is situated at a threshold: I am neither myself nor anyone else. Here one finds Pascal's figure of the between, which includes both persons while belonging to neither in particular.

> One does not display one's greatness by being at one extremity but by touching both at the same time, and by occupying the space in between.
>
> Pascal[20]

The space of transmission is an important meeting place where creation takes place. This transmission is creative, in that it is not limited to "conveying" gestures in a mechanical fashion but manages to endow the other with a taste for discovering its own gestures. It is a matter of getting him to discover his own body in a new way, of leaving him freedom of speech, of expression, in a dynamic of perpetual improvisation, even when already known "figures" are executed, "figures" which, however, can only exist and vibrate in the moment in which they are re-invented. In this way even the didactic art is situated in a practice both of research and of creation, since it is open to the unpredictable requirements of the moment. That's when the magic touch of *negligentia diligens* has to be added to artistic practices, a touch that makes the present instant unique and decisive, to the point of being registered in the memory of Time.

> I don't teach children to imitate my movements but to invent their own. I don't force them to study certain definite movements, I help them to develop those which are natural to them. Whoever observes the spontaneous movements of a small child who has not yet been educated can hardly deny the beauty of its movements. They are beautiful because they are natural.
>
> Duncan[21]

References

Anzieu, Didier. 1985. *Le Moi-peau*. Paris: Dunod.
Anzieu, Didier. 2000. *Les enveloppes psychiques*. Paris: Dunod.
Aristote. (around 350 B. J /1979) *Éthique à Nicomaque*, translated by *J. Tricot*, Paris: Vrin.
Bordier, Georgette. 1975. *Anatomie appliquée à la danse*. Paris: Amphora.
Bourdieu, Pierre. 1972. *Esquisses d'une théorie de la pratique*. Genève: Droz.
Bourdieu, Pierre. 1981. *Questions de sociologie*. Paris: Minuit.
Claparede, Edouard. 1929. *Comment diagnostiquer les aptitudes des écoliers*. Paris: Flammarion.
Dagognet, François. 1992. *Le corps multiple et un*. Le Plessis-Robinson: Delagrange.

---

[20] Pascal, Blaise (1678/2000). *Pensées* (28/575), in *Œuvres complètes*, vol. II, Paris, Gallimard, Bibliothèque de la Pléiade, Michel Le Guern (Ed.), p. 782. *Own translation*.

[21] Duncan, Isadora (2003). *La danse de l'avenir*, *op. cit.*, p. 57–58.

Damasio, Antonio. 1999/2002. *Le sentiment même de soi*, Paris: O. Jacob. Titre original: *The feeling of what happens. Body and emotion in the making of consciousness.*
David, Catherine. 2006. *La beauté du geste*. Arles: Actes Sud.
Denis, Michel. 2000. Image mentale. In *Grand Dictionnaire de la Psychologie*, ed. Henriette Bloch, 451. Paris: Larousse.
Dolto, Françoise. 1984. *L'image inconsciente du corps*. Paris: Seuil.
Dufrenne, Mikel. 1991. *L'œil et l'oreille*. Paris: ed. J.-M. Place.
Duncan, Isadora. 2003. *La danse de l'avenir*. Paris: ed. Complexe.
Faure, Sylvia. 2000. *Apprendre par corps*. Paris: La Dispute.
Gilson, Étienne. 1979. *Le thomisme. Introduction à la philosophie de saint Thomas D'Aquin*, Paris: Vrin.
Imberty, Michel. 2005. *La musique creuse le temps*. Paris: L'Harmattan.
Imberty, Michel. 2007. Introduction. In *Temps, geste et musicalité*, ed. M. Imberty and M. Gratier. Paris: L' Harmattan.
Mallarmé, Stéphane. 1887/2003. *Crayonné au théâtre*. In *Œuvres complètes*, Vol. II. Paris: Gallimard Bibliothèque de la Pléiade.
Merleau-Ponty, Maurice. 1945. *Phénoménologie de la perception*. Paris: Gallimard.
Merleau-Ponty, Maurice. 1964. *L'œil et l'esprit*. Paris: Gallimard.
Merleau-Ponty, Maurice. 1964. *Le visible et l'invisible,* text established by Claude Lefort. Paris: Gallimard.
Pascal, Blaise. 1678/2000. *Pensées*. In *Œuvres complètes*, Vol. II, ed. Michel Le Guern. Paris: Gallimard, Bibliothèque de la Pléiade.
Piaget, Jean. 1998. *De la pédagogie*. Paris: O. Jacob.
Pietragalla, Marie-Claude. 1999. *La légende de la danse*. Paris: Flammarion.
Ricoeur, Paul. 1984. *Temps et récit*. Paris: Seuil.
Ricoeur, Paul. 1990. *Soi-même comme un autre*. Paris: Seuil.
Saint Thomas D'Aquin. 1273/2002. *Somme théologique*, ed. J.-P. Torrell. Paris: Cerf.
Stefani, Gino. 1989. *Suonare uno strumento*. Cinisello Balsamo: Paoline.
Stern, Daniel N. 1985. *The interpersonal world of infant: A view from psychoanalysis and developmental psychology*. New York: Basic Book.
Stern, Daniel N. 1995. *The motherhood constellation*. New York: Basic Book.
Stern, Daniel N. 2004. *The present moment in psychotherapy and everyday life*. New York: Norton.
Trevarthen, Colwyn. 2008. The musical art of infant conversation: narrating in the time of sympathetic experience, without rational interpretation, before words. In *Musicæ Scientiæ*, 15–46. Special issue.
Valéry, Paul. 1957. *Œuvres*, Vol. I, ed. Jean Hytier. Paris: Gallimard, Bibliothèque de la Pléiade.
Valéry, Paul. 1973. *Cahiers*, Vol. I, ed. Judith Robinson. Paris: Gallimard, Bibliothèque de la Pléiade.
Valéry, Paul. 1974. *Cahiers*, Vol. II, ed. Judith Robinson. Paris: Gallimard, Bibliothèque de la Pléiade.
Vitale, Alessia R. 1997. *En corps chanté. Prospettive sulla voce*. Masters (*Tesi di laurea*) in Musicology and Semiotics, under the direction of: Professor Gino Stefani, University of Bologna.
Vitale, Alessia R. 2003a. *Dal silenzio al suono. Voce Corpo Sensi Memorie*. Roma: Borla.
Vitale, Alessia R. 2003b. *La leçon de chant*. Masters thesis in Psychology and Pedagogy of Music, University of Sorbonne – Paris IV, under the direction of Professor Michel Imberty and Professor Jean-Pierre Mialaret.
Vitale, Alessia R. 2004. La voix et ses gestes. In *Abstracts du Huitième congrès international sur la signification musicale*, ICMS, Costin Miereanu (Chairman of the congress and editor of the book of abstracts). Paris, 3–8 octobre 2004.
Vitale, Alessia R. 2006. The singing lesson. Learning and non verbal languages. In *Proceedings of the 6th Triennial Conference of European Society for the Cognitive Sciences of Music*, ESCOM, *and 9th International Conference on Music Perception and Cognition*, ICMPC, eds. Mario Baroni, Anna Rita Addessi, Roberto Caterina, Marco Costa, p. 347. Bologna (Italy): Bononia University Press, 22–26 August 2006.

Vitale, Alessia R. 2007a. The language of gestures in the singing lesson. In *Proceedings of the 3rd Symposium on Cognition and Musical Arts*, SIMCAM, ed. Diana Santiago, p. 415. Bahia (Brazil): Federal University of Salvador, 21–25 May 2007.

Vitale, Alessia R. 2007b. Gestes et mémoires dans l'apprentissage du chant. In *Temps, Geste et Musicalité*, eds. M. Imberty, and M. Gratier, pp. 101–125. Paris: L' Harmattan.

Vitale, Alessia R. 2007c. *La leçon de chant. Temps, espace, transitionnalité*. Doctoral thesis in Psychology and Pedagogy of Music, University of Sorbonne – Paris IV. Thesis director: Professor Michel Imberty.

Vitale, Alessia R. 2007d. *Transmission du corps – Corps de la transmission. Parallèles entre l'étude du chant et l'étude de la danse*, 34e Congrès international du rythme et de la rythmique, Geneva (Switzerland): Institut Dalcroze, 18–28 July (conference).

Vitale, Alessia R. 2008a. The singing lesson. Phenomenology of the non-verbal dynamics appearing in studying *l'instrument-voix*. In *Musicæ Scientiæ*, pp. 111–128, Special issue 2008.

Vitale, Alessia R. 2009a. Learning processes in the study of the singing voice. In *Proceedings of the 4th Conference of the European Network of Music Educators and Researchers of Young Children* (MERYC), eds. Anna Rita Addessi, Susan Young, pp. 443–452. Bologna (Italy): Bononia University Press, 22–25 July 2009.

Vitale, Alessia R. 2009b. From the vocal gesture to the writing of music. In *Proceedings of the 7th Triennial Conference of European Society for the Cognitive Sciences of Music* (ESCOM 2009), eds. Jukka Louhivuori, Tuomas Eerola, Suvi Saarikallio, Tommi Himberg, and Paivi-Sisko Eerola, pp. 567–572. Jyväskylä (Finland): University of Jyväskylä, 12–15 August 2009. https://jyx.jyu.fi/dspace/handle/123456789/20924

Vitale, Alessia R. 2009c. Towards a phenomenology of the *instrument-voix*. In *Analecta Husserliana*, Vol. 104, ed. Anna-Teresa Tymieniecka, pp. 407–425. Springer, 2009.

Vitale, Alessia R. 2009d. Temps et espace dans les processus d'apprentissage du chant. In *La coscienza di Gino: esperienza musicale ed arte di vivere. Saggi in onore di Gino Stefani*, eds. Dario Martinelli, and Francesco Spampinato, pp. 293–318. Helsinki: Umweb.

Vitale, Alessia R. 2010. For a grammar of vocal gestures. In *Proceedings of the 11th Conference on Music Perception and Cognition* (ICMPC), eds. Steven M. Demorest, Steven J. Morrison, Patricia S. Campbell, pp. 736–742. Seattle, University of Washington, 23–27 Aug 2010.

Vitale, Alessia R. 2011a. An analysis of the singing lesson as a 'space' both physical and psychical. In *Proceedings of the 9th Pan European Voice Conference* (PEVOC), eds. Antoine Giovanni, Nathalie Henrich, pp. 54–55, Marseille, Hospital 'La Timone', 31 Aug–3 Sept 2011.

Vitale, Alessia R. 2011b. Singing as a healing art. In *Proceedings of the 9th Pan European Voice Conference* (PEVOC), eds. Antoine Giovanni, Nathalie Henrich, pp. 129–130, Marseille, Hospital 'La Timone', 31 Aug–3 Sept 2011.

Vitale, Alessia R. 2011c. La voix comme geste – Le geste comme voix: Significations dans l'apprentissage du chant. In *Prismes no 15*. Revue de la Haute École Pédagogique de Lausanne, pp. 38–40. http://www.hepl.ch/files/live/sites/systemsite/files/prismes/prismes-numero-15-2011-hep-vaud.pdf

Vitale, Alessia R. 2012a. Voix, geste et habitus dans la didactique du chant. *Journal of EVTA no 14*, pp. 18–26. Switzerland: European Voice Teachers Association. http://www.evta.ch/Bulletins/j-14-10.htm

Vitale, Alessia R. 2012b. Apprendre "par corps": Corps de la transmission ou Transmission du corps? Le cas du chant et de la danse. In *Actes de la Biennale Internationale de l'Éducation*, pp. 1–15. Paris, 3–6 July 2012. http://labiennale.cnam.fr/medias/fichier/alessia-r-vitale-com-n-272-atelier36_1360078952176.pdf

Vitale, Alessia R. 2013a. Vocal gestures: Elements of a non-verbal grammar. A phenomenological, semantic and didactic approach to the transformation of the voice into a musical instrument. In *Proceedings of the 11th International Congress of Musical Signification* (ICMS), eds. Teresa Malecka and Małgorzata Pawłowska, pp. 90–108. Kraków (Poland), Academy of Music: Musica Iagellonica, 26 September–1 October 2010.

Vitale, Alessia R. in press-a. Les dynamiques spécifiques d'apprentissage de l'*instrument-voix* à l'interface du temps et de l'espace. In *Actes des Journées Francophones de Recherche en*

Education Musicale (JFREM), eds. Isabelle Mili, Pierre-François Coen, Geneva (Switzerland): University of Geneva, 11–13 November 2010.

Vitale, Alessia R. in press-b. The impact of vocal gestures within the dynamics of the learning process in singing and in the transitional process from oral musical culture to the writing.... In *Philosophies of performance*, ed. Eero Tarasti, Acta Semiotica Fennica XXXIV, Helsinki: International Semiotics Institute.

Winnicott, Donald Woods. 1957/1971. *L'enfant et sa famille. Les premières relations*. Paris: Payot. Original title: *The child and the family*.

Winnicott, Donald Woods. 1957/1972. *L'enfant et le monde extérieur*. Paris: Payot. Original title: *The child and the outside world*.

Winnicott, Donald Woods. 1971/1975. *Jeu et réalité*. Paris: Gallimard. Original title: *Playing and reality*.

Winnicott, Donald Woods. 1977/1980. *La petite "Piggle"*. Paris: Payot. Original title: *The Piggle*.

Winnicott, Donald Woods. 1986/1988. *Conversations ordinaires*. Paris: Gallimard. Original title: *Home is where we start from*.

Winnicott, Donald Woods. 1987/1992. *Le bébé et sa mère*. Paris: Payot. Original title: *Babies and their mothers*.

Winnicott, Donald Woods. 1996/1999. *L'enfant, la psyché et le corps*. Paris: Payot. Original title: *Thinking about children*.

Wiseman, Frederick. 2009. *La danse, le ballet de l'Opéra de Paris* (film).

Phenomenology and Archeology: Methodological Insights and Thematic Inspirations

Jaroslava Vydrová

Abstract Particular layers of re-constitutive analyses, the access to the structure of the core (*Kern*), the "unbuilding," as well as the "building-up"—there is a complex and structured notion of the movements of the phenomenological method that can lead us to challenges of that kind of investigation in which phenomenology and archeology overlap. We find several legitimate questions: Where does the concept of archeology occur in the phenomenological context? What kind of methodological tools does it use and what are the results of archeological discovery? The sphere where we can elaborate this kind of thinking represents methodologically anchored investigation and it leads phenomenology to its boundaries, to its borderline characterization as well. In this context we situate the question of origination and emergence. Our contribution will be based on Husserl's manuscripts collected in *Husserliana Materialien VIII* and *Husserliana XXXIV*, in which the problematics of reduction and time analyses are intertwined.

What does the phenomenological archeology as a part of the phenomenological methodology mean and which place does it occupy in the philosophy of Edmund Husserl? We start with a general characterization, that there are several possible ways in which we can grasp the phenomenological method in the meaning of its performance and impact. There are various manners in which it can be applied, uncovered or caught in its movements, without weakening its radical claim. And the method is determined by this achievement too. We shall pursue the issue in

This paper was created at the Institute of Philosophy, Slovak Academy of Sciences as a part of VEGA grant Nr. 2/0050/14. I would like to thank Radomír Masaryk, Peter Šajda and David Clayton for consultations and language editing of the English text.

J. Vydrová (✉)
Institute of Philosophy, Slovak Academy of Sciences, Klemensova 19,
Bratislava 813 64, Slovak Republic
e-mail: jaroslavavydrova@gmail.com

A.-T. Tymieniecka (ed.), *Phenomenology of Space and Time: The Forces of the Cosmos and the Ontopoietic Genesis of Life: Book One*, Analecta Husserliana 116,
DOI 10.1007/978-3-319-02015-0_15,

accordance with this dynamic (reciprocal) "logic" of reduction. At this point we choose three points of view which mean three different optics in the framework of our chosen problem.

The *first* circle will be associated with different forms of borderline phenomena. That which can be on the level of its appearing called "limit," can shift us also to other boundaries and limits: thus to the boundaries of the method and to its possibilities, as well as to the possible structure of the phenomenology. In other words, the first range of questions will be enacted on the line thematic—methodic.[1] What we subsequently find out about the character of the method then we develop further in the *second* point when we focus on the topics as discovering, originating, generating, transforming, which are by their nature also very difficult to be addressed and which are also very important in a methodological context. At this point we shall pay particular attention to the manners of phenomenological archeology as a special part of the phenomenological achievement. In the *third* point we shall outline how both of these methodological accounts can function in the context of intersubjectivity and communicativeness (of phenomenology/archeology). This presents a feasibility test of these borderline phenomenological lines of thought (from the thematic and methodological point of view).

These reflections will be based on the texts in which Husserl problematizes and radicalizes his methodological considerations. They belong to the 1926–1934 period and diverge in two directions—toward temporal analyses (Hua-Mat VIII) and toward phenomenological reduction (Hua XXXIV). As a supplement we shall use Eugen Fink's insight into phenomenology of phenomenology and Husserl's answers in the *Sixth Cartesian Meditation* (VI CM and Dok. II/1).[2] These selected texts are important and their scope is wide from the thematic and interpretational perspective, and they have been researched in depth in further phenomenological literature. We shall draw from these resources selectively with regard to the three chosen problematic areas of our interest.

[1] If under the phenomenological method we understand the steps which are not inserted to the system from outside but are its inherent part. The methodological possibilities of phenomenology in the texts of late Husserl were analyzed in the monograph by Jaroslava Vydrová *Cesty fenomenológie. Fenomenologická metóda neskorého Husserla* [The Ways of Phenomenology. Phenomenological Method of Later Husserl] (Pusté Úľany: Schola Philosophica, 2010). Cf. Georgy Chernavin, *Transzendentale Archäologie-Ontologie-Metaphysik. Methodologische Alternativen in der phänomenologischen Philosophie Husserls* (Nordhausen: Verlag Traugott Bautz GmbH, 2012).

[2] Eugen Fink, *Sixth Cartesian Meditation*, The Idea of a Transcendental Theory of Method, With textual notations by Edmund Husserl, trans. Ronald Bruzina (Indiana University Press, 1995). Fink's text will be quoted as VI CM, Husserl's as Dok. II/1. Orig.: *VI. Cartesianische Meditation*, Teil 1. Die Idee einer transzendentalen Methodenlehre, Texte aus dem Nachlass Eugen Finks (1932) mit Anmerkungen und Beilagen aus dem Nachlass Edmund Husserls (1933/34), ed. Hans Ebeling, Jann Holl and Guy van Kerckhoven (Dordrecht: Kluwer Academic Publishers, 1988).

The Method and Its Possibilities, Shifting of Boundaries and New Challenges

Each problem which stands aside the common range of investigation because of its specific nature is a kind of challenge which requires a distinctive approach, a specific "sensibility"[3] or a specific method. Husserl notices this in different points of his texts—although in his own work of exposing we can see that some problems come into the field of view earlier, some appear later, thanks to the change of optic or thanks to the deepening or broadening of the investigation and its tools. Not every phenomenon can be investigated in the same (phenomenologically common) way.

We can start with one distinctive and paradoxical case, which can in the end return us to the question of emergence, to the context where this question could be raised. "Only that is 'unthinkable' for me that I cease to exist transcendentally. The ending as a man in the objective world, dying, whereas others are bodily living further, that needs another interpretation, which does not belong here."[4] What is being suggested here? Thematically this opens up the sphere of other, "new" phenomena; and it is also a new level of investigation.[5] The development of this investigation can bring forward significant distinctions which in turn cast light on the possibilities of the method. The broader background of the problematic of death represents—as the phenomenon itself indicates—the temporal question of understanding the future, or the horizon of future. Its consequent thematisation is not appropriately available to static analysis and it is beginning to open up on the background of the question of the living present as a considerable phenomenological problem. The phenomena connected to the past and to the future have their specific nature partly linked to the characterization as possibility, ambiguity, lack of clarity, doubtfulness, almost nothingness—however, in both cases of past and future they are defined in specific ways, because the future is not accessible to investigation in the same way as in the case of the past.[6]

[3] As Róbert Karul indicates: "Sensibility, which is connected to appearing of being, is sensibility of intentionality of consciousness related to the openness of being…" Róbert Karul, "Subjektivita ako afektivita a trpnosť," [Subjectivity as Affectivity and Passiveness] *Filozofia* 51: 6 (1996), p. 387. This was analyzed in depth and pointed out in many works of A. J. Steinbock, to some of them we refer in the further text.

[4] Edmund Husserl, *Späte Texte über Zeitkonstitution* (*1929–1934*), Die C-Manuskripte, ed. Dieter Lohmar (Dordrecht: Springer, 2006), Hua-Mat VIII, p. 97. "Nur das ist 'undenkbar' für mich, dass ich transzendental aufhöre. Aufhören als Mensch in der objektiven Welt, sterben, während Andere leiblich fortleben, das bedarf einer anderen, nicht hierher gehörigen Auslegung."

[5] Which draws attention to itself by standing aside from the field of investigation, or it does not let the phenomena to appear in this investigation in the adequate manner.

[6] Cf. Husserl, *Späte Texte über Zeitkonstitution* (*1929–1934*), Hua-Mat VIII, C 4, No. 21 a – Husserl speaks about a different reconstruction of that which is past and future. This problematic is elaborated also by Anthony J. Steinbock in his text: "From Phenomenological Immortality to Natality," in *Rethinking Facticity*, ed. François Raffoul and Eric Sean Nelson (Albany: State University of New York Press, 2008), pp. 25–40.

The living present is a key, which opens and starts up this analysis in two directions as well as anchors it in *I* as a *Limes* of the beginning and the end.

Husserl asks: "*How* do we come to Limes or rather to ending?"[7] We need to come to the borderline which lets new options "enter" the problematic, the borderline that should "let through" the heterogeneity and thus make it thematised in a certain kind of way. We arrive to the borderline of experience and the borderline of its regularity. It is a problematisation of the question, how is this kind of differentiation possible within the continuity, sameness?[8] Let us mention three problematic spheres which can occur here: The example of stopping represents not only death, ending, but also sleeping, falling asleep, the unconscious, passivity, but also exhaustion, faintness, illness, which are slow forms of stopping, losing interest, decline of attention or ceasing any activity. There are different forms of "leaving." An example from the other part of the range of "beginning–end" is a birth, which stands aside the common range of understanding mainly by minimal overlapping of the past and the future[9] compared to the adult—who always orients with regard to his or her past and future horizons. The third sphere of borderline forms are animal beings with different degrees of intentional "distancing" (*Entfernung*), plants, different kinds of pathological anomalies, which thanks to their particularity in turn cast light on subjectivity and its normality—thus they co-determinate it.

Husserl indicates: "The original source of 'intuition' for all the possibilities of a transcendental subject lies always in myself, in the modifications of my inwardness. Possibilities through modifications occur also in the higher 'intuition' as borderline cases."[10] We could say, that these examples lead to the level of transformation, modification or in other words, of depth of that which is "psychic inward" (*Innerlichkeit*). The road there could be opened by the kind of method that is associated with acquiring the form of the world in the infinite opened progress of uncovering the horizon (as pointed out in the text C 6, No. 22).[11]

In these examples Husserl *on the one hand* continuously uncovers particular layers of the life of subjectivity, which cover the following: I and its life, different modes,

---

[7] Husserl, *Späte Texte über Zeitkonstitution (1929–1934)*, Hua-Mat VIII, p. 97 (italics J. V.).

[8] Husserl goes further: "Primal phenomenally belongs to the primal impression the difference (the unexplained) pertaining to sameness as well as not-sameness, of the sameness that is concrete self-same with itself (the unchanged), of the not-sameness that is continually and concretely similar with itself; further of the not-sameness as a leap, making diversion…" ("Urphänomenal gehört zur Urimpression der Unterschied (der unausgelegte) des Gleichmäßigen und des Ungleichmäßigen, des Gleichmäßigen als mit sich selbst konkret Gleichen (Unveränderten), des Ungleichmäßigen als mit sich kontinuierlich konkret Ähnlichen, ferner des Ungleichmäßigen als einen Sprung, <ein> eine Abhebung bildenden…") (Ibid., p. 98).

[9] Cf. Ibid., p. 101. The world of the child "begins in instinctive intentionality of the 'first childhood' in the body of the mother" (C 3, No. 17).

[10] Ibid., p. 105. ("Die Urquelle der 'Anschauung' für alle Möglichkeiten eines transzendentalen Subjekts liegt aber immer in mir selbst, in den Abwandlungen meiner eigenen Innerlichkeit. Möglichkeiten durch Abwandlungen ergeben sich auch in höherstufiger 'Intuition' als Limesfälle.")

[11] As a paradox (a very apt one) this is the point of intersection of the absence, the minimalism or the poverty of the borderline phenomena with its depth, or the depth it helps to unlock.

forms of its directedness, which create a special style or centralization of I; he discusses the higher degree of reflection and its pre-conditions, the constitutions of practical interests, the relationship between affectivity and activity, etc. *On the other hand* gradual deepening and layering enable us to open the theme of the horizon (the horizon of the world and the horizon of the situation), the foreground and background, implicit and the explicit. He thus brings into the game different possibilities of "the allowed," the acceptable or adequate thematisation. As we can see, some themes can fall out of the scope of investigation and are missing in its optic because they require another degree of approach or a different methodological index.

What do these examples suggest to the understanding of the phenomenological method with regard to the problematic of emergence? The thematic casts light on the methodological; the methodological on the other hand releases the thematic. On one hand, both spheres appear heterogeneous in their nature, i.e. they refuse simple or causal parallelism. On the other hand, they influence each other, they are intertwined. This double exposure is efficient and becomes obvious also as a result of the analysis of borderline cases. This analysis points out not only to limits of the method but also functions positively to enlarge its scope. We may relate this understanding to Georgy Chernavin's description who writes about different methodological strategies, potentialities and alternatives being "different perspectives of developing the phenomenological method."[12]

These methodological examples nevertheless need not be on the same level which moves us to yet another issue. This is the matter of a more complex outlook on the form, the build or the structure of phenomenology. As Husserl wrote, it is the case of *layering*. As we get more intense in investigating phenomena—and this goes hand in hand with problematizing of "how" and "if" we may explore them—there appears in the final analysis the problematic of phenomenology within the phenomenology of phenomenology. The layered nature of phenomenology is described by Eugen Fink in the *Sixth Cartesian Meditation*: (1) self-consideration is radicalized in the form of (2) the phenomenological reduction, bracketing, which leads to the transcendental onlooker and to the question of the constitution of the world. The next step is (3) the transcendental theory of elements (*Elementarlehre*) with the scope of examination of regressive and constructive phenomenology. And the highest level (4) is the transcendental theory of method.[13] Each layer has its own problem sphere of investigation. Other possibilities of structuring however offer different outlooks on the steps of reduction and epoché, as colorfully described and realized in particular texts by Husserl.

This approach which works with a certain kind of "classification" on one hand systemizes the process of investigation; on the other hand it frames the functioning of the method. Such framing can exclude a potential discovery of other contexts.

---

[12] Chernavin, *Transzendentale Archäologie-Ontologie-Metaphysik*, p. 11f.

[13] *Sixth Cartesian Meditation*, VI CM, p. 12f., cf. §§ 1, 2; cf. Roberto J. Walton, "The Constitutive and Reconstructive Building-up of Horizons," in *Epistemology, Archeology, Ethics. Current Investigation of Husserl's Corpus*, Continuum Issue in Phenomenology and Hermeneutics, ed. Pol Vandevelde and Sebastian Luft (London, New York: Continuum, 2010), p. 134.

Static phenomenology thus cannot see the genetic and the generative; the descriptive level should be adequately elaborated[14]; the analysis can enable the passage toward the interpretative, hermeneutic connections; or the basic constitutive analysis may not be sufficient for the deeper archeology, etc. Such enumeration points out the extent of achievement of the method. On the background of the notions of layering nature of phenomenology, there appears the question of new possibilities, plurality and pluralisation of the phenomenological method. Within the formalizing and systemizing approach there may also appear different characterizations of the method bringing forth something newer and more efficient—and the manner in which this happens shall be described in the upcoming example of phenomenological archeology as a possible access to the topic of emergence.

Movements and Transformations

The principle of plurality in relation to the nature of the method and its boundaries does not need to function only on the quantitative level (as snowballing of new possibilities) but also in the form of deepening, intensifying or radicalizing.[15] Thereby we can once again open the question of reflecting upon motivation with regard to the phenomenological method which functions as "a disturbance," a turning point in the common way of perceiving, living; reflecting its iteration, it is "the decision that is being decided each time, anew;"[16] and reflecting on its movements, movements of its performances. Husserl also uses the word "Windung," wrapping something around, winding something around (or also winding something off), which metaphorically suggests that neither the direction nor movements of the method are not straightforwardly linear but work within the variousness of layers. Similarly, in the relationship between attitudes (if we talk about the main differentiation between the natural and the phenomenological attitude; in the more narrow meaning of the attitude-thematic we may talk about the plurality of attitudes) it is not just parallelism that functions there, the transitions between attitudes do not occur along the same "avenues" from one to another and back. The focus of the phenomenological method—not only in the genetic phenomenology, but also in analyses of earlier texts—takes different courses and develops in different dynamics. The polysemy that occurs around the phenomenological method is in some later texts supplemented by the overlapping of epoché and reduction, or by some kind of liberation. This outcome is partly natural

[14] This was pointed out by Jagna Brudzinska in her contribution "On descriptive Methods." Cologne-Leuven Summer-School in Phenomenology: Methods of Husserl's Phenomenology (July 16th–20th, 2012).

[15] Cf. the work by G. Chernavin.

[16] Steinbock, "The Poor Phenomenon. Marion and the Problem of Giveness," pp. 129–130. Cf. Hua XXXIV (p. 194), where Husserl finds "the synthetic action of reflective iteration." Edmund Husserl, *Zur phänomenologischen Reduktion*, Texte aus dem Nachlass (1926–1935), ed. Sebastian Luft (Dordrecht, Boston, London: Kluwer Academic Publishers, 2001).

due to thematic variety. This specific development of the phenomenological method could be shown in Husserl's texts in the volume XXXIV that are focused on reduction: on one hand they are thematically divergent which on the other hand reflects their inner fixation within the methodological context.

The optic that captures the methodological variety is connected with the issue of layering, and phenomenological archeology brings forth the act of *de-layering*. The work of an archeologist—and by analogy also a researcher as well as a philosopher—functions as: progressing toward *Abbau*, walking backwards (*Rückgang*), the uncovering (*Aufdeckung*) of elementary structures. Behind these instructions (mentioned e. g. in the text C 6, No. 23) we can find Husserl's insights toward what he calls the kernel structure (*Kernstruktur*): "The primal hyle in its own temporalization is so to speak a core, alien to I, in the concrete present."[17] An archeologist proceeds by several steps by which he or she goes through different layers. What is uncovered on the first layer is my flowing present in the epoché (my past and my future), on the second layer we reduce the specific flowing present through *Abbau* "to primal impresional immanent presence-of-a-matter, to the 'alien to I,' i.e. the immanent hyle."[18] The movement through which we disclose these layers is indicated in the movement of Abbau, the movement back, *zurück*—through searching for different forms and layers of the (hidden) transcendence toward the original (pre-) impressional sphere, toward the hyletic core.

This "kernel" layers as if had two meanings. It is essential in that it creates the grounds or the "footwall" that we may examine, but we can go further from it and build up on it. In one of the key passages Husserl explicitly describes the archeological work:

> Phenomenological archeology, the excavation of constitutive building elements concealed in their structural members, the building elements of apperceptive sense-achievements that present themselves in their readiness as experiential world. The questioning and the laying bare of individual achievements that create the sense of Being all the way to that last, *archai*, to letting rise up again in the spirit the natural unity of variously founded validations of Being with their relative beings. As by the common archeology: reconstruction, understanding in 'zigzag'.[19]

Let us first look at the movement of this process. To move in the zigzag fashion means to go further but not in the linear direction but respecting layers of the "soil" through which we move. It provides a special dynamic to the phenomenological

---

[17] "Die Urhyle in ihrer eigenen Zeitigung ist der sozusagen ichfremde Kern in der konkreten Gegenwart" (Husserl, *Späte Texte über Zeitkonstitution* (*1929–1934*), Hua-Mat VIII, p. 110).

[18] "...auf die urimpressionale immanente Sach-Gegenwart, auf die 'ichfremde,' nämlich die immanente Hyle (Empfindungssphäre)" (Ibid., p. 109).

[19] "Phänomenologische Archäologie, das Aufgraben der in ihren Baugliedern verborgenen konstitutiven Bauten, der Bauten apperzeptiver Sinnesleistungen, die uns fertig vorliegen als Erfahrungswelt. Das Zurückfragen und dann Bloßlegen der Seinssinn schaffenden Einzelleistungen bis zu den letzten, den *archai*, um von diesen aufwärts wieder im Geist erstehen <zu> lassen die selbstverständliche Einheit der so vielfach fundierten Seinsgeltungen mit ihren relativ Seienden. Wie bei der gewöhnlichen Archäologie: Rekonstruktion, Verstehen im 'Zick-Zack'" (Ibid., pp. 356–357).

method and at the same time it casts light on the specific interconnectedness, on relationships uncovered in this manner.

So we already defined the shape of the trajectory of this movement and now comes the question of its direction, "from where" it goes and "to where" it progresses. To the passage from Husserl quoted above we could add the explanation by Roberto J. Walton: "On the one hand, it amounts to a restoration that advances in the opposite direction of unbuilding, and employs the elementary constituents that have been unearthed as a cornerstone for the reconstruction of what was unbuilt. It sets itself the task of 'letting rise up again'...what has been dismantled out of the *archai* provided by deconstruction ... On the other hand, building-up can be understood as an extension in the same direction into further depth dimensions." It is not just simple movement "from here to there," as Walton notices: "It is not a reversion but rather a continuation of unbuilding."[20]

Phenomenological archeology which at first sight seems like continuous disassembling to the level of elements incorporates thus—paradoxically—the question of creating, origination, emergence. What could archeology say about that? Let us note two moments: *The first point* is *orientational* meaning that if we want to capture ephemeral events of origination, creation, ontopoiesis, the deeper meaning and possibilities of modification, as a result of phenomenological archeology we can situate, anchor this problematic within the layers of phenomenological investigation. This point is essential for the methodological line of thought. With the help of Ronald Bruzina we would like to put this problem in the specific framework: "It is crucial, of course, to realize that this phenomenology of temporalization is not a rival to the cosmological account of the evolution of the physical universe." In the part of his analysis called "Limitations to the disclosure of origination" he also writes about a "careful theoretical critique of the limits in phenomenology's investigation into origins."[21]

Clarifying this problem from the methodological point of view and situating it in the specific phenomenological investigation is connected also with its development which represents *the second point*—we can call it *nodal*. Archeology "unlocks" this problematic on one hand in its center, on the other hand it then gets developed further, it is further thematisation. In other words, origination is here not only the theme as such, but it is developed in complex phenomenological connections. Husserl looks at it from different angles, more general or more partial, which are connected by the question of *modification*, *transformation* (*Umwandlung*)—in several philosophical differences as well as in inherently methodological matters.

---

[20] Walton, "The Constitutive and Reconstructive Building-up of Horizons," pp. 135–136. The problematic of phenomenological archeology is mentioned also by Angela Ales Bello, *The Divine in Husserl and Other Exploration*, Analecta Husserliana XCVIII (Dordrecht: Springer, 2009).

[21] Ronald Bruzina, "Phenomenology in a New Century: What Still Needs To Be Done," in *Analecta Husserliana CV*, ed. Anna-Teresa Tymieniecka (Dordrecht, Heidelberg, London, New York: Springer, 2010), pp. 58 and 65. "What is at issue in genetic phenomenology...is the origination of the real of *sense-sensuous manifestness* and of the way sense-manifestness *is for intentional experiencing*." (p. 58)

We shall present here two examples of two (possible) contexts of its development which serve as two cross-sections through this problematic:

1. In a broader sense origination is a matter of large steps of the phenomenological method: "The transcendental-phenomenological I (and then the transcendental we) comes in this action to a new *self-creation*, to a transformation of the natural I into an I of a radically pure self-consciousness, in radical and ultimate truth, and so to a radical and ultimate knowledge of the world and of everything."[22] It is a special kind of transformation, a radical turn in the natural way of life, which is connected with a series of consequences related to its methodological peculiarity. Therefore Husserl turns to the need of describing the original method (*Urmethode*) of phenomenology. As Ronald Bruzina writes: "The antecedency of transcendental origination is not temporal antecendency in the world; it is an antecedency of what can only be found as of that which it gives rise to...It would be, in other words, a sense of antecedency wholly specific to phenomenology, and drawn from its specific methodology."[23]
2. In a narrower sense the second type of analysis appears: "This determines the system of tasks: (1) We have to learn in the primal modal present and learn to understand it in its double or triple primal modal transformation, in the primal modal not-egoic transformation, in primal temporalization in which a hyletic quasi-world, alien to the ego, has its pre-being; (2) then the ego for which this pre-world is and through which or through the functioning of which, in affection and action, the proper world comes to creation, in a plurality of levels of creation, to which relative worlds correspond."[24] Such special insights show the thematic depth of the problem of origination. It appears in its explicit as well as implicit scope and contains not only transformation, modification, variation, creation but also the rise of unity and difference, diversity, otherness—the building-up of a horizon or "stages, in which the ego does not distinguish itself from the world,"[25] as described by Roberto J. Walton. They are exposed in the constitutive and re-constitutive sense, in the steps of archeological unbuilding, exposed in their basis, but also developed further within their efficacy or functioning. It is the

[22] Husserl, *Späte Texte über Zeitkonstitution* (*1929–1934*), Hua-Mat VIII, p. 126, italics J. V. ("Das transzendental-phänomenologisierende Ich (und nachher das transzendentale Wir) kommt in diesem Tun zu einer neuen Selbstschöpfung, einer Umschaffung des natürlichen Ich zu einem Ich radikal reinen Selbstbewusstseins, in radikaler und letzter Wahrheit, und damit zu einer radikalen und letzten Welterkenntnis, Allerkenntnis.")

[23] Bruzina, "Phenomenology in a New Century: What Still Needs To Be Done," p. 66.

[24] Husserl, *Späte Texte über Zeitkonstitution* (*1929–1934*), Hua-Mat VIII, p. 350. ("Von da aus ist das Aufgabensystem vorgezeichnet. (1) Wir müssen in der urmodalen Gegenwart uns zunächst umtun und sie in ihrer doppelten bzw. dreifachen urmodalen Wandlung verstehen lernen, der urmodalen nichtichlichen Wandlung, der Urzeitigung, in der eine ichfremde hyletische Quasi-Welt ihr Vor-Sein hat; (2) dann das Ich, für das diese Vor-Welt ist und durch dessen oder in dessen Funktionieren in Affektion und Aktion die eigentliche Welt zur Schöpfung kommt, in einer Vielheit von Schöpfungsstufen, denen relative Welten entsprechen.")

[25] Walton, "The Constitutive and Reconstructive Building-up of Horizons," p. 133.

sphere where unbuilding as well as building-up appears. "In the primal streaming standing present, in the absolute life of my I, we founded primal intertwining, unity, and primal difference (primal statement, primal un-unifying), which do not found unity, but rather distance and differentiation. Distance from the unity, manifoldness in joint action with intertwining, which brings about the unity."[26] This extends the analysis of subjectivity to other dimensions—which are located on its boundaries and could be discovered using a specific methodological approach—but it remains in frameworks that describe the tension as well as the primal blending of both spheres, the sphere of I and not-I.[27]

"What," "How" and "With Whom" of Communication

The next level represents the perspective of intersubjectivity,[28] which broadens both previous parts. It has however been already included in them. In which way? The series of borderline phenomena, as was mentioned in the first part, casts new light on the intersubjective scope of phenomenology, while it brings forth the specific *Miteinander* which is for example associated with a child, his or her world, shaped on the embryonic level especially by the body of mother (instinctive intentionality of the first childhood in mother's body—as described e.g. in the text No. 7). Their world constitution is different from that of an adult. We could develop the analogy with the constitution of animals (No. 55), too. Furthermore it is the world of others in a wider sense, the world of another nation, other traditions, cultures. The intersubjective level opens itself up in its "breadth"—with all variety of the other and possible layers of their investigation. When we look into the "depth," we could however find it in its very core, in the centre of constitution where *Miteinander* and *Ineinander* overlap. Husserl describes the fundamental level as "primordium": the level of the primordial where we can observe the layer of the primal experience as well as the primal alteration—"the alter immanent with the alter I, thus it alters for me the alter conscious world, alter-primal conscious"—the iteration of alteration, the other of the other.[29] The potentiality of archeology is applied also on the level of intersubjectivity, which is a part of unbuilding of fundamental construction layer of

[26] Husserl, *Späte Texte über Zeitkonstitution (1929–1934)*, Hua-Mat VIII, p. 76. ("In der urströmenden stehenden Gegenwart, dem absoluten Leben meines Ich, haben wir Urverschmelzung, die Einheit begründet, und Ursonderung (Urkontrastierung, Uruneinigung), die nicht Einheit, sondern Abständigkeit, Differieren begründet. Abständigkeit von Einheiten, Mehrheit im Zusammenwirken mit der Einheit schaffenden Verschmelzung.")

[27] Cf. "Konstitution von Seienden verschiedener Stufen, von Welten, von Zeiten, hat zwei Urvoraussetzungen, zwei Urquellen, die zeitlich gesprochen (in jeder dieser Zeitlichkeiten) immerfort ihr 'zugrundeliegen': (1) mein urtümliches Ich als fungierendes, als Ur-Ich in seinen Affektionen und Aktionen, mit allenWesensgestalten an zugehörigen Modis, (2) mein urtümliches Nicht-Ich als urtümlicher Strom der Zeitigung und selbst als Urform der Zeitigung, ein Zeitfeld, das der Ur-Sachlichkeit, konstituierend" (Ibid., p. 199).

[28] Cf. Walton, "The Constitutive and Reconstructive Building-up of Horizons," p. 146f.

[29] Husserl, *Späte Texte über Zeitkonstitution (1929–1934)*, Hua-Mat VIII, No. 85, p. 374.

(inter)subjectivity. In this sense, we are still in accordance with that which was already described in both previous parts. But the new dimension by which we want to extend this line of thought however means to *communicate* phenomenology, or in a specific sense to communicate phenomenological archeology.

How could this kind of analysis be communicated? To whom should this message be addressed? How and why should it be conveyed further? We can once again overlap the *Sixth Cartesian Meditation* with Husserl's texts from the volume Hua-Mat VIII. The sixth meditation points to one important dimension which is the appearing of phenomenology, the localization in the natural world, "making into a science," transforming it to a communicable science. This is the meeting point of both chosen texts, even the point where we may go beyond their scope or where we may try to move to *generative* phenomenology.

This generative broadening of the horizon anchors phenomenology in history, tradition, in how phenomenology becomes a concrete phenomenon, "cultural construct."[30] The output of phenomenology is a very complicated theoretical transition of its emergence in concrete historical situation; but we can observe also concrete everyday praxis, different activities, phenomenological achievements and work. As Eugen Fink writes about it: "...the phenomenological cognizer philosophizes as a functionary of the human community, he fits himself into the human generative habituality of philosophizing, he transmits, lectures, publishes, etc."[31] In this context we can speak about the idea of university, importance of research, sense and value of institution, about fellowship, loyalty, etc. Let us supplement this with the text C 16, No. 83: "When I practice with my co-philosopher a phenomenological world consideration, a layering of absolute traditions is uncovered for me and for us, a layering in which the world is already constituted, and is there for everybody and each possible communicative-social society with its formed special sense and horizon of possible development."[32] In communicating and applying, the phenomenologist turns mainly to other performers of reduction—and in our case in the scope of archeology. This creates the community of scientists. The place of co-philosophizing is here a natural sphere of communication. But its potential is however not exhausted by this. The alterity enters in play also here. The other can appear here not only as the other philosophizing subject, but—and this is essential—also as a non-phenomenologist, a mundane scientist, a non-scientist, a member of other cultural tradition etc. On this level such a person meets with a phenomenologist who can offer his or her knowledge as communicable, or even for further development.

---

[30] *Sixth Cartesian Meditation*, Dok. II/1, p. 214. Cf. Geo-historical analysis and wider generative analysis of A. J. Steinbock (*Home and Beyond*: *Generative Phenomenology After Husserl* (Evanston, Illinois: Northwestern University Press, 1995).

[31] *Sixth Cartesian Meditation*, VI CM, p. 145.

[32] Husserl, *Späte Texte über Zeitkonstitution* (*1929–1934*), Hua-Mat VIII, p. 370. ("Indem ich mit meinen Mitphilosophen die phänomenologische Weltbesinnung durchführe, enthüllt sich mir und uns die Stufenfolge der absoluten Traditionen, in denen Welt schon konstituiert ist und für jeden und jede mögliche kommunikativ-soziale Gemeinschaft mit ihrem ausgestalteten Sondersinn und ihrem Horizont möglicher Ausbildung.")

"The communicative surrounding as a field of communicative praxis has its intersubjective structure,"[33] that also applies to outcomes of the methodological process. As a result following the phenomenological archeology we can highlight three aspects in this regard:

1. Specific layers of investigation bring forth their own possibility of communication, verification, or further application of their processes. This makes it more complicated (limitary) for archeological research or borderline forms of givenness. This layer takes into account mainly the phenomenologist's investigation of "how".
2. On the other hand this is associated with the shift toward others, toward the question "with whom." On the intersubjective level there enters the diversity which influences the level of communication, where a phenomenologist communicates with an otherwise engaged actor, *et* vice versa.
3. Phenomenological archeology is at the same time interesting by moving the boundary of thematisation when bringing the specific "what," specific content of communication. In the context of Hua-Mat VIII phenomenological archeology brings specific deepening of investigating the layers of subjectivity, a certain in-depth egology, correlatively we could speak about the region investigation with regard to emergence (to which belongs the research on the boundary of I and not-I, research of temporality, history, or intersubjectivity), or some kind of "radicalized" uncovering. And it brings—this was our issue within the chosen topic—conclusion even in the area of methodology.

Concluding Notes

The three areas chosen to look into phenomenological archeology in this paper were referred to as the three optics through which we followed our problematic. The optic of borderline phenomena brought the opportunity to talk about variability and possibilities of the phenomenological method. The second optic followed movements of the method which brought us to the place where the themes of origination, emergence and archeology meet. The third optic focused on intersubjective and practical consequences.

Although we focused primarily on specifics of the method and particular levels of investigation where archeology occurs, in conclusion let us mention yet one more methodological note. We will do it with the help of Anthony J. Steinbock who remarked that "phenomenology as regressive or 'archeological' reconstruction, then, belongs to a static phenomenology: It is 'phenomenological' inasmuch as it inquires into the accomplishments of sense; but it is static because it questions back

[33] Ibid., p. 398. In text No. 90 Husserl shows, how the level of communication and praxis is connected to ground level, core: "Aber es liegt darin, <dass> in allem als weltlich Konstituierten vorweg ein Kern bloßer Natur steckt, der durch abstraktiven 'Abbau' aller Prädikate objektiven Geistes jederzeit hervortreten kann" (p. 402).

and then merely lays out the single sense accomplishments." On the other hand Steinbock admits the "possible genetic interpretation of reconstruction."[34] As we mentioned the differentiation of static, genetic and generative phenomenology, or other cross-sections/outlooks on the layers of investigation (for example through particular steps of the phenomenological method) they make some questions feasible yet avoid others. The strategy that we could borrow from phenomenological archeology is reflected in the peculiar "zigzag" movement, in transitions through individual layers (and in turn in exposing them to the reflection of their relationship), in the modification of the method, in a different outlook as we would have expected, and in asking what it would bring for us.[35]

References

Bello, Angela Ales. 2009. *The divine in Husserl and other exploration*, Analecta Husserliana, vol. XCVIII. Dordrecht: Springer.

Brudzinska, Jagna. 2012. On descriptive methods, Cologne-Leuven summer-school in phenomenology: Methods of Husserl's phenomenology (July 16th–20th).

Bruzina, Ronald. 2010. Phenomenology in a new century: What still needs to be done. In *Analecta Husserliana CV*, ed. Anna-Teresa Tymieniecka, 39–76. Dordrecht/Heidelberg/London/New York: Springer.

Chernavin, Georgy. 2012. *Transzendentale Archäologie-Ontologie-Metaphysik. Methodologische Alternativen in der phänomenologischen Philosophie Husserls*. Nordhausen: Verlag Traugott Bautz GmbH.

Fink, Eugen. 1995. *Sixth Cartesian meditation*, the idea of a transcendental theory of method, with textual notations by Edmund Husserl (trans: Ronald Bruzina). Indiana University Press. Orig.: *VI. Cartesianische Meditation*, Teil 1. Die Idee einer transzendentalen Methodenlehre, Texte aus dem Nachlass Eugen Finks 1932 mit Anmerkungen und Beilagen aus dem Nachlass Edmund Husserls 1933/34, ed. Hans Ebeling, Jann Holl and Guy van Kerckhoven. Dordrecht: Kluwer 1988.

Husserl, Edmund. 2001. *Zur phänomenologischen Reduktion, Texte aus dem Nachlass (1926–1935)*, ed. Sebastian Luft. Hua XXXIV. Dordrecht: Kluwer.

Husserl, Edmund. 2006. *Späte Texte über Zeitkonstitution* (*1929–1934*), Die C-Manuskripte, ed. Dieter Lohmar. Hua-Mat VIII. Dordrecht: Springer.

Karul, Róbert. 1996. Subjektivita ako afektivita a trpnosť, [Subjectivity as affectivity and passiveness]. *Filozofia* 51(6): 385–394.

---

[34] Steinbock, *Home and Beyond*: *Generative Phenomenology After Husserl*, pp. 90 and 284; cf. p. 79. This differentiation is also noticed by Walton: "On the one hand we have static relations in which the founding levels can be considered in themselves as finished formations without taking into account what is founded on them … On the other hand, the strata cannot be considered in themselves because the focus is on processes rather than results" (Walton, "The Constitutive and Reconstructive Building-up of Horizons," pp. 147–148).

[35] According to G. Chernavin different perspectives are possible: "in transcendental archeology (die *Rekonstruktion* der Urstiftungen, die *Abbau*-Analyse), in mundane ontology (die *Auslegung* der Vorgegebenheit) and in metaphysics of primordial facticity (die *Modalisierung* der Faktizität)," but as he notices: "it is the same experience, which is considered within one or another strategy" (Chernavin, *Transzendentale Archäologie-Ontologie-Metaphysik*, p. 117).

Steinbock, Anthony J. 1995. *Home and beyond: Generative phenomenology after Husserl*. Evanston: Northwestern University Press.
Steinbock, Anthony J. 2008. From phenomenological immortality to natality. In *Rethinking facticity*, ed. François Raffoul and Eric Sean Nelson, 25–40. Albany: State University of New York Press.
Steinbock, Anthony J. 2009. The poor phenomenon. Marion and the problem of giveness. In *Words of life: New theological turns in French phenomenology*, ed. Bruce Benson and Norma Wirzba, 120–131. New York: Fordham University Press.
Vydrová, Jaroslava. 2010. *Cesty fenomenológie. Fenomenologická metóda neskorého Husserla [The ways of phenomenology. Phenomenological method of later Husserl]*. Pusté Úľany: Schola Philosophica.
Walton, Roberto J. 2010. The constitutive and reconstructive building-up of horizons. In *Epistemology, archeology, ethics. Current investigation of Husserl's corpus*, Continuum issue in phenomenology and hermeneutics, ed. Pol Vandevelde and Sebastian Luft. London: Continuum.

Plotinus' "Enneads" and Self-creation

Ineta Kivle

Abstract The present study surveys Plotinus' philosophy within the context of self-creative activity. Owing to the fact that Plotinus' philosophy provides for a deep understanding of self-becoming in harmony with cosmic forces, I have examined the Plotinus notion of Soul and Intellect as well as explored such concepts, suggested by A.-T. Tymieniecka, as *ontopoiesis*, *Logos of life* and *self-becoming*. The article is divided into two parts:

In the first part I have tried to interpret the views by Plotinus, concerning the picture of Cosmos, and characterize his three Hypostases, namely, One, the Intellectual Principle, Soul and individual Souls. The question to be answered in the given passage, is the following: How independent and free is self-creative activity, taking into account that Soul rises and moves in the emanation of One, subsists as Hypostasis and is permeated by Logos?

In the second part I have surveyed self-creation and self-becoming within the context of fluxing wholeness, art and Cosmos. I have tried to show the differences and similarities between the philosophy which develops on the basis of phenomenological standpoints and Plotinus' cosmology which is rooted in the ancient understanding of Cosmos. I hold that self-creation is not only directed intentionally, but rather shows the place of the human in the world.

Plotinus' Picture of Cosmos

Plotinus' "Enneads" picture the harmony of Ancient Cosmos and influence further the Christian thought about the Trinity. Alongside with the contemplation of three Hypostases, namely, One (God, Good, the Divine Principle); the Intellectual

---

I. Kivle (✉)
National Library of Latvia, K. Barona 14, Riga, LV 1423, Latvia
e-mail: ineta.kivle@lnb.lv

A.-T. Tymieniecka (ed.), *Phenomenology of Space and Time: The Forces of the Cosmos and the Ontopoietic Genesis of Life: Book One*, Analecta Husserliana 116,
DOI 10.1007/978-3-319-02015-0_16,

Principle (Nous, Reason, Mind), Soul and individual Souls, Plotinus justifies a dependent self-creative activity which can be summarized by the following words: *creation of self from itself and by itself, but directed by Logos and Cosmos*. Creative activity refers both to Hypostases and to the self-seekings of Soul. Soul's inclusion in the light of the Divine Emanation and force of Logos determines its self-becoming and marks its flexible place in Cosmos. The order and forces of Cosmos insinuate not only to creative activity and self-becoming, but also provide for living in harmony alongside with Souls, being tended towards true existence.

In Plotinus' philosophy the order of Cosmos and its activity are illustrated by Hypostases and Logos where the former is viewed not as a separate Hypostasis, but as such which expresses the relationships present in Cosmos. "As to Logos, it is neither the Intellectual Principle, nor the absolute Divine Intellect, nor does it descend from the pure Soul alone. Logos is a radiation from two Divine Hypostases – the Intellectual Principle and the Soul. The latter, being preconditioned by the Intellectual Principle, engenders Logos which serves as a particular life holding measure of reason" (III. 1. 16.).[1]

Hypostases are non-spatial and cannot be diminished. Existing as non-separate levels of being, they remain unaffected by what they produce, however, they are always in connection with the process of emanation and Logos. Logos is considered to be an aspect of Intellect, Soul, and also nature. It expresses the order of each Hypostasis and makes a bridge between the intelligible and the sensible. Logos is dual: its activity shows an upward tendency towards Intellect, at the same time descending towards things and particulars. As Logos creates the visible World, administrates it and connects the principles of Cosmos with things and forms it is creative and connective. Logos displays itself as a creative activity, makes individual *Logoi* and insinuates in different qualities and things. Logos is neither *poiesis*, nor *praxis*. *Poiesis* is an aspect of contemplation and contemplative producing, whereas *praxis* is for contemplation, which involves deliberation and even physical instruments, needed for man's activity of creating real things. *Praxis* is inferior to *poiesis*, because *poiesis* transcends knowledge and is more directed towards the contemplation of true and intelligible realities. *Poiesis* as an activity of becoming is a flexible movement which is presented by Intellect and Soul. At the level of Intellect *poiesis* is compared to contemplation, whereas Soul contemplates Intellect in order to reach perfection.

Tymieniecka in her phenomenology of life states that *poietic* flow insinuates in life and in any other becoming being and that the *poietic* stream not only permeates in life, but also gives ground and order of existence for everything that exists and becomes. True existence, according to her, is a becoming being, engaged in *ontopoiesis*, or such a being who creates itself and makes the world. The process of making is woven together by life through creation and becoming. According to her opinion, the meaning of Logos can be recognized in a living being, in experiencing one's self by creative and *ontopoietic* activity, but not in cognitive acts.

---

[1] All references to the *Enneads* are in standard form. Thus (III.1.16.) indicates the sixteenth chapter of the first tractate of *Ennead* three.

Tymieniecka writes that Logos as an onward way, the same as Plotinus' contemplation of cosmic forces, directs to beings and in itself. Exploring the concepts of *Logos*, *poiesis* and *praxis*, one can see that the picture of Cosmos testifies to multilateral connections among things, nature, Soul and Intellect. Besides, it shows the realm of a self-becoming being and substantiates the wholeness of three Hypostases.

The highest Hypostasis One is beyond *ousia*, so, how to speak about One if it is higher than Intellect and Mind? How to reflect on One if it can be reached neither by creative, nor Intellectual Activity? Plotinus affirms that One can be reached by love and that Soul's union with One is a mystical experience of non-material light. In other words, it is an ideal case of self-love, which achieves culmination in irrational and ecstatic moments of life. Such expressions as *the One is from and through itself, it is tended towards itself, it makes or constitutes itself as a cause of itself, it is self-sufficient, One is before subsistence, One does not subsist* show that Plotinus comes to the core of the order of Cosmos by means of a particular contemplation on the Divine Power in his own Soul. Such a deep penetration shows that the order of life and heaven is predicted by being that is relatively beyond a human Soul but, at the same time, Soul's aspiration is a force which comes to the ground of life and architectonics of Cosmos. Consideration that One is like *energeia* which produces itself, but statements like *what is before subsistence, what makes itself to subsist and is the origin of every Logos, order and limit*, put forward a number of questions, among them: – How is self-creative activity possible in such a model of being? How does self-creation manifest itself in all three Hypostases? One creates itself from itself as a self-sufficient being which insinuates in other Hypostases and is always present. One preserves all things in being. It is more beautiful than Logos, it is of itself and roots Nous and Soul, but remains unidentified. The Union of Cosmos as an active fluxing being is determined by the Divine Light and One's emanation.

The Intellectual Principle, produced by One's Emanation is not a level of being, but something which lives according to Logos and apprehends itself. Separated from One, Nous represents the distinction and definition but it does not imply separation in parts. Intellect generates time, universe of thoughts and sensations through Soul. In apprehension of self, Nous activity concerns the whole architectonics of Cosmos, namely, One – Nous – Soul.

The existence of Soul is threefold. It comprises One as absolute Soul, Hypostasis as the Soul of the world and individual Souls. Being the Principle of sensations and emotions, Soul makes everything live. As Soul is an inhabitant of the world of the Divine intuitive Thought, its range has a cosmic dimension. Plotinus states that every living being is an intelligible universe and that we can choose on what level this living being is going to live: on the empirical level or on the ascending way towards One. These phases are not active in all Souls. Not always Soul has a tendency to transcend itself and reach the ecstatic experience of One. Every individual Soul marks its boundary from itself to itself and also from itself to other Souls by reasoning and desiring.

Soul manifests itself in different ways of life and is developing in various kinds of seeing. Its deepest self-knowing is inexpressible and inaccessible to any act of observation. Soul's knowledge about one's true Self is not in need of spoken word because it is like the awareness of being which is beyond all questions: "As speech is the echo of the thought in the Soul, so thought in the Soul is an echo from elsewhere. In other words, as any uttered thought is an image of the Soul-thought, so the Soul-thought images a thought above itself and is the interpreter of the higher sphere" (I. 2. 3.).

Plotinus begins *The First Ennead* with a description of Soul, then he searches for the *seat* of affections and experiences, writes about the living organism – *Animate*, and confirms that Soul uses the body as an instrument. *Animate* is a body, having life. It is a unity of Soul and body, where body's seat is the material but the body itself functions as a potential recipient of life. "The body is a brute touched to life; the true man is the other, going pure of the body, natively endowed with the virtues which belong to Intellectual Activity" (I. 1. 10).

In a downward or upward way through the spheres of Cosmos Soul adopts knowledge about things, the body, Intellect and Principles of Cosmos. Soul's experience of material things and the body is added to the essential being of Soul in its downward way. Unlike the omnipresent dwelling of Soul, the body is made up of single parts and is located in its own place. The true being of Soul is its own Beauty. In its way up Soul must get rid of everything that it has acquired while descending. Then it returns to its own true home by means of contemplation, thus reaching its origin and One. Plotinus compares the self-creation of Soul to the doings of a sculptor who is creating a statue.

> And if you do not find yourself beautiful yet, act as does the creator of a statue that is to be made beautiful: he cuts away here, he smooths there, he makes this line lighter, this other – purer, until a lovely face has grown upon his work. You also do the same: cut away everything that is excessive, straighten everything that is crooked, bring light to everything that is overcast, labour to make all one glow of beauty and never cease chiselling your statue, until the god-like splendour of virtue shall shine out on you from it, until you see the perfect goodness, surely established in the stainless shrine. When you know that you have become this perfect work, when you are self-gathered in the purity of your being, nothing remains that can shatter that inner unity, nothing from without, clinging to the authentic man, when you find yourself wholly true to your essential nature, so wholly that only veritable Light which is neither measured by space, nor narrowed to any circumscribed form, nor again diffused as a thing, void of term, but ever unmeasurable as something greater than all measure and more than all quantity – when you perceive that you have grown to this, you have now become the very vision: now call up all your confidence, strike forward yet a step – you need a guide no longer – strain, and see. (I. 6. 9.)

The moving of Soul from the bodily forms towards the utmost ideal, virtue and reason is characteristic of Soul towards its self-seeking and purification. Owing to its faculty of reasoning, Soul confirms its emanation from the Intellectual Principle and connection with the Divine Mind. The capacity of Soul to see the Divine Mind justifies its authentic existence, however, by dividing among living bodies, Soul presents itself as belonging to *Animate*. The true location of Soul is Intellectual Activity. In accordance with Plotinus' philosophy, Intellection is the highest phase of life which manifests itself both as an act of Soul and an act of the Intellectual Principle. Hence, purification as a necessary human quality depends on the capacity of Soul to raise itself above all passions and affections.

Self-creation in Plotinus' philosophy is a circular way home to the deepest layer of existence and back to the foundation of the universe. However, there are two movements: Soul makes a descent in order to create single individual Souls, the latter then having to find within themselves the way back to the One and to their origin. In the process of descending Souls are taking part in the work of creation, which proves to be an integrity of acts in that phase of the development of Soul when it is tended towards Ideal Principles. Individual Soul has its own self-will and an ability to make decisions. Besides, the true Self moves away from the influence of Matter. In other words, Soul gets free from bodily ties, separating itself spiritually. Soul, owing to the Power of Emanation, overcomes the lower world and evil. The ultimate goal of Soul is uniting with One that can be experienced within a short moment of life by Soul's liberation from the influence of the lower world.

Soul gets united to the Intellectual Principle by disrupting its ties with Matter on its way up. Soul's ascending towards One is a continuous self-activity which comes to an end when mystically uniting with One, namey, resting in ecstatic union with the Divine Light. A question arises, concerning the self-identity of Soul: Does Soul lose its identity, being in union with One? The experience of Soul uniting with One is characterized by a very special kind of awareness. By something like an erotic state and a particular experience of Self which shows that the deepest level of consciousness can be reached only uniting with cosmic forces.

According to Plotinus, Cosmos is a spiritual being where Soul always is in movement. Soul is linked with all other Hypostases by its own activity and predication of Cosmos. Self-creation is not only an activity, but also a vision, depending on the spiritual cosmic energy which Plotinus calls the "light or power". "Our vision is light or rather becomes one with light, and it sees light for it sees colours. In the Intellectual, the vision sees not through some medium, but by and through itself alone, for its object is not external: By one light it sees another not through any intermediate agency; a light sees a light, that is to say a thing sees itself. This light shining within the Soul enlightens it; that is, it makes the Soul intellective, working it into likeness with itself, the light above" (V. 1. 8.).

Soul is the seeker and conductor of rhythmic motions. Linking all heavenly activities with all living things, Soul unites the diverse heavenly system, its manifold existence outlines a directions of human activity. From Plotinus' point of view real existence and authenticity are observed in the life of Soul. Soul attains its genuine self-identity and authentic existence only after choosing and acting in ways exclusively appropriate for man's own higher nature. Plotinus suggests the following hierarchy of man's faculties: first – sensation and imagination; secondly – reasoning and Intellection. When Soul reaches the unity of intelligible world, it comes to rest in its native realm, however, the Intellectual Principle is not its ultimate resting place – the highest step is Soul's uniting with One.

According to Plotinus, Soul is a spiritual place which finds itself in movement by its own activity and by the emanation of One. The Divine Light insinuates in Soul, however, light is in Soul forever. During contemplation Soul sees the light and ascends towards One. In this spiritual motion Soul experiences its own spiritual place and adherence to Cosmos.

Self-creation

Plotinus acknowledges Self-creation by uniting two realms: spiritual cosmic Hypostases and the location of human existence and life. Self-creation as a becoming being unites the Intellectual Principle with the activity of Soul. It insinuates in cosmic directions as well as in individualities. Human personality as a variable entity, depending on the realm of existence he or she lives in, is dwelling between two ultimate areas – One and the body. In the direction of the body every Soul acquires inferior things, located on the lower levels, whereas, ascending towards Intellect and One, Soul acquires more elevated qualities. Human beings are of two kinds: for men of the lowest kind bodily power predominates, and on the contrary, men can achieve the transcendence of their body, becoming their truly authentic Selves. Plotinus distinguishes between two concepts – Soul and Self. Self is something that cannot be hypostatized and is not identical, as to its contents, with Soul. The concept of Self is mainly used by later interpreters of Plotinus' philosophy for a more clear understanding of man and the order of Cosmos. In Plotinus' "Enneads" we can find *itself, one-self, self-existence, self-directed, self-Intellection, self-centered, self-dwelling*. Self as the essence and the centre of human personality as an interior being is directed towards something. If Self concerns realm, relating to the man and his Soul, then the concept of Self can be compared to ontic Self. However, if Self is related to One, it is true Self.

Self-creation as an activity of Soul is directed towards the spiritual order of Cosmos, the Divine Mind and the Intellectual Principle. In uniting self-experience with cosmic energy human activity is acquiring likeness with his highest Self which manifests itself as a contemplative wisdom of everything that exists in the Intellectual Principle. Levels below Intellect can also be generated by contemplation, but – only as a causative-generative sequence. Plotinus holds that self-contemplation of Intellect is a self-directed activity which knows itself and thereby creates the visible world through Soul and *Logoi*. The Intellectual Principle is a self-directed and self-reflexive activity, thus, the Intellectual Subject is identical with its object and in such a way the object is located within Intellect and becomes part of it by focusing Intellect from within, but not from the outside, thus, self-becoming of Intellect always remains identical with itself. To put it differently, Intellect thinks of itself. Intellect and the object of thinking provide for a complex co-existence where Intellect is not filled with objects, but is fused together with images, with the subject and with the object. The same as being, motion and rest are apprehended as an integrity of Sameness and Otherness or as an activity which occurs simultaneously. Sameness and Otherness, self-identity and self-otherness are fused together like the subject and the object. To put it differently, Intellect grasps various concepts in Sameness by Sameness. Intellect considers *itself from itself and by itself* as such a being which does not create itself but, based on the movement of thought and images, creates art, philosophy, literature and other Intellectual and spiritual things.

Plotinus writes about art, music, mathematics, born lovers and philosophy. Art combines the mystery of creating with intelligible activity and constructive thinking, hence, the artist, producing true and genuine art, dwells on the upward trend. His journey to the Intellectual Principle is mediated by temperament and contemplation, thinking and creation. Artistic creation involves three aspects: (1) the source of beauty; (2) artistic creation; (3) the artist's answer to being.

The ultimate criterion of Beauty is presented as an idea, and total Beauty dwells in Hypostasis of the Intellectual Principle.

Plotinus highlights the mystical experience of Absolute Beauty without any shape and concreteness. Beauty manifests itself through the Cosmos by emanation and tends to unite with the Intellectual Principle. The source of Beauty is the highest realm of pure idea and wholeness, from which art derives its existence and such qualities as symmetry, proportion, unity, brightness and consonance of single parts.

The artist transfers the idea of Beauty in the representation of things. "No doubt, the wisdom of the artist may serve as a guide in his work. The artist himself goes back, after all, to that wisdom in nature which is embodied in himself; and this is not a wisdom, built up of theorems, but one totality, not a wisdom consisting of manifold detail, coordinated into a unity, but rather a unity, working out into detail" (V. 8. 5.).

The highest form of art comes close to the Intellectual Principle. Plotinus shows the way up, comprising several stages: lower life, the way to the sphere of the intelligible and to the third degree – the sphere of absolute Beauty. The artist shows himself from a spiritual place where he is and wherefrom he creates his work of art. If the artist is in contiguity with the material life, he is very far from the absolute form of true Beauty and Intellectual Cosmos. However, if the artist reaches his authentic existence by contemplative wisdom, he finds himself in the true spiritual realm of creation. These ideas of Plotinus show that the self-creative degree for any man depends on what that man thinks, sees, listens to and contemplates. Creating an artwork, the artist shows his own spiritual dwelling place in Cosmos. The aim of creation is to deal with being by true knowing. True knowing or contemplative wisdom is the highest point of creative Intellectual Activity which produces art and philosophy.

Plotinus holds that creative action is immanent to Intellect and Soul, that true reality of Intellectual World exists forever as the second Hypostasis of being. Intellectual World appears in created works and in everyday life as the realm of Otherness. Plotinus justifies the reality of Intellectual and conceptual World as the realm of Soul's self-seeking, as the union of Sameness and Otherness, as a movement that always possesses its objects where the contemplator and the contemplated are the same. Plotinus comes forward with fundamental philosophical questions: What are we? Where is our place of dwelling? His answer could be that our seat is in the wholeness of being, in connection of every realm where man lives and creates, in the movement between Intellection and sensibility, true Self and delusive Self.

Plotinus' writings are based on his own life experience and, being a philosopher of late antiquity, he uses the language and concepts of antique philosophy. Plotinus' philosophy shows notable philosophical and religious aspects of self–creation, but as the central axis I personally see the notion of the spiritual place of Soul. His description of activity of Intellect and Soul displays a deep self-understanding that enables us to analyze his philosophy in connection with self-knowing, self-thinking, self-becoming, etc.

The phrase *To think self* (*Intellection of self*), in the history of philosophy is analyzed from various aspects and in different modifications. This statement can also be interpreted and characterized as the origin of the philosophy of subjectivity. However, if Plotinus considers self-thinking as the activity of Hypostases as well as the activity of individual Souls, then the philosophy of subjectivity rises from "I." Since "I" is the central axis and beginning of every process of thinking, it postulates the following question: How do I think? In contradistinction to Plotinus' interpretation of Intellect, Husserl elaborates a subjective structure of consciousness as fluxing *noese* and *noema* correlations and shows that man knows object as a meaningful phenomenon that is grasped in the stream of consciousness. At the same time there is a common position for both thinkers – existence of objects is justified within Intellectual Activity. For Plotinus it obtains cosmic directions – the Intellectual Principle justifies that object is known in Intellect by Intellection and thus Intellect knows itself from itself, whereas for Husserl it is the realm of subjectivity – in structures of consciousness are grasped meaningful phenomena and the given structures of consciousness are illuminated by phenomenological reduction and reflection. Common for both philosophers is a standpoint that the faculty of knowing things is given *a priori*, difference appears in it what is contemplated. Plotinus interprets contemplation of One – Intellect – Soul in ancient cosmic directions where objects have not got any independent existence outside of Intellect. Husserl, in his turn, holds the opinion that we know an object as it is given in our subjectivity.

In some aspects Plotinus' philosophy is allied with such concepts in Tymieniecka's philosophy as *Logos of life*, *ontopoiesis and self-becoming*. Tymieniecka discovers ontopoietic Logos of life as a dynamic activity which pervades every level of being. In accordance with ontopoietic Logos of life each thing can find its proper place in the stream of flowing life, each thing can be characterized by ontopoietic position as becoming-towards-being, as a processes of self-creating. The concepts of *ontopoiesis* and Logos of life find their roots in pre-Socratic philosophy. Analyzing the Heraclitean interpretation of Logos, Tymieniecka writes: "We find, indeed, that the human Soul which grows "without limits" in its Logos is a microcosm interchangeable with the all-engulfing macrocosm. The human Soul, understood by Heraclitus as the centre of personality and as the one, caught in element transformation, is, actually, the measureless Logos. Seeking for one's own Self, one finds one's identity with the universe" (1. 5.).[2]

At the same time, Tymieniecka's concept of *ontopoiesis* shows a new perspective of human beingness nowadays, a way of better self-understanding. It is a new way of philosophical thinking in several aspects. I will mention some of them: The concept of ontopoietic self-becoming shows a unique approach to thinking that unites phenomenological standpoints with ancient thinking. So, what does phenomenology think about cosmic forces, if phenomenology arises as a philosophy of

[2]All references to citations, except *The Enneads*, are in standard form. Thus (1.5.) indicates the fifth page of the first reference in References list.

subjectivity, as a new method for philosophy, as such a philosophy that illuminates the structures of consciousness, the formation of meaning and intentionality? How can we reach the ground of cosmic order by method of phenomenological reduction and reflection, *epoche* and *empathy*? It provokes new and new questions. An assumption which is made on the basis of Plotinus' philosophy is, that Plotinus' reflection on Soul's movement to One resonates with transforming the subject into an object which can be measured not solely from the angle of capacity of the subject, but also from the angle of the Divine Force. In accordance with Tymieniecka's writings, the subject of perception transforms into an object and opens the sphere of the Divine in self-creative process. In such way, she shows the wholeness of a living being that opens to heaven and Divine Forces.

In comparison with the place of the human and his or her activity in the philosophies of Plotinus, Husserl and Tymieniecka we can identify some common positions and also differences. First of all, I would like to draw your attention to the notion of bilateral direction. The relations between Self and Other are considered in detail in the phenomenology and philosophy of existence. "I" is directed towards itself and also towards Other, Other can recognize "I" identity, and "I" life can be posited by Others. However these relations move through all living wholeness and identify the place where we dwell. In Husserl's philosophy the bilateral direction appears as a formation of meaning – the subject grasps an object as a meaningful phenomenon, where phenomenon shows itself from itself in intentionality. Also multiplicity of bilateral relations are justified in such concepts of Husserl's philosophy as *intersubjectivity* and *constitution of meaningful worlds*. These directions resonate with the activity of Soul in Plotinus' philosophy where Soul becomes performative, differentiating Self from itself, and together with it Soul performs the order of Cosmos. In such a way Soul illuminates its own spiritual place in Cosmos which depends on the intensity of contemplation. Soul's activity in Plotinus' philosophy shows the place of self-creation as the finding of one's Self in the cosmic order, and justifies that life is not only a subjective activity, but is also involved in the flux of being wherein man dwells and finds his true Self. According to Plotinus, self-creation is involved in the flux of being without being rooted in subjective intentionality as it is in Husserl's phenomenology. Husserl identifies three aspects of intentional contents: the intentional object of the act, the certain manner in which the object is intended in consciousness and the intentional essence of act. Later Husserl talks about *Lebenswelt* as a source of our experience, whereas Tymieniecka extends the meaningful horizon from *Lebenswelt* to the mode of Cosmos and *Logos of life*. Tymieniecka's concept of *ontopoiesis* provides for a new perspective for humanism in line with a conviction that philosophy comprises a deep understanding of Self through the wholeness of being, the latter involving nature, heaven, society, culture, etc. The ontopoietic interpretation of life justifies self-creation as an emphatic attitude and intentionality which is directed towards contemplation of the world as the wholeness of seeing, listening and thinking. It means that we are in dialogue with the World and the World is in dialogue with us. The World where we live is already preconditioned: according to Husserl, the meaningful worlds of theorizing, science and art are conditioned by intentionality and *Lebenswelt*, whereas from Plotinus we can conclude that the World is pre-conditioned by three Hypostases, Logos and self-activity. As for Tymieniecka, the World is pre-conditioned by the flux of being and Logos of life. Conditioned World is also a place for an endless process of self-realization, self-creation and self-understanding. Tymieniecka's philosophical approach reveals how intentionality develops in the wholeness of life and fluxing being. Life for her is more fundamental than particular activities, in other words, life is the initiator of philosophy, of thinking, of knowing, of unknowing, etc. Exclusively, the intentionality of consciousness does not concentrate on the illumination of life, art and philosophy as becoming beings. For Tymieniecka it is important to recognize that philosophical acting solely upon the subject does not solve the problem of human's place and time in the wholeness of life. The philosophy of subjectivity does not look upon the human from the metaphysical point of view as it is pictured in the philosophy of Plotinus, still I can agree that "Enneads" are created from Plotinus' self-seeking and self-becoming that prove to be a special kind of self-reflection and self-purification.

Focusing on Plotinus' contemplation of Soul, I admit that his self-reflection expresses his own deepest self-identity as Soul's mystical connection with One. Plotinus' philosophy confirms that self-being is not the subjective "I", but the kind of being that is involved in the order of Cosmos and is predicted by the Divine Emanation. Self-creation is an endless seeking for One, Good and God. In the meantime Soul becomes increasingly clear and beautiful in the process of going upwards to One.

References

Plotinus. 1992. *The Enneads* (trans: Stephen MacKenna). New York: Larson Publications.

Tymieniecka, A-T. (ed.). 2010. *Analecta Husserliana. The yearbook of phenomenological research: phenomenology/ontopoiesis retrieving geo-cosmic horizons of antiquity. Logos and life*, Vol. CX. Dordrecht: Springer.

Directing Anatoly Vasilyev, from Individual Creative Manner to the Method

Valery Kolenova

Abstract In S. E. Ozhegova's dictionary "direction" is defined as "a profession, activity". Efremova in his dictionary adds that "direction" is also a result of the director. And in Ushakov's dictionary, "direction" is a "successful or unsuccessful design" performance. However, it is important that the three authors of dictionaries agree on one thing: directing the possession of skills is directing profession. Vasiliev, director of special. He managed not only to raise the achievement of theater and performing arts to a new level, enriching drama school is not only a new methodology and new aesthetics. This unique artistic process played an important role in internally subjective feeling of the artist, figurative, metaphorical thinking, the ability to create complex shaped time and space, connecting fabric of a performance in a variety of genres and styles. We consider that this direction in the postmodern era has become an entirely new aesthetic quality. But to reach such a result seems to us impossible at once. Need a fine laboratory, experimental, artistic and creative work, leading author on the individual artistic style to your own author to the method.

The historical process, in its broadest sense, is the foundation of our lives, coupled with many of its phenomena, such as culture and civilization, the spiritual and material development of mankind, economics and politics, and other manifestations of human activity. Variety of levels of human existence, from the lowest to the highest in one way or another connected with the historical process, and to varying degrees, affect the latter. To understand that there is a historical process in all its richness – is to understand and correctly assess the patterns and characteristics of human

V. Kolenova (✉)
Chair of an Aesthetics and Art Culture, Vladimir State University, Vladimir, Russia
e-mail: kolenova0309@yandex.ru

A.-T. Tymieniecka (ed.), *Phenomenology of Space and Time: The Forces of the Cosmos and the Ontopoietic Genesis of Life: Book One*, Analecta Husserliana 116,
DOI 10.1007/978-3-319-02015-0_17,

development in the past and to understand the purpose of this development at present and know the basic directions of our future. Anatoly Vasilyev's personality is broad and unique. We will explore his journey as a director from personal creative manner to the method.

Let us turn to the heuristics of art. It is important that it starts with the comprehension of the artist's creative work. "The unique nature of the creative process is determined by the uniqueness of the individual artists, especially those tasks that he sets himself and decides as an author" (Bogdanova 2007: 229) For us, it is important that in laboratory of artist special role is played by spiritual means of his creativity and abilities: emotional, rational and subconscious. Recall that the emotional spiritual tools include emotion, inspiration, imagination, to rational: the talent, knowledge of life, observation, experience, outlook, hard work, intuition, insight, inspiration related to the subconscious spiritual means (Bogdanova 2007: 244). They are all one system, helping the artist to create and continually evolve in his own laboratory.

Anatoly Vasiliev was no exception. Starting to work in the theater in "the sixties", in the heyday of the creativity of Tovstonogov, Ephraim, Efros, he was the first of these names became more increasing to study the conflict of the person over time, using the theatrical modem. He begins to investigate the position of the person in reality (in the twentieth century it was engaged in the latest psychology: Freud, Jung, Erikson, Fromm). So, it opens the conflict in the hero, setting the type of "bifurcated man" and gradually comes to the famous conclusion: "The world does not determine a man and a man defines the world." How does this impact on the formation of a special artistic method Vasilieva, we will find out.

Recall that a creative manner "referred to in the aesthetics of a coherent and clearly expressed by individual art features art that distinguish one artist's works from the works of another" (Bogdanova 2007: 245). These features will undoubtedly affect all stages of artistic and creative processes of each artist. In an interview, "Look to the past and the future," Vasiliev talked about the stages of his artistic activities, and defined them as follows:

1. "The method of effective analysis," which he possessed in perfection, being a student M. Knebel.
2. "The beginning of self-analysis," Vasiliev describes his fascination with ritual, hexameters, epic texts.
3. Opposition of theater of fight to game theater became the following Vasilyevsky know-how. Hence, in his performances there is a special composition, which was later so-called "Basil".
4. At this stage, the artist engaged in the development of artificial methods of implementation of the word, i.e. study of verbal action, not with the psyche and physics, but only with the word. So, it opens the position of "inverted speech" and calls it a "solid intonation."
5. This stage develops the artist to this day. He named it chief of his creative life, because it has opened a special distance between actor and character, which allows one to control the action and energy (as an actor and viewers.)

Through an analysis of each of them, it becomes clear that the last of these periods is one of the most important in his artistic path. Since at this stage begins the laboratory experimental work of the artist in the theater "School of Dramatic Art" in Moscow. It was here that begins to develop a method of Anatoly Vasiliev. Artistic method in aesthetics – "category, which determines how and what the artist sees in reality." Thus, it is a deliberate selection, evaluation, synthesis, selection of the topics and ideas, variations and special forms. The problems of the creative method in aesthetics engaged Aristotle, Descartes, Zola, Belinsky, Chernyshevsky. "The basis of widely developed and used today to analyze the creative techniques of artists such as classicism, romanticism, sentimentalism, realism, critical realism, socialist realism, the realism of the Renaissance, and others, is *dihtomiya*: socio-historical conditioning installation wizard, as well as system of principles inherent in the direction of attitudes, styles, schools' (Bogdanova 2007: 245). In this dialectic of individual and common is the formation of the creative method in the art.

Formation of the Basil method was happening in the theater laboratory. Fine experimental work in the laboratory was the first to carry Polish director Jerzy Grotowski. Plays of amazing brevity and simplicity are combined with the semantic richness and diversity, filling the entire contents of a unique form of theatrical magic. Grotowski himself, commenting on his work, above all points to a process that takes place in the actor, the kind of freedom from his own ideal of the body's own talent, their own facial expressions, gestures, voice, tightly associated with certain socio-important task. The creative work of Anatoly Vasilyev also presents a partial removal of the artist's own will, desire, freedom, emotion, but in conjunction with the utmost concentration of plastic body. So, abandoning the idea of engagement, and along with entertainment, Vasiliev deliberately put himself in a situation of constant research and experimentation. This gives us, in our view, the theater an opportunity to be a moral and aesthetic body, united in an outburst of their civic tasks. We are interested to see how, through what means and mechanisms for the transformation of directing – the aesthetic ideas of Vasiliev, gradually acquired a unique creative style. How did they evolve, to become more and more permanent and recognizable? How, determining the angle of vision, depth vision, personal style has developed into a serious Vasilyevskaya, unique, somewhere in a revolutionary method of Anatoly Vasilyev?

Arguing about it, it seems impossible to trace the development of the artist, not defining art – an aesthetic perspective of the era, during which he works. For us, this is an important research task. Recall that Anatoly Vasiliev, before he started any serious work with the actors, the long period of time involved in the philosophy of art. Schelling defined the philosophy of art as "a real image of the ideal, contained in art" (Vasilyev 1995a/2003a: 53). It is known that the metaphysics of aesthetics is the core of any philosophy of art. These are two areas: Metaphysics as a classical aesthetic and metaphysical reality (aesthetic experience and artistic reality). We make the assumption that the deal with the philosophy of art Vassiliev pushes era itself, which is eclectic, unpredictable and at all levels is experimental.

Postmodernism – a "new classic", i.e., focus on tradition, where the change comes a whole new humanism anthropological. Citationality, collage, mosaic, – a game that's the nature of the attributes of postmodernism. The aesthetics of this era are not canonical, "she explodes from within traditional notions of wholeness, harmony, completeness aesthetic systems, there is non-classical interpretation of the classical tradition" (Mankovsky 2011: 103). N.B. Mankovskaya, talking about postmodernism, writes "... it is the diffusion of high-mass culture .... its main theme: *Interior-eksteriornoe*, nature, technology, masculine-feminine, the host-guest, the East-West, the man-animal." These topics are relevant to creativity itself Anatoly Vasiliev.

Exploring the aesthetics of the theater director Anatoly Vasilyev, we can not ignore the concept of an artistic image. In Bychkov, this "organic spiritually – *eydepicheskaya* integrity, expressing, presenting a kind of reality in the mode of larger and smaller isomorphism (similarity of shape), and realizing in the process of aesthetic perception of a particular work of art specific recipient" (Mankovsky 2011: 144). As a result, we have to compare the stages of formation the artistic image offered by modern aesthetics (e.g., the formation of artistic design or image, design, implementation of the plan in an artistic language or image-incarnation, the artistic perception of the finished product as a kind of invented authoring system) with the stages of laboratory and experimental Anatoly Vasiliev creativity through the idea of "relativistic" aesthetics of Grotowski E. Basil, and his own experiment by analysis of its directors, educational, theatrical and aesthetic ideas, as well as through the analysis of author performances of his students and followers Alexander Ogaryov and Vladimir Belyaykin.

Notes

1. Anastasyev A. Laboratory Hedgehogs Grotovsky [Text]/A.Anastasyev. – SPb.: Education, 2002.
2. Bogdanova P. Logic changes. Anatoly Vasilyev: between past and future. – M: UFO, 2007.
3. A.I.Estetika's drills: problems and disputes [Text]: studies. for higher education institutions on the specialties "Cultural science", "Philosophy", "Art Studies", "Musicology", "Philology", "Muzeology"/A.I. Drills. – M: Science, 1999.
4. Bull-calves V.V.Estetika [Text]: studies. for higher education institutions on the specialties "Cultural science", "Philosophy", "Art Studies", "Musicology", "Philology", "Muzeology"/ Century V.Bychkov. – M: Science, 2002.
5. Bull-calves of Century of Century. Russian esthetics of the nineteenth–twentieth centuries [Text]: [моногр.]/V.Bychkov. – M: Art, 1991.
6. Vasilyev A. Diary director [Text]/A.Vasilyev//Art of the film. – 1995a. – No. 10. 14. Vasilyev A. Anatoly Vasilyev. Plans [Text]/A.Vasilyev//Modern dramatic art. – 2003a. – No. 4 (second Agrarian Party of Russia.).
7. Vasilyev A. Time under the name "Ruins" [Text]/A.Vasilyev//Theatre life. – 2003b. – No. 7 (9 июл.).
8. Vasilyev A. Long ago it would be desirable to mix, destroy and forget everything that I am able [Text]/A.Vasilyev//Theatre life. – 1983. – No. 3 (12 March).
9. Vasilyev A. Diary Parisian director [Text]/A.Vasilyev//Art of the film. – 1995b. – No. 10 (23 окт.).

10. Vasilyev A. Martian chronicles [Text]/A.Vasilyev//Theatre life. – 2003c. – No. 7 (13 июл.).
11. Vasilyev A. To glance in the past and the future [Text]/A.Vasilyev//Theatre life. – 2009. – No. 8 (13 авг.).
12. Vasilyev A. The truth hunter in the lie markets [Text]/A.Vasilyev//GNII. – 2000. – No. (2.февр.)
13. Grotovsky E.Teatr awakes thought [Text]: [моногр.]/E.Grotovsky. – M: Vladis, 2002.
14. Kiray N. Anatoly Vasilyev. Theatrical fugue [Text]/N of Kiray. – SPb.: Science, 2006.
15. Mankovsky N.Fenomen of a postmodernism. Art and esthetic foreshortening [Text]/N Mankovsky. – M: Science, 2011.
16. Ogaryov A. Preacher of own concept [Text] A.Ogaryov//Culture. – 2011.-№ 5 (on May 13).

Part V

Teleology in Nature and Life-Transforming Art

Vladimir L. Marchenkov

Abstract The main goal of the paper is to argue, against the hypercritical treatment of teleology in poststructuralist writings, that the concepts of purpose and purposiveness are necessary elements in thinking about both nature and art. I begin with the examination of two contrasting views on teleology in nature and art: Immanuel Kant's treatment of the matter in the *Critique of the Power of Judgment* and G. W. F. Hegel's in his *Aesthetics*. The parallels and contrasts between them help delineate the problem and lead to a consideration of the role of art in modern thinking about nature, as well as in nature-transforming human activity. I end by arguing for a reexamination of teleology in the family of concepts dealing with the relation between praxis and poiesis, and for a substantive fleshing-out of its content in philosophical dialogue.

Introduction

One can speak of the cosmos today only with acute awareness of how problematic this concept has become for describing the currently prevalent view of nature. One can likewise speak of nature's *poiesis* only with a profound uneasiness about applying the notion of artistic creativity to natural processes. There was a time when cosmogony was perceived as the work of an artist, when God was the architect of creation. Modern thought destroyed this idea: God was first, in the beginning of the Enlightenment, demoted to the maker of the world mechanism who, having set it in motion, turned away in either boredom or disgust; and then, at the end of the Age of Reason, He was altogether "bracketed" as an "unnecessary hypothesis." God's place was taken by man, who was at first so filled with elation and fear that he wondered

V.L. Marchenkov (✉)
Ohio University, 192 East State Street, Athens, OH 45701, USA
e-mail: marchenk@ohio.edu

A.-T. Tymieniecka (ed.), *Phenomenology of Space and Time: The Forces of the Cosmos and the Ontopoietic Genesis of Life: Book One*, Analecta Husserliana 116,
DOI 10.1007/978-3-319-02015-0_18,

whether the world was merely his dream, but then got used to his new role and eventually began to chuckle at himself – without, however, giving up his pretensions to the title of the creator of all. René Descartes and George Berkeley were intensely concerned with proving the reality of nature, the ordered objective world, but in the end they could find nothing save faith in God, a remedy from an earlier time, to serve this purpose. The monumental edifice erected later by Immanuel Kant rested on a similar, if less openly acknowledged, foundation. These thinkers treated their task with utmost seriousness but could not solve it by philosophical means. (By contrast, David Hume, unable to decide whether or not empirical phenomena, "matters of fact," had any rational content, "ideas" – whether, in other words, the empirical world had a rational structure or was the domain of sheer chance where "it was not necessary that the sun would rise tomorrow" – consoled himself with facile witticisms.) G. W. F. Hegel did create a philosophical system that treated the relation between creative spirit and objective being without an appeal to a mystical Absolute but, apparently, his solution satisfied neither secular progressivists nor retrograde theologians – let alone mystics of various descriptions.

For Aristotle natural generation and artistic creation alike were purpose-driven, and there was no impassable gulf between *physis* and *poiesis*. In his *Physics* he stated, for example:

> It is absurd to suppose that purpose is not present because we do not observe the agent deliberating. Art does not deliberate. If the ship-building art were in the wood, it would produce the same results *by nature*. If, therefore, purpose is present in art, it is present also in nature. The best illustration is a doctor doctoring himself: nature is like that. It is plain then that nature is a cause, a cause that operates for a purpose.[1]

It was precisely purposive generation in nature that became the target for modern natural philosophy. The modern intellect does acknowledge partial, intermediary goals, but firmly refuses to allow an overall purpose in nature. When scientists muse about "the theory of all," their musings remain just that, musings, while their most concentrated effort is directed at discerning rational content (which necessarily evokes teleology) in every individual phenomenon. At present the debate about this issue – what is nature for? – seems to have degenerated into the conviction that it is a resource for humanity: to be *consumed* seems to be nature's destiny. Already Hegel noted, as he spoke of the dual, "amphibious" modern subject, that the latter "strips the world of its enlivened and flowering reality and dissolves it into abstractions, since the spirit now upholds its right and dignity only by mishandling nature and denying its right, and so retaliates on nature the distress and violence which it has suffered from itself."[2] The attitude that Hegel describes understandably provokes protest and yet there is little difference between those who continue to think of nature as a storage facility and those who advocate a more caring husbanding of its riches. Neither party is willing to give up the basic paradigm of modern history that can be called economic progressivism and projects a vision of an infinitely expanding

---

[1]Aristotle, *Physica* II 8 199b27-31, quoted from: *The Basic Works of Aristotle*, ed. Richard McKeon (New York: Random House, 1941), p. 251.

[2]G. W. F. Hegel, *Aesthetics*, vol. 1, trans. T. M. Knox (Oxford: Clarendon Press, 1975), p. 54.

material production. The voices of those who attempt to break out of this paradigm and propose a revival of a pre-modern, mythological attitude towards nature as a living, sentient interlocutor, are barely heard against the roar of earth-moving machines and the buzz of global networks. Soon everything will be drowned, it seems, in the most abstract current history has ever known: the flow of capital.

I believe that one of philosophy's most urgent tasks today is to return to the issue of teleology, both in nature and in art, and to subject to a constructive critique the two most ardently anti-teleological stances: modern progressivism, which claims to be rational but is riddled with irrationalities, and its patricidal progeny, postmodern scepticism, which would like to see the end of modern progress and does not mind being irrational.

Art and Teleology in Nature

Art "suspends," according to Theodor Adorno, "self-justificatory, teleological rationality."[3]Apart from resentment towards self-sustaining thought, there is in Adorno's words an echo of Kant's famous description of the aesthetic attitude as "disinterested interest" and of its object as "purposeless purposiveness." By suspending teleological thinking, as Adorno alleges, art enters the domain of disinterestedness and purposelessness. What is so often lost in evoking Kant's familiar phrases is the fact that in them the words "disinterested" and "purposeless" refer to a different sense of interest and purpose from the one that "interest" and "purposiveness" are struggling to call into existence. By saying "disinterested" and "purposeless" Kant wishes to point out the absence of a cognitive and pragmatic intent in a judgement of taste, whereas by using "interest" and "purposiveness" he strives to capture the genuine value and *telos* of art that has proved so elusive for the modern intellect. To call this *telos* "formal," as Kant himself does, is to surrender art to a hopelessly pedestrian agenda. What one should rather see in Kant's formulae is the starting point for exploring the dialectic of the pragmatic and the artistic, of *praxis* and *poiesis*, where the two are intertwined so closely that to detach them from each other is to destroy the new sense and the new vision of creativity that arises from their conjunction.

The school of phenomenology derives its name most directly from Hegel's *Phenomenology of Spirit* but Kant's *Critiques* should no doubt be counted among the most faithfully conducted phenomenological exercises in modern philosophy.[4] If one grasps Kant as a master-phenomenologist, then one can approach perhaps a correct view of his role in the construction of modern aesthetics. In other words,

[3]Theodor Adorno, "On the Contemporary Relation between Philosophy and Music," *Essays on Music* (Berkeley: University of California Press, 2002), p. 138.

[4]For the analysis of Hegel's concept of phenomenology see Wolfgang Bonsiepen's introduction to G. W. F. Hegel, *Phenomenologie des Geistes*, ed. by H.-F. Wessels and H. Clairmont (Hamburg: Felix MeinerVerlag, 1988), pp. IX–LXII.

rather than a metaphysician fantacizing about his object and rather than an obsessive legislator attempting to regulate the teeming world of art, Kant should be seen as a phenomenologist grappling with a new reality that had arisen in European culture by the middle of the eighteenth century: art as a human pursuit with its own intrinsic value, reducible to neither menial work nor cognition nor religious devotion nor moral indoctrination. Kant's *Third Critique* is an attempt to provide a detailed phenomenology of this reality, to comprehend its *eidos*. To appreciate the significance of this attempt one must firmly bear in mind that in the secular thought of the Enlightenment nature had been stripped of its pre-modern mantle of God's handiwork, a development that simultaneously voided nature of its teleological dimension. Kant understood very well that without this dimension, without a purpose, nature could not become an object of either rational cognition or aesthetic judgement. "It is in fact indispensable for us," he wrote, "to subject nature to the concept of an intention if we would even merely conduct research among its organized products by means of continual observation; and this concept is thus already an absolutely necessary maxim for the use of our reason in experience."[5] He was bold enough to reverse the empiricist's earthbound wisdom – a reversal made more difficult by his own empiricist inclinations – and to propose that, in order to see any natural phenomenon as a rational whole rather than a melange of sensory stimuli, one must approach it *as if* it were a work of art. As he himself put it, "We cannot conceive of the purposiveness which must be made the basis even of our cognition of the internal possibility of many things in nature and make it comprehensible except as a product of an intelligent cause."[6] Further on he linked this "intelligent cause" specifically with art:

> Physical teleology certainly drives us to seek a theology, but it cannot produce one, however widely we may scrutinize nature through experience and however much we may supplement the nexus of ends discovered in it with ideas of reason...What help is it, one may rightly complain, to ground all these arrangements on a great and for us immeasurable intelligence, and have it arrange this world in accordance with its final aim, without which, however, we can form no common reference point for all these natural ends, no teleological principle sufficient for cognizing all the ends together in a single system as well as for forming a concept of the supreme intelligence, as the cause of such a nature, which could serve as a standard for our power of judgment for reflecting upon nature teleologically? In that case I would, to be sure, have an **artistic intelligence**, for various ends, but no **wisdom**, for a final end, which, however, must really contain the determining ground of the former.[7]

Thus "artistic intelligence," in Kant's opinion, may be responsible for the organized and therefore purposeful *individual* phenomena (even if it is incapable of bringing them under *one* universal purpose, which is the task of "wisdom"). And yet the talk of this "artistic intelligence" that must be presupposed behind all phenomena should always be correlated with what Kant says about it in the First Introduction to his *Critique of the Power of Judgment*: "[T]he representation of nature as art is a

---

[5] Immanuel Kant, *Critique of the Power of Judgment*, trans. Paul Guyer (Cambridge: Cambridge University Press, 2001), §75, p. 269.

[6] Ibid.

[7] Ibid., §85, p. 307.

mere idea, which serves as a principle, merely for the subject, for our investigation of nature, so that we can where possible bring interconnection, as in a system, into the aggregate of empirical laws as such, by attributing to nature a relation to this end of ours."[8]

Kantianism is often regarded as a variety of Platonism, but this instance highlights its extremely complex interaction with Plato's thought. For Plato the world was a priori a cosmos, the perfect artefact; hence the couch that the painter had before him in Plato's most notorious example from the *Republic* was already a superior artwork, admitting only of inferior copying.[9] Kant lives, by contrast, in an entirely different world: infinite, shapeless, and as far removed from intentional creation as possible. The relation between the couch and a painting of the couch is reversed in it and the painting now must serve as a model for the object that it depicts – if this object is to become anything that rational thinking can grapple with.

There is logic in this, a logic that exonerates teleology and shows how myopic Adorno's rebellion against it was. Like Voltaire, who thought that the senseless death of thousands in the Lisbon earthquake was sufficient proof against the divine design in the world; and like Hume, who drew similar conclusions from the imperfection of creation; Adorno rejected objective teleology out of moral outrage at the holocaust.[10] Neither Voltaire nor Hume nor Adorno, though, noticed that in pronouncing their judgements they usurped absolute moral authority and made themselves supreme judges over the entire order of things. To be able to deny divine design one must assume a divine vantage point. Such gestures are self-contradictory in their very essence, a circumstance that has not prevented large numbers of modern philosophers from repeating them *ad infinitum*.

Kant's discussion of teleology culminates in the idea of man as the pinnacle of creation and, further, in presenting "the moral man" as the highest avatar of humanity.[11] Kant's "ethicotheology" is the end of his *Third Critique* and thus it is clearly implied, in typical Enlightenment fashion, that the purpose of art is the moral betterment of man. Still for Kant the imposition of a teleological stance borrowed from art upon presumably non-teleological natural phenomena was a purely subjective operation (hence *as if*), not intended to argue that teleology inheres, after all, in nature herself.[12] He did not dwell overmuch on explicitly admitting the contradiction between claims to rational cognition of natural phenomena in a purposeless universe, on the one hand, and the realization that they become rationally comprehensible only under a teleological view of things, on the other. But his difficulties

[8] Ibid., p. 10.

[9] Plato, *Republic* X 597–599, in Edith Hamilton and Huntington Cairns (eds.), *The Collected Dialogues of Plato Including the Letters* (New York: Bollingen Foundation, 1963), 821–823.

[10] See Theodor Adorno, *Negative Dialectics*, trans. E. B. Ashton (New York: Seabury Press, 1973), pp. 142–143.

[11] "[I]t is only as a moral being that the human being can be a final end of creation…" (Kant, *Critique of the Power of Judgment*, §86, p. 309).

[12] This is made abundantly clear in §58 of the "Critique of the Aesthetic Power of Judgment" (Kant, *Critique of the Power of Judgment*, pp. 221–225).

negatively show that without teleology, construed along the lines of artistic creation, a rational view of nature becomes impossible. The relation between art and nature, in other words, is far more intricate than the model underlying Plato's example suggests, i.e., the one copying an already available reality of the other. It turns out that, in order for the kingdom of nature to become available to human reason at all, it must first become the kingdom of art.[13]

This dialectic did not escape Hegel who argued in his lectures on aesthetics that the beauty of art is, in fact, prior (and superior) to the beauty of nature. Hegel's response to Kant is clear and uncompromising. He calls the theory that "art has to serve as a means to moral purposes" a "false position." Moral dilemmas themselves, the urgent need to resolve them, he remarks, push us along "to a higher standpoint." The main thrust of his objection is against the instrumental approach to art that denies it its own intrinsic value and makes it merely a tool for achieving extraneous ends. Instead, he states:

> [W]e must maintain that art's vocation is to unveil the *truth* in the form of sensuous artistic configuration, to set forth the reconciled opposition [between the universal and the particular, abstract law and individual phenomena, sensual and spiritual, spirit and flesh, duty and feeling, inner freedom and the necessity of external nature, the dead inherently empty concept and the full concreteness of life, subjective thinking and objective existence] and so to have its end and aim in itself, in this very setting forth and unveiling.[14]

And yet this did not mean that Hegel identified the teleology inherent in natural phenomena with the one found in artistic creation. The teleology of nature – objective spirit, as Hegel referred to it – is not of a conscious variety; things do evolve in a rationally comprehensible way but the universal design does not enter the consciousness of these evolving beings themselves. Only subjective spirit – humanity – has the self-consciousness that makes rational intent possible, and Hegel defined art as a strictly human activity. Kant used to call the kingdom of art "second nature"; in Hegel's case it would be misleading to use "nature," however metaphorically, to describe art. For him, art is rather a total transformation of immediately given natural existence into a product of self-conscious spirit. What emerges from this transformation is no longer natural in any sense, but is thoroughly reworked and transmuted by the spirit into its own actuality, where the *telos* is pursued with full awareness of its significance. (The natural, i.e. instinctual aspect of genius, so important for Kant, is played down as much as possible in Hegel's treatment where

[13] I leave aside the matter of the Kantian "as if" surviving the mind's first encounter with its object and permeating *all* knowledge of the objective world that derives from this encounter. The constant presence of this "as if" in the generation of modern knowledge renders the latter thoroughly hypothetical, locks it in within the domain of the hypothesis as surely as the original sin locked the medieval person within the domain of the fallen world – yet without the corresponding hope of ever breaking beyond it.

[14] G. W. F. Hegel, *Aesthetics: Lectures on Fine Art*, vol. I, trans. T. M. Knox (Oxford: Clarendon Press, 1975), p. 55.

the pride of place is given to the artist's intellectual development and purposeful mastery of technical skill.)[15]

In Hegel's philosophy temporal reality – both natural and human history – and its universal *telos*, the Absolute, find themselves flowing continually into each other. The temporal is the inverted image of the Absolute and the Absolute is in turn constituted by nothing other than the temporal flowing into it and finding its entelechy in it. The farther the spatiotemporal world finds itself away from the final fulfilment of its purpose, the less pronounced its teleological dimension. However, in Hegel's thought the end is not simply *posited* as rationally necessary, as it was in Aristotle's and Aquinas' cases, but developed into a category that is dynamically upheld within the network of constantly interacting fellow-concepts. To remove it, as Adorno did in constructing his so-called "negative dialectic," is to render both the concept of the goal and the whole to which it belongs irrational and fragmented at a fundamental level. "Negative dialectic" is an oxymoron and no amount of talk about "determinate negation" can save it from its own self-induced collapse. Adorno and his followers remain prisoners of modern infinitism, of blind, unphilosophical faith in progress without end. I have argued elsewhere that this wilful silencing of philosophical reason is driven by a moral agenda and it bears repeating here that a morality which wishes to dictate to reason turns into its own opposite.[16]

It goes almost without saying that, as I have already pointed out, Hegel's thoroughgoing rationalism has proven equally unacceptable to all those who wish to set limits to philosophical thinking on behalf of various mystical authorities, be they of a religious or secular variety. A rational comprehension of the universal *telos* of history is antithetical to the notion of inscrutable, ineffable beginnings – or endings, for that matter. From Friedrich Nietzsche's voluntarism to Carl Jung's psychological mysticism (to say nothing of fundamentalisms of all sorts), with the aesthetic mysticism of the romantics mediating between them, generations of post-Hegelian thinkers have laboured to exorcise the ghost of rational teleology from modern consciousness and to confine it to oblivion.

But teleology refuses to sink into oblivion, and for good reasons. As I explained above, these reasons were well known to Kant – even though he could not get a full view of them. Moreover, to the extent to which anti-teleologists themselves wish to persuade by rational arguments, they are forced to accept teleology – as Nietzsche certainly did in his doctrine of the new human being, the Overman, and in his poetic-philosophical imagery of the "new dawn." Teleology remains as (an unwelcome) trace in the writings of his postmodern heirs: when Gilles Deleuze and Felix Guattari argue for a new, rhizomic humanity, they definitely engage in goal-setting, try as

---

[15] It should be noted, however, that Kant did propose that taste must set limits to genius in certain respects and at the very least is necessary for the education of genius. (Kant, *Critique of the Power of Judgment*, §50, pp. 196–197).

[16] See my "Instead of an Afterword," in Vladimir Marchenkov (ed.), *Between Histories: Art's Trajectories and Dilemmas* (Cresskill, N. J.: Hampton Press, 2013).

they may to get away from purpose-driven, hierarchical thinking.[17] When Susan Buck-Morss attempts to articulate a sense of universal history distinct from both Hegelianism and postmodernism, she resuscitates the typically modern endless pursuit of a goal that is forever slipping away.[18] When they do become conscious of the internally contradictory nature of their stance, they blame reason, rational thinking – and propose that its reach be curtailed. In this they join hands with the obscurantism of all ages that likewise saw in rational thinking a threat to the foundations of its mystical-moral edifices.[19]

The Emerging Telos

What kind of sense is struggling to emerge from the paradoxical antinomy of "interest" and "disinterestedness," "purposefulness" and "purposelessness"? Translating, as I did above, these conjunctions into the language of *praxis* and *poiesis*, helps one view their antinomies in a way that lets both sides find their justification. The teleological defect of *praxis* consists in the fact that it operates only with proximate and intermediary goals and remains oblivious of the need for a final and universal one. The technical problem of *poiesis* consists in the fact that its final goals can be realized only within the strictly defined boundaries of time and space – be that a temple, museum, or the concert stage; it is incapable of embracing existence as a whole. In other words, *praxis* does not reach a final goal because it is

[17] See Gilles Deleuze and Felix Guattari, *A Thousand Plateaus: Capitalism and Schizophrenia*, trans. Brian Massumi (Minneapolis: University of Minnesota Press, 1987), pp. 3–25 ("Introduction: Rhizome").

[18] Susan Buck-Morss' argument in *Hegel, Haiti, and Universal History* is a recent example of advocacy for this infinitism. Inevitably its defenders find themselves forced to resort to metaphors rather than logically defensible reasoning. Buck-Morss borrows, for example, from like-minded authors the metaphor of the "motley crew" (of various marginalized "others") to propose an allegedly new collectivity – after the embarrassment suffered by the *citoyen* of the French Revolution and the Marxian proletariat – as the agent of history that will push progress further (Buck-Morss, *Hegel, Haiti, and Universal History* [Pittsburgh, PA: University of Pittsburgh Press, 2009], pp. 104–107 and 151). It is logically necessary, in other words, that theorists embracing modern infinitism should abandon philosophical thinking at the most crucial moment and reach for non-philosophical tools to give their cause a semblance of coherence.

[19] Incidentally, Buck-Morss is mistaken when she equates Hegel's Absolute with the mystical absolutes of traditional religions (e.g., Buck-Morss, *Hegel, Haiti, and Universal History*, p. 115). Hegel's treatment of the concept of mystery in general is quite negative and his interpretation of the Holy Trinity is clearly intended as non-mystical (G. W. F. Hegel, *Lectures on the History of Philosophy: Plato and the Platonists*, trans. E. S. Haldane [Lincoln and London: University of Nebraska Press, 1995], pp. 418 and 440ff; and *Lectures on the History of Philosophy: Medieval and Modern Philosophy*, trans. E. S. Haldane and F. H. Simson [Lincoln and London: University of Nebraska Press 1995], pp. 4–5). When he does acknowledge mystery in a positive light, as he does in his ruminations about the religion of ancient Egypt, it is still regarded as only a phase in the evolution of spirit, a stepping stone that must eventually be left behind (Hegel, *Lectures on the Philosophy of Religion*, Volume II Determinate Religion, trans. R. Brown et al. [Berkeley, Los Angeles, London: University of California Press, 1987], pp. 365, 369–70, and 531).

incapable of articulating it for itself, while *poiesis* fails to reach it because it lacks practical means to achieve it.

In the history of modern technology we see a constant striving to embrace the entirety of existence, to subjugate all that is to *praxis*. We see also, in close conjunction with this expansion of *praxis*, the desire to assert as rationally defensible only intermediary goals, while condemning final goals as irrational, fictional, and destructive. *Poiesis* has not been entirely innocent in this regard either. Acutely aware of its own limited nature, it wished either to dissolve itself entirely in *praxis*, thereby turning art into pure technology, or, plunging into the opposite extreme, locked itself in an ivory tower and declared its particular goals the only worthwhile ones, treating what remained outside as fertilizer for the rose garden of art.

Such are the deformities created by the irrationality that is partially contained already in the Kantian antinomies. This irrationality can be expressed as follows: a means without an end (*praxis*, narrowly understood) and an end without a means (*poiesis*, likewise narrowly understood). The aesthetic is called to life, on the one hand, by the dream of humanity's power over the entire universe, over external being as such, and, on the other hand, it assumes its specific shape under the pressure of humanity's weakness and limitations. As they collide, these two forces give birth to aesthetic activity in which humanity displays its full control over external existence – but only within a strictly delimited, *ludic* space. In its pure form aesthetic activity is quite indifferent to what material it exercises its power over: its main purpose is to show the artist's omnipotence ("the free play of imagination and understanding") over the chosen material. In this sense aesthetic activity is purely formal – albeit even its formal character expresses its essential content *qua* aesthetic. But our constantly expanding power over nature and our evolving technological prowess increasingly endanger precisely the ludic character of aesthetic activity. As long as simulacra – pure products of human fantasy – comprised the peripeteias of Oedipus or of Madam Butterfly, our art could preserve its ludic character. But when global landscape, climate, and the conditions for the preservation of human race on planet Earth become pliable material in the hands of man-the-technologist, then aesthetic activity can no longer remain purely ludic and its content – what precisely is, after all, done with the chosen material – becomes no less and perhaps even more important than the idea of our power over it.[20]

In the past several decades the thesis about the end of history once again became a subject of debate among philosophers, including aestheticians. Echoing Arthur Danto, I should like to propose that what is coming to an end is not history in general, but only a certain phase of it, namely the one that has come to be known as *modern history*. Its end does not mean that the future is devoid of historical events. Against postmodern historical despair, we must realize exactly the opposite: the end of modern history opens the opportunity for us to begin a *genuine* history. To the extent to which it was the result of purposive action, modern history was a history not of reason, but of the instrumental intellect, what Hegel called "the

[20]A parallel process in *political* life must be examined in a separate work but it is worthwhile to point out here that in politics this set of problems is present in an even more acute form than in ecology. The popularity of the cliché according to which the contemporary political process has degenerated into show business is only one symptom among many.

understanding" (*das Verstand*); it was not so much rational as rationalistic. Its end marks the exhaustion of the understanding rather than reason, and in the final analysis it has led to the triumph of the rationalistic intellect rather than genuine rationality.

One must carefully distinguish, however, between the working of the understanding and that of reason in modern history, or rather to understand the dialectical relation between these two agents. History in general is driven by the self-unfolding of reason but reason itself is, in turn, something historical. It presents itself as divided in modernity: it recognizes itself mostly in its own truncated form, namely that of the understanding. (The story is more detailed at a closer look but I am simplifying it for the sake of brevity.) The actions to which the modern instrumental intellect induces humanity can be summed up in the notion of *the infinite progress of immanent humanity*, i.e., humanity construed as devoid of a transcendent dimension. This formula – and especially the fantastical construct of immanent humanity – is shot through with contradictions and, left to its own devices, leads to irrational outcomes. And yet there is a rational element in it, too, there is logic in its rise as a reaction to medieval transcendental eschatology, and there is, finally, in it the possibility to resolve its internal contradictions and to deduce a rational concept of history, humanity, and its goals.

All this is directly connected with the question of teleology in nature and life-transforming art. Nature as an inexhaustible reservoir of resources fated to be consumed by a human being that is embarked upon an endless external expansion is part of modern mythology in which anthropocentric voluntarism and rationalistic intellect are fused together. Modern art, which precariously teeters between *l'art pour l'art* and industrial technology, is the child of the same forces. This mythology must be surmounted and must yield its place to a more rational relation between nature and art where the aporias of the abstract intellect can be resolved without doing violence to nature and without reducing art to something existentially irrelevant.

References

Adorno, T. W. 1973. *Negative Dialectics*, Trans. E. B. Ashton, New York: Seabury Press.
Adorno, T.W. 2002. *Essays on music*. Berkeley: University of California Press.
Aristotle. 1941. *The basic works of Aristotle*, ed. Richard McKeon. New York: Random House.
Bonsiepen, Wolfgang.1988. Einleitung. In *Phenomenologie des Geistes*, ed. G.W.F. Hegel., H.-F. Wessels., and H. Clairmont, IX–LVII. Hamburg: Felix MeinerVerlag.
Buck-Morss, S. 2009. *Hegel, Haiti, and universal history*. Pittsburgh: University of Pittsburgh Press.
Deleuze, Gilles, and Felix Guattari. 1987. *A thousand plateaus: Capitalism and Schizophrenia*. Trans. Brian Massumi. Minneapolis: University of Minnesota Press.
Hegel, G.W.F. 1975. *Aesthetics: Lectures on fine art*. Trans.T.M. Knox. Oxford: Clarendon Press.
Kant, Immanuel. 2001. *Critique of the power of judgment*. Trans. Paul Guyer. Cambridge, UK: Cambridge University Press.
Plato. 1963. *The collected dialogues of Plato, including the letters*, ed. Edith Hamilton and Huntington Cairns. New York: Pantheon Books.

The Creative Potential of Humor

Anna Malecka

Abstract The paper focuses upon creativity as the essential trait of philosophically understood humor and the comic.

Arthur Koestler's bisociation theory is referred to, which interprets humor as one of the forms of human creative activity, consisting in the ability to perceive phenomena simultaneously in traditionally separate and incompatible planes of reference. True humor escapes routine, presenting familiar situations from multiple angles and thus in a new light. The akin potential can stimulate creativity also in other spheres of human activity, including art or science.

Anna-Teresa Tymieniecka's phenomenology of the human creative act, which is shown to possess its universal cosmic foundation is also applied to analyze humor as one of the manifestations of the sources of life.

In the present author's opinion, humor can be recognized as a symbolic – though perverse – expression of the creative potential inherent in human spirit as well as in life itself.

In philosophy, and especially in aesthetics, creativity has traditionally been understood as an activity bringing forth novelty in the field of culture and social life. In Medieval times, in particular, creativity was interpreted as an imitation of the Divine act of creation *ex nihilo*. Accordingly, in the most radical and metaphysically fundamental meaning, human creativity in the sphere of vastly understood culture was recognized as a modest imitation of God's act of summoning to being something out of nothingness. In an alternative meaning, in light of Jung's psychoanalysis

A. Malecka, Ph.D. (✉)
AGH University of Science and Technology, Faculty of Humanities,
Department of Culture Studies and Philosophy, ul. Piastowska 3, Krakow 30–211, Poland
e-mail: amm@agh.edu.pl

A.-T. Tymieniecka (ed.), *Phenomenology of Space and Time: The Forces of the Cosmos and the Ontopoietic Genesis of Life: Book One*, Analecta Husserliana 116,
DOI 10.1007/978-3-319-02015-0_19,

and Mircea Eliade's studies in the origins of religions, the mythical creation of the world itself is presented as an archetype of all human creative gestures.[1]

Poiesis, understood as a process of "making" original artefacts and summoning the fictional worlds into existence, was in ancient Greece reserved for poetry which was seen as an unrestrained activity, and thus excluded from the domain of art governed by strict principles. Freedom, originality, extraordinariness and novelty – these notions have proved inseparable from the concepts of creativity interweaving the history of Western culture, regardless of the differences in the applied terms. However, from the nineteenth century on the differentiated aspects of novelty have been emphasized, not necessarily inherent to the act of "creating out of nothingness", but rather consisting in various transformations, new combinations, novel associations and juxtapositions, supposed to intimate the hitherto unknown horizons of fresh interpretations within the domain of culture. It should also be noticed that aesthetically defined creativity is closely related to axiological postulates: a novelty must convey value.

The main question which will be posed in the paper is whether true humor is creative in its essence, and if so, what type of creative potential it involves in comparison with other human faculties and acts, particularly those related to artistic endeavours. The term 'humor' will be primarily used to designate the subject's approach and activity, finding expression predominantly (though not always) in laughter or smile, and evoking amusement on the part of the spectator. All in all, the notion of humor will be dealt with as superior in relation to the totality of akin phenomena, such as comicality and wit.

As Maurice Charney pertinently remarks, the essence of the creativity of humor (and jokes) does not lie in the thematic materials but rather in the daring new combinations.[2] This vital aspect of the innovativeness of humor was philosophically developed and discussed in detail by Arthur Koestler in *The Act of Creation* (1964).[3] Koestler analyzes creativity as manifested in the three spheres of human performance: humor, art, and science, finding its common pattern in the so-called *bisociation of matrices*, i.e. the ability to perceive an event or idea simultaneously in two previously unrelated contexts and planes of reference, in two matrices governed by completely different sets of rules. In view of this theory, the engagement in the bisociative thinking, i.e. in bringing the concepts or perceptions together in an entirely original way is the essential aspect of creative humor.[4]

---

[1] Cf. Mircea Eliade, *The Sacred and The Profane: The Nature of Religion,* trans. Willard R. Trask (New York: Harper Torchbooks, 1961); C. G. Jung, *The Archetypes of the Collective Unconscious*, in *Collected Works of C. G. Jung* (Princeton: Princeton University Press, 1968), Vol. 9, Part 1.

[2] Maurice Charney, "Comic Creativity in Plays, Films, and Jokes", in *Handbook of Humor Research,* ed. Paul E. Mcghee and Jeffrey H. Goldstein (New York, Berlin, Heidelberg, Tokyo: Springer-Verlag, 1983), p. 34.

[3] Arthur Koestler, *The Act of Creation* (New York: The Macmillan Company, 1967).

[4] Cf. Don L.F. Nilsen, *Humor—Other Views* http://science.jrank.org/pages/9718/Humor-Other-Views.html (access: 10/15/2012).

It was Immanuel Kant who pointed out that the source of jokes lies in the assimilation of heterogeneous juxtapositions which, according to the principle of association, are usually remote, and evoke laughter when put together. As such, this capability of startling assimilations implies the freedom of disposition.[5] The idea was psychoanalytically reinterpreted by Sigmund Freud for whom humor – resulting from the dynamic of the conscious and the unconscious and rooted in the suppressed thoughts – finds expression in the free associations of *der Witz*.[6]

Koestler's understanding of humor as a creative phenomenon is also anchored in the tradition of the so-called incongruity theories. These theories with their multifarious modifications are generally based on the assumption that humor can be explained in terms of discrepancy between the actual outcome of a certain situation or thought and the schematic expectation related to the development of a given situation or idea.[7] Such incongruous stories or events are amusing because they disobey conventional expectations and introduce paradoxical new links between ideas.

As I have already pointed out, for the author of *The Act of Creation* the heart of humor lies in startling clashes of incongruous planes of interpretation of the same ideas or perceptions. Each of such planes of reference is governed by a different code and a specific logic. Thus, the simultaneous and brilliant juxtaposition of the two incompatible connotation frames evokes a certain rational amazement which finds its outlet in a spontaneous laughter. The originality of the comic guarantees the effect of surprise.[8] Consequently, the sense of humor can be equalled with the gift of creativity, consisting in the invention of a shockingly incompatible pair of contexts in which a given phenomenon can be approached from diametrically different angles and explained through different logics perspectives. However, the comic requires that the conflicting rules of the two contextual matrices be hidden and only implied, because the amusing effect would be destroyed if they were made explicit. Thus, the audiences are led to expect a certain outcome compatible with a particular matrix (e.g. introduced by a story); but in a punch line we have a replacement of the original matrix with an alternative one. The overall comic effect seems to be based on swift intellectual oscillating from one matrix to the other.

In Koestler's theory, it is this act of *oscillating* between two competitive matrices of reference that constitutes the essence of creativity inherent in humor. Thinking confined to a single matrix has its obvious limitations and clichés, while escaping such a confinement by a one-track routinized dictate of commonsensical scheme opens the so far not intimated horizons, introduces new forms of outlook that contrast with the familiar ones. The tension and conflict between discrepant connotative planes are inspiring but also bewildering.

---

[5] Immanuel Kant, *Anthropology from a Pragmatic Point of View* (Cambridge: Cambridge University Press, 2006).

[6] Cf. Sigmund Freud, *The Joke and Its Relation to the Unconscious* (New York: Penguin Classics, 2002).

[7] Cf. John Morreall, "Introduction", in *The Philosophy of Laughter and Humor*, John Morreal, ed. (Albany, NY: State University of New York Press, 1987), p. 6.

[8] Marcella Tarozzi Goldsmith, *Nonrepresentational Forms of the Comic. Humor, Irony, Jokes* (New York: P. Lang, 1991), p. 30.

In his book *Laughing and Crying* originally published in 1941, Helmut Plessner pointed to the situations of crisis of expression as a source of laughter (and crying).[9] In a similar vein, Arthur Koestler analyzes the emotional tension that accompanies the intellectual tumbling between the conflicting sets of rules, which finds its outlet in laughter. In his view, in certain situations our emotions are "incapable of following these acrobatic turns", "of keeping step with our reason and become divorced from reason".[10] Thus, the sudden switch of ideas to a different type of logic or a new rule of the game results in this specific "crisis of expression". In so understood a crisis, the comic effect will occur provided that the narrative carries the right kind of emotional tension, and relief of tension in laughter is somehow pleasurable, or at least satisfies our specific need for the comic.

Laughter itself signals a rebellion against the single-mindedness and habitual perception of phenomena that the non-creative mind is used to. Humorous creativity would mean balancing on the line of intersection of two planes which are traditionally mutually excluding, conducted in an amusing mood. The discovery of the intersection lines, of brilliant and meaningful *bisociation* planes requires imagination and innovativeness. The element of surprise and aesthetic satisfaction with novelty of the so-far unknown associations lies at the core of philosophically analyzed humor.

Following Koestler, we can refer to the scheme of humor analysis as follows: first, one should discover "the type of logic, the rules of game which govern each matrix" (usually only implied and hidden); then, find "the 'link' – the focal concept, word or situation which is bisociated with both mental planes; last, define the character of the emotive charge and make a guess regarding the unconscious elements that it may contain".[11] In this context the Freudian idea of humor expressed in jokes satisfying our unconscious and suppressed desires may be again called in.[12] For Koestler, the emotional charge of humor is usually of an aggressive character, though laughter typically alleviates the destructive emotions.

The bisociative clash finds its expression in emotive laughter, thereby marking the breach with – and release from – the cultural and familiar *status quo*. As Lawrence Kimmel puts it: "(…) laughter is manifestly a release, however brief, from time, from law, from obligation, from sense and sensibility, from suffering and death, indeed from the felt burdens of life (…). It begins in the sweeping release from all that has preceded, from everything that has built up into whatever constrains and restrains…"[13] Defined as a release from the dictate of the familiar and superficial, laughter is an expression of the free spirit. Accordingly, humor opens the way to a deeper

[9] Cf. Helmut Plessner, *Laughing and Crying: A Study of the Limits of Human Behavior*, trans. J. S. Churchill and Marjorie Grene (Evanston, IL: Nortwestern University Press, 1970).

[10] Arthur Koestler, op. cit., p. 56.

[11] Ibidem, p. 64.

[12] Cf. Sigmund Freud, *The Joke and Its Relation to the Unconscious*, op. cit.

[13] Lawrence Kimmel, "Philosophy, Literature, and Laughter. Notes on an Ontology of the Moment", in *Enjoyment: From Laughter to Delight in Philosophy, Literature, the Fine Arts, and Aesthetics*, ed. Anna-Teresa Tymieniecka (Dirdrecht/Boston/London: Kluwer Academic Publishers, 1998), pp. 176–177.

existential experience of the so-far unfelt core dimension of life as such, hidden for the commonsensical and restrained eye.

Considering the problem of a criterion by which the aesthetic "depth" and value of the comic are to be measured as contrasted to the tragic, Anna-Teresa Tymieniecka notices: "Undoubtedly, it is in experience of the receiver, first, and of the author second, that we have to anchor our inquiry into the profound roots of this contrast (…). In the spread of various kinds of comedy (…) it is not to reach the deepest feelings and to crystallize them in a personal subliminal experience (…), but it is to slide over the surface of things, selecting those which are most striking in order to present them to us in an enjoyable, entertaining way. It is enjoyment, mirth and laughter that the spectator expects from a comedy".[14] This interpretation allows us to analyze the whole comic context with its creative impact in the aesthetic terms proper, which would take into account the author, "the work of humour/the comic" itself, and, finally, the recipient – the "subject of humorous/comic perception". In other words, we shall attempt to employ the scheme of Ingardenian aesthetic situation to phenomenologically interpret the phenomenon of humor as an example of human aesthetically valuable creative performance and its reception.

The author of a comic presentation/a joke can be described in terms of the Ingardenian artist as a creator, who in his intentional act expresses physically (in speech, writing, drawing or gesture, treated as material carriers) an original bisociative juxtaposition of ideas or perceptions. The bisociative situation should be, in turn, "deciphered" on the part of the recipient within his specific experience, in another intentional act, characterised by the aesthetically comic enjoyment. Thus the "comic or humorous performance" should posses special "humorous" artistic qualities that within the experience of the spectator should be actualized as aesthetic values of the comic, effectively satisfying his need for amusement. The actualization of the "humorous" aspects of a given performance would require a kind of "a comic competence" or "a comic sensitivity" as well as an enjoyable mood. Such a specific attitude differs significantly from the one in which tragedy is perceived. As Anna-Teresa Tymieniecka remarks, the comic is more appealing to the public at large than tragedy, so it can be more easily received as the so-understood aesthetic phenomenon, and does not require so much subtleness as tragedy does.[15]

The roles of the author and the recipient are likewise described in the Koestlerian theory, according to which "the comic derives from a creative act of the mind, reflected and recreated in the hearer or spectator".[16] The acts of humor, although basically rooted in the creative mind of the individual, exist – similarly to the works of art – within social space. They are expressed physically, and evoke the active engagement of the recipient. They are designed to induce a special type of amusing

[14] Anna-Teresa Tymieniecka, "Introduction", in *Enjoyment: From Laughter to Delight in Philosophy, Literature, the Fine Arts, and Aesthetics*, op. cit., p. xi–xii.

[15] Ibidem.

[16] Marcella Tarozzi Goldsmith, "Nonrepresentative Forms of the Comic. Humor, Irony, Jokes", in *Enjoyment: From Laughter to Delight in Philosophy, Literature, the Fine Arts, and Aesthetics*, op.cit., p. 29.

perception and re-creation within the experience of the spectator, while the character of the comic perception is to be affected by the spectator's own sensitivity and knowledge, as well as his cultural awareness and – last but not least – the mood in which the act of perception is carried out.

There is, however, another aspect of the comic reception. Modifying Coleridge's famous formula, Maurice Charney points out that it is "the involuntary suspension of disbelief that constitutes the comic faith".[17] The indulgence in the "humorous situation" which commonly finds expression in laughter is here linked to the overcoming of the natural defences against the absurdity of the presented *bisociation*. Thus the indulgence in the shocking and unusual combinations of the comic would be strongly associated with the suspension of commonsensical disbelief. A strict single-minded and sober approach would make it impossible to perceive the brilliant novelty of the apparently unbelievable comic. Creativity is here related more to comedy as a literary genre and the comedian's performance, and is considered in terms of various illusions created by comic characters or comedians (omnipotence, unbounded energy, autonomous language, and perfect timing).[18] Yet, it can also be associated with any aesthetic situation in which a specific amusing fictional situation is constituted by the author of a joke.

In the Ingardenian theory, the aesthetic concretization is determined by the personality of the recipient, the cultural atmosphere of the epoch and, above all, by the work of art itself. Linking the spectator with the work of art, it necessarily transcends the work itself, becoming the aesthetic object only when it is revealed in concretization.[19] So similarly "the work of humorous performance" assumes its aesthetic and meaningful values only when it is concretized within an individual experience of the receiver, in which places of indeterminacy become filled with the content compatible with "the humorous work" itself, yet at the same time bearing the unique trait of the recipient's imagination. Such a concretized "aesthetic comic object" is transcendent in relation to both "the humorous work" or "the humorous performance" and to the experience of the aesthetic subject of a comic perception. Thus concretizations of the same amusing performance could differ one from the other, being at the same time adequate to the original, or at least rendering the essential comic fundamentals of such performance.[20] It is only when the concretization is faithful to the original that the artistic values of the perceived performance within "the comic situation" are manifested in the form of aesthetically amusing values in the experienced (and constituted) object of "the comic experience". All in all, concretization involves creative acts on the part of the spectator, though within certain limits delineated by the perceived "work of humor" itself.

[17] Maurice Charney, "Comic Creativity in Plays, Films, and Jokes", op. cit., p. 34.

[18] Ibidem.

[19] Roman Ingarden, *The Literary Work of Art,* trans. George G. Grabowicz (Evanston, Illinois: Northwestern University Press, 1973), § § 64, 68.

[20] Ibidem, § 64.

The phenomenology of life emphasizes the deeper existential message of the comic as conveyed by the comic enjoyment. Anna-Teresa Tymieniecka notices: "Life enjoyment carries the dynamic unfolding of life (...) in all circuits of its creative unfolding (...)".[21] In this vision, the comic creativity and its reception points to its cosmic foundation with its creative processes. In the symbolically understood comic, enjoyment reveals its graded perspectives of creativity – also in the process of unfolding the life-rooted bases: "from the most primitive, physiologically conditioned laughter, through mirth, joy, amusement etc. to the 'smile of the soul' and the swing of the human spirit towards the sublime".[22] The role of humor and comic enjoyment would have its part in the vast process of vigorous and creative unfolding of life forces, though with various degrees of intensity and sublimity.

The expression 'smile of the soul' orients the discussion towards the crucial aspects of the highest ranks of humor. This type of 'smile' would imply involvement of the most subtle and intrinsic human faculties in the creation and reception of comic situations in terms of symbolic and even metaphysical meaning. Within such an approach, even the seemingly trivial can carry metaphysical intimations.

Let us refer to the Ingardenian model of the aesthetic situation again. Ingarden reflects on the so-called "metaphysical qualities" that can be found in the higher ranks of artistic phenomena, including the most valuable comic ones. The problem of the specific metaphysical burden of humor requires a deeper insight, and the aesthetic values related to the comic as well as the creative originality of humor clearly play an important role in the process of such manifestation of metaphysical qualities. In a similar way, the aesthetically valuable qualities enable the revelation of metaphysical qualities through the work of art within the Ingardenian model.

Referring to the category introduced by the author of *The Literary Work of Art*, we can conclude that metaphysical qualities can also accompany specific "acts of deep humor" as their specific "areola". The artistically presented situations themselves would require adequate concretization by the observer in order to become fully revealed. Metaphysical qualities as such disclose that which for the commonsensical mind remains inaccessible and mysterious; moreover, they themselves lie at the source of life and being.[23] Revealing themselves within a given situation, such qualities are at the same time a way of disclosing the symbolic truth about the world. The humorous situations may be then treated as revelatory, transcending their own limits, and in an *à rebours* way referring to life and being as such. However, as it is primarily the case with art, the experience of metaphysical qualities of the comic endeavours differs from their experience in real life. First of all, humor allows for a distance towards the clashes of bisociations that are intimated as symbolising existential or metaphysical contradictions lying at the

[21] Anna-Teresa Tymieniecka, "Introduction", in *Enjoyment: From Laughter to Delight in Philosophy, Literature, the Fine Arts, and Aesthetics*, op. cit., p. xiii.

[22] Ibidem.

[23] As an example of such metaphysical quality Ingarden cites the grotesque. Cf. Roman Ingarden, *The Literary Work of Art*, op. cit., § 48.

core of life's creativity. If viewed as distanced in the artificially evoked situations of the comic, such contradictions are not painful, and their distant revelation (or constitution) enables one to experience both intellectual and emotional puzzlement and enjoyment. On the other hand, if perceived as existing intentionally in a given humorous situation, they can enrich our exuberant experience of life with its mysteries.

The deepest meaning of the humorous amazing acrobatics lies in its existential and even metaphysical implications, which themselves are rooted in semi-artistic originality, creativity and novelty. In conclusion, we can repeat, after Lawrence Kimmel, that laughter (alongside humor) means "a purity of acceptance and unerring affirmation of existence (…)" in which we are "loosed from the fetters of time".[24] The momentous act of creativity beyond time would bear its obvious – though much lighter and perverse – parallel to mystical glory. The affirmation of existence rooted in the freedom from a commonsensical and superficial layer of living, with its creative manifestations of novelties and brilliances, outlines the unlimited philosophically intimated horizon of humor in a sudden clash of contradictory planes of existential reference. Such a clash can bring forth a modified and joyful mood, creative transformation of our attitude towards life – even if lasting in its fullness only for one glorious moment of amazing comic insight.

---

[24] Lawrence Kimmel, "Philosophy, Literature, and Laughter. Notes on an Ontology of the Moment", op. cit., p. 181.

Educational Paradigm Shift Towards Phenomenological Pedagogy

Kiymet Selvi

Abstract The main aim of the educational system is to help individuals improve their own learning to deal with searching suitable ways of improving learning within the educational context. Learning is one of the most important issues in development of individualistic and societal bases. The current education system is based on a positivist paradigm which states that objective knowledge is the best knowledge of our lifeworld and it must be transmitted to younger generations. It is known that the objectivist ideas introduce only the scientific methods of searching for and reaching meaning about a phenomenon. The current educational system also mainly focuses on cognitive development of individuals as if it is the only characteristic of them. These are creating some barriers to free search for and creation of authentic knowledge. In the current education system, there are lots of limitations such as taking attention only to social constructivism and social learning, cognitive development, learning outcomes that rely on an exam-based performance evaluation system for students. However, besides cognitive development, individuals need to develop as a whole, that is as a social being, a creative being, an emotional being, a physical being and an authentic being.

Since the current paradigm has many limitations, a shift in the paradigm is likely to emerge and a new paradigm will be important in the near future. A new paradigm should take into account individualistic intuition, perception, imagination, creation and construction of knowledge. Moreover, a new paradigm should take attention to support the individual as a social being and an authentic being as well as an intellectual being. In this context, phenomenological pedagogy will provide some explanations about educational paradigm shift. Thus, a new educational paradigm needs to be discussed very broadly in many aspects, especially in institutional education contexts by educators, teachers, students and interested people.

K. Selvi (✉)
Faculty of Education, Anadolu University, Eskisehir 26470, Turkey
e-mail: kselvi59@gmail.com

A.-T. Tymieniecka (ed.), *Phenomenology of Space and Time: The Forces of the Cosmos and the Ontopoietic Genesis of Life: Book One*, Analecta Husserliana 116,
DOI 10.1007/978-3-319-02015-0_20,

Introduction

This paper proposes discussion of paradigm shift in education. First of all, the question of "what is the meaning of paradigm shift in education?" needs to be answered in this context. This question is very important in understanding the meaning of the concept of paradigm, the current paradigm of education and paradigm shift in education. The concepts of paradigm and paradigm shift must be clarified at the beginning of this discussion. The concept of paradigm should be explained briefly to understand the necessity of paradigm shift in education. Main features of the current education paradigm should be analyzed to comprehend the problems of it. Then the reasons or roots of paradigm shift in education should be discussed in detail. The paradigm shift in education must be discussed extensively so that the whole education system should be regarded.

It is known that a paradigm includes main approaches and acceptances or viewpoints related to any system or field as well as science, social system or individualistic base. Preferences determine main paths for not only a paradigm but also individuals and systems. A paradigm covers many aspects such as beliefs, values, rules, concepts, feelings, viewpoints, preferences, tools, ways, technics, methods, approaches and theories related to any field. A paradigm preferred by someone or some systems helps establishment of common interest about any topic or field especially. A paradigm introduces the same pathways for questioning, understanding and explaining phenomenon. It is very close to philosophy and many philosophies turn to paradigms in many cases. Philosophies of education project the direction of an educational system. It means that the paradigm of education is mainly clarified by means of philosophies of education. Thus, the paradigm of education is based on philosophies. Philosophies discuss every aspect of education such as learning, teaching, importance of knowledge, acquisition of knowledge, learning environments, learning-teaching processes and the main feature of students, learners and teachers.

It is necessary to clarify the concepts of paradigm shift and reform studies in education. Some reform studies are referred to as paradigm shifts, but reform studies cannot be referred to as paradigm shifts. For example changing students' performance system in education may be defined as a reform in education. It may be a radical change but may not affect the whole system. If reform studies affected the whole system, they could be mentioned as paradigm shifts. Otherwise, they could only be called changes in education. Thus, a reform in education is a kind of change, and does not correspond to the meaning of paradigm shift. Although a paradigm shift in education is based on reformist ideas, it does not have the same meaning with a reform in education.

Main Features of the Current Education Paradigm

The current educational system mainly focuses on the cognitive development of individuals by means of applying behavioral constructivist and social constructivist approaches in the learning-teaching process. Constructivism has been the main approach in education context over five decades. This approach is called constructivism but may not allow students behave as real constructivists while studying phenomenon. Students behave constructively just in terms of reconstructing current ready-made knowledge. Therefore, it is known that individuals have some problems while creating and constructing their self-knowledge in the current constructivist approach. The current system attaches importance to transmission of knowledge to younger generations. Thus, it has some limitations to free search and creation of authentic knowledge by individuals. Some examples of limitations in the current education system are only paying attention to social constructivism and social learning, cognitive development, learning outcomes depending on exam-based performance evaluation system for students' learning. The current system regards social constructivist ideas rather than creation and construction of individualistic meaning. According to the social constructionist point of view, individuals can create meaning by means of interacting with other individuals (Rimmel and Diedrich 2000). It means that individuals create meaning under living conditions in their social environments. But, it is known that constructivism is a very complex issue that consists of internal and external dimensions. There are lots of questions related to creation and construction of knowledge in current education system. For example, which knowledge comes from outer and inner worlds of individuals? How can individuals create knowledge? How does the individual mind perform while creating and constructing knowledge?

There are many gaps in the current education system and they create many problems. All problems of the current education system cannot be discussed in this paper. But, the main features or characteristics of the current education system and most of the problems connected with these features are mentioned very briefly in this paper. The main approach and main acceptances of the current education paradigm are still based on the teacher-centered approach. The current education system tries to apply the learner-centered approach in an education context. Nevertheless many problems emerge while applying the learner-centered approach in the learning-teaching process because the main tools of the learning-teaching process such as learning-teaching methods and techniques, teachers' roles and responsibilities, number of the students in the classrooms, textbooks, and information and communication technologies are suitable for teacher-centered approach rather than for the learner-centered approach. The current education paradigm basically aims at transmitting knowledge to students. The current education paradigm is based upon three important features. Firstly, the current system prefers teaching to learning. Secondly, it approves teacher-centered approaches instead of learner-centered approaches. Thirdly, it is related to transmission of ready-made knowledge to younger generations. In other words, the current education system prefers the topic-centered approach rather than free creation and construction of knowledge through self-inquiries of individuals. The current paradigm must change beliefs and knowledge related to the concepts of learning and teaching by means of changing curriculum developers, instruction designers, teachers, students, parents, textbook writers, and the mass-media. It should also change the whole education system including pre-school, primary school, and high school and higher education systems. A reform only in secondary level education is not a paradigm shift. Paradigm shift must cover the whole educational system.

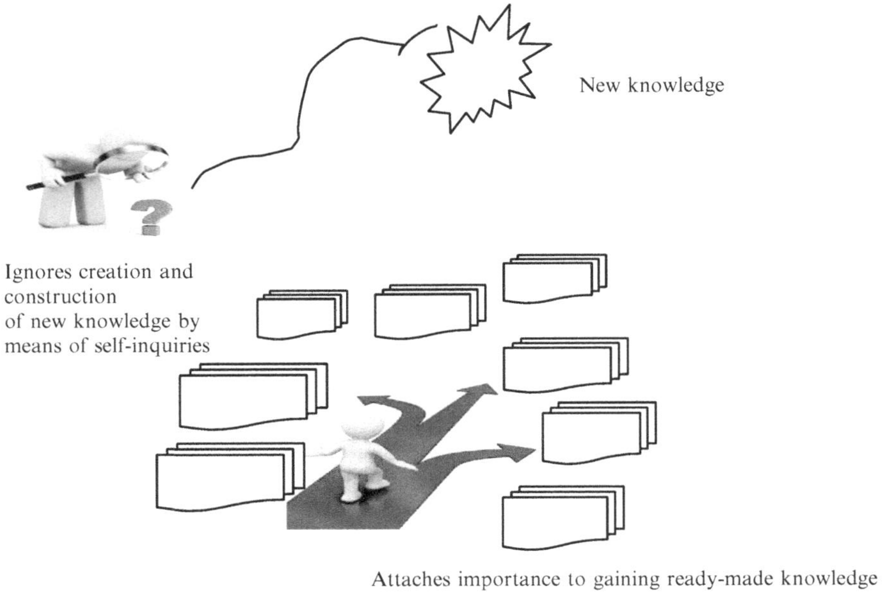

Fig. 1 The current education paradigm prefers reconstruction of meaning by means of learning ready-made knowledge

In sum, the most important feature of the current education system is that it ignores creation and construction of new and authentic knowledge by individuals. The current system is based on reconstruction of ready-made knowledge by individuals. It means that the current paradigm mainly attaches importance to gaining knowledge as seen in Fig. 1.

In Fig. 1, it is clearly seen that the current education paradigm is mainly based on learning ready-made knowledge. This paradigm ignores creating and constructing new knowledge.

Reasons for Education Paradigm Shift

Current educational paradigm shifts are due to many reasons such as demands of scientific and technological developments, demands of individuals and demands of society. Educational paradigm shift has certain indicators. In this paper, nine important indicators are mentioned that relate to the main reasons for an education paradigm shift. The *first important indicator* is related to the dominant pedagogical approach based on transmission of ready-made knowledge to younger generations. The *second indicator* is based on perfection of the current system without perfection of self. The *third indicator* is the fact that the current system defines an individual as a knower despite individual known at the same time. *The fourth indicator* is that the current system approves instruction or teaching rather than learning. The *fifth important indicator* is related to crises of scientific research methods. The *sixth indicator* is that rapid development of information and communication technologies accelerate the education paradigm shift. The *seventh indicator* is that the current system attaches importance to institutional education. But, lifelong learning becomes important besides the institutional educational paradigm. The *eighth indicator* is related to curriculum design in education. The *ninth indicator* is about teachers' roles connected with students' learning. All of these indictors may trigger a paradigm shift in the educational system.

The first important indicator is related to the dominant pedagogical approach in the education system and based on transmission of ready-made knowledge to younger generations. With regard to students' learning, ready-made knowledge may create many crises. For example, memorization of ready-made knowledge, pacifism of students in the learning-teaching process and textbooks become important tools for learning and teaching in school whereas mature ideas and positivist research results based on these ideas become the main sources of knowledge. The main ideas of the current education paradigm can be created by expert or mature people or geniuses and intelligent people. Some educational institutes become main sources of creation and construction of new knowledge. This idea refers to disciplinary knowledge that is mostly positive and created and constructed based on positivist research methods. Institutional sources and disciplinary bodies of knowledge ignore individuals as knowledge creators by means of individualistic search for authentic meaning. The current educational paradigm approves individuals' passive reception of knowledge served by some sources for them. However, a human naturally becomes a creator of knowledge rather than a passive receiver of it. They feel the need to behave freely while searching for new meaning and to attempt freely while creating, constructing and acquiring a disciplinary body of knowledge and their own authentic knowledge. According to Von Glasersfeld (cited in Phillips 1995: 8) "the notion that knowledge is result of a learner's activity rather than that of passive reception of information or instruction… ". Similarly, Dewey (1907) stated concept of learning as activity of learners.

The individual can behave as a creator at the beginning of a search for meaning of phenomenon as seen in Fig. 2. He/she must start a new inquiry about learning topics. At the beginning phase of a search for meaning, the individual ignores ready-made knowledge. The cycle of a search for meaning begins with a new search of individual, continues with individual creation and construction of new knowledge, and then reconstruction of new knowledge based on ready-made knowledge follows. This cycle must follow this path so that the individual becomes a creator of new and authentic knowledge as seen in Fig. 2.

It means that the current paradigm must shift from teaching to learning. The new paradigm must be based on learning rather than teaching.

What is indispensable is the knowledge of the circumambient world of life to which individuals have adjusted in their life courses along with self-recognition of personal talents, capacities, and predilections. Knowledge can be a strategically important power for individuals who live in complex and changeable living

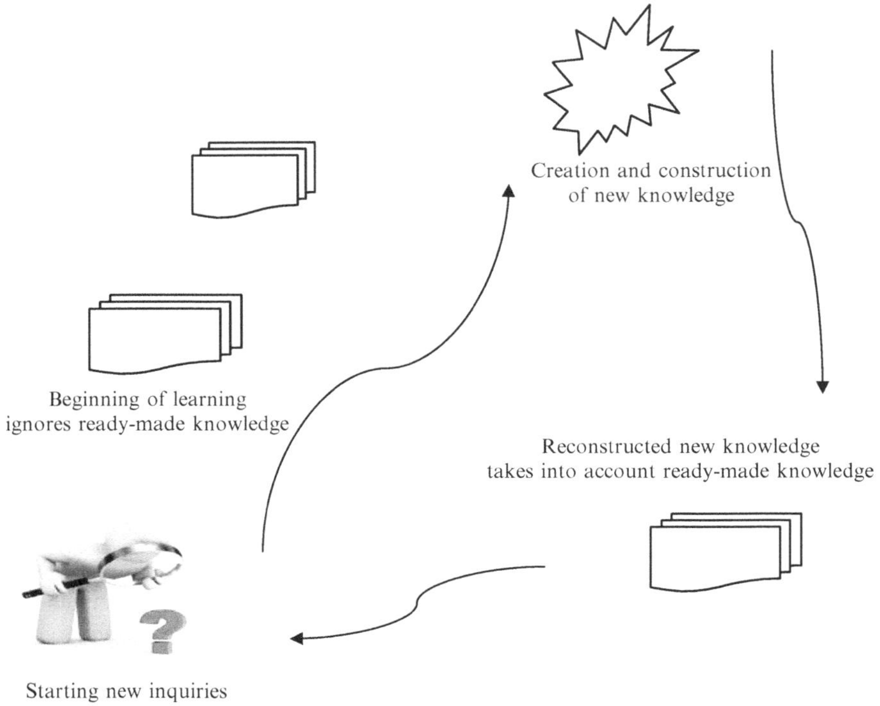

Fig. 2 Paradigm shift to creation and construction of knowledge by means of self-inquiries

conditions, acquire roles, responsibilities and signification of life. Knowledge refers to those individuals who become creative and authentic beings by means of creating and constructing their own knowledge through behaving as meaning makers. Individuals not only gain ready-made knowledge but they also create knowledge based on their own individualistic experiences. It is clear that individuals' learning may be seen and defined as just acquiring knowledge from others such as teachers, books, and mass-media. It is necessary to discuss and answer following questions, such as "what we experience as learners", that create inherent risk of human learning (Smeyers and Hogan 2005). The question is directly related with education paradigm and has many possible answers is connected with your understanding of the concept of "experience." Because of learning experiences, the learning-teaching process and the whole educational system can be designed depending on an understanding of the concept of "experience." According to the current paradigm, the concept of experience corresponds to mature experiences and ready-made experiences of others. Individualistic experiences can be classified into two groups as "imaginative experience" and "real-life experience" (Selvi 2006). It is clear that in a formal learning system, both types of experiences from creativity studies at school cannot exist.

The second indicator of a shift in the current educational system is based on perfection of the current system without perfection of the self. Knowledge which comes from inner worlds of individuals and forms important reasons for crises of individuals' learning is neglected in the current educational system. "It may well be as a result of this lack of understanding of the inner world that peace and happiness are still a far cry from manifestation in the world we see around us" (Grassom 2008: 80). A long period of human education has been based on knowledge only about outer world. It means that knowledge from the physical and material world has formed a persistent and sincere attempt for all educational activities. Ignoring knowledge of the inner world may create many barriers to actualization of self as a whole being. Self becomes happier if she/he as a self has got a chance for full development of her/his capabilities. The self cannot be happier in the education system because education ignores development of her/his full capabilities. However, for a long period of time educational studies have focused on educating individuals depending on perfection of educational system. Nevertheless, it is obvious that the educational system cannot reach perfection through perfection of curriculum, tools, methods, techniques without perfection of individuals. In other words, the focus must be perfection of individuals in the education system. Following perfection of individuals, educational system gradually becomes perfect. The current educational paradigm is established upon perfectibility of a system without perfectibility of individuals. This creates crises related to current educational paradigm. A paradigm shift in education must take into account the whole development and perfection of individuals.

The current educational paradigm mostly pays attention to behaviorist and social constructivist approaches. In the current education system, these approaches may cause lots of limitations, such as only paying attention to behaviorism and social constructivism, cognitive and social learning, learning outcomes, exam-based performance evaluation, and teachers' evaluation system for students' learning. The social constructivist approach may not be fully implemented in the current education system. However, it helps students and teachers to change their understanding of learning and teaching that corresponds to acquisition of ready-made knowledge of others. This can be interpreted as the first steps of change in educational paradigm already in use. Partial implementation of social constructivism in the education system is not enough for students' learning. Social constructivism is important in terms of creating new meaning while studying at school. For more than three decades, social constructivism has come into prominence in the formal education system and insists on learner-centered implementations in education context. Demands of implementation of social constructivism must become signs of current paradigm shift in education because social constructivism focuses on search for meaning by means of individualistic attempts in the learning-teaching process. It is known that social constructivism, which claimed that individual experiences might lead to acquiring subjective and authentic meanings of phenomenon, has become dominant in individuals' learning. Individualistic base mainly comes out of individuals' inner speeches connected with their authentic meanings. Social constructivism gives individuals a chance to bring out and catch authentic meaning of phenomenon.

The third indicator of paradigm shift in the current system defines the individual as a knower. But the individual projects himself/herself as "knower" and also identifies himself/herself as "known" in many cases. When the individual knows about phenomenon, he/she as a knower and a known includes outer and inner worlds of him/her. However, the current educational paradigm approves the individual only as a knower. In this case the individual may accept only the knowledge received from the outer or physical world. But, it is known that the individual becomes both a subject and an object while gaining knowledge about phenomenon. Individual as a knower means that a student must be responsible only for acquiring knowledge from the outer world. Individual as a known has capabilities and features and must know to design teaching-learning in education system. Students' learning, needs, expectations, intentions, views, thoughts, preferences, ideas, creativities, imaginations must be taken into account during the learning-teaching process. These all refer to creative capabilities and conditions of individuals. Individuals creative conditions and powers (Tymieniecka 2000) must be taken into account in the education system as they become known in education. If students become known in the education system, they learn by means of their own search and learn by doing. They also discover their own creative power from their inner worlds. Approach of the known may become a balance between outer and inner worlds of individuals. According to Grassom (2008: 84), "we can help redress the balance of outer and inner that is displaced by overemphasis on other…". This idea must be put into practice in an education context but the current educational paradigm ignores the individual as a knower.

Individualistic creative acts of a human include experiences of the individual as a knower. It is clear that the current education system neglects the individualistic dimension and attaches importance to the institutional dimension. Individuals need to freely reflect their own creative experiences but the current education system cannot regard their free reflections. Tools and ways of reflection allow only students' reflections related to learning outcomes of ready-made knowledge acquisition. Answering teacher questions is the main tool of reflecting students' learning. Authorities designed and decided about the way of students' reflection and students mustn't choose or design tools of reflecting their own experiences. In the last decades, some alternative ways for free reflection of students have been tried to be alternated in an education context. However, this attempt seems to be unsuccessful because of general acceptance about students' reflection.

The fourth indicator about paradigm shift is that the current system approves instruction or teaching rather than learning. Herman Miller (2006) in higher education undergoes a paradigm shift from instructional paradigm, in other words teaching paradigm, to learning paradigm. Herman Miller discussed the shift at the higher education level but, the whole education system and all levels include this shift. Paradigm shift in learning promoted some new discussions about concepts such as lifelong learning, learning to learn, learning styles and strategies, learning autonomies and self-directed learning. All these redefine learning and also attach importance to learning instead of teaching. Teaching refers to teacher-centered implementations; content-based learning, institutional teaching and a learner becomes only a learner for the teaching process. The teacher must organize a whole teaching-learning process and instruct main tools for teaching and learning.

Another important indicator of educational paradigm shift is related to crises of scientific research method and promotes paradigm shift in scientific research methods. The positivist research method has been mainly applied by means of quantitative research methods for long time. Scientific research has provided immense development during this period but, reaches a limit that is the end of rapid development phases. Yet, many questions cannot be replied to by means of the quantitative research method. Quantitative research methodology to some extent restrains research about phenomenon. Phenomenon needs to be analyzed more deeply to discover new knowledge about it. During this analysis, individuals must follow an authentic pathway and also need to behave creatively while creating and constructing new knowledge about phenomenon. But, it is certain that quantitative research method must follow certain research steps. Otherwise, research results must be questioned and may be rejected based on positivist ideas. This tradition should be changed to reach and reflect individualistic meaning of self. An individual can reach true meaning of phenomenon by means of his/her authentic experiences.

Strict positivist scientific research methods such as the quantitative research paradigm get close to the qualitative research paradigm. New paradigm in scientific studies is changing from the quantitative research method to the qualitative research method. Turning point of research paradigm begins when that qualitative research method becomes dominant in scientific studies. This change refers to those positivist paradigm shifts towards naturalistic approaches and these approaches become dominant at present and in the near future. It means that nature of individual such as thinking, perceiving, questioning of phenomenon should reflect the results of qualitative research of individual mind. Furthermore, individuals' inner powers or capabilities such as intelligence, motivation, intuitions, inspiration, and perception can affect creation and construction of knowledge. Authenticity of individual becomes important for scientific studies related to individualistic meaning about phenomenon. For more than three decades, two coherent research paradigms have become popular within scientific researches. In order to accelerate improvement of scientific studies, instead of borders, new horizons are necessary.

The sixth *indicator* is that rapid development of information and communication technologies accelerate a paradigm shift in education. Intelligent technologies are improving and robots have been doing many works perfectly. Technological development affects not only human life but also the whole system of the universe dramatically. For example, changes in seed technology also change food sector and farming. These changes affect atmospheres, earth, seas, health, and so on. Technological improvements directly affect education and lead to tangible results. It means that implementations such as computer-based learning, distance learning, virtual learning and blended learning come out. They create new implementations in education and roles of education also change. The frame of learning-teaching process, responsibilities and roles of students and teachers change depending on use of information and communication technologies (ICTs) in education. ICT leads to revaluation in teaching-learning process, learning environments, learning autonomies, self-directed learning and evaluation of students' performance. Nevertheless, using ICTs in the current education system cannot reach real performance. Although use of ICTs is in a rapid development, access to ICTs is not enough. Using and reaching ICTs cannot be easy for everyone because it is expensive. Moreover, it provides many facilities but also creates many problems in education. Development and implementation of ICTs need to be reanalyzed with regard to current paradigm.

The seventh indicator related to the current paradigm shift is that the current system attaches importance to institutional education. But lifelong learning is becoming important besides the institutional educational paradigm. The concept of lifelong learning is a very important development in education context. Lifelong learning constitutes main roots for individuals, societies and many different areas such as schools, firms, social services, work places. Learning in life process through gaining new competencies, qualifications and skills can be defined as lifelong learning. Lifelong learning competencies of teachers help students to develop self-learning methods. In educational contexts, teachers' lifelong learning competencies can create positive impacts on developing students' lifelong learning skills. Thus, teachers need to acquire lifelong learning competencies (Selvi 2011). In educational context, lifelong learning gives a chance for self-actualization of an individual. Lifelong learning refers to a paradigm shift in education. Learning occurs in a life process and cannot be bound with institutional learning. Lifelong learning creates a "culture of learning" within society. The main problem of institutional learning in human learning is that it only focuses on certain period of an individual's life. But it is clear that learning exists as reality in an individual's life process. The other problem of institutional learning is that students have to study certain topics in a certain time. This is also a very important problem in human life because students may not need and want to study certain topics in certain periods of their own lives. Institutional learning is weakening despite strengths of lifelong learning understanding.

The eighth indicator of paradigm shift in education is related to curriculum design in education. Curriculum design must be based on students' learning needs rather than other sources and powers such as preferences and admirations of teachers, decision-makers and textbook writers. Students' needs must be defined so broadly that they include full development of individuals. Herman Miller (2006: 2) stated that "curriculum design is based on an analysis of what student needs to know to function in complex world rather than on what the teacher knows about how to teach." It is known that teacher's capabilities and responsibilities form the main force for curriculum design. Two-level curriculum designs like macro-level curriculum and micro level curriculum must be taken into account. Macro-level curriculum refers to general planning of curriculum such as primary school curriculum at the national level. Macro-level curriculum must be designed flexible in order to meet students' needs. Micro-level curriculum refers to redesign of macro-level curriculum by teachers. In other words, teachers design micro-level curriculum while redesigning applied macro-level curriculum at school. Micro-level curriculum is similar to the school-based curriculum understanding. Students' needs, learning-teaching environment at school, features of families, living conditions and possibilities and resources of environment are very important in school-based curriculum implementations. Teachers must prepare their own micro-level curriculums based on macro-level curriculums. It means that the macro-level curriculum should be reanalyzed and reorganized by teachers to facilitate students' learning.

The ninth *indicator* of paradigm shift is related to teachers' roles in students' learning. Teachers turn out to be facilitators rather than instructors. It is known that teacher's role is defined as an instructor in learning. And also students' responsibilities in learning, learning only by means of memorizing learning topics, comprise the beginning of profession of teachers. Teachers' and students' responsibilities in the current learning-teaching process seem to be the end of the period. According to new approaches about teachers' roles and responsibilities in students' learning, teachers' competencies mainly refer to teachers become facilitators. Teachers' new competencies such as lifelong learning competencies, ICTs competencies, environmental competencies, research competencies, curriculum competencies must be discussed and redefined (Selvi 2010). Teachers' competencies, roles and responsibilities in students' learning are becoming important for future teachers. Students' roles and responsibilities in their own learning is also a turning point in terms of students becoming self-directed learners. Thus, teachers as micro level curriculum designers must help students to specify and decide about their own goals and learning activities to enable them to become self-directed learners.

Conclusion

Some of the common *indicators* and reasons discussed in this paper trigger a paradigm shift in educational system. All indicators about paradigm shift in education aim at improving individuals' learning. Knowledge and learning become very important agenda for individualistic and societal development. The main problem about the current education system is connected with the approach to learning and teaching and also individuals' ways of learning. It is clearly seen that the main approach and main acceptances of education paradigm must shift and need to transform learner-centered approach. However, actualizing paradigm shift in education is very hard because paradigm covers many aspects such as beliefs, values, rules, concepts, feelings, viewpoints, preferences, tools, ways, techniques, methods, approaches and theories related to education still based on teacher-centered approach. Teachers, students, decision-makers, scientists, politicians, parents have their own beliefs, values, habits and preferences about the current education system. They insist on applying their paradigms and they do not change their paradigms about education. These kinds of individual resistances become very important barriers to shift and change in any system. Many indicators refer to paradigm shift in education but human beliefs, values, habits and preferences may not lead to this shift. Thus, the current education system remains incapable and creates many problems for individuals' learning. For this reason, the current education system undergoes a shift in its own paradigm.

The paradigm shift in education should move from teaching to learning. New learning approaches also insist on changing students' learning habits and preferences. But, the current education paradigm still follows and applies the same paths about learning and teaching at school. This situation leads to conflicts between learners and learning ways approved and implemented in the current education system. Harmen Miller (2006) stated that a new student in an old education system may lead to conflict in institutional learning system. It means that instructional learning follows the same ways while studying with students who are equipped with new learning skills and have new preferences. But, it is clear that instructional learning may not manage adaptation to developed structure of learning and teaching. For example, a learner-centered approach is to be applied in the current educational context. But, a learner-centered approach creates many problems when it is applied in the current learning-teaching process. This is because the main tools of the learning-teaching process such as learning-teaching methods and techniques, teachers' roles and responsibilities, number of students in classrooms, textbooks, and information and communication technologies are designed for a teacher-centered approach rather than a learner-centered approach. All of the sub-systems in the current education system are designed according to a teacher-centered approach and topic-centered approach.

In educational context, most discussions about learning are related to the claim that students' preferences must be at the center of the learning-teaching process. This claim means that subjectivity, individuality and freedom of learning must lead all educational activities at schools. However, teachers may not be able to manage how students' subjectivities come out and are reflected during the learning-teaching processes. Students' individualistic learning preferences should be the main criteria while designing and applying curriculum at school. Thus, a new paradigm must consider self-directed learning and self-actualization in an educational context. Schools must aim at teaching individuals through creating new ways of creating and constructing new knowledge. In this sense, phenomenological pedagogy may provide individuals with methods of acquiring some skills such as imagination, inner-speeches, creativity, self-directed learning and self-organization.

Verducci (2008: 23) stated that "It would seem that the task of education is no longer simply to articulate and develop a paradigm of humanity, known and shared in its essentiality…". Thus, the current paradigm prefers cognitive development of self rather than self-actualization of individuals. It is needed that the new educational paradigm aims at full development of individuals and they must also be meaning-markers while studying in the educational system. In the self-actualization process individuals must feel free to search for meaning, to catch meaning of phenomenon and to be self-interpreters of their own experiences. New paradigm in scientific studies is changing in the direction from quantitative research method towards qualitative research method. It means that the nature of human thinking, perceiving and questioning should reflect the results of qualitative research about individualistic experiences of self. The paradigm of institutional education must change and it should design individualistic perspective that pays attention to self-directed learning and learning to learn. According to Ryan "learning is about learning across life and requires a new perception of education" (2007: 195). Education must put emphasis on developing individuals' skills of learning to learn, that is skills for becoming lifelong learners. Learning about learning or learning to learn is mainly discussed on a theoretical base and it may not be applied to operate self-directed learning for individuals. Self-directed learning is based on individuals' ability of learning to learn. This change is related to crises of positivist paradigm and it is going to be the turning point that naturalistic approaches will be dominant at present and in the near future.

It is necessary to deeply discuss the direction of paradigm shift in education. Paradigm shift requires a shift in philosophies, beliefs, values and preferences. Paradigm shift in education must begin with discussion based on philosophical fundamental that improves awareness of individualistic bases. In order to establish a strong philosophical background of paradigm shift in education, different philosophical approaches must be regarded. It is known that knowledge is getting immense day by day and also these questions are becoming important for everyone. It is necessary to ask a question about what individuals experience while trying to gain knowledge from a persuasive teacher-centered system. It means that the educational system should aim at improving individuals' awareness of creation, construction and reflection of knowledge rather than transmission of ready-made knowledge by a passive receiver. Thus, directions of paradigm shift in education also need to involve new education paradigms such as phenomenological pedagogy. Ikere stated that "phenomenological approach may open new avenues of investigation and serve as a source of inspiration for twenty-first-century educators" (2008: 9). The phenomenology introduces new horizons for the new humanism that is a kind of new enlightenment of human history. Individuals may easily capture knowledge from phenomenon if they are aware of their own learning and freedom of learning. Phenomenological approach includes humanistic approach, anarchic approach and naturalistic approach. Thus a new paradigm must be based on phenomenological pedagogy.

References

Dewey, J. 1907. *School and society*. Chicago: The University of Chicago Press.

Grassom, B. 2008. Beyond knowledge: Questioning the episteme through art's alterity. In *Analecta Husserliana*, The yearbook of phenomenological research, vol. XCV, ed. Anna-Teresa Tymieniecka, 79–86. Dordrecht: Springer.

Herman Miller. 2006. Paradigm shift: How higher education is improving learning. Retrieved from http://www.hermanmiller.com/content/dam/hermanmiller/documents/nresearch_summaries/wp_LearningParadigm.pdf. Accessed 7 May 2012.

Ikere, Z. 2008. Human being as a creative differentiator of the logos of life. In *Analecta Husserliana*, The yearbook of phenomenological research, vol. XCV, ed. Anna-Teresa Tymieniecka, 9–22. Dordrecht: Springer.

Phillips, D.C. 1995. The good, the bad, and the ugly: The many faces of constructivism. *Educational Researcher* 24(7): 5–12.

Ryan, J. 2007. Exploring 'lifewide learning' as a vehicle for shifting pre-service teachers' conceptions of teaching and learning. In *Handbook of teacher education*, ed. T. Townsend and R. Bates, 193–204. Netherlands: Springer.

Rimmel, G., and A. Diedrich. 2000. Knowledge as-action and knowledge development. Paper presented in International conference managerial knowledge between globalization and local contexts, Rome, Luiss University, 15–17 June 2000.
Selvi, K. 2006. Learning and creativity. In *Analecta Husserliana*, The yearbook of phenomenological research, vol. XCIII, ed. A.-T. Tymienecka, 351–369. Dordrecht: Springer.
Selvi, K. 2010. Teachers' competencies. *Cultura: International Journal of Philosophy of Culture and Axiology* VII(1): 167–175.
Selvi, K. 2011. Teachers' lifelong learning. *International Journal of Curriculum and Instructional Studies* 1(1): 61–69.
Smeyers, P., and P. Hogan. 2005. The inherent of human learning. *Educational Theory* 55(29): 115–119.
Tymieniecka, A.-T. 2000. The orgins of life: The origins of the existential sharing-in-life. In *Analecta Husserliana*, The yearbook of phenomenological research, vol. LXVII, ed. Anna-Teresa Tymieniecka, 1–12. Dordrecht: Springer.
Verducci, D. 2008. Education and the Ontopoietic conception of life: A noeanthropological perspective. In *Analecta Husserliana*, The yearbook of phenomenological research, vol. XCV, ed. Anna-Teresa Tymieniecka, 23–37. Dordrecht: Springer.

Human Soul, Body and Life Horizons

Maija Kūle

Abstract Phenomenology of life develops an essential transformation of the positioning of life, human being, soul and life horizons. Human soul reflects the passions of the earth and of the skies. There are two directions characteristic for the soul – upward and downward. Life horizons are closely connected with life forms, styles of living. It has been described in rather different cultures including post-modern culture. This paper deals with A.-T. Tymieniecka's ideas about the New Enlightenment and critique of too narrow an explanation of human subjectivity and body and discusses the need for balance between soul's directions and decreasing of materialistic, consumerism life form, orientations to primitive feeling of the world and Cosmos.

Concept "Soul" in Phenomenology

Concept "soul" is difficult to explain because it is multi-meaningful. From ancient times in old Greek language *soul* means – *psyche.* But it does not mean that today – when we are using the term "psyche" in the psychological sense – it is the same as the philosophical concept "*psyche* as soul". Soul is described as living essence or inner force of living beings, as the distinguishing feature of all living creatures, as the subject of emotional states, force that is responsible for acting, planning, doing. Brittanica online describes "soul" as the immaterial aspect or essence of human being. Soul can be explained as a bearer of moral qualities because ancient poets and philosophers spoke about courage souls, wet and dry souls, responsible souls, rightful souls and so on. The Italian philosopher Angela Ales Bello characterizes

M. Kūle (✉)
Institute of Philosophy and Sociology, University of Latvia,
Akademijas laukums 1, Riga, Latvia
e-mail: maija.kule@gmail.com

A.-T. Tymieniecka (ed.), *Phenomenology of Space and Time: The Forces of the Cosmos and the Ontopoietic Genesis of Life: Book One*, Analecta Husserliana 116,
DOI 10.1007/978-3-319-02015-0_21,

soul from the phenomenological point of view: " [..] phenomenological analyses tend to consider the soul not as a monolithic unit, but as a complex terrain of acts and operations that have also different qualities; some of these constitute the psyche, which has to be referred to everything that we find within ourselves by way of impulses, tendencies and spontaneous assumptions of position that cannot be eliminated, though they can eventually be controlled by a series of free and voluntary acts; since the latter enable us to take decisions, they have peculiar characteristics and therefore form part of a different sphere that is defined as spirit. The psychic and spiritual complex is different from corporeity (bodiliness) and, wanting to use a unitary term, can be called soul."[1]

There are different aspects how soul is characterized at the history of philosophy and culture:

1. As power of life, expression of being alive;
2. As core of sensations and psychological functions;
3. As form of body (Aristotle);
4. As immaterial substance which stays alive after subject's death, it differentiates body from corpse and represents an essence of living [human] being;
5. As God's implementation into the human's body;
6. As consciousness, reason; result of the combined action of the sensual and the spiritual force; soul is an intermediary between physical and the spiritual;
7. As person's essence, expressed by philosophical concept "*me-ness*";
8. As historical and cultural background for changing identity;
9. As poetical metaphor but not something real;
10. As phenomenon of imagination;
11. As empty concept, mistake of philosophy and theology.

Anna-Teresa Tymieniecka in the phenomenology of life recognizes that soul exists, it isn't an empty category but the concept which plays a very important role in the system – Cosmos, Logos, human being. Body is animated by a psyche and enlivened by the spirit. Soul requires the body for its natural existence. An important step for the explanation of soul is attributing to it intellectual activities and ethical values. The philosophical idea of soul helps to unifies mind with body, cognitive with emotional; to see close relationships between perceptions, feelings, reflections, doings; it unifies theoretical thinking with practical doing; thinking with will and evaluation. Soul tends to include all the functions of human beings and represents human development on the basis of going upward, it means – to the higher values and self-development. Soul is associated with feelings like hate and love, joy and grief, shame and anger, honour and dishonour, it means – moral qualities and virtues. Soul stands between life and death, wrote Plato. The same idea we can see today when children are playing computer games – heroes have souls and possibilities to lose them, but today games give to players non-possible potentialities – many souls; never before people had an imagination that the number of their souls can depend on their artificial, playful wish.

---

[1] A. Ales Bello, "The Study of the Soul between Psychology and Phenomenology at Edith Stein", in *Cultura. International Journal of Philosophy of Culture and Axiology*, 8/2007, p. 107.

Phenomenology of Life: Three Movements of Soul

Phenomenology of life is to develop the extended understanding of soul. Tymieniecka develops an essential transformation of the positioning of life, human being, soul and life horizons. The concept "soul" alongsite with the concepts "Logos", "unity-of-all-there-is-alive" and "creative act" are main concepts of Tymieniecka's life philosophy. For phenomenology of life is understandable: human soul belongs to the process of life development on the Earth and self-individualization of human persons. Human soul reflects the passions of the Earth and of the skies. It is not a narrow and subjective characteristic of human subjectivity but involves human beings in the large cosmic circle of creativity. Human existence in the orbit of life is the main problem at the phenomenology of life. She includes soul into the context of Cosmos and demonstrates development of rather different strata of souls as the creative development of the Universe under the expanding of the potentialities of Logos. Tymieniecka shows that the logos of life is an intelligent design of all things and intelligence itself, measure and proportion of all things, in itself – logic, insight, intuition, awareness, sentience. The Imaginatio Creatrix transforms the more primitive stirrings of the human soul into subliminal passions of human existential significance. "Imaginatio Creatrix proceeds from the womb of life and depends on it."[2]

Tymieniecka describes three sense-bestowing functions of Imaginatio Creatrix:

(a) Aesthetic, poetic sense;
(b) Objectifying sense;
(c) Moral sense.

Moral sense is not the product of reasoning, but rather the result of subliminal passions that acquire their moral dimension guided by the benevolent sentiment.

Tymieniecka is interested in the lived soul, including its virtualities, its *élan vital*, soul's structure and unity of besouled human person. Kathleen Haney characterizes the specific understanding of soul in Tymieniecka's philosophy: "Tymieniecka's own examination of the life of the soul focuses on the affective dimension."[3] Affective dimension is expressed in the explanation of soul as power of life; as individualization of Logos; as core of sensations; as representation of human being in the Cosmos. In the book "*Logos and Life: the three movements of the soul*" Tymieniecka describes the extended phenomenology of soul characterizing the soul as the "soil" of life-forces, as the subliminal "soil" of individualized life. Soul as person – it is meaning endowing complex. "This complex (which, in its manifestation we call the soul) is a germinal soil in which the play of the primeval life-forces within the life-schema

[2] A.-T. Tymieniecka, "Creative Imagination in the Converting of Life's Sensibilities into Full Human Experience" in *Phenomenology of Life—from the Animal Soul to the Human Mind*, Analecta Husserliana, Book 2, *The Human Soul in the Creative Transformation of the Mind*, vol. XCIV (Springer, 2007), p. XVII.

[3] K. Haney, "The Three Movements of the Soul in Tymieniecka's Philosophy" in *The Passions of the Soul in the Metamorphosis of Becoming*. Islamic Philosophy and Occidental Phenomenology in Dialogue, vol. 1 (Kluwer Academic publishers, 2003), p. 49.

enters, into generative context with the virtualities of Human condition."[4] Olga Louchakova-Schwartz comments on her understanding of Tymieniecka's philosophy, that mental health is the ontopoietic functionality of the soul. Emergence of the soul creates a differentiation of Self and Other, realizes principle of unity-of-all-there-is-alive and principle of individualization. Soul allows human being to reflect about his or her trans-empirical experience, striving for the beautiful, truthful, just, infinite and the Absolute.

The theory of soul has been developed in the context of the New Enlightenment, it means new understanding of human being, his or her reason, unifying imagination and creative aspects. Reason therefore is inseparable from life process. New ontology for the phenomenology of life means unity of knowing and being. This understanding counteracts the post-modern loss of meaning. New Enlightenment emerges in the post-post-modern time, it involves networks of life; ontological self-poiesis of life, unification of reason and intuition. Tymieniecka recognizes that postmodern Self is endangered Self with disintegration of identity, based on self-fragmentation and disappearing of the unity of soul. She does not focus on self-fragmentation, for her central category is LIFE, not existence; but beingness as process and time; logic ontopoiesis, not time as structure which categorizes existence. For her life times itself; constitution of worlds and others is an opportunity for the creativity. Logos is directly intuited within a phenomenological horizon of life.

Soul: Life and Death

Sixteenth-century French essayist Michel de Montaigne wrote: "As we are born we die, and the end commences with the beginning. All the whole time you live, you purloin from life, and live at the expense of life itself. The perpetual work of your life is but to lay the foundation of death. You are in death, whilst you are in life, because you still are after death, when you are no more alive; or, if you had rather have it so, you are dead after life, but dying all the while you live; and death handles the dying much more rudely than the dead, and more sensibly and essentially. If you have made your profit of life, you have had enough of it; go your way satisfied."[5] Recognizing the uniqueness of birth and death people have always been baffled by the body.

In contemporary Europe, however, there are but few who would consider death a part of life. People try to avoid the concept "soul". Artists, novelists, poets, priests and philosophers might form the exception.

Traditional Europeans admit dying of diseases and do not see the connection between soul, attitude to death and life. Thus, people blame doctors for death and

[4] A.-T. Tymieniecka, *Logos and Life: the Three Movements of the Soul: the Spontaneous and the Creative in Man's Self-Interpretation-in-the-Sacred* (Springer Verlag New York, 1988), p. 8.

[5] M. de Montaigne, Essays. Vol. 1, http://oll.libertyfund.org/simple.php?id=107#chapter_20794 20794

diseases and do not take them to be part of life and its final stage. Every achievement of medical science gives rise to a new variety of death, a new mutation. Death adapts and changes just like a virus. The moral philosopher Hans Jonas admits: "Death does no longer seem to be an innate necessity of live nature, but rather an erroneous organic creation that could be avoided – certainly to be discussed and depicted in detail."[6] The connection between soul – as essence of living beings– and death in many post-modern cultures has been lost dramatically.

People usually avoid thinking about the grievous questions of life and death, or else they just believe religious doctrines. Life, death, evaluation of one's lifetime that are essentially philosophical and ethical problems in life are more and more often treated with soulless practicality, shameless greed and industrial functionalism. The reasons for it are to be sought in a distant past – when life started to be regarded as a sum total of separate parts and human being – mainly as a physiological mechanism and a social role player; when economy, finances and politics came to the fore in the social life of society. It does not mean that the New Enlightenment has come. The other way round – we can almost with certainty state that life is no longer sacred, belief in soul's existence has been lost, death does not inspire reverence and man's life can quite often find no satisfaction. Body, soul, spirit have lost unity. In the contemporary secular life form death does no longer mean the beginning of life after death or the meaningful summary of life, but an inability to plan something in the future. There is not a concept of soul. In value degradation conditions sufferings become more multiform and longstanding; increase egoistic, insensitive and indifferent attitudes towards the life of others.

Value degradation comes to light most clearly in the attitude towards the great mysteries of being – birth, life process and death. If death is not regarded to be a tragically sacred, unique event (if there is no thought that the death of every man is death of a whole world), but just statistics, an everyday occurrence, then in the morgue corpses can be mixed up and relatives handed out a stranger, as has already been the case in some developed states. Then victims of an accident can be calmly robbed because they are taken to be people just like all others, only they do not move and cannot slap the thief's hand. Idea of soul disappears.

It is strange, but for the majority of people views on death have simplified thus far that in their eyes death does not accord a different status – inviolability, eternity and liberation from any social roles or relationships. In Soviet socialistic countries it was fashionable to remove the burying places of the soldiers killed in World War II: the remains were dug out by excavators and then dropped in the earth again. Killed soldiers continued to play social roles after death and became part of ideologized parks. Nowadays there is also a wish to settle accounts with the dead people as if after the death they were continuing to perform their political and military mission. The social, political and military role outlives man. But the idea about human soul does not. That is the paradox of post-modern society.

---

[6] Hans Jonas, *Das Prinzip Verantwortung*, (Frankfurt am Main: Suhrkamp, 1984), S.48.

Death has assumed the meaning of a socially practical, financial and medical phenomenon losing to a great extent the existential and life-completing sense.[7] It is an unfortunate event, as it were, (one cannot exclude medical errors, of course), an unexpected moment (how could it, in principle, be unexpected if everybody knows the absolute truth that man is mortal?). People have lost feelings of the dialectics of life and death. Ancient and classical philosophers closely put together life and death based on the concept of soul. Today living is interpreted as self-sufficient phenomenon for long time because lifetime in the post-modern opinion depends on new projects in medicine and technologies which will stop aging.

In post-modern societies living life means that you repeatedly find yourself concentrating on unessential, separate events, a succession of obscure processes in which man's birth and death are not accorded a fundamental meaning. Death is connected with hospitals, medicines and body weakness but not with the life one has lived being ensouled.

Most people avoid and experience fear of those who are on the deathbed. The usual tactic is to send an old and ready-to-die man to a medical establishment for care and cure. In old age man is much more than in youth at the mercy of medical manipulations. Honour and praise to medicine undertaking to conquer death! However in its deeper sense it is unconquerable and demands its share of love. In modern Europe death comes to hospital wards, in loneliness, without the intimacy with children, without summing up one's life, without soul.

It is only the Christian church that still keeps valid an invitation to confession, the nearness of the priest, deference to death. In worldly life deference to death has been lost.

Why is death taboo in contemporary life? Nobody wants to speak or write about it, its image appears either in a commercial or a perverse form (as a mishap or a killer's victim). Love has lost its spiritual magic and also its tragic component. It is mostly reduced to sex as a technology of life or to entertainment. Medical technology produced pregnancy prevention means enable one to separate sex from the life continuation mission.

Contemporary philosophy is looking for an answer to the question why in the classical life form people could have deeper and more durable feelings (and also conflicts and contradictions). Now deep feelings and durable faithfulness arouse suspicions: how come man cannot comply with the style of the age, what is the reason for it – an illness, "an old-fashioned upbringing" or stupidity. Sympathy and help without recompense is usually regarded as an exceptional case and reasons are sought to understand what has made people act like that.

Death is being commercialized. Just think of the number of firms, offices, enterprises that gain profit from death! Selling places for burial is profitable and of late a still better business has appeared in European Union countries – selling small coffins, burial places and small monuments to pets. Death that is not profitable and does not give rise to a journalistic intrigue is of no interest.

---

[7] See: J. Dollimore, *Death, Desire and Loss in Western Culture*, (London and New York: Routledge, 2001).

Commercialization of death and its direct submission to medicine raises apprehensions as to one-sidedness in the understanding of life. People's death becomes a commodity in modern industry. My country neighbour says: "Now you can earn more making coffins than making beds." And he is quite right.

Death is manipulated with, especially in respect to those who are inferior in power, significance and the ability to oppose. The interconnection between enlived soul and end of life has been lost in post-modern cultures. It has been reflected in many contemporary philosophical trends which do not reflect upon the concept "soul".

Soul and Body: Life Horizons Upward and Downward

The seventeenth-century French philosopher Blaise Pascal says: "Who would not think, seeing us compose all things of mind and body, but that this mixture would be quite intelligible to us? Yet it is the very thing we least understand. Man is to himself the most wonderful object in nature; for he cannot conceive what the body is, still less what the mind is, and least of all how a body should be united to a mind. This is the consummation of his difficulties, and yet it is his very being. *Modus quo corporibus adhaerent spiritus comprehendi ab hominibus non potest, et hoc tamen homo est.*[8]

Unity of material body and spirit has been analyzed as mind-body problem at the history of philosophy. It is one of the main problems in the cognitive sciences and philosophy of mind. History of philosophy demonstrates turn from the Cartesian dualism of body and mind. The turn has been directed to the recognition of ensouled body, as Maurice Merleau-Ponty recognizes, to the intentionality of body.

Man's being in a meaningful cultural world affords human dimensions to the sensational level. The body experiences sensations. The bodily sensations are grasped by the mind. People on the basis of their experience, upbringing and culture are able to evaluate the information the body provides. Sensations provide the roughest contact with the world; feelings are spiritually on a higher level reflecting life forms. Feelings can be refined, lasting or primitive, rash and transitory. Merleau-Ponty thinks of the body as "symbolism of the world". He writes: "My body is not one of what is grasped, it is the measure of all, the zero point of all the dimensions of the world."[9]

Every one of us grasps the world based on one's own body as the centre. The body is an entity grasping all the other objects around him being in the "centre". The body as a centre of a reading system forms the parameters of perception "nearer", "further off", "up", "down", etc. Bodily sensations are at the basis of many world myths. Space from the point of view of human relations is grasped through the body.

---

[8] Blaise Pascal, *Thoughts*. No.72 http://etext.lib.virginia.edu/toc/modeng/public/PasThou.html

[9] Maurice Merleau-Ponty, *The Visible and the Invisible* (Evanston: Northwestern University Press, 1968), pp. 248–249.

Thus, the body is the "point of vision" from which we look at the world. "My body is not an object, sooner I am my body."[10]

The body is the basis for communication because it acts as a mediator in order to grasp another I. A caress of a loving hand sends a message to the other of the tenderness of their mutual relations; a blow conveys hatred. Another I we perceive in his bodily shape and ourselves in the flesh, so to say. Characteristically, I perceive my body very personally because my grasp is accompanied by my understanding of it. Another does not feel the pain in this or that part of the body the way I feel it. The body we grasp as an object cannot be my body – it seems impossible, as it were, if we wish to be human beings, not some degraded biological clots.

In the classical life form *the body as flesh* is the carrier of the spirit of another man, a mediator for understanding and cooperation to appear. In this sense the body cannot be grasped only as some sign.

In all life forms the human body has been an important entity included in the value hierarchy. In the life horizon *upward* the body is not at the top of the ladder because matter, decay, unconscious passions and desires weigh it down. Spirit is undoubtedly superior to the body because it rises upward to God. Since the time Europe accepted Plato's view of the world it has been divided into the sensuous and the trans-sensuous world. The trans-sensuous part is superior to the sensuous one because it can grasp the eternal and the imperishable. The body is born and dies; that is why in the Western classical culture the body is valued lower than the immortal soul.

In the classical age there are different ways of interpreting the relations between the body (flesh) and the soul, between the physical and the spiritual spheres. If superiority of the spiritual sphere is accentuated, the material body acquires the role of a lackey that is jokingly called "brother donkey". The flesh can be starved; its passions can be suppressed and exploited. Europe is familiar with different religious ways of controlling passions: asceticism implying complete refusal of all worldly goods, life in a convent by way of separation from the temporal world, a life of a hermit, piety – a spiritual trend that started in the seventeenth-century Lutheranism and denotes strict devoutness in everyday life.

The soul and the flesh are not separated from each other. Christianity teaches that the flesh influences the soul; it is carried away by passions, grief, anxieties and apprehensions. The soul may be sinful and it is not only the flesh that is at fault. In Augustine's opinion, a sinful soul spoils the flesh. That is why the soul should cling to the upper rank constantly purifying itself from sins and not submitting to the flesh. Even such an extremely active bodily sphere as sexual relations is devoted in the first place to the continuity of one's life in children, not to sensual enjoyment. In the life horizon *upward* it is clearly manifested in the restriction of passions, a drive for spiritual values, immortality. If the flesh strives upward, then it is redeemed flesh.

[10] Gabriel Marcel, *The Mystery of Being*, vol.1, trans. G.S. Fraser (Chicago: Henry Regnery, 1960), p. 123.

Some psychologists consider that the soul and the body are only theoretically separated because there is actually no such division. Carl Gustav Jung admits that he doubts whether this separation of the soul and the body is not only a way to investigate one and the same fact we effected division into two concepts illicitly affording them an independent existence. The English philosopher Bertrand Russell thinks that it is with Christianity that the tendency to separate spirit from matter is closely connected because Christianity tries to separate the soul from the body. Christianity is averse to the body as such. That is what influences the negative attitude towards the body in the life horizon *upward*.

With the change of life forms in no other sphere is there such a drastic overturn of hierarchy as in the relations of the body and the soul. In the post-modern secular age the word "soul" is to be found in the wastepaper basket of old-fashioned words. The words "spirit" and "spirituality" in many languages drag out a miserable existence reminding us of former Christian values. The English word "spirituality" occurs mainly in religious contexts while the German "*Geist*" is not a politically correct word if people remember its role in the Nazi ideology.

In the post-modern age the body is grasped in its meaningful liveliness and the spiritual sphere – in a much closer affinity with the body. In the contemporary life form it is not the soul that tries to save the body, but the flesh embodies the soul. The body is ascribed everything that is valuable that was formerly ascribed to the soul. The body lays claims, raves, enjoys and puts on masks. The body thirsts for immortality, but now it is promised as a life prolongation programme as far as technology and pharmacology are able to insure. If at the beginning of the twentieth century bodily challenges were regarded to be an attestation of liberation, then in the contemporary life forms emancipation of the flesh is over. Twenty-first-century post-modern culture demonstrates games with body without soul, not human liberation.

In the life horizon *upward* feelings are regarded to belong to a lower rank. Understanding and the mind are above sensations. Sensations supply the material the mind processes. The life form characteristic for post-modern people turns it upside down: sensations can express feelings. Feelings are reduced to the level of sensations, the higher – to lower. Life horizon turns downward.

In the classical culture the merger of feelings with sensations is inadmissible. Taste does not give rise to feelings because the lower is in no way able to create the higher. A fine taste for coffee can change your mood for the better, but it will not directly affect your spiritual feelings, your soul. Classical culture takes care of developing feelings, looks for grasping the world in a spiritually refined way, soul tends to go upward, to spirituality, in German philosopher Max Scheler's words, to the fourth strata of feelings.

In the post-modern age the refined cultural layer that decreed inadmissibility of subjecting feelings to senses is shrinking. The American philosopher Robert Solomon analyzing the understanding of emotions nowadays admits that there is mythology and ideology of emotions.[11] Emotions develop man's self-understanding

[11] Robert C. Solomon, *The Passions. Emotions and the Meaning of Life* (Indianapolis: Hackett Publishing company, 1993), pp. 21–23.

and determine his relations with others, in some sense it is ideology of interpreting of "soul". The emotions that form our world create ideology – desires, hopes, demands, expectations that have to be satisfied. Mythology of emotions is an unconscious orientation; it turns into ideology of emotions when man starts clearly and definitely submitting to emotions; lives, acts and evaluates events under their impact.

Characteristically, in the post-modern life form there arise all kinds of emotion-based ideologies that are not formed as a result of refined upbringing, but aiming at satisfying one's bodily requirements. It isn't spiritually ensouled body but functional body. One of those is that sexual needs are at the centre of man's life. Sex is spoken of as a totality of technologies, a source of sensations of pleasure and physical energy. Emotions accompanying sex are the simplest ones – lust, desire to subject and use.

Another emotional ideology of today refers to glamorized appetite or gastronomy. In affluent countries with no hunger problems a cult of eating with gusto has sprung up. An enormous number of cookery books leave in the shade all the other types of literature. Delicacies, finesse and originality in garnishing – all to please your eye and thrill your palate when you relax in delight after an aesthetically prepared meal. Wine experts in the eyes of society are more eminent than poets who write about souls, for instance. Gastronomy even manages to become poetic and "ensouled". Eating a lot and making love without measure, man satisfies his or her elementary needs, but he or she does not grow in eminence. At times exactly the opposite is the case – emotions evoke degrading sensations and diseases. Overdrawn elemental bodily feelings lead to collapse of personality, neurosis, obtuseness, excessive infatuation with the eating cult leads to vegetative diseases and an aversion to food.

Man's freedom is turned into perverse freedom. Man's activity is limited by his or her unwillingness not by impossibility or inaccessibility. However unwillingness is not given naturally, it has to be cultivated. The human body freed from the control of the spirit desires every moment of gratifying his physiological urges. To stop the triumphant march of the body one needs a different value system from the one prevailing in the life horizon downward to lowest feelings. However, the classical feelings of duty, abstinence and spirituality are unfortunately moved aside as outmoded.

The body is like a sign on which the consumer culture inscribes its codes. Soul has been lost or enlivened by marketing. Human spirit has been changed into social role and trends. Personality has been broken into parts, ideas of permanently changed identity have been developed in social sciences. No recognition of soul as a core of human historical transitions. Unity of body, soul and spirit becomes only illusion. Peter Sloterdijk tells when the famous French prostitute Arletty was accused of having had sexual relations with members of German occupation forces, her answer is said to have been: "My heart is French, but my backside is international."[12] Her rude but apt answer is typical of a person who feels divided, as it were, in spirit and body, it means to be a broken post-modern personality.

[12] Peter Sloterdijk, *Critique of the Cynical Reason* (Minneapolis: University of Minnesota Press, 1987), p. 148.

However, body is not as capable as the mind, it is unable to grasp interconnections, it is unable to reflect. Paradoxically, but the organs of senses can perceive objects affecting them, but they cannot perceive their own selves. Sight can see things, but it cannot see seeing. The organs of sense lack something that is inherent in the mind, namely, reflection. That is why sensations cannot create anything higher – feelings. Sensations cannot replace soul. The contemporary post-modern life form exaggerating the experience of senses reduces human being to a lower level of existence in comparison with the life form *upward* that anticipates feelings that are not connected with one's body (the feelings of sanctity, for instance).

The *Body Art Movement* coined the slogan – truth belongs to those who experience physical pain. It does not mean redemption. That spells an essential difference from the standpoints of the classical age. In the classical standpoint truth is connected with sufferings that can raise human being to a higher spiritual level, soul strives upward – to God.

In the post-modern standpoint the aspiration *upward* does not exist, sensations are used only as one of the many designators. It evokes shock, fear and a sense of emptiness in the audience. Post-modern artist Gina Pane thinks that alongside her performances "physical suffering is no longer a personal problem; it becomes a language problem. The body itself becomes an idea while formerly it was nothing else but a way to transfer of ideas."[13] However, on the other hand, the body does not become an idea, but a thing of no higher value than other things.

Turning the body into a thing is something contemporary philosophers have been warning against. Thus, even blood loses its carnality, its corporeality and turns into a threat, a sign of disease or viewed from another stance – into a manifestation of endangered sexuality, not into the basis of life.

At present attested are not corporeal values, but sooner lack of their value. The diseased as such is a fact in our view, but the idea of his and every one's including our own immortality, finality and vulnerability is generated in us, the living ones.

Changing the body into a thing tends to get closer and closer to the body which is not ensouled body. Consumer society gathers speed to cultivate sensations and trivialize feelings. Getting the world of things man lost soul. Pascal writes: "The victory over death". "What is a man advantaged if he gain the whole world and lose his own soul? Whosoever will save his soul, shall lose it."[14]

Phenomenology of life emphasizes the need for balance between soul's directions and decreasing of materialistic, consumerism life form, orientation to primitive feeling of the world and Cosmos. Phenomenology of life is right that postmodern Self is endangered Self, it is characterized by disintegration, broken identity, based on self-fragmentation and disappearing of the unity of soul. Restoring of harmony between human body, soul and spirit is the fundamental basis for developing of contemporary culture.

---

[13] Francesca Alfano Miglietti, *Extreme Bodies. The Use and Abuse of the Body in Art*. (Milan: Skira editore, 2003), p. 28.

[14] Blaise Pascal, *Thoughts*. No.782. http://etext.lib.virginia.edu/toc/modeng/public/PasThou.html

Part VI

The Unity of Eastern and Western Thought Traditions in Anna-Teresa Tymieniecka's Phenomenology of Life

Salahaddin Khalilov

Abstract Phenomenology is traditionally thought to be a product of Western thought. However, Buddhism is also usually considered as one of the sources of Husserlian phenomenology. The close relationship of this teaching with Islamic philosophy, particularly with Ishraqism (Illuminationism) has been researched under the leadership of A.-T. Tymieniecka in the last decades. Although the existence of this relationship was first revealed by Henry Corbin, the wider analytic comparison of these two different philosophical research directions has been made in the works of A.-T. Tymieniecka.

In the Islamic (Illuminationist) philosophy, the knowledge gained by means of the senses is not considered as true knowledge (as truth). On the contrary, what is required for achieving truth is to be free from the effects of sensory experience. In fact, it reminds us of the phenomenological induction.

The epistemological system in Islamic philosophy has been studied better and more comprehensively than Husserlian phenomenology. Merely Western philosophy, which was unaware of it, had to rediscover all these notions. However, Husserlian phenomenology can only one-sidedly explain the problem. It does not take into consideration the variety and multi-stageness of the structure of the soul. Reason in Islamic philosophy is only one aspect of the soul. Perhaps for this reason, A.-T. Tymieniecka establishes her teaching not only on reason and logic but also on the world of soul which is a richer world of consciousness.

The formation of intuition is considered by Tymieniecka in the context of creativity which is a wider notion. She thinks that people only live when they create.

The idea 'I create, therefore I am' is more precise than Descartes' thesis 'Cogito Ergo Sum (I think, therefore I am) '. If we consider the fact that creativity is also an experience and life then it could be said that the philosophy of Tymieniecka includes

S. Khalilov (✉)
Azerbaijan University, Mirzaga Aliyev 247/6, Baku, AZ1014, Azerbaijan
e-mail: ssx@azun.baku.az

A.-T. Tymieniecka (ed.), *Phenomenology of Space and Time: The Forces of the Cosmos and the Ontopoietic Genesis of Life: Book One*, Analecta Husserliana 116,
DOI 10.1007/978-3-319-02015-0_22,

in itself the synthesis of Husserlian phenomenology and Existentialism. However, Tymieniecka goes a step further, and by considering the achievements of Islamic philosophy in this field, she rises up to a higher state of synthesis.

Every man lives out his own ordained life; a life that has been predestined for him. However, after the human being reaches a certain mature age, he begins to weigh up, and if he reaches a higher mature age then he calculates his own life. Namely, this life is a conscious life perceived in the first stage, but man still cannot build his life on his own. Consciousness here is posterior to events and life. In the second stage, man chooses his life on his own; that is to say, what is of concern here is a consciously built life.

The cognition of man of his own life and his building his future life consciously become possible only due to philosophical thinking. However, not every man can rise to the level of philosophical thinking and the vast majority live an unconscious and spontaneous life.

The aim is to convey the illumination and self-cognitive practices of a limited number of human beings, who have perceived themselves and the world, to others and to present it to the use of everyone. Is it however possible? For the present, we can only transfer knowledge. Namely, the practice of the transfer of inner illumination and mental energy among people is not yet known. For this reason, everything can be transferred to others only after arranging them into logical patterns and bringing them into the form of knowledge, and as if they go into space-cosmos that is common for everyone, and only after this they become useful for being accepted by others; or more precisely, the transfer becomes possible by means of language. A man can transfer his condition as well as his feelings through scientific, philosophical and artistic languages.

It is impossible for every man, of course, to be a philosopher. However, can philosophers, besides perceiving the world and the life they live, also find the methods of conveying it to others?

Self-cognition, that is, man's understanding of his personal life as if by observing from outside, and his using the dialogue and mutual agreement between him and "Ego", which he has pushed it aside, in his life experience, differ from the abstraction and depersonalization of gained knowledge and carrying them out of personal life and as if entering these types of knowledge, which belong to man and to his relations with the world, into the system and creating from them a theoretical teaching.

However, in the next stage, this system of depersonalized and abstracted knowledge – philosophy – must re-descend into life and illuminate the lives of non-philosophers.

A specialized man, of course, in a certain meaning, is asymmetric and sometimes even degenerative, that is, in spite of his poor development in all other fields, he has great unexampled knowledge and skills in a certain field. For this reason, he has left the entire harmonious human image and has possessed a specific and different image.

However, it does not mean that the germ of completeness in his activities has disappeared for ever. The image of that completeness, which has lagged behind development, is probably continuing to remain as a germ in that man. When life "treads on his corns" and when the world "twinkles" in front of his eyes, then as if that completeness within him becomes revived, and by gaining completeness for an instant, man understands the great, true meaning of life. Literature and the arts undertake the mission of providing the harmony for such specialized people and causing them to experience the beauty of the complete image of the world. Literature forms a virtual world for us and shows the completeness and another aspect of the world to specialized but yet incomplete men; and the real world becomes completed with this figurative world; and thus man reaches his completeness.

Unlike the universe, which is infinite, a stone that weighs a gram, a small part of the world is finite in measure. Air, which weighs a gram, spreads all over the world and includes the whole world.

Reason, as thickened and patterned senses in the form of knowledge, has lost its opportunity to spread over the world by being re-thinned. Senses, in turn, are active and agile like air. If "a drop" of a sense "evaporates" then it can re-conquer the whole world; or more precisely, with our senses we are determined to include the entire world. How many times should we gather, with our knowledge, different local beings, which have been detached from the world, together so that we can re-create an image of the world?

Literature can make man complete just because it excites feelings. Science takes man to pieces just because it has been increasingly localized in the small parts of the world. In this respect, unlike science, literature is formed due to inspiration and fervency rather than rational thinking. For this reason, literature is closer to philosophy in terms of its mission and duty. It is not a coincidence that Tymieniecka speaks of the contest between them: "Philosophy and literature are caught in a constant contest as each attempts to absorb each other's task".[1]

The aim of teaching philosophy to people is, in fact, to return them their completeness and give them a chance of living a perfect and complete life. It is very difficult. Music and poetics can accomplish this function better. Philosophy, in turn, attempts to establish that complete image, not on feelings and senses, but on reason itself. The reunification of the separated.... However, each one builds a different world from the parts of this "constructor". They build and thus every time establish themselves once again.

There cannot be a philosopher who does not have senses. That is, only by means of senses philosophy can gain completeness. This is a specific sense. This has been always called "love" in the history of philosophy. Love is a connective sense. As Abu Turkhan says: "Reason separates, love unites".

The personal sensory experiences and life of every man are, first of all, the realisation of senses. The life of the body, to wit, as it is accepted in biology,

[1] A-T. Tymieniecka, *Logos and Life. The Passions of the Soul and the Elements in the Onto-Poiesis of Culture*. Book 3: the Life-Significance of Literature, Dordrecht-Boston: Kluwer, 1990, p. 16.

"life"-breathing, the metabolism with nature and the activities of every organ of the body like circulation of blood, heartbeat, etc. – all these are usually the same processes for all people. Namely, the model of the body and its functionality cannot characterize personal life. Considering this fact, philosophers do not seek human life in the context of the body's functionality. However, the body is an instrument and means for the practical activity of the human being. The activities realized by means of it, reveal the intention, aim and also the knowledge and skills of the human being. Man becomes "revealed" in society as a social being. For this reason, there are some inclinations towards seeking human life at social plane and in its social functioning.

Nevertheless, what kind of actions does the human being do? What kind of functions does he perform and which position does he represent as "Self" in relations with others?

When seeking the answer to these questions we have to return to the wishes and ideals of the human being as well as to his "functioning" programme and to his knowledge and skills. However, there is a need here to differentiation and identification. Namely, although knowledge is a main index of human life, only quantitative indices could be shown as its criterion. That is to say, the contents of knowledge are usually the same for all people who study in a certain field. There are few differences and many common aspects. Less or more knowledge cannot be a major factor that characterizes a person. What distinguishes people are rather their goals of life and what they look forward to. There is a saying: "Where there is a will there is a way". However, the human being does not have a single intention and goal in his life. There are small and big plans in life; and the achievement of a certain goal paves the way for a greater goal.

It would be naive to think that the whole activity of the human being is consciously planned by him and he chooses his way of life on his own. The outside impacts, social norms, traditions and tendencies, moral values, religious canons, prejudices, temporary fashions and the impacts of the local social environment and collective that a person has accidentally entered, etc. – all these have serious impacts on man's daily life and sometimes on his whole life.

In fact, there are people who, despite all outside impacts, choose their lives themselves and consciously determine it, but they are minorities. In any case, we mean these very minorities when we speak of "personal life" and the "meaning of life", because the lives of the rest might be suggested not by themselves but by social environment.

The meaning of human life, in fact, manifests itself not due to the absolutization of any of man's rational, sensual or practical functionings, but at their crossroads and in the whole, which includes all his aspects. Tymieniecka writes: "Man is there present under all his aspects: natural man who fulfils physiological functions, social man in the bosom of the human community, as well as the man who feels, who loves, or who hopes and who, so doing, transcends himself and wins his freedom by breaking the chains of natural man. This system which knots harmonious bonds between man and the world does it on a different basis, issuing from man's conquest of himself".[2]

R. Descartes, E. Husserl and other followers of the line of "Cogito Ergo Sum" treat reason and rational thinking as a central element. Nevertheless, what is the role of reason in choosing personal life and in living it? We have mentioned above that the contents of knowledge are usually the same for all people who study in a certain field. Then what about reason? Can the human being be characterized by the knowledge that he has obtained not from his sensory experiences but from rational thoughts? With what does the thought of a man differ from that of another man? At least, logic is the same for all people as well as mathematics. Yet the degree of adopting and using them can be different. However, all adopters adopt a same thing and a same content. If all people keep to the "norms" of thought then they will be the same in this aspect; or if a cosmic mind, the logos of the world, etc. are taken as sources of reason then it should be taken into consideration that this is also the same for everybody. In other words, reason can only be a means for individual life, but cannot be life itself.

Then what can individualize man, characterize his personal life, cause him to know himself and reveal 'Self'? If we return to sensory experience once again, we will find there the experiences that are common for everybody and are usually repeated. Nonetheless, there are some individual, dear and native experiences and besides being the most striking pages of human life, they are also his personality and indicator of his level. Such pages are related not to the current, traditional and "normal" life of man, but rather to his creative activities. For in the act of creativity, the real nature of man and his essence becomes unveiled. Paying attention to this very necessity Tymieniecka writes: "First and foremost the discovery of human creative experience allowed us access to the logos of life, for it is reflected in human creative experience in its manifold radiation."[3] Afterwards, Tymieniecka shows the way that leads to revealing the "Ontopoiesis of Life": "We found a definitive station (platform) and our compass not in cognition but in the human creative act, which enters the sphere of becoming-individualizing life. We thus interpret in its original nature the becoming that reveals the logos of life within pristine nature. With only one step further (but what an infinite step!), the entire field of the becoming of life, of the ontopoiesis of life, lies open."[4] "The great enigmas of the Universal Logos"[5] become unveiled due to Tymieniecka's conception of "Ontopoiesis".

---

[2] A-T. Tymieniecka, *Logos and Life. Creative experience and the critique of reason Book 1, Analecta Husserliana*: v. 24, Dordricht-Boston, 1988, p. 58.

[3] Anna-Teresa Tymieniecka, *The Fullness of the Logos in the Key of Life. Book 1. The Case of God in the New Enlightenment*. Springer, 2009. p. xxxiii.

[4] Ibid.

[5] Ibid.

The Husserlian phenomenology is not sufficient for the identification of individual life. The points, which are put forward here, are not for individual and personal lives, but for an abstract "common man". However, the window to man's native world can be open only from within his life.

"Ego" stands at the centre of the world and the goal of philosophy is to create a world model that Ego stands at its centre. One of the main goals of Tymieniecka is to lay stress on human life, look at the world from the prism of the life values of man and see, as a central element in the architectonics of the world, not a cognitive process, knowledge, concrete senses, etc., but life itself: "Life" as such means primordially differentiation and constructiveness. It means individualization from within a circumambient realm, an individualization which simultaneously transforms that realm into one's own milieu".[6]

Life, in turn, is related to the factors, which have an important place in the daily activities of man and are native to him, excite him and encourage him to live, rather than science and rational cognition, which are "alien" to him and do not charge him emotionally.

No doubt Tymieniecka does not reach this conclusion all of a sudden. If we follow the evolution of her thought we do not encounter any rectilinear trajectory. This way is zigzagging. Till reaching this final and decisive formulation of her philosophy, Tymieniecka had continued her studies in many directions and more than once she "walked up and down" in philosophical space. However, because the history of philosophy is "in sight", if we draw this way linearly in retrospect, we would see the line of Leibnitz – Ortega y Gasset – Tymieniecka.

José Ortega y Gasset thinks that phenomenological thinking must be based on a phenomenon, which is an independent system, and this system must be human life. He writes that he "abandoned Phenomenology at the very moment of accepting it. Instead of withdrawing from consciousness, as has been done since Descartes, we become firm in the radical reality which is for every one his [or her life]".[7]

Maybe Tymieniecka has directly benefited from Ortega y Gasset; because he himself started his studies from Leibnitz. However, when we look back from the twenty-first century, we can determine the shining and deflation periods and places of this brilliant idea. This is may be a virtual model. However we stand on this very line of Leibnitz – Ortega y Gasset – Tymieniecka.

Ortega y Gasset brilliantly anticipated that phenomenology and in fact future philosophy would follow this way. However, this idea has not been developed by him. He emphasized the necessity of the application of the phenomenological reduction in the phenomenon of life, but did not work on the realization of this idea. The attempts of Tymieniecka, who is a representative of phenomenology and, besides the line of Leibnitz, has also represented the line of Descartes-Husserl-Heidegger

---

[6] A-T. Tymieniecka, *Logos and Life. Creative experience and the critique of reason. Book 1. Analecta Husserliana*: v. 24, Dordricht-Boston, 1988, p. 327.

[7] José Ortega y Gasset, *The Idea of Principle in Leibnitz and the Evolution of Deductive Theory*. New York: Norton, 1971. Sect. 29. http://plato.stanford.edu/entries/gasset/

for a time, are of prime importance in this issue. As if she unites different sources and directions in phenomenology in one large way. Her works have been formed in near retrospect from the syntheses of the line of Husserl and Ortega y Gasset, and in far retrospect, from the syntheses of the Western way of thinking and Eastern thinking, mainly medieval Islamic thought and in particular Sufi and Illuminationist (Ishraqi) thoughts.

Heart and inspiration are praised more than mind and reason in the East. Although reason is accepted by Islamic Peripatetics as the highest stage in the structure of the soul, divine love is considered higher than anything else in Sufi philosophy.

According to Suhrawardi's classification, there is a great emotional-spiritual spectrum between animal and reasoning souls in the structure of the soul and it basically implies aesthetic emotions. Unlike Western Sensualism, which overrates the role of the five senses in cognition and unlike Western Realism that insists on the importance of the reasoning soul, Suhrawardi pays attention to the active role of all three steps in the cognitive process, and this, in turn, becomes possible on account of the synthetic functions of aesthetic emotions. For the reason that beauty is related to both instinct and truth, the way to reach the truth goes through the artistic-aesthetic life. A poet needs to be filled with enthusiasm as well as a philosopher cannot embrace truth without ecstasy (*wajd*).

In her article called "The esoteric passion for space" Prof. Tymieniecka tries to explain the relationship between aesthetic and voluptuous feelings of human being and the space factor, which is a more fundamental-existential factor than others. Here we remember al-Farabi's notion of "the soul of the Earth". This notion means that without being dependent on the human approach to them, different places have esoteric power and attractiveness. Uniting with the vital process of humans or animals and even plants, this attractiveness creates an opposite relationship. The idea of the unity of the world appears not only ontologically but also subjectively – as a coherence and unity of emotional experiences. The unity of man and world is put forward in the emotional perspective of space.

From what necessity do the expressions such as "the soul of the Earth", "the esoteric passion for space", "the reviving of idea" and other non-traditional and metaphoric and even mystical expressions originate?

Although the word "aliveness" is mainly used as a synonym of "having the soul" there are other shades of meaning too. The verbs "to be alive" and "to breathe" coincide with one another in most languages.

However, there is one more term of aliveness. The living being is born from another living being. In other words, all living beings come to the world through genetic inheritance and increase in generation and, after remaining as living beings for a period of time, they pass away from this world. Life is established between birth and death.

The dead do not return to life.

However, the word "reviving" is also used in a broad sense. Namely, if any being has not yet died and favourable conditions have been created (or cured) then it can become revived. What is of concern here is the reviving of a living one. That is to say, the movement, which is directed from life to death, changes its moving direction again towards life. Why do the plants become withered? Is it because of the lack of water, light and fertilizer? If a withered plant is watered or is lightened then it becomes revived. The main point is that it should be taken into consideration what and how much the plant needs. The potential opportunity is within the plant. The material (or energy) needed for its realization is taken from outside. The reviving becomes possible only when the opportunities of the environment meet the needs of the central factor.

Things are divided into living and non-living ones. The reviving of non-living things is considered impossible.

However, we will view "reviving" as a universal category. Namely, we will investigate the terms of the reviving of non-living things.

The living world has been programmed. However, the objects of the non-living world also contain certain information; they have certain forms and structures. These forms are the copies of an idea. The reviving of this copy requires energy from outside. The energy source for the reviving of idea could be obtained only due to the intellectual potential of men. The attention of the human being is directed to the object and thus a certain image of the object is formed. Whose reviving do we mean in this case?

For the reason that the idea in the object (its structure and form) is a copy, it is not capable of being revived. It can only take part in "fermentation". Only the human being, who has access to the real McCoy of these ideas, is capable of giving birth to idea.

The human soul also does not keep all ideas alive in itself. As a rule, they settle in the passive fund. At the result of the contact with the copies, which are in the material world, man's ideas, which are reserved in the passive fund, can be activated. As if this is a process of "fermentation" and, as a result of it, passive ideas can become revived.

The revived ideas are those phenomena that a special teaching based on them has been established in the West.

At the heart of the controversy is what attitude do phenomena have towards the relation between objects and human mind? To what extent is their connection to the object as well as their adequacy to it? For Kant, the 'thing-in-itself' is incomprehensible. What do we comprehend then?

According to Husserl, what consciousness is directed to are phenomena. Namely, what is of concern is not the object, but its meaning. On the other hand, what we take as being is phenomenon, and the existence of the object and its perception as a whole stays out of cognition.

The different aspect of Hume's agnosticism is that what it talks about is the image of sensory impression. Namely, what is perceived is not the object itself, but the mental image of sensory impression. The problem of to what extent this image corresponds to the object is out of cognition. The model is the same. Hume, Kant and Husserl take, as an object of cognition, not a concrete object, which is considered as material reality, but its image, form and appearance. However, in Hume this image is formed through organs of sense, while in Husserl it is accepted as a pure meaning purified of senses. The purification of this image of the senses and the liberation of reason from all kinds of psychologism and the adoption of meaning purely is one of the important problems in Husserl.

However, the senses also can be completely at different levels. The cognitive senses serve the perception formation of the knowledge of an object gained though organs of sense. This sensory image resembles the object in terms of its external signs, or more precisely, we think so. Nevertheless, there can be another way of being affected by the object. Namely, we do not form an image considering its colour, form, sound and other physical indicators; on the contrary, by being liberated from this sensory information, we suffice only with an impression formed within us through non-apparent ways; more precisely, the external signs of the object recede into the background and we take into account only the sensual experience and emotional state formed by it. This experience, however, is incomprehensible for us. As if an abstract image gradually differentiates and the secret relations between our sensual condition and the object become revealed. For the reason that such clarification mostly does not happen, we become unaware of our senses and their sources.

Unlike a sensory image, an image of idea is related to our primary knowledge. Leibnitz says: "...the knowledge that our ideas give us, for ideas are the only things that knowledge has anything to do with".[8]

* * *

Just as different things, every plant and all living beings have been programmed, and just as the whole future life has been taken into account in every germ as a project, likewise the whole nature and world can be viewed as a programmed system. Individual programs are included within the macro-programme in the context of nature.

Every man as well as society as a whole has inner development regularity and inner motive impulses as an organism.

Just as every human thought has a certain relative independence, likewise the world of reason is also autonomous and it has its own inner regularity and objective development ways. Merely, sometimes man follows one of these ways and the way leads man. Husserl writes: "According to the guiding ideal of the Renaissance, ancient man forms himself with insight through free reason. For this renewed "Platonism" this means not only that man should be changed ethically [but that] the whole human surrounding world, must be fashioned anew through free reason, through the insights of a universal philosophy".[9]

However, on the one hand the human being has to consider himself in the context of the experience of humanity, but on the other hand as a part of nature. Nature itself, besides being a manifestation of an idea, is also a reality of a certain

---

[8] G. W. Leibniz. New Essays on Human Understanding. Book III: Ideas, p. 177.http://www.earlymoderntexts.com/pdf/leibnew4.pdf

[9] Edmund Husserl, *The Crisis of European Sciences and Transcendental Phenomenology. An introduction to Phenomenological Philosophy*, Evanston, 1970. p. 47.

experience. Leibnitz writes: "space is only an order of coexisting things".[10] Order, in turn, is a manifestation of logos. In this sense, space itself, on the one hand, is a carrier of divine idea and, on the other hand, a carrier of reason. Thus the mental relief of space, together with the ideals of man, is a manifestation of "collective sensory experience" (Carl Jung) and subconscious memory in a different perspective.

In her studies, Tymieniecka pays a special attention to the subconscious desires of man and their roles in human life. This sense, which is called by her "esoteric passion", is presented as a result of the hidden attempts of human reason.

In the analysis of "the esoteric passion for space", which is related to the cosmos and imagined extraordinary space, Tymieniecka applies to the ontopoietic revealing of life and ascribes it, in fact, to its essence that arises from the individual cosmic beginning of life. She writes: "Can there be a more fundamental grounding, a firmer and more indicative point of departure than life itself? *I submit that the living being recognizes itself as "himself" or "herself" not by a cognitive act but by "being alive"* – by experiencing itself within its milieu of beingness, directing its instincts and appetites, recognizing the elements of the circumambient world in their vital relatedness to itself, and lastly but foremostly, by recognizing that one is *the acting center of the existence,* as a self-sustaining agent who directs within this universe of existence through experience, observation, reflection, and deliberation his or her own course and who, finally, endows that course with moral and aesthetic values, and upon the wings of the spirit seeks to understand the reasons of it all and soars to the metaphysical and spiritual realm above, carrying within a thoroughly felt self-aware conviction that to be is to be alive."[11]

Tymieniecka mentions that under the wings of creative imagination, man's subconscious passions enable him to rise to higher positions and thus man becomes capable of moving away from the bounds of his personal existence.

Which inner aspect of man does Tymieniecka mean when speaking of the "passions of the Earth" or "the esoteric passion for space"?

The human being has conscious searches as well as, together with a space that his mind reaches and his knowledge make it possible to describe it, he has also an unbelievable broadness, which cannot be comprehended by ordinary reason, and the inconceivable world of imagination which is revealed by a spiritual will directed to eternity. This eternal world and cosmic expanse has a mirror reflection as well as it has a projection to the inner world of man, to the material environment to which he is directly related, and to the "esoteric homeland". In fact, before Tymieniecka, a number of great philosophers, including Sufi thinkers, have spoken of the projection of the whole sea in a drop of water and of the manifestation of the whole world in every particle of the world, or modern nature scientists speak of the existence of the code system of the whole organism in each cell of the organism.

[10] G. W. Leibniz, *New Essays on Human Understanding*, Book II: Ideas, p. 102.http://www.earlymoderntexts.com/pdf/leibnew2.pdf222

[11] Anna-Teresa Tymieniecka, *The Fullness of the Logos in the Key of Life. Book 1. The Case of God in the New Enlightenment*. Springer, 2009. pp. xxxi–xxxii.

Tymieniecka's contribution is that she attempts to conceive of every individual as a projection of the cosmic expanse, which corresponds to human life and which is codified to this life. Namely, the human being seeks his life and his native material place directly in the image of the cosmic expanse which could be revealed, in fact, by individual sense and reason. What brings her creativity closer to Sufism are mostly these searches. For this very reason, in the last period of her creativity, she increasingly refers to distinguished Islamic thinkers. It is not coincidence that Tymieniecka acknowledges that her "phenomenology of life considers with the metaphysics of Mulla Sadra on the matter of a 'radical metamorphosis'".[12]

There have been written many works on the comparison of a man with the whole world and universe and on revealing the sameness of them. Ibn Arabi's *al-Tadbirat al-Ilahiyya* (التدبيرات الالهية- Governance of the Human Kingdom) is dedicated completely to this problem. According to Hurufism, the secrets of the universe are manifested in man and man, in turn, expresses his secret in letters. In this sense, the soul and letter are considered the same. For the reason that all these searches have not found their systematic continuation in following philosophical investigations, it is very hard to put them in the context of modern philosophical thinking. Nevertheless, the studies of Tymieniecka seem to be a continuation of these ideas. Although she is a western person, her closeness to the eastern spirit means the transformation of Western philosophical thinking from spiritlessness to spirit and the determination of the return of the thinking way existed since Plato and Neo-Platonism to its beginning.

In Eastern philosophy, in particular in Sufism, this problem has been symbolized as a relation of "drop" and the "ocean". Mawlana Jalaluddin Rumi says:

> Listen, O drop, give yourself up without regret,
> and in exchange gain the Ocean....
> Give a drop, and take this Sea full of pearls.

"The drop and the ocean" is a poetic and figurative expression of the serious philosophical problems like finite and infinite, death and eternity, soul and spirit, dark and light, individual and society, quantity and quality.

And once again two poles are of concern; but the poles of sameness, and the essences that are more sharply expressed in these poles; or the manifestation of the same content to minimum and maximum extents, and the change in quality and essence due to quantity difference.

These poles could be taken at different perspectives and scales. Tymieniecka speaks of the different poles of living being: "We may see living beingness as a filigree, a microcosmic counterpart of the great macrocosmic horizon".[13]

This world is between two infinities: the infinite small and the infinite big are the opposite poles. Opposites are in unity and can be transformed to one another. This, in turn, leads to the conclusion that space is inclined and the lines of the world are circular. The fact that the drop in the example of "the drop and the ocean" is, in fact,

---

[12] A-T. Tymieniecka, "The Unveiling and the Unveiled", *The Passion of the Soul in the Metamorphosis of Becoming*, Edited by A-T. Tymieniecka, Dordrecht-Boston, 2003, p. XLIII.

[13] Ibid, p. XXIX.

infinitely divisible leads us to the questions put in Zeno of Elea's *Aporia.* Although the drop is finite in terms of its final measure, it is infinite in terms of its complex inner structure. It is impossible to find the correlation of infinities in mathematics, or it is a special discussion.

In this case, the correlation of the drop to the ocean is zero. Namely, only the ocean exists, and the drop is only an apparent being. Speaking with the terminology of Plato, it is **the shadow** as well as it is **appearance** (*Schein*; *Erscheinung*) in Hegel and **fana** (annihilation), **non-existence** and **nothingness** in Sufism and Existentialism.

If we take it in the spiritual context, it could be said that in return for the Divine or Absolute Soul, the individual soul (human soul) is nothing, and it needs to join to the Absolute for existing. The wish of the drop to become, by being united with the ocean, eternal, arises from this need. Only in this case it can prove its existence. The reason why individual "existence" is abandoned is because it is temporary and mortal.

Transcendental Morphology: A Phenomenological Interpretation of Human and Non-Human Cosmos

Bence Peter Marosan

Abstract In this present study I would like to propose the following thesis: at the core of the phenomenon of the world we find the phenomenon of the *cosmos*. In order to demonstrate this idea, I will expound in detail on the philosophical interpretation of the problem of world in the works of three authors: Husserl, Heidegger and Eliade. Husserl interpreted the phenomenon of the world as a transcendental, intersubjective constitution. In Heidegger's interpretation the world was first and foremost "being-in-the-world" and as such the existential structure, the mode of being of human existence or "being-there" ("Dasein"). Eliade, in his comparative religious studies, accepted the standpoint of Heidegger's existential phenomenology. For him, the "sacred" and the "profane" were two distinct fundamental modes of "being-in-the-world". Eliade conceived of the ancient man's world essentially as the "cosmos", as a well-formed *order* of sacred and profane matters. I will try to show how these three conceptions coincide with one another, forming a single coherent phenomenological structure. I will call the method whereby we investigate the ideal laws of this structure "transcendental morphology".

Abbreviations

GA	Heidegger Gesamtausgabe
Hua	Husserliana. Edmund Husserl's Collected Works
Hua Mab	Husserliana Materialienband

B.P. Marosan (✉)
Budapest Business School, College of International Management and Business, Budapest, Hungary
e-mail: bencemarosan@gmail.com

A.-T. Tymieniecka (ed.), *Phenomenology of Space and Time: The Forces of the Cosmos and the Ontopoietic Genesis of Life: Book One*, Analecta Husserliana 116,
DOI 10.1007/978-3-319-02015-0_23,

Introduction

In his book *The Fundamental Concepts of Metaphysics – World, Finitude, Solitude* (winter semester, 1929/1930), Heidegger drew a threefold distinction between the three levels of existence: the stone is worldless, the animal is poor in world and man is a world-forming being, (GA 29/30: §42). Heidegger later abandoned this idea; in his 1935 seminar entitled *Introduction to Metaphysics,* he said that "the animal has no world, nor any environment" (GA 40: 48, English [2000]: 47). With this gesture Heidegger placed the human being at the heart of Being – him and him alone. According to Heidegger, man as being-there was alone in the clearing of Being. Thus Heidegger continued the old tradition of anthropocentrism in Western metaphysics. Heidegger was not radical enough in this respect – and because of this perception of him, he was criticized by his student Hans Jonas.[1] Elsewhere I proposed that an anti-anthropocentric, that is a nature-centric turn would be needed in contemporary philosophy (particularly in phenomenology).[2] But I think that the world-analyses we could find in the notes of Heidegger's 1929/1930 seminar could provide some crucial points in articulating such a viewpoint in phenomenology.

By the word "world" I basically mean "cosmos", in the sense Mircea Eliade used the term in his book *The Sacred and the Profane* (as well as other writings on comparative religion). With this term Eliade referred to ancient man's structure of *existential experience*, to the particular way the pre-modern person lived through his or her world. Eliade pointed out the dual meaning that "cosmos" had in the Ancient Greek language: it meant both *order* and *world.*[3] In Eliade's interpretation, for the ancient man the world presented itself as a set of fixed, binding laws and rules, as a cosmic order. But Eliade also used a concept that is coterminous with "cosmos": he treated the concept of "chaos" as a sort of "counter-world", other-world (we could say: "Unwelt" in German), a domain that lacked the familiar, well-known laws and rules of the home-world, which was chaotic, and in this sense was not a world at all, (Eliade, 1987: 29, ff.).[4] In the present paper, drawing on Eliade's terminology I will use the word "cosmos" to denote everything that is *constitutive* of the "world" understood in its phenomenological meaning, by "chaos" in turn I will refer to those things that are *destructive* for the world, the cosmos of the subject.

[1] Cf. Marosan, "The Primal Child of Nature – Toward a Systemic Theory of Eco-phenomenology", 2011. Forthcoming.

[2] Loc. cit.

[3] Κοσμος = (amongst others:) (1) order, arrangement, composition, composure, (2) world-order, universe, world.

[4] Mircea Eliade, *The Sacred and the Profane*, (New York: Harcourt Inc., 1987), p. 29. "One of the outstanding characteristics of traditional societies is the opposition that they assume between their inhabited territory and the unknown and indeterminate space that surrounds it. The former is the world (more precisely, our world), the *cosmos*; everything outside it is no longer a cosmos but a sort of "other world", a foreign, *chaotic* space, peopled by ghosts, demons and «foreigners» (who are assimilated to demons and the souls of the dead)".

One might reasonably raise the following question: Why do we speak about the "cosmos of subject"? Why not simply say "human cosmos"? One reason for this particular phrasing is the intention of overcoming Western anthropocentrism and of moving towards a nature-centric phenomenology. Another reason is that we could develop a more adequate view of the phenomenological content of the cosmos (world) if we approach and treat it as the cosmos (world) of an indeterminate subject rather than as the cosmos of subject of a particular kind, namely a human (or animal) cosmos. It is the subject and its subjectivity that constitutes a world, and it is the subject that constitutes itself as a human, animal or non-human being (if we take cases of science-fiction) belonging to a human, animal or non-human world. In philosophy, only the acceptance of *phenomenological reduction* could justify such an interpretation. Pursuant to the aforementioned methodological approach, introduced and elaborated by Edmund Husserl, the subject appears as a world-constituting being, and the human, animal or non-human existence belongs to the special mundane meaning, determined by the particular way in which the subject in question constitutes its world.

I claim that the real phenomenological content of the cosmos phenomenon can only be revealed by way of Husserlian phenomenological reduction. In order to see the real meaning and depth of this concept, one must also realize that the "human" and "animal" (as well as the "human world" and the "animal world") are constituted meanings, and can only be properly understood in relation with the transcendental process of world-constitution. The proper meaning of "being-a-human" or "being-an-animal" could be disclosed by an analysis of constitution. These are constituted meanings and validities.

Transcendental-phenomenological reduction presents everything as the result of a constitution. The phenomenological concept of *constitution* does not mean "creating", but rather to unfold or disclose something as a complexity of meanings or validities.[5] Despite his radical criticism of Husserl, Heidegger also borrowed heavily from Husserl's view of constitution and the transcendental procedure of sense-bestowing.[6] *Being and Time* was essentially a transcendental project. In the following paper I will accept phenomenological reduction as an initial methodological step of philosophy, and try to unfold the concrete meaning of the world as cosmos from this point of view.

---

[5] See: Robert Sokolowski, *The Formation of Husserl's Concept of Constitution*, (The Hague: Nijhoff, 1970), pp. 1ff, 10–11. There Sokolowski refers to constitution as the *meeting* or *coincidence* of something subjective and something objective. Cf. further: Sokolowski, *Introduction to Phenomenology*, (Cambridge: Cambridge University Press, 2000), pp. 92–93. Dermot Moran, *Introduction to Phenomenology*, (London and New York: Routledge, 2002), pp. 164–166, especially: 165.

[6] Cf. László Tengelyi, *Der Zwitterbegriff Lebensgeschichte*, (München: Wilhelm Fink Verlag, 1998), pp. 140–146, especially: pp. 145–146. See also: Moran, "Heidegger's Transcendental Phenomenology in the Light of Husserl's Project of First Philosophy", in Steven Crowell and Jeff Malpas (ed.), (*Transcendental Heidegger*, Stanford, California: Stanford University Press, 2007), pp. 135–150. Moran, 2002: 197–198.

We might say with good reason that in the case of an animal being one can certainly not speak about meaning or validity in the *same* sense of the terms as in the case of a human being, only – at best – in an *analogous* manner. Phenomenological reduction reduces things to pure subjective immanence, and in the phenomenological attitude one faces the phenomenon of the world in this immanence. Elaborating his *genetic phenomenology*, the late Husserl tried to unfold the genetic origins of meaning and validity. In the course of this project he sought to unfold *proto-meanings* and *proto-validities* in the purely passive sub-layers of subjectivity, in original affectivity and non-ecological spontaneity. Husserl, during the later periods of his life, attempted to approach the problem of animal subjectivity in an indirect manner, providing a phenomenological description of animal behavior and dismantling the different layers of subjectivity in general.

We shall see that in their own peculiar way both Husserl and Heidegger pursued a sort of "phenomenological archeology",[7] that is they tried to dig down to the most fundamental layers of the world-phenomenon. Husserl did so with his transcendental phenomenology, Heidegger chose the way of existential phenomenology. In the course of the analysis below we shall see that the transcendental and the existential way of phenomenology merely describe different aspects of the same phenomenon. The main thesis of this paper is that what lies at the core of world-phenomenon is the phenomenon of the cosmos. Ultimately, it is the phenomenon of the cosmos that could shed light on any phenomenon in the world, be it environment, encircling ring ("Umring"), home-world or any other. In elaborating the phenomenological concept of the idea of world, I will rely heavily on Eliade's abovementioned analyses of this phenomenon. I will refer to "transcendental morphology" as the phenomenological discipline whose task is to unfold, describe and analyze the laws of formation, development and inner life of the cosmos, (because the term "transcendental cosmology" would be dangerously ambiguous).

This paper contains four major parts: (1) the concept of world in Husserl, (2) Heidegger's existential analysis on the phenomenon of world and being-in-the-world, (3) the phenomenon of the cosmos as a structural part of religious experience in Eliade, and, finally (4) the relationship of world and cosmos and the fundamental layers of the cosmos-phenomenon.

The Concept of World in Husserl

Husserl developed the proper sense of the phenomenon of world using the methodological means of phenomenological reduction. The concrete conception of reduction appeared to him at around 1905, in his work-notes and lectures about the problem of time-consciousness. The first signs of this notion appeared as early as

[7] Cf. Nam-In Lee, Edmund *Husserls Phänomenologie der Instinkte*, (The Hague: Kluwer Academic Publishers, 1993), p. 77.

1896, in Lectures on Logic, (Hua Mab 1: 7–8).[8] In these lectures Husserl employed the Cartesian concept of doubt to show that no matter how extensive a latitude we grant skepticism, the most radical doubt is *self-refuting*. We cannot doubt that we doubt. Husserl claimed that our mental acts *qua* acts possessed apodictic self-evidence. The existence of the transcendent content or "target" of these acts might be doubtful, but the acts as such aim at those things which have a peculiar resistance against any form of doubt (loc. cit.).

In his later lectures on time-consciousness, which began in 1905, Husserl introduced the methodological step of excluding the objective, physical time, and focusing the philosophical interest only on the immanent, purely subjective aspect of time.[9] It is the central core of the act of reduction: *reduction to immanence*. Husserl wanted to switch off the phenomenon of nature's objective time-flow because he thought that we would only be able to clarify the proper meaning of time only on the basis of immanent time-consciousness, and that we could only understand the real meaning and origin of objective time with regard to this subjective basis (Hua 10: 4–8). In the following years Husserl worked out a systematic, methodological architecture of reductions. He developed two basic types of reductions: the *transcendental* and the *eidetic*.

Husserl presented his "transcendental transformation" of phenomenology to a wider audience for the first time in his 1907 lectures entitled "The Idea of Phenomenology" (Hua 2). As stated before, the reductions followed two basic lines: the reduction to pure immanence (transcendental-phenomenological reduction, Hua 2: 45) and the reduction to the a priori, to the necessary and essential structures of experience (the eidetic reduction, Hua 2: 50–59, 68ff.).[10] It is important to distinguish these two fundamental types of reductions because in Husserl's opinion eidetic reduction is in principle possible without transcendental-phenomenological reduction. The eidetic reduction is nothing else than the methodologically conscious elaboration of the eidetic view or view of essences ("Wesenserschauung", "Wesensanschauung", see e.g. Hua 3/1: 14–18, 20–22, etc.), which is inherent in both, the natural attitude and natural perception.[11] Husserl argues that scientists, in their theoretical efforts of clarification, also try to determine essences and essential structures. Outside of phenomenology, these essences and essential structures make up regional, mundane ontologies, in correlation with the widest fields of the natural, positive sciences (Hua 3/1: 1–9). In Husserl's opinion, the phenomenology, as the ontology of transcendental essences (of the essential structures of transcendental subjectivity), is more original than these mundane ontologies; transcendental phenomenology is the foundation of the regional and mundane ontologies, as well as of the ontology of formal essences (Hua 34: 24–28, 264–278, B I 5 III).

---

[8] Cf. especially, Peter Andras Varga, *The Formation of Husserl's Notion of Philosophy*, 2012, Manuscript.

[9] See: Inga Römer, *Das Zeitdenken bei Husserl, Heidegger und Ricœur*, (Dordrecht, London, Heidelberg, New York: Springer Verlag, 2010), pp. 28ff.

[10] On the problem of reduction see in detail: Moran, 2002: 124–163. Lina Rizzoli, *Erkenntnis und Reduktion*, (Dordrecht, London, Heidelberg, New York: Springer Verlag, 2008).

[11] Cf. also: Sokolowski, 2000: 177–184.

Husserl, in his first main work, the *Logical Investigations*, reinterpreted the meaning of phenomenon as defined by his mentor, Franz Brentano. While Brentano drew a distinction between "physical" and "psychical phenomena", in his above-mentioned book Husserl rejected this conception and talked about "phenomena" in general.[12] For Husserl everything that appeared in the consciousness was a phenomenon, and the basic types of phenomena were determined by the sort of consciousness in which they appeared. In the *Logical Investigations*, phenomenon was defined as the aim or target ("Meinung") of intentionality. The term "phenomenon" acquired its proper meaning – which was then established as a standard throughout the rest of Husserl's life – in Husserl's second main work, his *Ideas Pertaining to a Pure Phenomenology and Phenomenological Philosophy* (1913).[13]

Ideas marked the first time that Husserl presented his transcendental phenomenology in a publication. After 8 years of intellectual development, he published the fully elaborated complex methodology of phenomenological reductions in this book. In *Ideas,* he redefined the meaning of phenomenon as follows: "*all real unities are unities of sense*" (Hua 3/1: 120, English [1983]: 128). All phenomena are unities of sense, ("Sinneseinheiten") – this is an extremely important definition, which served as a point of reference for all the following members and generations of the phenomenological movement after Husserl, up to the present day. But Husserl added a *second* definition to this first one, and it was no less crucial or essential with regard to his phenomenology than the former. According to this second definition, "[u]nities of sense presuppose [...] *a sense-bestowing* consciousness", (loc. cit., English [1983]: 128–129). In the subsequent periods of the history of phenomenology this second definition emerged as highly problematic. Though phenomenologists after Husserl accepted that the normal meaning of phenomenon is that it is a "unity of sense", they discovered certain kinds of phenomena that apparently did not presuppose the activity of a sense-bestowing transcendental consciousness. In terms of Husserl's original definition, these later discovered types proved to be hyper, paradox or nonsensical phenomena (in Merleau-Ponty, Michel Henry or Paul Ricœur). As Tengelyi put it, the history of phenomenology could be interpreted as the "metamorphoses of phenomenon".[14]

In the following we will be mindful of the meaning of the phenomenon of world that Husserl put forth in his *Ideas*: the world, under the phenomenological reduction, appears as a structure of sense and validity.

---

[12] Cf. Moran, *Edmund Husserl: Founder of Phenomenology*, (Cambridge: Polity Press, 2005), pp. 18, 123.

[13] Herein after referred to as *Ideas* – the author.

[14] Cf. László Tengelyi, Hans-Dieter Gondek, *Neue Phänomenologie in Frankreich*, (Berlin: Suhrkamp Verlag, 2011), pp. 9–34.

The Static Notion of World

In his famous book on Husserl's phenomenology of instincts, Nam-In Lee, expressed his opinion that the real caesura in Husserl's lifework is not in the so-called realist phenomenology of the *Logical Investigations* and the transcendental phenomenology of the *Ideas*, but between static and genetic phenomenology.[15] Though one could find signs of the conception of genetic phenomenology in the unpublished second and third book of *Ideas*,[16] the real, developed and mature notion of genetic phenomenology was born around 1916/1917.[17] He elaborated the notion of "transcendental genesis" in detail for the first time in his "Bernauer Manuscripts" on time-consciousness (cf. Hua 33: xlvi). Static phenomenology addressed the problem of "validity-foundation" and "validity-origins", while genetic phenomenology investigated the temporal genesis of these structures. Static phenomenology is not responsible for exploring the temporal dimension of the formation of these sense- and validity-structures, but – in contrast – genetic phenomenology is responsible for discovering the a priori structure of the temporal genesis of validity-structures.[18] Husserl did not entirely drop the idea of static phenomenology; he emphasized its pre-eminent importance even in the later stages of his life (in the 1930s), but he insisted that it is only a preliminary stage ("Vorstufe") for genetic phenomenology, which is to say for the deeper layers of the realm of phenomena.[19]

Thus according to Husserl static phenomenology investigates timeless structures of sense- and validity-foundations, but in genetic phenomenology we treat the constitution as a temporally extended process, and we study the different temporal phases of this constitution with regard to the a priori structure of temporal genesis and constitution. Static phenomenology gives fixity to a complex of sense and validity. Genetic phenomenology strives to unfold the temporal substructure and subsoil (*Unterboden*) of this complex, and at the same times tries to give fixity to the a priori structure of the temporal dimension of the process of constitution. What can we say about the phenomenon of world from the perspective of static constitution or static phenomenology? Husserl developed his notion of world on the basis of his concept of *horizon*. He elaborated the concept of horizon during his Göttingen Period.[20] The term first appeared in his works around 1908; we can find it in the lecture notes of *Old and new logic* 1908/1909, (*Alte und neue Logik. Vorlesung 1908/1909*, Hua Mab 6: 4f). But we can also discover certain preliminary forms of this notion in earlier works; it manifests in the term "*fringe*" (he used the English word), for

[15] See: Lee, 1993: 18.

[16] See: Moran, 2002: 166–168.

[17] Lee, 1993: 24.

[18] Lee, 1993: 17–28.

[19] Lee, 1993: 20.

[20] Cf. Hua 39: xxxvii, in footnote 1.

example, which appears in the *Logical Investigations* and which Husserl borrowed from William James (cf. also: Hua 6: 267).

What does the word "horizon" mean in Husserl? He defines it the following way: "the horizons are "predelineated" possibilities" ("Die Horizonte sind vorgezeichnete Potentialitäten", Hua 1: 82, English [1999]: 44). A horizon is a direct or indirect *indication* of possible intuitions or meanings. A horizon is a horizon of indicated possibilities. In the *Ideas,* there are two fundamental meanings of the term "world" for Husserl: the world of natural attitude, which we are compelled to exclude with phenomenological reduction (Hua 3/1: 56 ff., 103 ff.), and the transcendental meaning of the world, which we gain after performing the reduction. The core of this transcendental meaning is the phenomenon of horizon; the indication of possible future intuitions, actions, meanings, etc. The things around us indicate other things, and the way we use them, the structures of everyday and theoretical practice. The immediate system of these indications makes up our immediate environment, our familiar surrounding ("Umwelt"). The ultimate phenomenological meaning of the world is that it is a horizon of every possible horizon, it is an absolute and encompassing horizon ("Totalhorizont" or "Urhorizont").[21] There are two basic types of horizon in Husserl: *external* and *internal* horizons. External horizon refers to the relationship of things amongst each other. Internal horizon is everything that might be discovered about the very same thing, treated in isolation, only by itself.[22] The widest context of these horizons is the phenomenon of world. The world is the *hyper-context* of every possible context. The ultimate meaning of world-phenomenon in Husserl is that it is an infinite open system of indications.

In the phenomenological attitude, from the standpoint of static phenomenology the world-phenomenon not only possesses the meaning of a horizon, but, in strong intertwinement with it, also the meaning of a systematically articulated and ordered *region*: a realm or domain of *contents*, along with the relations between those contents. That is how one can speak about the spiritual and cultural world, the world of nature, etc.

The Genetic Origins of the World

Static phenomenology presents the phenomenon of the world as a total-horizon of indications. The constitutional *origins* of the particular and concrete types of world, as well as the origins of the hyper-context of all these worlds, could only be unfolded from the perspective of genetic phenomenology. Since from the standpoint of static phenomenology the world appeared as a system of indications, genetic phenomenology will show us the formation and genesis of this system, or the particular subsystems of the total system. We speak about particular worlds in the meaning of

---

[21] Cf. Rolf Kühn, *Husserls Begriff der Passivität,* (Freiburg/München: Karl Alber Verlag, 1998), pp. 318 ff.

[22] Kühn, 1998: 320.

particular contexts, "subcontexts" in a total-context (the world of different cultures, subcultures and communities, as well as the world of mathematics, the world of mathematicians, the world of philosophers, etc.). The formation or genesis of such a particular system of indications is based on the phenomenon of *motivation*. According to Husserl the basic law of the physical world is causation, and "the fundamental law of the spiritual world" is motivation (Hua 4: 211, English [1989]: 223).

A system of motivations serves as a foundational basis for any formation and genesis of a system of indications. Husserl speaks about "motivational systems" and "motivational connexion" in this context (Hua 1: 109, 144, English [1999]: 75, 114). This system, this connexion provides the basic level for the birth, development and crystallization of a system of indications. The indication indicates things and their relationships in the world. The indication is a vehicle of the worldly, mundane practice. The world in its entirety as well as its particular environments, always has a certain *style* or certain styles. It means that it always possesses certain familiar features that we can rely on. The world (and its particular surroundings) has some characteristic ways in which it manifests itself to us. We are accustomed to these ways and they might be referred to as the styles of the world. The style is the familiar and accustomed order of indications.

The system of indications, the particular styles of the world are born from the *passive* constitutional processes of association, habitualization[23] and sedimentation. For Husserl, the "principle of passive genesis" is *association* (Hua 1: 113, English [1999]: 80). He takes this to be a transcendental-phenomenological principle and draws a sharp distinction between the meaning of the term from an empiricist viewpoint and its psychological interpretation (loc. cit.). "*Association* is a general name for a set of laws determining why it is that one experience points forward to something similar. It is, then, a primitive feature of all sense-bestowal".[24] Association is the original, foundational event of the birth of meaning or sense. When a child "understands for the first time the final sense of scissors", it is due to the transcendental operation of association (Hua 1: 141, English [1999]: 111). *Habitualization* is the appropriation of certain patterns of activities (Hua 1: §32). For Husserl, the ego is a system of possible activities, it is a space of possibilities of "I can" ("ich kann", Hua 1: 128, English: 96–97). The "I can" does not necessarily only imply bodily movements, but also extends to the direction of one's attention and other ego-activities. Sedimentation is the sinking of once-established complexities and formations of sense into the depth of historicity and tradition. They are concealed in the flow of cultural and spiritual history, but under the surface they affect the present stage of history, our general attitudes in everyday life, and they also impact other fields of life, such as science, philosophy, arts, etc. But these sedimented layers of

[23] "Habitualisierung", Hua 6: 471, also: "habitualities", "Habitualitäten". Cf.: Klaus Held, "Husserl's Phenomenology of the Life-World", in Donn Welton (edited), *The New Husserl*, (Bloomington and Indianapolis: Indiana University Press, 2003), pp. 52f., Anthony Steinbock, *Home and Beyond: Generative Phenomenology after Husserl*, (Evanston, Illinois: Northwestern University Press, 1995), pp. 47, 205.

[24] Moran, 2002: 168.

meaning and sense can be unravelled and reactivated. The dual movement of sedimentation and reactivation lends a peculiar rhythmic, pulsing character to the movements of tradition.[25]

The transcendental processes of association, habitualization and sedimentation take place on the basis of a complex system of motivations. But in the late Husserl's view the level of motivation is not the deepest layer in the life of subjectivity. As it is well-known (at least since Nam-In Lee's book of),[26] the late Husserl's thought took an important turn towards the central importance of *instincts*. In his later research manuscripts, especially in the so-called C-manuscripts, Husserl recurrently dealt with the notion of a systematic phenomenology of instincts (Hua Mab 8).[27] The first appearance of the idea of a phenomenology of instincts can be found in Husserl's so-called "St. Märgen Manuscripts" from autumn 1921.[28] In his St. Märgen-Manuscripts (A VII 13), Husserl spoke for the first time about the concepts of "instinct-intentionality" and "instinct-association" ("Instinktintentionalität", "Instinktassoziation"). Pursuant to these conceptions, the deepest level of subjectivity is a system of different instincts (but the higher levels of consciousness also have correlative instinctive backgrounds), and motivations are based on this system of instincts from the very outset.

In the late Husserl's opinion the most original, most fundamental instinct is that of self-preservation ("Selbsterhaltung", cf. Hua 15: 518ff, 590ff). The instincts of feeding and defecating are the most primitive forms of instincts that seek to preserve the subject as an individual; there are also intersubjective instincts, as are the instincts of sexuality and mother-child relationships, which strive to preserve the species. In certain places Husserl takes the mother-child relationship to be the most original, "primordial intersubjective" relation.[29] Every higher type and system of instinct is a more sophisticated and more complex manifestation of the basic instinct of self-preservation. Thus Husserl spoke of the "instinct of reason" ("Vernunftinstinkt", "Vernunfttrieb"), the "drive of cognition", ("Erkenntnistrieb"), the "instinct of culture", the "transcendental instinct", etc.[30] All of them are instincts on a higher plane than the original instinct of self-preservation. In Husserl instinct means the primordial connection between the subject and her/his world.

As a matter of course Husserl did not treat the phenomenon of instinct as a biological fact but rather tried to unfold its transcendental meaning. He described it as a necessary connection between the subject and her/his world, a connection

---

[25] Here I would like to refer to Andrea Carroccio's excellent paper on "Edmund Husserl on Tradition"; presented at the "61st International Congress of Phenomenology", Istanbul, Friday, 1st of July, 2011.

[26] Lee, 1993.

[27] Cf. also: Hua 15: xxxix, ff., Inga Römer, 2010: 86–112.

[28] See also: Donn Welton, *The Other Husserl: The Horizons of Transcendental Phenomenology*, (Bloomington, Indiana: Indiana University Press, 2000), pp. 201, 221; Steinbock, 1995: 37, f.

[29] See: Dan Zahavi, *Husserl's Phenomenology*, (Stanford, California: Stanford University Press, 2003), pp. 113f.

[30] Cf. Lee, 1993: 186ff.

that manifests itself in increasingly complex forms. Instinct is the ultimate original passive basis for any conscious, egological activity. Through instincts the subject and her/his world are disclosed as two necessary moments of the same structure, in mutual and inseparable correlation. It is the instinct that reveals the world to the subject in certain necessary ways.

Life-World: Home-World and Alien-World

According to the late Husserl, the ultimate context and horizon of every practical, theoretical and philosophical (ultimately of all transcendental phenomenological) activity and achievement was the hidden underground life-world ("Lebenswelt"). In terms of the workings of transcendental subjectivity the life-world is an achievement of constitution, as is everything else. But the lowest layer of transcendental subjectivity, as we saw above, is the instinct. Hence the constitution of the life-world is an instinctive process. The life-world has its roots in the instinct-life of the subject. It is important to see that the life-world is a composition of cultural, spiritual and intellectual achievements, and does not primarily stem from the world of nature. Ernst Tugendhat tried to maintain that the life-world must be understood as the world of nature.[31] In my view, Nam-In Lee was right to criticize him for this interpretation; I am of the opinion that Lee showed perfectly that we should understand the life-world as a collective *cultural* and spiritual achievement of transcendental *inter* subjectivity, and the sense of purely physical nature first emerges only from this original intersubjective basis.[32]

The life-world is first and foremost a cultural world, the everyday of subjective and relative validities. The rigorously scientific achievements of natural sciences come from this original soil of subjective-relative certainties. In other words, the problem of the life-world can be found in Husserl as early as the time of *Philosophy of Arithmetic* (1891, in Hua 12). The numbers that we use in exact, formalized and symbolic mathematics come from our pre-scientific and pre-theoretical natural life. The authentic numbers (low positive cardinals, up to circa 10 or 12) find their fulfillment in direct intuition. According to the young Husserl the inauthentic, symbolic representations of numbers are based on these original and intuitively concrete representations of authentic numbers. That is a notion that also crops up in Husserl's last works: in *Crisis* and *The Origin of Geometry* he developed the concept that in their roots, scientific idealizations go back to the pre-theoretic and pre-scientific world of everyday, intuitively based practice (Hua 6).

In the early and middle periods of his work, the precursory concept of the life-world in Husserl is the world of pre-theoretical and pre-scientific everyday life, which appeared as "natural attitude", ("natürliche Einstellung") in the *Ideas* (Hua 3/1: §27).

[31] Cf. Ernst Tugendhat, *Der Wahrheitsbegriff bei Husserl und Heidegger*, (Berlin: Walter de Gruyter & Co, 1970), pp. 240–245.

[32] See: Lee, 1993: 149.

The term "natural attitude" can also be found in the *Logical Investigations*, but it is a subsequent insertion in the second edition (1913) of said study. Heidegger, in his Marburg Period, harshly criticized Husserl's notion of natural attitude because in Heidegger's opinion Husserl treated this phenomenon through the paradigm of modern scientific world-view, and consequently distorted it. In Heidegger's interpretation, Husserl suggests that we see "physical bodies" and "animated beings" in the natural attitude, and we have to "suspend" all of these under phenomenological reduction. But in Heidegger's opinion this is a fundamentally mistaken approach towards natural attitude: in our everyday life we do not encounter anything like "physical, three-dimensional objects" and "bodies", or "animated, organic, biological beings", etc. According to Heidegger these are all scientific abstractions that have nothing to do with natural life. In the natural attitude we see the table in front of us, we hear the neighbour's dog barking, we meet our friend on the street, etc. (GA 20: 131ff, 148ff, English [1992]: 92ff., 117ff.).

But if we take a closer look at what Husserl himself said about natural attitude, we find that his conception of it was in fact very similar to Heidegger's, and in reality there is no sign of the "scientific distortions" of which Heidegger accused Husserl. Husserl wrote the following on natural attitude:

> Immediately, physical things stand there as Objects of use, the "table" with its "books," the "drinking glass," the "vase" the "piano," etc. These value-characteristics and practical characteristics also belong *constitutively to the Objects "on hand" as Objects*, regardless of whether or not I turn to such characteristics and the Objects. Naturally this applies not only in the case of the "mere physical things," but also in the case of humans and brute animals belonging to my surroundings. - They are my "friends" or "enemies", my "servants" or "superiors," "strangers" or "relatives," etc.
>
> (Hua 3/1: 58, English [1983]: 53).

In the *second* book of *Ideas* it turns out that this level of familiar, accustomed things is not even the lowest one. The lowest level is made up of our bodily existence and our bodily movements through which we are embedded in our surrounding world. Thus, in my opinion, Husserl's and Heidegger's conception of natural attitude were in fact quite close to one another (as we shall also see in the following section of this paper).

The life-world is the world of the pre-theoretical, and pre-scientific natural attitude. The expression "life-world" appeared in Husserl's writings as early as 1917, in a manuscript (Hua 4: 374). The young Heidegger, during his early Freiburg Period between 1919 and 1923, used it as an integral part of his terminology (GA 56/57: 18, GA 58: §§15–20).[33] Their common source was – in all likelihood – the Neokantian Georg Simmel, who used this term in his 1912 book, *On Religion* (*Die Religion*).[34] Both in Husserl and Heidegger this term designated the subjective,

[33] Mathias Jung, "Die frühen Freiburger Vorlesungen und andere Schriften 1919–1923. Aufbau einer eigenen Philosophie im historischen Kontext", in D. Thomä (ed.), *Heidegger-Handbuch. Leben – Werk – Wirkung*, (Metzler, Stuttgart, Weimar, 2003), pp. 13–22. o.

[34] See also: Hans-Helmuth Gander, *Selbstverständnis und Lebenswelt. Grundzüge einer phänomenologischen Hermeneutik im Ausgang von Husserl und Heidegger*, (Frankfurt am Main: Klostermann, 2001), p. 110.

relative familiar surrounding world of everyday life, which thereby served as the original foundation for every scientific, philosophical and abstract achievement. But it would be hard to deny that Heidegger's vehement criticism of Husserl (which Husserl was exposed to in 1929, in *Being and Time, Kant and the Problem of Metaphysics* and Heidegger's lecture on *What is Metaphysics?*) played a crucial role in the fact that in the 1930s Husserl made the problem of the life-world central to his phenomenological investigations.

The life-world is an intersubjective constitution. We can understand the essential structures of the life-world by considering the basic level of intersubjective genesis. The most fundamental level of intersubjective genesis is *analogizing appresentation* (cf. Hua 1: §§50–55). Analogizing appresentation is a particular way of association, namely the most basic level of intersubjective association by way of "pairing". By means of analogizing appresentation I "learn" to recognize the other person's body and my own body as a "pair", as both being animated, organic *bodies*, as incarnated subjectivities. The working of analogizing appresentation involves *empathy*, which is not an artificial, conscious and psychological operation, but the original manifestation of the intersubjective character of my existence.[35] Empathy is a transcendental operation that unfolds an original and essential connection between another and myself, presenting both of us as members of an intersubjective community.

In the course of analogizing appresentation I experience both the sameness and the difference between myself and the other. Both of us are subjects: I have my own sphere of "ownness" ("Eigenheit", Hua 1: §44), and the other person also has her/his own. The sphere of ownness or *primordiality* designates an exclusive access to my own experiences. If my sphere of ownness were to reach directly into the other person's ownness, then I and the other would not be two different subjects but one and the same (Hua 1: 139, English [1999]: 108–109).[36] I cannot experience the other person's experience in the direct, intimate and personal way she/he does, and neither can she/he experience my experiences as I do in my sphere of ownness. We both have *indirect* access to each other's sphere of private subjectivity, and this indirectness involves an ineliminable moment of "strangeness" ("Fremdheit") or alterity.[37] By virtue of its very essence, the communication with the other necessarily implies this dialectic of sameness and alterity.

According to Husserl, this dialectic returns to the higher levels of intersubjective constitution, on the constitutive level of the life-world. We experience the styles of our intersubjective surrounding, about the *normality* and *abnormality* of our

[35] See: Marosan, "Apodicticity and Transcendental Phenomenology", in *Perspectives: International Postgraduate Journal of Philosophy*, Vol. 2. (Autumn 2009), pp. 89, 91, 92ff.

[36] Cf. loc. cit. "If it were, if what belongs to the other's own essence were directly accessible, it would be merely a moment of my own essence, and ultimately he himself and I myself would be the same."

[37] Cf. Sokolowski, 2000: 154f.

intersubjective relationships.[38] We have our common norms of behaviour and communication in our common intersubjective praxis. But these norms, which make up the particular normality of our community, primarily characterize the peculiar life-world of *my* cultural community. This normality constitutes the "primordiality" of my intersubjective community, and thus makes it my intimate, familiar world, my home-world ("Heimwelt"). The world of another cultural community, with its strange, unfamiliar habits and norms constitutes an "abnormal" case of intersubjectivity for me. With their strangeness, the life-worlds of subjects belonging to another cultural community appear to me as "alien-worlds" ("Fremdwelt").

The home-world may be treated as a sphere of a higher order primordiality.[39] In his late research manuscripts, especially in the 1930s, Husserl analysed in great detail the problem of communication between different cultural worlds.[40] In these texts Husserl's performed investigations might almost be called hermeneutic. Alien culture is a radical anomaly for us that could easily serve as a source of *conflicts* (Hua 29: 42).[41] We see the other culture as a manifestation of irrationality that we try to eliminate. At the lower level of history we try to do so by means of *force* and *violence*, (Hua 29: 42ff), while at the higher historical levels of intersubjective rationality, the different cultures increasingly try to bridge the differences and gaps in mutual understanding by means of communication. They try to create a common open space for intercultural communication. In Husserl's view history has a tendentious direction toward mutual rationalization and *harmonization* between the domestic and the alien. According to Husserl, the process of world-history is a process of universal harmonization between different cultural communities, during which the different cultures learn to recognize what they have in common with alien cultures and in which regard they are the same. The progress of history points to the idea of a common planetary life-world, the common home of every different culture. *This is a sort of the Kantian ideal of "eternal peace"*.

The World as Existential Structure in Heidegger: Being-in-the-World

As is well-known, Heidegger attempted to undertake a hermeneutical transformation of Husserl's transcendental phenomenology.[42] Heidegger's attempt realized a fairly different type of phenomenology, which at first glance appears to be fundamentally

[38] See amongst others: Klaus Held: "Heimwelt, Fremdwelt, die eine Welt", in *Phänomenologische Forschungen* 24 (1991), pp. 305–337; Steinbock, "Generativiy and the Scope of Generative Phenomenology", in Donn Welton (ed.), *The New Husserl,* (Bloomington and Indianapolis: Indiana University Press, 2003), pp. 289–325, especially: 296ff.

[39] Held, 1991: 308.

[40] These texts can be found in the 15th, 29th and 39th volumes of Husserliana.

[41] See also: Held, 1991: 323f.

[42] Cf. Moran, 2002: 197.

different from the Husserlian way of phenomenology. The striking differences between them cried out for bridging or mediating between the two. Contemporary authors[43] as well as later interpreters – up to the present day[44] – have been analyzing the common features and differences of Husserl's transcendental philosophy and Heidegger's existential phenomenology.

Heidegger's criticism of Husserl's transcendentalism encompasses a wide and complex range of issues, with many layers and embranchments. A thorough analysis of this problem definitely exceeds the limits of this paper; here we can only refer to the most basic points.[45] Up until the very end of his life, Heidegger's relationship with his former teacher was marked by ambivalence. This ambivalent relationship haunted Heidegger even in his last seminars, which he gave in Zähringen in 1973 (in GA15: 372ff.). Heidegger claimed that Husserl gave him "eyes to see", which is to say that Husserl's phenomenology enabled him to articulate his philosophical problem adequately for the first time, (GA63: 5).[46] It was Husserl's *Logical Investigations* that enabled him to delineate the philosophical foundations of the question of the meaning of Being ("Seinsfrage") (GA20: 34, English [1992]: 27).[47] Though Heidegger strongly criticized Husserl's second main work, the *Ideas*, as "distorting the question concerning the meaning of Being" (GA20: §11), he also owed a lot to that volume of his mentor, since the conception of regional ontologies provided him with an essential inspiration in the elaboration of his own conception of phenomenology as the fundamental ontology of Being.

There was also another side to Heidegger's ambivalent relationship to Husserl: Heidegger thought that Husserl had not been able to fulfil his promise; he was unable to "return to the things themselves". Instead, he followed in a very uncritical way the modern natural scientific paradigm, and – strongly connected to the previous thought – the Cartesian paradigm of immanent consciousness and subjectivity. What was the "thing itself" ("die Sache selbst") for Heidegger? At the specific level, it was the concrete factual life of man. More generally, it is the question concerning the meaning of Being. In Heidegger these two things are closely intertwined: the factual existence of man is an open clearing of Being, and also a particular way in which the Being exists.

We would like to highlight only three crucial points in Heidegger's arborescent and extremely complex critique of Husserl. (1) In Heidegger's view Husserl's interpretation of the natural attitude was fundamentally distorted by the paradigm of natural sciences. (2) Husserl's main concern was the question of the structure of transcendental subjectivity. But thereby Husserl distorted the human existence as such, in its very roots. Consciousness is only an abstract aspect of human existence, and Husserl took this abstraction to be the essence of man. (3) Husserl spoke about

[43] See: Georg Misch, *Lebensphilosophie und Phänomenologie*, (Bonn: F. Cohen Verlag, 1930).

[44] Amongst others: Tugendhat, 1970, Buckley, 1992, Hopkins, 1993, Keller, 1999, Crowell, 2001, Moran, 2000, 2002, 2007, Tengelyi, 1998, 2011, just to name a few.

[45] As regards the most essential point of Heidegger's criticism, see: Moran, 2002: 226–238.

[46] Cf. also: Moran, 2002: 228.

[47] See also: Römer, 2010: 121ff.

the "Being" ("Sein") of subjectivity as a general concept, but he never clarified what we should mean by "Being", (GA20: §11). He failed to raise the most fundamental question, the question about the meaning of Being. But Heidegger also added that this "omission" ("Versäumnis") is not something that Husserl could be blamed for, since it is something "fateful" ("Schicksalhaft"). The "oblivion of Being" ("Seinsvergessenheit"), in Heidegger's view, is the innermost fate of Western philosophy.[48]

Just as for Husserl, Heidegger's lifework may be broken down into several periods and subperiods, each of which has their own peculiar centercenters and emphases. The most important turning-point in Heidegger's life was "the Turn" ("die Kehre"), that is the shifting of attention from the Being of being-there ("Dasein") as existence to the Being as such, increasingly apart from particular entities.[49] In this second period Heidegger proposed the "mildness of letting be" ("Gelassenheit der Milde"), which meant that the human existence does not try to unfold or bring out the Being violently, but takes a *passive* attitude towards it, and allows the Being to unfold itself, to let the Being itself speak. The essence of human freedom in this second period is that it can attune itself to the voice of Being, it can listen to the call of Being.[50] There are debates about the exact date of Heidegger's Turn,[51] but according to his own account the Turn first appeared in his 1930 lecture "On the Essence of Truth" (GA9: 177–202, English [1998]: 136–154).[52]

Before the Turn, the phenomenon of world was an existential structure of being-there in the form of being-in-the-world for Heidegger, but after the Turn the world become the clearing for the manifestation of Being. In my interpretation it is quite the same, but from a radically different perspective. Both these periods had different stages. Though Heidegger's thinking has certain main directions, the focus of his interest and the emphases changed from time to time during the main periods of his work, both before and after the Turn. Before the Turn, Heidegger's work had three major stages: (1) the early Freiburg period as the assistant of Husserl; (2) the Marburg period until the completion of the manuscripts for *Being and Time* (1926); (3) his efforts and projects concerning a systematic elaboration of the question of Being ("Seinsfrage") and the fundamental ontology he sketched in *Being and Time*. This third stage also encompassed the second half of his Marburg professorship, as well as the first few years of his second Freiburg period before Turning (1926–1930).

---

[48] Römer, loc. cit.

[49] See: Moran, 2002: 198f., 208f.; Römer, 2010: 10f., 205f., 220f.; Peter Trawny, *Martin Heidegger*, Frankfurt/New York: Campus Verlag, 2003: 15f., 68, 99f.

[50] Cf. Römer, 2010: 214f., 220f.; Moran, 2002: 213, 218; Robert J. Dostal, "Beyond Being: Heidegger's Plato", in Christopher E. Macann (ed.), *Martin Heidegger – Critical Assessments*, (London: Routledge, 1992), pp. 76–77.

[51] See: Moran, 2002: 198; "Some time after the publication of *Being and Time*, and probably around 1930, though exactly when is a matter of much debate". See also: Laurence Paul Hemming, "Speaking Out of Turn: Heidegger and *die Kehre*", in *International Journal of Philosophical Studies* 6(3) (October, 1998), pp. 393–423.

[52] Cf. Römer, 2010: 220.

1. Heidegger's fascination with the problem of Being began with Brentano's book, *On the Several Senses of Being in Aristotle*, which he received as a present from a friend of his father, the Catholic priest Conrad Gröber, in 1907.[53] "By way of Brentano Heidegger came to Edmund Husserl".[54] Heidegger's thinking was determined by phenomenology, until the end of his life. In his early period the psychologist debate was a central topic of his interest. The psychological interpretation of logic and mathematics argued that meanings, logical and mathematical laws and relationships are nothing but the products of our mental activities. They are psychological laws and entities. Heidegger connected the critique of this stance to the question of Being. Formal structures and relationships all belong to the Being itself, meanings and senses are the manifestations of Being. Heidegger, in his own interpretation, arrived at this insight with the help of Husserl.[55]

   Heidegger's exclusive, central concern was the problem of Being. The primary philosophical approach by which he tried to articulate and elaborate it was phenomenology. But over and above phenomenology, he absorbed and synthesized many other intellectual streams and tendencies, such as scholastic philosophy, theology, hermeneutics (Dilthey), philosophy of life (Bergson),[56] existentialism (Karl Jaspers) and many others.[57] From these studies and tendencies Heidegger gained his essential life-long interest in *factual life*. For him, phenomenology was in part the hermeneutics of factual life. The phenomenological analysis of factual human existence was an essential part in the elaboration of the question of Being for the following reasons: a. man is the being who has a relationship with his own Being; b. man is the being who has the capacity to raise questions and to elaborate these questions; c. with regard to the human being, the stake is her/his own existence. The question of Being concerns man both in her/his pre-theoretical and theoretical attitude, as the most fundamental level of every possible theoretical concern and question.

   Heidegger did not hide his critical remarks and doubts concerning the philosophy of his spiritual mentor from Husserl. The following note is dated July 19th, 1919: "In the usual Saturday discussion Julius Ebbinghaus, Martin Heidegger and Gerda Walter criticized Husserl's conception of the pure ego".[58]

[53] Cf. William J- Richardson, *Heidegger. Through Phenomenology to Thought*, (The Hague: Martinus Nijjhoff, 1963), p. xi. Moran, 2002: 26f, 2001f.1.

[54] Rüdiger Safranski, *Martin Heidegger. Between Good and Evil*, (Cambridge, Massachusetts: Harvard University Press, 1998), p. 25.

[55] Safranski, 1998: 25f., 71–88. See also George Heffernan, "A Study in the Sedimented Origins of Evidence: Husserl and His Contemporaries Engaged in a Collective Essay in the Phenomenology and Psychology of Epistemic Justification", in *Husserl Studies* 16 (1999): 141–157.

[56] Safranski, 1998: 49–53.

[57] See: Istvan M. Feher, "Religion, Theology, and Philosophy on the Way to 'Being and Time': Heidegger, the Hermeneutical, the Factical, and the Historical with Respect to Dilthey and Early Christianity." In *Research in Phenomenology* 39:(2009/1): 99–131.

[58] Karl Schumann, *Husserls Staatsphilosophie*, (München/Freiburg: Alber Verlag, 1988), p. 235.

Even in this early period Heidegger stated that if we were to speak about a pure or transcendental ego, then we must conceive of it as *historical*.[59] But Heidegger worked out the details of his criticism of Husserl only during his professorship in Marburg.

2. Through Husserl's help, Heidegger was offered the position of an "extraordinary professor" ("professor extraordinarius") at Marburg University in 1923. The distance in space helped him to become more independent from his mentor in terms of his thinking. The notes of the last course he held in Freiburg, "Ontology: The Hermeneutics of Facticity", as well as the so-called "Natorp-Bericht" ("Phenomenological Interpretations to Aristotle"),[60] can be considered part of the Marburg Period. Both of these works can be considered as groundwork directly preceding the philosophical enterprises he engaged in while at Marburg University.

   His main efforts there aimed to systematically articulate the phenomenology of human facticity. During these years he laid down the foundations of *Being and Time*. The first drafts of his first main work were written in the winter of 1923–1924.[61] In the following years he finished the first part of *Being and Time*, whose manuscripts he presented to Husserl on the 67th birthday of his master, 8 April 1926.
3. As is well-known, the actually published text of *Being and Time* is only one third part of the planned volume. Based on Heidegger's own account he finished the entire book, but – partly due to the criticism of his friend, Karl Jaspers – he abandoned the intention to publish the last parts of his work and destroyed the unpublished sections of the manuscript (cf. GA66: 413ff.). In the opinion of some observers, Heidegger included the unpublished parts of his main work in his later works, foremost "The Basic Problems of Phenomenology" (summer semester, 1927, GA24) and in *Kant and the Problem of Metaphysics* (GA3).[62]

   What is certain in any case is that Heidegger, after finishing the first part (in fact the first third) of *Being and Time*, tried to elaborate the systematic ontology he was speaking about in that book. His main work included an existential analysis of being-there. The existential analysis is the road towards the elaboration of the question of Being, that is: to the ontology of fundaments. In "The Basic Problems of Phenomenology" he tried to work out the temporality of Being in general (GA24: §§20–21). In the Leibniz-lecture ("The Metaphysical Foundations of Logic", GA26) he sketched the outlines of a universal ontology

[59] Cf. Otto Pöggeler, "Die Krise des phänomenologischen Philosophiebegriffs", in Christoph Jamme und Otto Pöggeler (ed.), *Phänomenologie im Widerstreit. Zum 50. Todestag Edmund Husserls*, (Frankfurt am Main, 1989), pp. 255–276; in particular: 257f.

[60] *Phänomenologische Interpretationen zu Aristoteles. Anzeige der Hermeneutischen Situation*, Stuttgart: Reclam Verlag, 2003, ("Natorp-Bericht"); "Ontologie: Hermeneutik der Faktizität": **GA63**.

[61] Cf. Moran, 2002: 205–206.

[62] Safranski, 1998: 171f.

that encompasses both human and non-human nature, both beings with existence and entities of mere presence-at-hand. This universal ontology he called "metontology", the ontology of the fundaments of fundaments.[63]

The World of Anonymous Publicity

In the following three subsections we will discuss two conceptions of the world in Heidegger: we encounter one in *Being and Time* and in his 1929/1930 winter semester lecture, "The Basic Concepts of Metaphysics", and another after the Turn in the 1935 lecture entitled "Introduction to Metaphysics". The most important difference between the two conceptions is that the pre-Turn interpretation of the world conceived of different levels of world. Specifically, it referred to pre-human or subhuman world, as a mere animal surrounding ("Umring"). The post-Turn interpretation of the world conceived of it purely and exclusively as human and cultural ("geistige") world. In both interpretations the world was an existential structure of being-there, a horizon of indications, accesses, and connections of things and purposes. Heidegger, in defining the most essential characteristics of his conception of the world, joined Husserl in taking the "horizon" as the basic structural feature of world, and conceiving of it as the existential structure of the human being. (In Husserl "horizon" was the structural moment of transcendental subjectivity. Since for Husserl it was the transcendental subjectivity itself that constituted things according to the structures of a horizon.

To what extent can we regard Heidegger as a "transcendental thinker"? My opinion is that as regards his pre-Turn period, we can definitely consider him a transcendental philosopher, something he himself explicitly acknowledged. The elaboration of the question of Being is essentially a transcendental project. "Every disclosure of Being as the *transcendens* is *transcendental knowledge*. Phenomenological truth (disclosedness of Being) is *veritas transcendentalis*" (GA1: 38, English [1996]: 34).[64] My view is also that after the Turn Heidegger's thought clearly retained certain transcendental motifs in his account of the history of Being ("Seinsgeschichte"). He sought the description of certain fundamental structural features of Being as such until the end of his life. We can refer to these efforts to unfold the basic structures of Being, and conditions of possibility in the history of Being, as transcendental. Every philosophy that aims to unfold the a priori necessary conditions of possibility for Being may be defined as transcendental with reason, and in light of the foregoing we should call both Heidegger's and Husserl's philosophy transcendental.

---

[63] Cf. Römer, 2010: 206–231. The term "metontology" acquired its name from the Greek expression "metabole" ("overturning") because this ontology implied the turn of ontology away from the ontology of being-there to the ontology of Being in general, of universal nature as such ("das Ganze").

[64] Concerning the transcendentalism of Heidegger, see: Tengelyi, 1998: 140ff., Moran, 2007: 135ff.

In *Being and Time* the human being as being-there appeared essentially as being-in-the-world and being-*at*-the-world. Thereby Heidegger sought to solve an old epistemological mystery: how can an isolated subject reach beyond the sphere of its immanence to the transcendence of a similarly isolated object (GA1: 62, English: 58)? According to Heidegger, this is a fundamentally mistaken question; neither is the "subject" isolated from the object of its immanence nor is the "object" separated from the subject through its transcendence. Instead, they make up a single unit in the structure of being-in-the world. The so-called problem of the "external world" is the product of an abstraction. The human being is a being-at-the-things and a caring-about-the-things. Every illusion, dream and hallucination, as a deficient mode of existence, is made possible by this original ontological connection with things. Heidegger borrowed this stance of his from Husserl.[65] Heidegger understood the Husserlian notion of intentionality as an original openness to things, as transcendence towards the world, and an ontological connection with them.[66]

Heidegger tried to develop the phenomenon of worldliness starting from the problem of signs and indications (GA1: 76ff, English: 71ff). We are surrounded by a web of signs, indications and references. This web makes up the "substance" of our being-in-the-world. The things around us are organized according to a system of references and functional utilities. These things are immediately one with their functions in the everyday practice in which we use them. They are completely dissolved in the flow of practice. We are in such close acquaintance with these things that they fail to capture our attention. We concentrate only on the task at hand, which facilitates the unperturbed course of our daily duties. We are one with the pulsating stream of this practice, we regularly do not differentiate ourselves from the *care*.

Heidegger calls this stream of original everyday practice *care* ("Sorge"). The things call our attention to themselves only if they fail to fulfill their special purpose, if they stop the flow of care. They stop this flow in their deficient modes of conspicuousness, obtrusiveness and obstinacy, when they "refuse" to work as they normally should (GA1: 74, English: 69). The being-there must take care of her/his everyday matters. His/her practice is referred to as care when it aims at things. But the being-there is an intersubjective being, a being-with ("Mitsein"). When her/his practice is aimed at the other human beings, it is called "concern" ("Fürsorge"). The being-there *cares* for the things that exist in the mode of "readiness-to-hand" ("Zuhandenheit"), but she/he *is concerned with* her/his social matters with her/his fellow-beings. The being-there lives in and lives from this web of practical, instrumental connections and her/his social relationships. These connections and relationships constitute the substance of her/his existence.

In her/his everyday life the being-there exists in a social world, in a world shared with her/his companions. Her/his world is first and foremost a shared world, which is not even her/his. It is the "one" ("das Man") who owns the public world. The

---

[65] Cf. Moran, "Heidegger's Critique of Husserl's and Brentano's Accounts of Intentionality", *Inquiry* 43 (2007), pp. 39–66. See also: Sokolowski, 2000: 14f.

[66] Cf. Römer, 2010: 121ff.

"one" is everybody else: the "they", the "others". I do not live my own life, I do not make decisions on my own: the others live in my stead, the others make my decisions. In my being, I depend entirely on the being of others. I am a dependent, *inauthentic* being. I am inauthentic because my decisions and choices are not authentically mine, as they are made by *others*. We have to *fight* for our real possibilities, opportunities and chances, we have to *fight back* against the dictatorial pressure of publicity. We have to create our independent, authentic existence; or, to put it better, we have to *discover* our independent, authentic self, and we have to let this independent self express itself through us. According to Heidegger we are ripped out from the anonymous world of the others by the mood of *anxiety*.[67] Anxiety confronts us with the fact that my death is mine alone, and nobody else could take over my own death, substituting me in my death. What is more: I may die at any moment. "Death is the possibility of the absolute impossibility of being-there.[68] Thus death reveals itself as ownmost, non-relational possibility, not to be bypassed" (GA1: 250, English: 232).

This fundamental experience of the inevitability and omnipresent possibility of one's own death alienates the being-there from the whole of the public world. Heidegger goes as far as to talk of "existential solipsism" in this context (GA1: 188, English: 176).[69] On the one hand, anxiety concerning one's own death completely isolates and secludes the being-there. On the other hand, it confronts the being-there that the *opportunity* to make her/his own decisions is too *valuable* to let the others decide on her/his behalf. The *conscience,* in the solitude of her/his anxiety, calls upon to the *authentic practice*; the conscience shows the alienated, isolated and secluded being-there a way back to the public world. In the public praxis, the being-there moves in the double movement of isolation and absorption. In the last paragraphs of Being and Time Heidegger refers to the possibility of *collective breakout* from inauthenticity, but does not elaborate this idea in this book.[70]

World and Encircling Ring: The World of Man and Animal

The public world deprives me from my independence and places me in the anonymous publicity of the "one" ("das Man"). This is one of the most essential consequences of the worldliness-analyses of *Being and Time*.[71] "The Fundamental Concepts of Metaphysics", written between 1929/1930, has different emphases and treats the animal existence in extensive details. The animal cannot alter its world in

[67] See: Moran, 2002: 241f.

[68] Translation altered – BPM.

[69] In Pierre Keller's interpretation Heidegger replaces Husserl's methodological solipsism with an existential solipsism. Cf. Pierre Keller, *Husserl and Heidegger on Human Experience*, (Cambridge: Cambridge University Press, 2004), p. 145.

[70] Cf. Safranski, 1998: 167f, 225ff.

[71] See also: Hubert Dreyfus, *Being-in-the-World*, (Cambridge, Massachusetts: The MIT Press, 1991), especially: pp. 140–162.

terms of said world's structural features (for Heidegger the creations of the spider web or of the beaver dam do not really count as genuine changes in the "world" of the animal). The animal habituates itself into its world; it is one with its surroundings; it does not differentiate itself from its surroundings. It cannot distance itself from its surroundings. The world of man consists of more than its mere surroundings. Man has the capability to alter the basic structures of her/his world. Man has the power to form and enrich her/his world.

In "The Fundamental Concepts of Metaphysics" Heidegger situated the animal existence between the mere presence-at-hand and being-there. The animal is "poor in world". What does that mean? Heidegger emphasized that the "poorness" of the animal world does not imply a comparison between it and the human world, which is "rich" in some sense. Heidegger observed that "poverty" is an essential and positive feature of the animal's existential structure. It refers to the fact that the animal is *immediately directly one with its surroundings*; one might say it *exists* by virtue of its surroundings, that it is interwoven with its environment itself from said environment. The animal does not have an access to the things in the world ("Zugangslos"). When a lizard basks in the sun, lying on a stone, it has a relationship with neither the stone nor the sun. It merely follows its *instincts*. The instincts simultaneously articulate and impoverish the environment of animal. The animal instinctively seeks special signs and indications in its surrounding. It does not recognize these signs and indications as such. In Heidegger's interpretation the animal does not recognize anything at all. The animal is pure instinct. The signs and indications affect the animal's system of instincts, they set this system into motion in the form of instinctive reactions. The animal is an existence based on response to stimuli.

The "substance" of the human being is its existence. The human existence is a relationship with itself and the world. The animal being lacks the capacity of relating itself to anything, because it lacks distance to itself, to the world and to the things around itself. The "substance" of the animal being is the *instinct*. The instinct directly seeks to sustain the animal. It drives the animal toward food, water, potential partners for reproduction, to seek the closeness of its parents and the companionship of its own species and to avoid potential enemies, etc. These things as such do not exist for the animal. It does not recognize them. There are only signs and indications which govern the purely instinctive existence that we refer to as animal.

The existence of the animal is entirely determined by this instinctive spontaneity. The instinct discloses and encloses the animal's surrounding. The animal existence is defined by the instinctive movements of its own environment. Heidegger cites the works of contemporary biologists Emanuel Radl and Jakob von Uexküll to illustrate his thesis regarding the purely instinctive character of animal existence. Heidegger cites the following example: when we put a bee on the edge of a dish full of honey, the bee sucks as much honey as it can handle. If researchers were to remove the tiny organ from the body of the bee which "informs" the animal that it is full, then the bee would be unable to stop sucking the honey at a given point and would instead ingest so much from the sweet liquid that it would be killed as a result. Rather than sustaining the animal existence, in this scenario the instinct, simply kills it.

Hence we cannot speak of behavior in the same manner in the case of animal as in the case of man. The totally instinctive behavior ("Benehmen") of animals is different from human comportment ("Verhalten"), which gives man the capacity to place herself/himself at some distance and to engage in a relationship with the around her/him. Heidegger emphatically distinguished between the animal environment and the human world. The animal does not have language, understanding or selfhood; thus it lacks everything that makes existence especially human. The Being cannot make itself manifest to an animal. The world of man discloses and opens these things. It is a clearing of Being. The environment, or rather encircling ring ("Umring"), encloses the animal in its system of instincts. The human world is disclosure, the animal environment is enclosure. Thus even in these lectures a very strong type of Cartesianism and anthropocentrism is present in Heidegger, as Matthew Calarco points out.[72]

What does it mean that man is a world-forming ("weltbildend") being? First and foremost it means that the human being possesses the distance to the world that animal existence lacks. She or he has a relationship to things, to herself or himself, to other human beings, to the world and to the Being as such. In her/his life her/his own existence is at stake, and she/he is aware of it. Being-there possesses the capacity to disclose a world, and to disclose the things in the world. But man does not only disclose her/his world; with her/his presence she/he constantly changes and shapes her/his world as well. We have an effect on the world and on our relationship to others. No human life vanishes without leaving *traces* in the world.

But Heidegger pinpoints the essence of man's world-forming capacity in two fundamental characteristics: the *as*-structure and temporality.[73] An animal lacks both. Being-there discloses the world and things according to the *as*-structure ("das hermeneutische "Als""). This means that being-there conceives the things always "as this and this thing specifically". I conceive of a thing as a scissor, a spoon, a fork, a writing-table, a house, somebody else, etc. I disclose things as concrete things, attributing concrete objective meanings to them. Heidegger adopted this conception of as-structure from Husserl; this is a transcendental structure that is based on the transcendental procedure of projection ("Entwurf").[74] The being-there projects things *as* concrete somethings. This procedure is a *temporal* process. The other fundamental existential dimension of being-there that distinguishes her/him from animal existence is her/his existential temporality.

The animal lives in the present, but in a way that we cannot even call it present in the same manner as we speak about the present or now in the case of human beings. The present acquires its special temporal meaning in the context of past and future. The present has an ecstatic meaning, just as the future and the past. The ongoing present of animal existence lacks such an ecstatic character. It is the

---

[72] Matthew Calarco, "Heidegger's Zoontology", in Matthew Calarco and Peter Attenton (ed.), *Animal Philosophy*, (London and New York: Continuum, 2004), p. 24.

[73] Cf. Graham Harman, *Heidegger Explained – From Phenomenon to Thing*, (Chicago: Open Court Publishing Company, 2007), pp. 84–91, especially: 90f.

[74] Cf. Tengelyi, 1998: 143ff.

continuous present confined within the bounds of instinct. In a strict sense, this is an atemporal present. The temporal dimensions of human existence presuppose the distance and ecstatic character that we cannot find in the animal. Each temporal dimension (past, present and future) is a movement towards the world, the existential movement of ecstasy and transcendence. In conformity with this threefold structure of human temporality, according to Heidegger, the world-forming capacity of being-there also has three basic structural modes: "The being-there in man *forms* world: (1) it brings it forth; (2) it gives an image or view of the world, it sets it forth; (3) it constitutes the world, contains and embraces it" (GA29/30: §68, English [1995]: 285).[75] These are different phases of projecting a world. This projection shapes and forms the world; and every new projection reshapes and re-forms the world of being-there.

The Spiritual World: The Late Heidegger on the World

"The Fundamental Concepts of Metaphysics" is the first text in which Heidegger addresses the problem of the *body* in great detail. The body is always a body of an existence. The world is articulated by our bodily being. But the human body and animal body are completely different in existential terms, despite their organic similarities. Heidegger, in his later writings and notes, returned at different times to the problems of the body and the animal. Heidegger is ambivalent on the question of the animal. We might as well say that Heidegger's philosophy is haunted by the spirit of the animal. In his 1935 lecture "Introduction to Metaphysics", as we already noted, he denied that the animal has any environment at all. Later he was not quite as categorical concerning the animal existence, but a definite Cartesian motif persisted in his philosophy until the end of his life. In his "Letter on Humanism" (1946) he wrote:

> Because plants and animals are lodged in their respective environments but are never placed freely into the clearing of being which alone is "world," they lack language. But in being denied language, they are not thereby suspended wordlessly in their environment. Still in this world "environment" converges all that is puzzling about living creatures. In its essence, language is not the utterance of an organism, nor is it an expression of a living thing. Nor can it be ever thought in an essentially correct way in terms of its symbolic character, perhaps not even terms of signification. Language is the clearing-concealing advent of Being itself.
>
> (GA9: 326[158], English [1998]: 248–249).

The animal's "worst sin" is that it lacks language. Language in Heidegger is essentially human language, an existential structure of being-there. Language is the capacity to unfold and disclose the depths of Being, in such a way that is by its very essence closed off from the animal. Language is the capacity to develop the hidden regions of Being and to *listen* to its voice. The animal cannot speak and cannot listen, and is therefore unable to reach the clearing of Being. In the "Zollikon Seminars" Heidegger

---

[75] See also: Harman, 2007: 90–91.

emphasized again the capacity of language as the most essential difference between man and animal: "Kant once said that man distinguishes himself from animals by the fact that he can say "I". This assertion can be formulated still more radically. The human being distinguishes himself from animals because he can "say" anything at all, that is to say, because he has a language" (GA89: 113–114, English [2001]: 87).

For Heidegger language, is not merely an instrument of communication, but the fundamental way we understand Being, and could be the means whereby we can understand our Being-together ("Mitsein"). Language is the fundamental way in which Being could be there. The event and appearance of Being, according to Heidegger, takes place in the medium of language. "Language is the house of Being. In its home human beings dwell. Those who think and those who create with words are the guardians of this home" (GA9: 313[145], English [1998]:239). Language, for Heidegger, is essentially spiritual ("geistig"). The appearance or revelation of Being is thus essentially also a spiritual event. The world, which is the disclosure of Being, is a linguistically understood, spiritual one. That is why in "Introduction to Metaphysics" Heidegger denies that an animal could have a world (or an environment) at all. Then and there Heidegger wanted to grant the unique capacity to reveal and disclose Being only to man. In regard of his capacity to disclose and reveal, man is alone in the world. With regard to this anthropocentrism, we can call Heidegger a Cartesian thinker.

The world is essentially spiritual. We live in a spiritual world. But what does "spirit" and "spiritual" mean according to Heidegger? He treats this problem in details in §15 of the "Introduction to Metaphysics". In that text, Heidegger addresses those conceptions of the spirit ("der Geist")[76] that he thought to be fundamentally improper and false. He analyzes these misinterpretations and misconceptions one after the other, and tells us why he thinks them to be essentially incorrect. He also tries to uncover the origins of the distortion of the proper sense of spirit. The spirit is: (1) not intelligence; (2) not an instrument of practical goals; (3) not culture; and (4) definitely not a collection of "showpieces and spectacles". These misinterpretations, according to Heidegger, follow from one another.

The "spirit" came by its original meaning during the period of "classical German idealism". The intellectual age after this period (the age of positivism in the second half of the nineteenth century) was not "strong enough" to endure and sustain the inherent power of this conception, (1) so it degraded the spirit to *intelligence*, to practical *cleverness* and *techniques*. (2) Practical intelligence became instrumental, a "serviceable tool" of practical goals. (3) Thus practical and instrumental intelligence organizes the domain of *culture* based on its practical considerations. The different fields and regions of culture must be ordered according to practical interests and instrumental reasons.[77] (4) The instrumentally and practically organized domain of

[76] See: GA40: 48–53, English [2000]: 47–53.

[77] "As soon as this instrumental misinterpretation of the spirit sets in, the powers of spiritual happening – poetry and fine arts, statescraft and religion – shift to a sphere where they can be *consciously* cultivated and planned. *At the same* time, they get divided up into regions. The spiritual world becomes culture, and in the creation and conservation of culture the individual seeks to fulfill himself", GA40: 51, English [2000]: 50.

culture finally emerges as the "showpieces and spectacles" of modern human existence. In this conception of culture the spirit loses even the last remainders of its original meaning.[78]

In Heidegger's interpretation these are all misconceptions, which on the one hand result from the previous misinterpretation, and which are on the other hand necessary outcomes of the decaying stage in which the spirit finds itself. In some sense, according to Heidegger, these misconceptions are all ways in which the spirit misunderstood itself. But what would be an authentic and more proper conception of the notion of spirit?

> "Spirit is neither empty acuity, nor the noncommittal play of wit, nor the understanding's boundless pursuit of analysis, nor even world reason, but rather spirit is originally attuned, knowing resolution to the essence of Being" *(Rectoral Address*, p. 13). Spirit is the empowering of the powers of being as such and as a whole. Where spirit rules, beings as such always and in each case come more into Being (*wird... seiender*).
>
> (GA40: 53, English [2000]: 52).

According to Heidegger the spirit, as spiritual world is nothing but the disclosure of Being, in its double movement of self-revealing and concealment. For Heidegger, being a spiritual being means confronting the self-revealing and self-depriving movement of Being, regardless of what it keeps in store for us. It is intellectual (spiritual) bravery to confront the revelation of Being, even if these experiences of Being are inconvenient, or even terrible and disastrous for us. We must face the dangers or defeat in the event of Being. The appearance of Being in its double movement is entirely pervaded by the spirituality of language. Thinkers and poets must help the Being in its appearance by intensifying and deepening the event of disclosure.

The Concept of the Cosmos in Eliade

In the previous sections of this essay, following the analyses of Husserl's and Heidegger's writings on the problem of the world, we were able to uncover some fundamental characteristics of this phenomenon. The world is essentially a horizon. This horizon has a fixed structure; the world is always arranged in a certain order. The world is always an ordered world. The world is first and foremost the familiar order of everyday instrumental and interpersonal practice; this means that primarily the world is an articulated surrounding of acquainted and accustomed things, indications, habits, rules, roles, norms, people, etc. In brief, the world is first and foremost our pre-theoretical and practical home-world. The world is the primal home of our existence. Furthermore, the world is a projection. Its character is

---

[78] "The spirit as intelligence in the service of goals and the spirit as culture finally become showpieces and spectacles that one takes into account along with many others, that one publicly trots out and exhibits as proof that one does not want to deny culture in favor of barbarism", GA40: 52–53, English [2000]: 52.

necessarily temporary. The world arranges itself in this projection; it is a self-arranging, self-ordering horizontal projection of the way we originally inhabit the world. There are several possible modes of habitation, which all fit in a coherent system of being-in-the-world; the latter is a shared-world ("Mitwelt") with other people, who live this world in ways very similar to my own. Habitation is always essentially *cohabitation* with others. The inhabited world is a *cohabited world* for a community of people or subjects. The projection of the world is a common, intersubjective achievement, always accomplished jointly with the others. The different modes of habitation and cohabitation with our more or less familiar fellow-subjects make up the home-world of my immediate community; they make up our immediate, more intimate and acquainted life-world.

In this ordered and self-ordering projection of our original habitation we continuously experience the phenomenon of familiarity and strangeness ("Fremdheit"), normality and abnormality. Both familiarity and strangeness, as well as normality and abnormality display certain *schemes* and *patterns*. At the very basic level of subjectivity, the patterns of strangeness represent the *alien*, i.e. the other subjects as such. In the context of our collectively inhabited home-world, they present the different communities and groups ("different" as compared to my family or closer community, for example), "subcultures", etc. In the general context of the life-world as such, the structures and patterns of strangeness and abnormality indicate alien life-worlds (different peoples and cultures), that is to say: alien-worlds ("Fremdwelten"), and alien forms of subjectivity. Alien in the latter case means: alien to human subjectivity, that is to say: animal or non-human subjectivity.

Subjectivity is essentially an incarnation; the subject is always an incarnated subjectivity. The life-world of subjects as an essentially intersubjective world is constituted by an incarnated intersubjectivity: the collective corporeality of the cultural community in question. "The "we" has its collective corporeality" as Husserl put it, ("Das *Wir* hat seine *kollektive Leiblichkeit*", Hua 39: 181). Due to its corporeal nature, the subject and the intersubjectivity are embedded into the body of nature.[79] The subject and the community of subjects take the *energies* and materials to sustain their bodily, natural and corporeal existence from the universal body of nature. The world, as the structure of human and subjective existence, continuously needs certain sources of energies to renew and reserve itself in its existence. Without energy it starves, withers and finally perishes. It is very important that in this context we must not construe the concept of energy in a naturalistic, material sense, but we need instead to render a phenomenological account of this concept. We will only be able to fulfil this task in the fourth section of this paper.

In the present section I will try to provide a brief account of Eliade's interpretation of the phenomenon of the cosmos. This will also enable me to present here the main thesis of this essay: in my view it is the phenomenon of the cosmos that gathers and holds together all the other phenomena of world investigated in the earlier sections; this phenomenon fits all those other phenomena – as dependent structure-moments – into a coherent and original structural whole. The primal

---

[79] Cf. Marosan, 2011.

structure whole is the phenomenon of the cosmos; it is the cosmos which serves as the heart at the depth of any possible world-phenomena; no matter whether it is a human, animal or whatsoever type of world. The key to the problem of the world, in my view, is Eliade's conception of the cosmos of pre-modern man.

The Sacred and the Profane

According to Eliade, ancient man experienced the world fundamentally differently than modern man. The world of modern man is a secularized, homogenous, and infinite horizon of space and time. The space and time of modern man is a system of points in which every single point is equivalent to the other; no point has special significance or importance as compared to the other ones. Ancient man had a fundamentally different experience of the world, namely that he/she experienced two essentially different dimensions of the world: the sacred and the profane. Modern man today apprehends the world as something basically profane, as lacking in sacredness. The world of ancient man was ordered and articulated into a cosmos by the sacred. The profane space and time is homogeneous, it is like a coordinate system: every point is equal to the other; no one point is more special than any other. The sacred space and time has points that are more important than the others, it has some central points.

It was the sacred that organized the world and life of ancient man and gave it a firm structure and order. The world of ancient man was a de-centered world: the center of his cosmos was not himself, but the sacred center. In his life it was not himself that was really important for him, but rather the sacred in his life and world. For the ancient man, the world as world was born with the experience of the sacred. The sacred unfolded his world as the cosmos. The eminent sense of world in the experience of pre-modern man was that of the cosmos. Beyond the borders of the cosmos, the world ceased to exist. There is a strong interrelation between the sacred and the cosmos. The sacred founded a cosmos, which for the ancient man was identical with the world itself. The opposite of the cosmos was chaos, which was the negation and absence of the world as such. Chaos was non-world ("Unwelt"), the lack of laws and order.

Eliade described the fundamental structure of religious experience as *hierophany*. Hierophany was the manifestation of the sacred. He made a distinction between hierophany, epiphany and theophany. *Epiphany* was an immediate union with divine and supernatural forces, an "unio mystica", an insight into the divine. *Theophany* was the manifestation of a god. Eliade favored the concept of hierophany in describing the fundamental features of religious experience and ancient man's relationship to his world, and he did so with good reason: *theophany*, compared with hierophany, was too specific a concept for him. In the ancient man's world-view the status of sacred also characterized universal cosmic laws, such as the law of *tao* (or *dao*) in Confucianism, Taoism and Zen Buddhism, or *karma* in Buddhism and Hinduism.

In the sacred ancient man experienced a transcendent power, an immense force of higher order. He realised his own mortal, mundane and finite nature against these supernatural forces that ruled, controlled and articulated the entire cosmos, or at least general domains of the cosmos. (Aphrodite, the goddess of love, naturally did not rule the entire cosmos of ancient Greek man, but she had potential power over every Greek man and woman in the particular form of love, and she also had control over the laws of love). The sacred represented an infinite and invisible power of higher order. (*Invisible*: the Greek man could only experience his obsession through the presence or power of Aphrodite. He could not do anything against this power, he was simply helpless and powerless against it.[80] *Infinite*: In the form of love, Aphrodite had potential power over any possible or actual Greek mortal. Higher Order or *transcendent*: Aphrodite belonged to the divine world of Olympic Gods). In brief: the experience of sacred in a certain form was the experience of *infinite*.

The sacred was first and foremost the manifestation of infinite: because (1) it was not bound to a certain object, figure or visible shape and it could make itself manifest everywhere and in every form; (2) it represented an absolute, ultimate power against any mundane and mortal power, which could exert and impact anywhere in the cosmos; (3) it gave universal structure and universal laws to the world, articulating it in the form of the cosmos. The sacred and the profane are two fundamental modes of "being-in-the-world" ("in-der-Welt-sein"), says Eliade, with explicit reference to Heidegger.[81] The sacred unfolds a dimension of the world for the man that remains essentially hidden in the context of the profane attitude. Eliade interprets these two attitudes (sacred and profane) as fundamental existential structures of the human being as such.[82] He argues that the sacred has some ineradicable manifestations in the modern world, too; thereby he tries to demonstrate that the sacred is really an ontological-existential structure of human existence, and not a mere empirical-psychological fact.[83] The cult of the Leader ("Führer") in totalitarian regimes and the cult of celebrities in consumer societies are both manifestations of the original existential structure of what is sacred in man. Atheist interpretations of world-history, like Marxism, fully adopted the eschatological view of the world proffered by Judeo-Christian religion.[84] There are universal, fundamental values and norms in which atheists also believe; atheist scientists never cease admiring the order and structural complexity of the universe.[85] Even modern world and modern society knows several forms of *initiation*. In Eliade's opinion, all this is proof that

[80] Cf. Karl Kerenyi, *Hermes, the Guide of Souls*, (Dallas: Spring Publications, 1996).

[81] Eliade, 1987: 14ff.

[82] Eliade, 1987: 15, 188ff.

[83] Eliade, 1987: 202ff.

[84] Eliade, 1987: 206f. See also: Karl Löwith, *Meaning in History*, (Chicago: Chicago University Press, 1957), pp. 33ff., Jacob Taubes, *Occidental Eschatology*, (California, Stanford: Stanford University Press, 2009), pp. 184ff.

[85] Cf. Richard Dawkins, *The God Delusion*, (Boston: Houghton Mifflin Books, 2008), pp. 31–41.

the human being has a profound tendency to seek the sacred in the world, and the tendency toward the sacred cannot simply be torn out of human existence.[86]

The sacred broke into the world of ancient man as a radically alien power from a transcendent domain. The sacred obsessed, inhabited the things, places and people in the world. In the context of totemism it was not the stone, the river, the tree or an animal itself that counted as sacred for ancient man, as opposed to the notions embraced by some who did not engage in the study of religion thoroughly enough. It was the sacred power that made the stone, the river, the tree etc. special. That thing, animal or person was obsessed by the sacred. As Eliade wrote:

> The modern Occidental has certain uneasiness before many manifestations of the sacred. He finds it difficult to accept the fact that, for many human beings, the sacred can be manifested in stones or trees, for example. But, as we shall soon see, what is involved is not a veneration of the stone itself, a cult of tree in itself. The sacred tree, the sacred stone is not adored as stone or tree; they are worshipped precisely because they are hierophanies, because they show something that is no longer stone or tree, but the sacred, the completely different.
>
> (Eliade, 1987: 11–12).

The sacred pointed to special places in space and time, it pointed to things amongst things, people amongst other people. The special places articulated the spatial world of ancient man, whereas the special occasion structured his time. The manifestation of the sacred in space and time implicated some more concrete structures of the experience of the sacred, which on the one hand also belonged to the existential structure of human beings, but which were on the other hand also highly contextualized by the socio-cultural history and experience of the people or community in question.

Sacred places, such as temples, churches, sanctuaries, sacred groves, oracles in ancient Greece and Rome, medieval pilgrimage destinations, etc. were centers of pre-modern life. Ancient man knew several centers, among which there was always a highest center: an *axis mundi*, a *world-axis*. All sacred places served as a channel between the mundane and the divine world, but there were places which served as a center for all other centers, and where the presence of divine or transcendent was especially intensive and strong. This *axis mundi* connected the world of mortals, gods and the dead with one another. These three regions belonged to the cosmos of ancient people: they made up three domains of the same cosmos. During the early stages of the history of religion this was meant in a very literal spatial manner. The divine world belonged to the sky, the human world was that of the earth, and the world of the dead was the domain under the surface of the earth, it was the underworld. There were special places where ancient man could journey to these different parts of the cosmos (journeys to the world of gods and world of the dead); tales of these journeys were preserved in accounts of journeys to the Otherworld (such as Odysseus' journey to Hades, or Dante's descent into Hell and ascension to Heaven).

---

[86] Eliade, 1987: 211f.

The special points in time were the religious *fests*.[87] Whereas sacred places signified a spatial connection to the divine or transcendent, sacred occasions and fests created a temporal connection to the transcendent domain. In and through the fests ancient, fundamental events from the history of the cosmos, gods and people returned. The fests of ancient peoples were usually connected to *myths of origin*. During these fests archaic people did not only *commemorate* fundamentally important ancient events from the history of the people in question. That was Jung's interpretation of fests and rites, but Eliade criticizes this conception. According to Eliade, premodern man experienced himself through repeating the event in question. The fest was not simply a form of commemoration, but an existential experience concerning the *repetition* of the mythical event itself. He called this existential structure "eternal return" – he transformed the Nietzschean idea of Eternal Return into a structure-moment of the complete existential structure of sacred.[88] This meant that the ancient Christian did not simply commemorate the event of the birth of Christ at Christmas, but that in his mind, the Saviour himself was reborn every Christmas. For him, Good Friday was more than a commemoration of the crucifixion of Jesus Christ; to his mind, Jesus was crucified every year on Good Friday. The New Year Fest amongst ancient peoples was the festival of the birth of the world, a festival of the myth of origin. They experienced every New Year fest as the re-genesis or rebirth of the world itself.[89]

Sacred time intertwined with the problem of myths of origin. The cosmo-genesis was the event when the world – both in its spatiality and temporality – was born. The myths of origins also represented the origin of time. In the beginning the world was born with its most fundamental structures, with the community's ancestors and ancient heroes, and with the norms and laws that bound both nature and society. The principal moments of the creation of the world, the most important events of a community's pre-history, returned again and again in the sacred fest, and they were also evoked by the rituals, religious practice, the *rite*. Through special places, times, tools, instruments, materials, procedures and also special persons (the sacred king, priests, shamans, the chosen heroes, etc.) premodern man could gain access to the transcendent, otherworldly regions of the world. But for the ancient man even the otherworld had many different types.

The Concrete Structures of the Ancient Man's Cosmos

Ancient man, just like his modern descendant, was surrounded by things and people. They made up his world. Some of those things and men were special; they were possessed by the sacred. Archaic man was born into a world of objects and people, and through the 'sacred' this objective and intersubjective world was formed

---

[87] Eliade, 1987: 68ff.

[88] Cf. Eliade, *The Myth of Eternal Return*, (New York: Harper & Brothers Publishers, 1959).

[89] Cf. Eliade, 1987: 105f.

and articulated into a 'true' world, that is to say into a 'cosmos'. The sacred made itself manifest in certain places, times, people and things. The sacred could also abandon those things and people which it had previously possessed. For ancient people usually there were sacred *cosmic laws*, which controlled these events and processes of possession and abandonment. Thus for example from time to time, the sacred kings of certain tribes had to fight for their position with a rival.

As I mentioned above, the world of ancient man was synonymous with the cosmos itself. Beyond the cosmos the world ceased to exist. Ancient man experienced the world as the cosmos. This cosmos was highly structured and articulated, with several centers that were ordered into a hierarchy with the main spatial, temporal and communal center at the top, which is the sacred feast, sacred place or sacred leader; a center of world, a center of time and a center of the community (in certain cases, in certain cultures there was not one but many of these highest centers). The factual details of the religious beliefs and practice of the particular community – that is to say the factual manifestation of sacred in a cultural community – were determined by the soil and particular history of the life-world of the community in question.

As was noted previously, the cosmos had different regions, namely one for each, the world of mortals, the deities and the dead. The three regions were, the world of earth, the world of *above* (skies, heavens) and the world of *below* (underworld, world of shadows, hell, etc.). In the beginning, these spatial directions were taken in a very literal sense. During the later periods of the history or religion of a culture this spatiality became more and more metaphorical. There were universal laws that governed and ruled all these regions of the cosmos. These rules and laws also possessed the status of the sacred also, and the cosmos gained its ultimate form and structure from them.

These cosmic laws bound both the natural and social world; they functioned as laws of nature and society at the same time. Ancient man simply did not distinguish between laws of nature and laws of society. In pre-modern societies the breaking of community laws counted as an offence against nature. Laws of nature and society were mundane laws, and deities stood above these laws. But there were laws which bound even the gods. In monotheistic religions the one and only God appeared as the source of every law. His will was the only, ultimate law. In cosmic oriental religions (such as Confucianism, Taoism, Buddhism)[90] there was but one universal law (and its different levels and many concrete specifications) that bound everything.

Just as the world of gods and world of the dead, the world of the sacred belonged to the archaic cosmos. They were, so to say, domesticated forms of the sacred, which in making the immediate world of ancient man feel more familiar and better known to its inhabitants. The cosmic forms of the sacred had firm and well-known *rules* and laws. Beyond the border of the cosmos these rules ceased to apply. Beyond the world, the ultimate negation of the world began: the chaos, the non-world or counter-world ("Unwelt"). The *chaos* was the domain of the uncertain, it was a

---

[90] Cf. Eliade, *History of Religious Ideas*, (Chicago: University of Chicago Press, 1981, 1985, 1988), Vol. I–III.

blurry, lawless, ruleless terrain. The cosmos was the world of men. Beyond the cosmos everyone counted as non-human, as infernal, demonic creatures. Greeks considered non-Greeks subhuman barbarians whose speech cannot be understood. For the Romans, the world of civilization ended at the borders of the Roman Empire. The conquering Roman emperors understood the task of conquest and extending the borders of the Empire as a sacral responsibility; it was the duty of *civilization* (of civilizing the uncivilized), of transforming the chaos into the cosmos.

The cosmos was the closer, familiar, acquainted life-world of ancient man; those who did not belong to this life-world counted as potential and actual enemies. The oath, the given word was only binding in the case of a member of the tribe, of the same cosmos. Oaths made to members of another could be broken because said persons were not men at all. When Odysseus and Diomedes caught the Trojan spy, Dolon, in the Greek camp, Odysseus made a "strong oath" that they would spare Dolon's life if Dolon told them everything he knew. Dolon did as Odysseus had requested, but in taking Dolon's head Diomedes broke the "strong oath" Odysseus had made, arguing that an oath counts nothing in the case of a Trojan.[91] In ancient societies such breaking of the given word was not counted as a form of dishonesty or villainy, as long as it was performed against someone who did not belong to the given tribe or community. Odysseus' and Diomedes' deed was treated as a cunning and clever action.

The otherworld contained several different fundamental types for the ancient man. Basically, it had two fundamental types: (1) on the one hand it had what we might term a "culturally domesticated" form that was fit into the cosmos of the particular cultural community in question; (2) on the other hand there was a "wild form" of the otherworld that began beyond the borders of this cosmos. The world of the unknown, the world of alien cultures was the form *par excellence* of the otherworld, of chaos. Fighting against the chaos was a form of world-creating activity. *The* battle against the unknown aimed at the transformation of chaos into the cosmos. The world of one's own gods, the world of the dead of one's community represented a familiar form of the otherworld. The unknown was a chaotic, demonic and infernal world. The basic line of rupture that hides under the difference between the cosmos and chaos is the difference between *known* and *unknown*. The world was *per definitionem* known; the domain of unknown was the negation of the world.

No laws of the home-world were valid in the domain of chaos; neither human nor natural, not even divine law. In the beginning of Hesiod's *Theogony* Chaos created a rupture in the *order* of the world, of cosmos itself. As long as this gap was open it was possible for all the species of mixed, chaotic and demonic creatures, such as

[91] Homer, *The Iliad*, 445–455; see, Homer, *The Iliad,* New York: Oxford University Press, 2011: 162. ""I warn you, Dolon do not put thought of escape in your heart/your news may be good, but you have fallen into our hands,/and if we not accept a ransom or let you go free/you will surely return some days to Achaeans's swift ships,/either to spy on us or to fight us, matching strength to strength/But if you are beaten down by my hands and lose your life,/you will never after this be an affliction to the Argives"//So he spoke. Dolon was about to touch his chin in entreaty,/with his brawny hand, but Diomedes lunged with his sword and/drove it through the middle of his neck, severing both tendons;/and his head rolled in the dust while he was still speaking."

Cyclopes, Giants, Titans, Centaurs, etc. to be born. Once this rupture was closed, the age of Chaos was over, and species like them were born no more. In the terrains of the otherworld the familiar laws of the home-world changed fundamentally, if they did not cease to function at all. I mention the example of tales of journeys to the otherworld as a final illustration of this thesis. In the *immrama* (singular: *immram*, "journey to Otherworld") of Celtic folks and peoples the well-known laws of human and natural laws ceased to exist. When in ancient Irish mythology Oisín (Ossian) followed the fairy-queen Niamh to Tír na n'Óg, to the "Land of the Young", to the island of the fairies, during his journey one step counted as a hundred miles in the man-world, and 1 year in the land of fairies was a 100 years in the land of men.[92] In the case of medieval pilgrimages, when a man received permission for his journey from the landlord, or from his ecclesiastic superior if he was a friar, he proceeded to collect some devotional objects or relics to protect himself because he knew that beyond his well-known familiar home-world there was the unknown, the beyond the otherworld, where – as he knew full well – the customary laws were not valid anymore, and thus anything could happen.[93] The crusaders, as they went to the Holy Land, asked in every village they passed: "Is this Jerusalem yet?" because the prior laws of space and time were not valid after crossing the borders of their closer home.

When assessing things from the first person plural vantage points of his community, ancient man found himself in the perspective of "we". This first person plural perspective was the original basis of that which was known, and also the basis of the human/non-human distinction, and thus – consequently – the good/evil distinction as well. The first person plural perspective served as a basis for ethnocentrism: my group is the center of the world; my group is *ab ovo* good; the other group is potentially an enemy, potentially evil, and – in case of extreme conflicts – not even human.[94] This distinction between my group and the other group as good and evil, as human and non-human, left a deep trace in human history. As Reinhart Koselleck put it, this distinction served as a basis for *asymmetric counter-concepts*.[95] But in history there was also the possibility of harmonization and mediation between different cultures and different cultural perspectives. The names of gods and deities can be translated into different languages because in most cases gods had certain functions that had an analogy or parallel in other pantheons. This fact opened the way for something Jan Assmann referred to as "intercultural translatability".[96] This translatability created the possibility of a relatively peaceful cultural co-existence

---

[92] Cf. Patrick Weston Joyce, *Ancient Celtic Romances*, (London: Parkgate Books, 1997), pp. 385–399.

[93] Cf. Aron Gurevich, *Categories of Medieval Culture*, (London: Routledge and Kegan Paul, 1985), pp. 35ff. 72, 74ff.

[94] On ethnocentrism see: William G. Sumner, *Folkways*, (New York: Ginn, 1906).

[95] Cf. Reinhart Koselleck, *Future Past: On the Semantics of Historical Time*, (New York: Columbia University Press, 2004), pp. 155ff.

[96] Cf. Jan Assmann, *Moses the Egyptian: The Memory of Egypt in Western Monotheism*, (Cambridge, MA: Harvard University Press, 1998), pp. 1–22, 44–54.

and cohabitation in ancient empires, and finally the multicultural, multi-religious world of the Roman Empire.

What raised a serious problem for the multicultural and multi-religious environment of empires were monotheistic "counter-religions", as Assmann calls them.[97] These posed a problem because counter-religions defined themselves as the only true religion, as standing in opposition to any and every other actual and possible religions and religious cults. In the beginning, this view of the world excluded any possibility of harmonizing different perspectives (save for the instance of converting someone and abandoning the previous religion), and later it made harmonization very difficult. From the perspective of a counter-religion, all other monotheistic religions and cults counted as heresies and every polytheistic cult was idolatry – it regarded both cases as manifestations of infernal forces. The ultimate solution to this conflict-generating character of counter-religion was the abandonment of the character of "counter-religion" as such, that is to say to abandon the view that the particular religion in question is the one and only possessor of the one and only truth. This process began only in modern Europe, and it developed very slowly only since the seventeenth century, since the spread of secularization in Europe. It was the slow and painful process of recognizing that there is not only one "good" and "true" perspective, and that we have to learn how to mediate and communicate between different perspectives.

Transcendental Morphology

It is necessary to devote more detailed analysis to some parts of Eliade's comparative investigations of the history of religions in order to see the fundamental features of his conception of the phenomenon of the world. He treated this phenomenon in a deeply phenomenological manner, by reconstructing modern man's perspective through the phenomenology of pre-modern human existence. The phenomenological analysis of pre-modern existence could shed light on some phenomena much more intensively than in the case of the modern human being (as Heidegger already suggested in *Being and Time*, in the case of "primitive being-there"; cf. GA2: 50ff., English [1996]: 47ff.). Eliade's analysis of the ancient man's experience of world as the cosmos displayed some features that invite us to treat the phenomenon of the cosmos as the seed of the phenomenon of world as such.

The most important features of Eliade's investigations of religious experience are (1) the sacred as an existential structure, and (2) the cosmos, as an ordered, articulated collective perspectivity, which was bound to the life-world and cultural-historical practice of a community, and which was defined by its own rules and norms. The *sacred* appeared as an invisible, omniscient and omnipresent power of higher order (or at least a power that had a much more thorough knowledge of the world than any mortal can possess, and whose spatial manifestation was not limited

[97] Assmann, 1998: 44ff.

by the boundaries that apply to human beings). The sacred was experienced by the ancient man either as a personalised power, such as deities and other supernatural beings, or as an impersonal, cosmic power, in the form of a cosmic law, which works through an ultimately binding and inescapable normativity. The *cosmos* was experienced by the ancient man as a normatively and structurally ordered horizontal perspective, whose boundaries simultaneously designated the end of the aforementioned norms, laws and structures; that is how the negation and absence of the world began beyond those borders resulting in *chaos*, the complete absence of any rule, norm or law.

What makes the ancient experience of the cosmos *special* in comparison with the previously discussed interpretations of the world advanced by Husserl and Heidegger? The special reason on account of which I sought to highlight Eliade's conception of the cosmos-chaos duality is its sensibility concerning two fundamental features of the world: it is extremely *fragile* and extremely *flexible* at the same time. It is fragile because the flowing in or breaking in of any new experience could shake the fundaments of the fixed and crystallized form of the cosmos. The rupture of chaos into cosmos shakes the fundamental structure of the cosmos, but after any and all harm and damage it experienced from chaos, the cosmos tries to reorganize, re-establish and regenerate itself. The cosmos is a very stable, tough and enduring phenomenon: it is very difficult to fully destroy it. As long as the minimal "physical" conditions exist for it to preserve and restore itself from its damage and injuries, it strives to do so. (In the phenomenological sense, we must use the word "physical" with great caution; from a phenomenological perspective we can obviously construe this word only in an analogous sense). In my opinion, these two features of the cosmos are very apparent when analyzing the world-experience of ancient man.

I gave preference to the word "cosmos" over the term "world" because the former expression, in light of Eliade's comparative religious analysis, on the one hand precisely expresses the ordered, articulated and structured character of the world, ruled by rules and norms, with its inherent perspectivity, but on the other hand also signifies its flexibility and fragility. It was only the juxtaposition of cosmos and chaos, the permanent *threat* of chaos to cosmos that made it possible to properly emphasize these characteristic features. Every event that preserves, nourishes and enriches a cosmos, in other words has a positive effect on it, counts as *cosmic*; every event that in one way or another harms the cosmos, halts or encumbers its proper and adequate functioning, and ultimately threatens its bare existence, counts as *chaotic*. What helps, supports and increases (collective) self-preservation is cosmic, whatever constitutes a threat to it is *chaotic*.

We could now provide a preliminary definition of the phenomenon of cosmos: a cosmos is a horizontal disclosure that has an inherent form, order and structure. It is ruled by norms and rules, it strives to assimilate the different, the non-cosmic, the chaotic, but it can also accommodate itself to this difference. From this latter feature arises its essentially ambivalent nature: the cosmos is flexible and fragile at the same time. Any chaotic, non-cosmic event could inflict harm on it and damage its texture, but it is capable of achieving even the toughest, most unbelievable forms of adaptation in order to just remain in existence, to not perish. *The discipline of transcendental morphology deals with the laws and structures of changes, accommodation and injuries, birth and death, starvation and flourishing in the context of the cosmos.* Chaos is only a parasitic form of being in the cosmos; the laws of chaos ought to be investigated with regard to the laws of the cosmos.

The phenomenon of the cosmos clearly shows the structural moments of *flexibility* and *fragility*, as well as the *horizontality*, which is articulated in the form of an order with either a single highest point or several *central points* that organize this horizon. Among these structural moments one should also include the cosmos' internal relation to those forces that – actually or at least on the level of possibility – threaten the order and existence of the cosmos, namely its essential relationship to *chaos*. Now, at the end of our study, we have to have a look at three more important questions concerning the problem of the cosmos: (1) the problem of embodiment, (2) the phenomenon of energy, (3) the question of the sacred and its relationship to the phenomenological account of the cosmos.

The Body of the Cosmos

In the Husserl- and Heidegger-section we repeatedly mentioned the problem of embodiment as a fundamental feature of human existence and, as such, a phenomenological characteristic of fundamental importance with regard to the world itself. Moreover, in the context of our common, public world not only our particular bodily nature or our own embodiment has peculiar importance, but also the embodied nature of intersubjectivity, that is the collective embodiment of a community.

It is crucial to clarify in what sense we shall understand this embodiment. Especially in the French phenomenological tradition the issue of embodiment was one type of argument against Cartesianism, and particularly the solipsism of an isolated mind. The Hungarian philosopher János Tőzsér engaged in a debate with Sean Gallagher, in which Tőzsér argued against the necessity of the embodied experience of mind. He observed that some out of body experiences do not support this presumed necessity of embodiment. The mind can very well experience and conceive of itself as a pure perspective, as something completely bodiless.[98] In Tőzsér's opinion the body is just another phenomenon amongst many other phenomena.

This view illuminates a very important fact: we *must not* naturalize the phenomenon of embodiment, and we *must not* conceive of it as being in a *direct* relation with the physical reality itself. The body is just a system of phenomena amongst other phenomena – without a doubt, Tőzsér had a profound reason to say so. Husserl shared this opinion. To put it another way, the presumption of a direct relationship with the physical reality would have been a break with the concept of phenomenological reduction. Indeed, the body, even the proper body, is just a system of

[98] János Tőzsér-Balogh Zsuzsanna, *Much Ado about Nothing: The Discarded Representations Revisited'*, Manuscript.

appearances or phenomena. But it is not without any consequence to other types and systems of phenomena: Husserl stated that these systems of phenomena exhibited mutual influence on one another, and there were some phenomenal systems that showed a more fundamental determination of or influence on other phenomenal systems. The own body belonged exactly to those types of phenomena. The phenomenal system of the own body determined in a most fundamental way the phenomenal system of the world, it determined the way in which the world could appear to us. The phenomenal system of the collective embodiment of our community determined the way in which the intersubjective, common world could appear to us.

All these phenomenal systems showed inner layers and were themselves organized into layers, which interpenetrated each other on the one hand, and at the same time also determined the functioning of the other, underlying systems. The cosmos (and also the chaos) is a phenomenon that appears through the determination of the own body and collective corporeality. But the own body and the "collective body" are in all cases systems of phenomena that determine the manifestation of other systems and layers of phenomena, and which should not be naturalized. However the body demands *energy* from which it can sustain itself – also from the phenomenological point of view.

The Phenomenology of Energy

The phenomenon of energy, like the phenomenon of the body, should not be naturalized either, but must be kept in phenomenological reduction. Thus we lack the physical terms with which we would be able to define it otherwise; we must seek other ways of definition. Energy is something from which a cosmos could nourish and preserve itself, and without which it starves and finally dies. How could we make this phenomenon manifest without turning towards materialistic and naturalistic notions?

We must concentrate our attention exclusively on the phenomenal-phenomenological aspect of the processes of nourishing the cosmoses. At first sight, the phenomenon of energy is connected to certain patterns of events. We find that every cosmos-like formation lives according to a certain *rhythm*. The existence itself has a rhythm. We sleep and awake, eat and drink, defecate, etc. The animal must find something to eat and drink, a place where it can sleep. Plants demand water, good soil and sunlight. If a pattern of events is broken, there arises confusion in the particular functions of the cosmos, in its life and existence. If a living being (also us, human beings) cannot get food and drink for a very long period, it starves and then dies. If a plant does not get water for a certain time it withers away. If a living being cannot sleep at all for a longer time, it dies.

The case is quite the same in the situation of every cosmos-like being. Take a shop, for example, a grocery: it needs customers to sustain itself – customers who pay for the goods they buy there. Their money "feeds" the grocery. If there are fewer customers at that grocery, then the shop begins to "starve". The shopkeeper then tries to find ways in which he can spare expenses that he considers unnecessary. Just like plant which in the times of drought "withdraws" humidity from the leaves to the inner parts of branch. But after a longer time of drought even these "techniques" of survival are not enough; the plants will inevitably die. Just as in the case of our shopkeeper. He can enact these cuts and savings (he could even drastically reduce the price of his products and goods) for a while – but if there are too few customers to sustain his grocery (e.g. because there is an economic crisis), then the store will necessarily die, it will go bankrupt. In the case of this grocery, if it were treated as a "cosmos", the event of economic crisis would be a chaotic one, a manifestation of chaos.

In all these cases we see patterns of events that in one way or another display a correlation with the phenomenon of energy. In the absence of these events, the cosmos perishes, maybe slowly, maybe more quickly. But these events do not represent the phenomenon of energy. Not even the pattern they make up. We say that someone has great supplies of physical energies. He sleeps little and he is always active, doing everything he does with great enthusiasm and impulsiveness; we never see him get tired, we see him moving incessantly as if he was a perpetual motion machine. We also say that somebody has immense intellectual energies. He writes and speaks with great impulsivity, he never ceases writing new essays, studies and books, he has one new idea after the other – and we do not stop wondering: how can he have such great supplies of intellectual energy? In this case the term "energy" obviously refers to some *capacities* and *potentialities* of the person in question.

These people can also *exhaust* their supplies, even in a dual meaning of the word. First, they become *very* tired after a longer period of especially intense activity; they require some extra time for recreation. Second, if they live permanently with an extreme intensity, their capacities to "pool" and "restore" energies – either physical or intellectual – continuously decline. The phenomenon of energy has something to do with capacity or potentiality – which makes the situation a bit complicated, since it uncovers the fact that the phenomenon of energy, by its very essence, is an *indirect* or *invisible phenomenon* – something that we can only localize between two visible types of phenomena: the patterns of events, which from a certain point of view "feed" and "nourish" the cosmos, and the functioning of the cosmos itself, its visible activities and manifestations.

The phenomenon of energy can therefore only be made manifest in an indirect way. We could also speak about the increase or decline in the capacities for physical or intellectual activity (to stick with the example of a specific human being), which is to say an increase or decrease in the supply of energy which man uses for physical and intellectual activities, and which can only be observed in their manifestations in the latter. There are some events that increase these capacities and there are others that decrease them. We could speak in this case about something very similar to what Spinoza discussed in his *Ethics*, when he wrote that "happiness" increases the *conatus* of man, while "sadness" decreases his *conatus*. We could observe in our own case that some sad events or failures can drain all our energies, and some happy, fortunate event or success can give us extra energy.

The Relationship to the Sacred

As we already mentioned in previous sections of this essay, Husserl and Heidegger found an ingenious solution to the challenge of solipsism: they made man's relation to the world and to other people his existential (in Husserl: transcendental) structure. Husserl, in his late period, embedded this conception into his theory of transcendental instincts. For him, instinct was the original relation to the world and to the other. He based every intentional tendency upon the original instinct of self-preservation, and presented every such intentional tendency and activity as an achievement and manifestation of self-preservation at ever higher levels of complexity. Husserl was anxious not to naturalize the conception of instinct in his theory of instincts, but to reflect it with the method of transcendental reduction.

For Husserl, it is of course impossible to prove the existence of natural world and the other as a natural being. But one can apodictically show up the intentional, instinctive tendency toward an objective world and toward an objective, independent other. In addition to these two fundamental types of relations (to an objective world and to the other), in the late Husserl there appears – though only in occasional remarks – a third basic type of relation: the "divine instinct" ("der göttliche Instinkt", A VI 21: 102). For Husserl there is an instinctive relation or tendency towards the divine region of the world. This was an ingenious conception that could be traced back to Descartes' views on the Idea of Infinity. This conception in Husserl stood in parallel to the other two descriptions concerning the inherent, instinctive, intentional relation to the world and to the other; there is but one possible way of proving the existence of a divine transcendence: if we make the relation to such transcendence a structural moment of human subjectivity. In Husserl this instinctive tendency towards a divine region of the world was a tendency toward the unfolding of a universal, divine teleology that embraced the totality of monads.

Eliade – as well as Max Scheler[99] – did something very similar when he tried to clarify the relationship between the sacred and the existential structure of man. I believe that there is no other possible way to find a passage to divine transcendence but this one. Eliade described the concrete structural moments of this phenomenon of sacredness: an invisible, omnipresent, omnipotent and omniscient transcendent power. When the phenomenon of the sacred took the form of personal transcendence, the sacred made itself manifest as an unlimited, infinite subjective force. The task of transcendental morphology with regard to the experience of the sacred is that it needs to describe the ways in which the transcendent in general and the concrete forms of the sacred make themselves manifest in the human cosmos.

[99] Max Scheler, *Vom Ewigen im Menschen*, (Berlin, Der neue Geist, 1933).

Conclusion

We analyzed three phenomenological accounts of the problem of the world. I tried to argue for the standpoint that at the core of these three accounts there is a single phenomenon that is most readily apparent in Eliade's comparative studies on the history of religions. It is the phenomenon of the cosmos, which shows *horizontality* (as the world did in Husserl), which – *in case of the human cosmos* – is both *spiritual* and *cultural* (as the late Heidegger stated in his investigation concerning the world), and which displays a fixed *order*, held together by some centers (as in the case of ancient man in Eliade), and which exhibited the fundamental features of flexibility and fragility at the same time.

I also asserted that the pre-form of the human cosmos, its seed-like form, can also be demonstrated in the lower levels of subjectivity, as in animals. I called the phenomenological discipline whose task it is to describe the a priori laws of the birth and death, the enrichment and starvation, and the development and stagnation of a cosmos *transcendental morphology*.

The Outside's Inside: The Phenomenology of the External World in Hedwig Conrad-Martius' Thought

Ronny Miron

Abstract *On the Ontology and Doctrine of Appearance of the Real External World* (1916) is the first publication from a vast corpus of writings by Hedwig Conrad-Martius (CM), a phenomenologist from the Munich School. The phenomenon of "the real external world" encloses within itself everything that "exists outside" (*Daraußenseinde*) and that is "of the external world" (*Außenweltliches*). The "self-presentation" that deeply characterizes the sensory givenness is an essential foundation in the phenomenon of the reality, to the extent that it distinguishes it from everything that "lacks a Being- for-itself" and thus misses what might be presented externally. Although sensory appearance is not itself the totality of the external world, CM determines that the pure observation of what the sensory appearance presents by itself and in itself, and not of what is above and beyond it, provides the "framework for the whole" of the research, since by sensory presentation "the book of the real world is being opened". The paper proposes a critical explication of both constitutive phenomena of the sensory givenness, "feeling's givenness" and "manifest appearance givenness", and suggests a metaphysical interpretation that explicates them in terms of the relation between immanence and transcendence that seems to be a key to the understanding the phenomenology of reality that that unifies the entirety of CM's writings.

Preface: Intuition of Essence

The issue of the existence of the world as an independent reality and as one's object of reference and context of experience occupies modern philosophy, especially since Kant. In *On the Ontology and Doctrine of Appearance of the Real External World* (1916)

R. Miron (✉)
Professor, Head of the program of Hermeneutics and Cultural Studies,
Interdisciplinary Unit, Bar Ilan University, Israel
e-mail: mironronny@gmail.com

A.-T. Tymieniecka (ed.), *Phenomenology of Space and Time: The Forces of the Cosmos and the Ontopoietic Genesis of Life: Book One*, Analecta Husserliana 116,
DOI 10.1007/978-3-319-02015-0_24,

(*Doctrine of Appearance*), Hedwig Conrad-Martius (CM) (1888–1966) explores a phenomenological-realistic conception of the real external world.[1] She characterizes this phenomenon as enclosing within itself everything that "exists outside" (*Daraußenseinde*) and that is "of the external world" (*Außenweltliches*), which as such is real (386). As in the rest of her writings, so also in *Doctrine of Appearance*, CM was committed to "intuition of essence" (*Wesensfassung*), which she shared with the early phenomenologists of the Munich Circle, which in her opinion is applicable in any thinkable object sphere (355, n. 1).[2] Decisive in this approach is the observation that searches for the most primordial characteristics of things, or alternatively the essential "what" thanks to which they became these specific objects, while putting aside any previous theories or assumptions regarding them. The emphasis, typical of analytic discourse, about the formal and logical aspects of the issue under discussion, is explicitly rejected by the early phenomenologists in favor of

---

[1] Hedwig Conrad-Martius, "Zur Ontologie und Erscheinungslehre der realen Außenwelt" [1913], in Jahrbuch für Philosophie und phänomenologische Forschung 3 (1916), pp. 353–354. References to this book are given in the body text. Bolds in citations follow the original. In the works of CM, the year mentioned first is the year of the work's writing, while the second year denotes the year of publication. Archive materials are taken from the Munich Estate Archive, Die Nachllässe der Münchener Phänomenologen, Die Bayerische Staatsbibliotheck, München (BSM, Nachlass). CM explains that the title Doctrine of Appearance refers to an object domain that is between "nature" and "life's essence" in which the human subject is CM, Zur Ontologie und Erscheinungslehre der realen Auβenwelt", pp. 345–542, n. 1). However, the aspect of the nature as such remains in this book as a latent layer that will be revealed only in her later writings. See in particular: Hedwig Conrad-Martius, Der Selbstaufbau Der Natur, Entelechien und Energien (Munich: Kösel-Verlag, 1961). *Doctrine of Appearance* is an exploration of the first chapter in her first essay (Hedwig Conrad-Martius, Die erkenntnistheoretischen Grundlagen des Positivismus [1912] (Bergzabern: Heinrich Müller, 1920). pp. 10–24), that received an award from the department of philosophy at the University of Göttingen. The subtitle "associated with a critique of positivistic theories", as well as the debate with positivism throughout the text (CM, "Zur Ontologie und Erscheinungslehre der realen Auβenwelt", pp. 345–347; 352; 357–358; 361–365; 378; 382–386; 390–391; 398–400; 423; 425) clearly indicate its roots in the first essay. In 1912 Alexander Pfänder recognized *Doctrine of Appearance* as a Ph.D. thesis in the University of Munich (Ursula Avé-Lallemant, "Hedwig Conrad Martuis", in Jahrbuch der Evangelischen Akademie Tutzing XV (1965/1966), p. 212). In 1913, the expanded chapter of the award-winning essay was printed and submitted as a dissertation, in a version almost identical to *Doctrine of Appearance*. In the epilogue to the special print in 1920 (Hedwig Conrad-Martius, Die erkenntnistheoretischen Grundlagen des Positivismus, pp. 130–131), CM referred to this fact and explained that she left behind the direction of criticism of positivism in favor of an ontological direction. Indeed, the plan to elaborate the rest of the chapters has never been carried out. Eberhard Avé-Lallemant, CM's assistant, estate curator, collector, and editor of her published writings told me (conversation, Munich, July 4th, 2003) that the publication of *Doctrine of Appearance* in 1913 was well received at the time.

[2] The Munich Circle included a group of intellectuals and philosophers from Munich, the first generation of the phenomenologists, whose prominent members included: Alexander Pfänder, Johannes Daubert, Moritz Geiger, Theodor Conrad, Adolf Reinach, Dietrich von Hildebrand, Maximilian Beck, Max Scheler, Jean Hering, Alexander Koyré, Roman Ingarden, Edith Stein, and Hedwig Conrad-Martuis. For further reading about this circle see: Eberhard Avé-Lallemant, Phänomenologie und Realität: Vergleichende Untersuchungen zur 'München-Gittinger' und 'Freiburger' Phänomenologie (Habilitationsschrift) (Munich, 1971), pp. 19–38.

"pure description of the given, which is carried out by pure observation".[3] Jean Hering determined that in phenomenology we are not troubled about the specific use of logic, since "it is incapable of matter-of-fact (*sachlich*) recognition".[4] CM explains the reservation of this method from a definition-oriented approach as follows: "definitions deliver always the tiny measurement – mostly the external – of what must be available and thus the matter [*Sache*] is in place. Yet in genuine essence analysis, the matter is grasped by its middle-point and by the origin-point belonging to it. Hence, we must have the matter as such within our gaze".[5] These words clearly follow Husserl's appeal "to return to the things themselves"[6] that wished to focus only on what "delivers itself through observation", i.e., "the given" or "the givenness".[7] Following Husserl, the early phenomenologists where convinced that perceived objects and the modes in which they become known were

[3] Alexanader Pfänder, cited from Eberhard Avé-Lallemant, "Die phänomenologische Bewegung. Ursprung, Anfänge und Ausblick", in Hans Reiner Sepp (ed.), Edmund Husserl und die phänomenologische Bewegung (Freiburg/Munich: Alber, 1988), p. 69.

[4] Jean Hering, cited from Franz Georg Schmücker, Die Phänomenologie als Methode der Wesenerkenntnis, unter besondere Berücksichtigung der Auffassung der München-Göttinger Phänomenologenchule (Dissertation) (München: 1956), p. 32.

[5] Hedwig Conrad-Martius, "Naturwissenschaft und Naturphilosophie" [1950], in Schriften zur Philosophie, 3 vols. (Munich: Kösel-Verlag, 1964), Vol. 2, p. 4. Also in *Doctrine of Appearance*, CM argues that no matter how we characterize the phenomenological investigation, it will never define in advance its mission and the essence belonging to the matter under discussion (CM, "Zur Ontologie und Erscheinungslehre der realen Außenwelt", p. 354, n. 1). See also Edmund Husserl, Ideen zu einer reiner Phänomenologie und phänomenologischen Philosophie, Husserliana III (The Hague: Martinus Nijhoff, 1950), §63. For the realist phenomenologists the main point is that definitions indeed do not meet the object itself. For further discussion of the differences between a defending approach and "intuition of essence", see: Gerhard Ebel, Untersuchungen zu einer Realistischen Grundlegung der Phänomenologischen Wesensschau (Dissertation) (Munich, 1965), pp. 19–15.

[6] Edmund Husserl, Logische Untersuchungen II (Hamburg: Felix Meiner Verlag, 2009), p. 19, p. 22. This saying is widely discussed, see for example: Josef Seifert, "Was ist Philosophie? Die Antwort der Realistische Phänomenologie", in Zeitschrift für Philosophische Forschung 49/3 (1995), pp. 92–98; Helmut Kuhn, "Phänomenologie und Realität", in Zeitschrift für Philosophische Forschung 23/3 (1969), pp. 397–399.

[7] The early phenomenologists were inspired by Husserl's struggle in "Logical Investigations" against psychologism, relativism, and varying reductionism (Edmund Husserl, Logische Untersuchungen I, p. 81, p. 117), in particular by his principle that it is possible to observe consciousness' condition apart from the thinking subject (Edmund Husserl, Logische Untersuchungen I, p. 240). See also: Franz Georg Schmücker, Die Phänomenologie als Methode der Wesenerkenntnis, unter besondere Berücksichtigung der Auffassung der München-Göttinger Phänomenologenchule, p. 31; Gerda Walther, Phänomenologie der Mystik (Olten: Freiburg im Breisgau, 1955), p. 190. CM admits the influence of "Logical Investigations" on her (CM, "Zur Ontologie und Erscheinungslehre der realen Außenwelt", p. 355, n. 1). The principles of the object's oriented observation were phrased by Hering (Jean Hering, "Bemerkung über das Wesen, die Wesenheit und die Idee", in Jahrbuch für Philosophie und phänomenologische Forschung VI (1921), p. 496). For a detailed discussion of this observation in regard to the Munich Circle, see: Eberhard Avé-Lallemant, Phänomenologie und Realität: Vergleichende Untersuchungen zur 'München-Göttinger' und 'Freiburger' Phänomenologie, pp. 89–105; Schmücker, op. cit., pp. 3–8.

established upon the lawfulness of essence, which is independent of consciousness and of the subject in general. The method of "intuition of essence", based on this conviction, is characterized as a pure description of the givens, free of particular and contingent features,[8] and as "mental clinging to givenness".[9] Adolf Reinach regarded this method as fundamental for philosophy in general, although not the only one, since it enables one to reach the things themselves, while its sole leading criterion is the matter itself.[10] Wilhelm Schapp highlighted phenomenology's fundamental bond to reality, since in it "the world is not formally imposed but present according to its genuine mode of givenness but is present in the originality of its givenness' mode from the beginning till the end and in each stage of the investigation".[11] In the same spirit, CM too understood Husserl's early focus on objects as an attempt to follow the way they reveal themselves and not as a description of the subject's constitutive relations to objects, of laws of thinking, or of the way objects are given to consciousness. She pinned the knowability of objects on the objective logos inherent in the things themselves that is capable of declaring its own existence,[12] i.e., on what she will later designate as the "kósmos noetós",[13] and the subject's power for transcendence that poses one simultaneously in his/her own place and in the "foreign world" (408).

In *Doctrine of Appearance*, CM explicitly expresses her identification with the method of "intuition of essence" and designates it as a "genuine philosophical mission" (348), responding to which might open knowing's possibilities and whole objects' domain that cannot be accessed otherwise, in particular not by psychology and biology of consciousness (346).[14] She devotes *Doctrine of Appearance* to "a sui generis idea of 'real being' surrounding factually existing being" (365), or alternatively "for proving of **phenomena that are capable of**

[8] Alexandra Elisabeth Pfeifer, Hedwig Conrad-Martuis, Eine Phänomenologische sicht auf Nature und Welt (Würzburg: Orbis Phenomenologicus, Königshausen & Neumann, 2005), p. 15; Eberhard Avé-Lallemant, query "Hedwig Conrad-Martius (1888–1966): Phenomenology and Reality', in Herbert Spiegelberg (ed.), The Phenomenological Movement: A Historical Introduction (3rd ed., The Hague: Martinus Nijhoff, 1982), p. 212.

[9] Walther, op. cit., p. 21.

[10] Adolf Reinach, Was ist Phänomenologie (Munich: Kösel-Verlag, 1951), pp. 71–73.

[11] Wilhelm Schapp, Beiträge zur Phänomenologie der Wahrnehmung (Göttingen: Kaestner, 1910), p. 12.

[12] Hedwig Conrad-Martius, "Über das Wesen des Wesens" [1956], in Schriften zur Philosophie, 3 vols. (Munich: Kösel-Verlag, 1965), Vol. 3, p. 347.

[13] Hedwig Conrad-Martius, "Phänomenologie und Spekulation" [1956], in Schriften zur Philosophie, 3 vols. (Munich: Kösel-Verlag, 1965), Vol. 3, p. 377.

[14] For further reading about the method of "intuition of essence", especially in the realistic school of phenomenology, see: Adolf Reinach, Die apriorischen Grundlagen des bürgerlichen Rechtes (2nd ed., Halle: Max Niemeyer, 1913), pp. 1–163; Alexander Pfänder, "Zur Psychologie der Gesinnung", in Jahrbuch für Philosophie und Phänomenologische Forschung 1 (1913), pp. 325–404; Pfeiffer, op. cit., pp. 27–35; Schmücker, op. cit., pp. 1–33; Ebel, op. cit., pp. 1–25.

being perceived concretely" (390).[15] The idea designates here the separateness from consciousness, the bearing of absoluteness of its own, essentiality, and substantial unity per se. However, as will be revealed later, being independent of specific essence, the "phenomenon" involves within itself also aspects of consciousness (353).

At the center of the present article stands *Doctrine of Appearance*, in which CM wishes to achieve "independent and matter-of-fact objectivities" regarding the modes of givenness of the real external world. It is an extremely loaded, compressed, and complicated essay that creates considerable difficulties for the reader. However, already in this early book, one encounters her peculiar phenomenological investigation that enfolds within itself a realistic approach and a profound study of transcendentalism.[16] Indeed, the importance and value of *Doctrine of Appearance* becomes evident especially in CM's later writings, in which she continues exploring fundaments that often appeared in this essay in the raw, without the necessary explanations. Devoting a wide part of this article to the presentation and explication of the perceptions that first appeared in *Doctrine of Appearance* wishes to fill the vacuum in the relatively inadequate literature on CM, in which this essay is almost never mentioned, despite its decisive importance for the understanding of the entirety of her work.[17]

From the 'Concealed' to the 'Primordial'

Two fundamental questions posed at the outset of *Doctrine of Appearance* guide CM's study of the external world: The first is in which real mode are essences given to us? The second is where do we encounter essences in their concrete realization,

---

[15] Similarly to CM, Friz Heinemann also wrote about the affinity of phenomenology to the concrete being. He mentioned another essay by CM ("Phänomenologie und Spekulation") but surprisingly not "Zur Ontologie und Erscheinungslehre der realen Auβenwelt", in which she established this theme.

[16] CM hardly mentions other philosophers who dealt with the theme of the external world. Besides her, also Roman Ingarden, another realist phenomenologist, has discussed this issue. See: Roman Ingarden, Der Streit um die Existenz der Welt, vol. 1, Existenzialontologie, (Tübingen: Max Niemeyer, 1964). On his approach to the essences, see: Roman Ingarden, "Essentiale Fragen", in Jahrbuch für Philosophie und phänomenoligische Forschung 7 (1925), pp. 125–304. Franz Josef Brecht criticised the realistic orientation of the Munich Circle for not dealing with the problem of transcendentalism. See: Franz Josef Brecht, Bewußtsein und Existenz: Wesen und Weg der Phänomenologie (Habilitationsschrift) (Bremen: J. Storm, 1948), p. 42 n. 2. However, this judgment cannot be addressed to CM, who dealt with it early in 1916.

[17] This argument is valid also for the research that dealt with CM's understanding of the world (see for example: Angela Ales Bello, "The Controversy about the Existence of the World in Edmund Husserl's Phenomenological School: A. Reinach, R. Ingarden, H. Conrad-Martius, E. Stein", in Analecta Husserliana LXXIX (2004), pp. 97–115). Exceptions to this are Schmücker, who mentioned it as a "decisive breakthrough" (Schmücker, op. cit., p. 39, n. 1) and Ebel, who mentioned it four times in his dissertation (see: Ebel, op. cit., p. 16, n. 42; p. 17; p. 22 n. 48; p. 42). Still, neither of them delivered a systematic and comprehensive discussion of the issues at stake in this essay.

and is it indeed impossible to doubt their real concreteness in general as well as in a special case? (356). These questions initiate the problematization of the given, and thus pose a clear limit between the immediate appearance of the external world and its real being, which in her opinion are not identical (427). CM contends that sometimes there exists only a "semblance of real presence-being that does not correspond to the actually present being" (356), or more specifically "that the uncovered beings-position (*Seinsstelle*) of the concretely given does not always hold what actually appears in it" (358).[18] Accordingly, CM distinguishes between "phenomenal beginning-material" (*phänomenales Anfangsmaterial*) and "genuine phenomenon" or "primordial phenomenon" (*Urphänomen*). The phenomenally given serves as a starting point in the philosophical study of the entirety of objectivities of possible consciousness (351), while the essence is inherent there in a "cover and distance" manner. Yet, this condition is due to the nature of the object itself, that does not completely expose its essence, and not because of aspects that refer to one's consciousness' capabilities. In her opinion, "the genuine and specific phenomenological-philosophical work progresses from the still concealed, yet as such already visible "primordial phenomenon', to the 'pure primordial phenomenon'. This progress "demands first the specific phenomenological stance (*Haltung*) and then in it the totally direct and undeterred in its heading gaze towards the phenomenon in its pure 'what'" (352). Helmut Kuhn well described this as follows: the things towards which the gaze is directed are always known in advance, we do not start at a null point. They show themselves to us, but they are concealed. They are standing up against us as known but also as mysterious, and impose on us the distinction between what things are in their beginning and the essence that is uncovered by penetrating observation.[19] The "uncovering" demands discounting of all contingents that appear only up against me, and are only a "certain side" of the phenomenon, while "the substantial totality that bounds it is under darkness". Only when the phenomenon "steps out in a complete absolute objectivity" is the philosophical work ended (353) and the "primordial phenomenon" comes to light.

An approach that treats the givenness that appears up against it as concealed is likely to lead to skepticism and dogmatism – two consequences that should be included in what CM designates as "going against the given" (358), because they do not enable the careful and restrained observation of the phenomenal givenness and of the world's phenomena in general. First and foremost, one faces the gap between skepticism and dogmatism and CM's approach from watching the slow and prolonged observation, to the point of exhaustion, typical of her discussion in *Doctrine of Appearance*. Moreover, one can find in this essay an implicit answer to these approaches. At the outset, one notices the widespread use of the word "believe" (*glauben*) that

[18] For the "semblance of reality" typical for perception's objects, see: CM, "Zur Ontologie und Erscheinungslehre der realen Außenwelt", p. 380.

[19] Kuhn, op. cit., p. 399. See in this context Husserl's argument that alongside the grounding of the value of the original givenness there is also an acknowledging that things are given to us under restriction (Husserl, Ideen zu einer reiner Phänomenologie und phänomenologischen Philosophie, I, β24, "The principle of all principles").

clearly indicates the adoption of the typical certainty that stood at the foundation of the Husserlian phenomenology.[20] A more direct opposition to the skeptical approach is apparent in CM's argument that the determinations that were reached by the method of intuition of essences are not relative or conditioned by specific circumstances (349), because every genuine primordial-phenomenon responds in an essential mode to an idea that can be discerned by intuition of essence (353).[21] An answer to the dogmatic stance is given in the fundamental determination that although the search for essences is in itself independent of factual existence in the real world, in *Doctrine of Appearance*, CM assumes the original belonging of this essentiality to a certain phenomenal state of affairs (349).[22] It transpires, then, that not everything belongs to the phenomenal beginning-material or alternatively not every given is rooted in a primordial phenomenon (351). Thus, the phenomenology of the external world is not exhausted in the study of phenomenal appearance, but is anchored in a most fundamental certainty. However, "matter-of-fact" or objective rootedness is not real unless it is realized within a concrete reality (351). Therefore, a careful observation of the concrete phenomenal appearance must always be the starting point of the phenomenological investigation of the external world.

Realism and Transcendentalism

In the first place, CM wishes to establish the understanding of the external world as real existence (386), having a self-constant-Being (*Seinselbstständigkeit*) (391), autonomous and absolute in its being (392), closed in itself, and transcendent to human consciousness and spirit (424).[23] In this context, she regards as a "self-evident and comprehensible" issue that "every real being has being 'for-itself'"

[20] See in particular CM, "Zur Ontologie und Erscheinungslehre der realen Außenwelt", pp. 355, 370, 398, 407, 413, 418, 423, 446, 496, 500, 513. Husserl regarded skepticism as a denial of apodicticity, i.e., necessary and universal truths that are essential for any theory to make sense. He distinguished between three forms of skepticism: "logical", "noetic", and "metaphysical". See: Husserl, Logische Untersuchungen, I, Chap. 10, §57–§61. As for Husserl, so also for CM in *Doctrine of Appearance* the metaphysical skepticism that denies the objective knowledge of the real world is the most problematic. For further discussion, see: Brice R.Wachterhauser, "Introduction: the Shipwreck of Apodicticity? ", in James M. Edie and Brice R. Wachterhauser (eds.), Phenomenology and Skepticism: Essays in Honor of James M. Edie, (Evanston: Northwestern University Press, 1996), pp. 1–62, 227–238. Regarding Husserl's certitude, see: Leszek Kolakowski, Husserl and the Search for Certitude (New Haven and London: Yale University Press, 1975).

[21] CM criticizes skepticism in the context of positivism, see: CM, "Zur Ontologie und Erscheinungslehre der realen Außenwelt", p. 398, p. 358.

[22] Regarding dogmatisms, see also: CM, "Zur Ontologie und Erscheinungslehre der realen Außenwelt", p. 347.

[23] CM explains that it is a mistake (quite common in positivistic approaches) to identify "existence's independence of consciousness" with the "real external world", see: CM, "Zur Ontologie und Erscheinungslehre der realen Außenwelt", p. 391.

(472). In the "natural point of view", the world of external space is assumed to be unconstrained within itself and can be expressed only by negative speaking, while within the sphere of observation, i.e., where reflective consciousness is involved, positive speaking about this world becomes possible (392).[24] In any event, CM determines that our accidental perception of the external world cannot define it, and the possibility to look into the space's reality as such exists above and beyond all obstacles (393).[25] Hence, in her opinion, the realistic grounding of the external world demands a separation between the sphere of representation (*Vorstellung*) and that of perception (*Wahrnehmung*) (371) – the first has contact, yet not identity or complete overlapping, with the realm of reality that in itself imposes limits upon observation (383). The second happens in the sphere of consciousness. Yet CM argues that "the genuine and original aspects of the consciousness of reality do not fall in line with the grasping of something" (382).[26] It appears that the sphere of presentation enables accessibility to the "phenomenal beginning-material", as well as to the "primordial phenomenon" in which it is founded. Thus, CM determines that the analysis of the external world must bypass the perceptive attitude, which in her opinion has no function in our consciousness of reality (364). This determination is grounded on two fundamental distinctions:

The first differentiates between two types of objects: the manifest, which are characterized as "uncovered self-emerging" (*unverhüllte Selbsthervortreten*) and the concealed, which are depicted as "covered presentiveness" (*verdeckte Anschaulichkeit*) (371). The emphasized aspect in each of these characterizations

[24] The distinction between the natural and reflexive approaches recalls Husserl's division between the natural attitude and the phenomenological one (Husserl, Ideen zu einer reiner Phänomenologie und phänomenologischen Philosophie, I, pp. 3–15), except that Husserl directed the reflexive approach to the transcendental subject and founded upon it the phenomenological reduction. However, similar to the early phenomenologists (see: Helmuth Plessner, "Bei Husserl in Göttingen, in H.L.Van Breda and J. Taminiaux (eds.), Edmund Husserl 1859–1959: Recueil commémoratif publié à l'occasion du centenaire de la naissance du philosophe (The Hague: Martinus Nijhoff, 1959), p. 38; Otto Pöggeler, "Eine Epoche gewaltigen Werdens", in Otto Pöggeler, Ernst Wolfgang Orth et al. (eds.), Die Freiburger Phänomenologie (Freiburg/Munich: K. Alber, 1996), pp. 15–17), CM's approach was directed towards the object, and later on she will explicitly reject the phenomenological reduction. See: CM, "Seinsphilosophie" [1931], in Schriften zur Philosophie, 3 vols. (Munich: Kösel-Verlag, 1963), Vol. 1, p. 17; CM, "Die transzendentale und die ontologische Phänomenologie" [1958], in Schriften zur Philosophie, 3 vols. (Munich: Kösel-Verlag, 1965), Vol. 3, pp. 394–402.

[25] Like CM, Spiegelberg also argues that genuine phenomena are not influenced by theoretical or other interpretations, while untrue ones collapse as soon as their falsification is uncovered. See: Spiegelberg, op. cit., p. 164. Spiegelberg's ideas in this essay closely resemble those of CM in *Doctrine of Appearance*. Obviously he was familiar with her work, but surprisingly neither *Doctrine of Appearance* nor any of CM's later writings are even mentioned in his essay. However, Spiegelberg provides the lacking but important background and explanation of CM's principles of realism. I will point to the main similarities in the notes below.

[26] This argument is supported by a detailed discussion of two types of representation (CM, "Zur Ontologie und Erscheinungslehre der realen Außenwelt", pp. 361–378). For further reading, see Heinemann's discussion of the relation between "experience" (chaotic in its concrete manifestations) and "appearance" (the dynamic dimension in the static, the unshaped that becomes shaped) (Heinemann, op. cit., p. 188).

mirrors a consistent principle in *Doctrine of Appearance*, i.e., with regard to the manifest objects, whose accessibility to observation is immediate, to emphasize their independence of the observing subject. Thus, the foundations are laid for the demand for the I to be pulled back into itself and prevent its spiritual gaze (409), or alternatively the request for the spirit "to keep constant in eliminating its continuous effort for representation" in order to be able to maintain the self-imposed objectivity "as external world reality" on itself (386–387). With respect to concealed objects, which are not immediately available, the emphasis is put on the presentiveness, which as such assumes the involvement of the I (by means of the senses and consciousness) in validating the reality of the concealed object, though this reality itself is considered as undoubted and independent of such validation. One way or another, the realistic argument is completed by the I.

However, the status of the two types of objects is not equivalent in the study of the external world. CM determines that only 'uncovered self-emerging', later to be designated as "self-announcement" (*Selbstkundgabe*), provides a descriptive and immediate guarantee out of the factual suchness and self-existence of the object (371).[27] Hence the possible and required effort to make "perceptible as sui generis" the moment in which the object is being represented (364), namely not by reasoning that is taken from the ideal patterns of consciousness, but by study of the real modes of its appearance, cannot but be based on the investigation of "uncovered self-emerging" of the sensually given manifest objects. However, as a metaphysical disposition, the realistic approach cannot be exhausted in the plane of externality, hence the reality of the concealed objects must be confirmed as well.

The distinction between manifest and concealed objects and the choice to focus the study of the external world on them enable us to figure out the depth and complexity of *Doctrine of Appearance's* constitutive movement from the "concealed" to the "primordial". Obviously, there is no identification between the manifest objects and the "concealed" layer of appearance or between the concealed ones and the 'primordial'. Additionally, *Doctrine of Appearance* does not present a move that leaves behind the manifest and the accessible in favor of the concealed inside the appearance – otherwise the perceptive attitude would have been preferable to the representative. It is especially the insight that the "primordial" is already present in the manifest objects that means the study of the external world can and must focus on them. Yet, the duality between the manifest and the concealed is so profound that it is present in both the manifest and the concealed object, although settled differently in them: the manifest show their concrete aspects, while the involved dimensions of

[27] In her later writings, CM continued to deal with affinity between the suchness of the object and its substantial being, see: CM, Das Sein (Munich: Kösel-Verlag, 1957), p. 57. See Gerhard Ebel's criticism of the realistic direction in phenomenology, including CM's, for not being able to produce a genuine realism, and instead turns reality into a sheer "phenomenon" of reality, which is therefore especially not real (Ebel, op. cit., p. 2). Ebel admits that CM brought to the fore aspects unnoticed by the realistic school. Yet in his opinion these are insufficient (ibid, p. 42). For supportive evaluation of this school for suggesting the suchness-experience alternative, see: Seifert, op. cit., pp. 97–98. Like Seifert, Heinemann also speaks for the value of phenomenology's focus on appearance, see: Heinemann, op. cit.

the I are cloaked. Regarding the concealed objects, it seems that their concreteness is covered, and accordingly their way of appearance brings to the fore especially the aspects of the I.

The second distinction refers to the two dispositions of the I: "perceptive presentiveness" (*wahrnehmungsmäßiger Anschaulichkeit*) and "representative presentiveness" (*vorstellungsmäßiger Anschaulichkeit*) (382). In the first, the I is located at an "unobservable distance" from its object. The perception that is enabled from this disposition is depicted as "filling" with objects that are equipped with only a "semblance of reality" and "habits" of real objects, whereby they "hover" in the face of the spiritual gaze, but actually carry within themselves "non-reality" (*Unwirklichkeit*) (379). However, when the I is posed in "observable closeness" typical of representation, the space world is accessible to direct and immediate observation, and thus concretely given outside the position of perception. Observing an object from proximity seems as unveiling the covered presentiveness and thus bringing to light the external world as "uncovered self-emerging" or as "uncovered lucidity". While within perception objects are given to consciousness' judgment as distant and separate from the I, in the context of representation objects, as well as the I as real body-soul unity, are present in the same firm sphere of being. Since one's consciousness is not constitutive for the sphere of presentation, with the shift from the attitude of perception to the realm of representation no new sphere of observation is being established. Moreover, the presentive possession of the space sphere in which the I exists for itself does not suddenly collapse with the cessation of perception. Whereas the perceiving I "fills" its observed space sphere with content and meaning granted by its consciousness, the representing I is depicted as present in the same sphere where objects are given "in an unique mode which is 'void of content' but presentatively full" (380–381). Although CM did not point to that, the affinity between the two types of objects and the two dispositions of the I is apparent, i.e., the "uncovered self-emerging" objects require the representive I, while those of "perceptive presentiveness" need the perceiving I. Finally, it is clear that the focus on the study of representation is a continuation of the dominance granted to the manifest objects over the covered ones, and it is rather understandable in an investigation of the external world.

The study of the two dispositions of the I, whose start is already within the grounding of the reality of the external world as an independent Being, serves as a base for the second achievement of *Doctrine of Appearance*, that is heading for achieving a new understanding of transcendentalism. Long before Heidegger established his philosophy of existence in *Being* and *Time*,[28] CM achieves a new

[28] A comparison of CM and Heidegger in *Being* and *Time*, is crucial for the understanding of the novelty of CM's idea of transcendentalism in *Doctrine of Appearance* and later in her entire writings. Yet this exceeds the scope of this article. CM criticized Heidegger in several contexts. See: CM, Die Zeit (Munich: Kösel-Verlag, 1954), pp. 13–31; CM, "Heidegger 'Sein und Zeit', Metaphysische Quellpunkte" [1930], in BSM, Nachlass, AIII6a, pp. 1–42; CM, "Heidegger 'Sein und Zeit' " [1932], in Schriften zur Philosophie, 3 vols. (Munich: Kösel-Verlag, 1963), Vol. 1, pp. 185–193. For further reading, see: Wolfgang Behler, Realität und Ek-Sistenz, Auseinandersetzung mit der Konzeption Martin Heidegger in Konfrontation mit der ontologischen Schriften von Hedwig Conrad–Martius (Dissertation) (Frankfurt am Main, 1956).

ontological-existential conception of the spirit in which, as Schmücker put it, "for the first time the subject is released from Kant's prison",[29] meaning it becomes possible to refer to the I not by means of its cogitations. She regards as obsolete the idea that representation of the external world is independent of the I (364). She criticizes positivism for its "setting outside" (*Außenstellung*) the contents of perception but "then somehow mysteriously the I is 'projected outside' into its appropriate phenomenal position" (383–384). Moreover, by setting outside the appearance, the real and the non-real are grasped in the same way (353), i.e., according to the way they refer to consciousness. Unlike this approach, designated as contradictious and absurdist (384), CM discovers from the study of the external world a new understanding of transcendentalism, according to which the human spirit holds two fundamental situations: it is simultaneously "underhand" and "covered" (386), or alternatively "nearby itself and out of itself". Yet since consciousness cannot be totally against itself, we are dealing here with an artificial situation that might be achieved if the spirit is able to depart itself a little from itself, in order to enable itself to experience an objective appearance, that as such demands separateness from the perceiving disposition of consciousness (470–471). Against this background, CM describes the stance of the spirit toward real existence as "playing in the face of the matter itself" and as "hide and seek". Even the tiniest loosening in the effort to keep this duality will lead to loss of objectivity, or, alternatively, to the grasping of objectivity as something that is created or carried by spirit (386–387). As will transpire from the following discussion, the issue of realism is much more dominant in *Doctrine of Appearance* compared to transcendentalism. Nevertheless, the essentiality of the I for the realistic grounding of the phenomenon of the external world is unequivocal.[30]

The Sensory Givenness

The phenomenological investigation of the external world in *Doctrine of Appearance* focuses on "the sensory given". CM wishes to demonstrate that it is a sort of "sensory givenness" (*sinnliche Gegebenheit*) and not the other way around (399), namely: the focus is on the object of the sensory givenness as distinct from the

[29] Schmücker, op. cit., p. 39. The intensive dealing with senses and their relation to consciousness apparently raises the expectation for a dialogue with Kant and other philosophers who dealt with the issue. Yet Kant is mentioned only one time, see: CM, "Zur Ontologie und Erscheinungslehre der realen Außenwelt", p. 487.

[30] CM deals with the problem of the subject in several contexts, see: CM, "Die Problematik des Subjekts" [1932], in BSM, Nachlass, AIII8a, pp. 1–35; CM, "Existentialle Tiefe und Untiefe von Dasein und Ich" [1934], in Schriften zur Philosophie, 3 vols., (Munich: Kösel-Verlag, 1963), Vol. 1, pp. 185–193; CM, "Dasein, Substantialität, Seele" [1932], in Schriften zur Philosophie, 3 vols., (Munich: Kösel-Verlag, 1963), Vol. 1, pp. 194–227; CM, Das Sein, pp. 118–141; CM, Die Ziet, pp. 13–31; CM, "Heidegger 'Sein und Zeit', Metaphysische Quellpunkte"; CM, "Heidegger 'Sein und Zeit'".

sensory itself as an experience in which subjects are involved.[31] CM justifies taking "sensory givenness" as the starting point in the investigation of the external world by what she designates as "the totally ancient and specific nature of the sensory given" (398), meaning the characteristics of the sensory given that in her opinion enables "real contact" with the external world (423).[32] Firstly, among all the existences of the external world, it is possible for the 'sensory given' to approach me as a content of givenness, due to its capability to make obvious its real being in its "here and now", and to make itself felt from the itself and outwardly. This also true in the face of the I that keeps constant in its circle of being, and as such lacks the necessary distance that is crucial for confirming the reality of the world external to it (412–413). Secondly, the sensory given as such raises from itself the argument regarding its own real and factual existence. In her words: "to the 'face' of the sensory appearance belongs in a factually indivisible manner a worldly and external existence of real being" that announces itself to me personally and thus brings its own existence to presentation (422).[33] These characteristics enable the sensory given to mediate between me and the being and the factual arrangement of the real external world. Indeed, the sensory given is the only means that ensures for me the external world in its time-space factuality, since its essence is to "present" the external-world's Being. This given is not verified, and in principle epistemological arguments might demonstrate that sensory "semblance" misleads, and thus, according to CM, to some extent rightly, raises the need for investigating whether the phenomenal state of affairs is capable of validating the demand of our knowing of the external world.[34]

[31] Herman Krings, who is admittedly influenced by CM, explains that the focus on object as a real existing being is not simply equivalent to the inversion of the Kantian beginning in which the I directs itself to consciousness. Here we assume that there is a real relation between the existent and the essence referring to it. Yet this assumption does not contain an argument about the possibility of knowing this existent, see: Krings, op. cit., pp. 193–195.

[32] Like CM, Spiegelberg also provides a justification for relying on sensory givens within a realistic approach. He argues that a critical and phenomenological inspection of the immediate phenomena of reality will remove the most frequent objections to the reliability of perceiving that is mediated by senses. See: Spiegelberg, op. cit., p. 153.

[33] Spiegelberg explains that a phenomenon and reality do not exclude each other, namely: what is real exists within itself and can be presented to us in its very existence out of itself. This means that real things in the world can remain exactly as they are, including in case of being presented and having relation to us. He designates the phenomena in which subjects are involved "subjectival", not in the sense that they are not real or that they mislead us, but as objective parts of subjects and of their world. See: Spiegelberg, op. cit., pp.134–135. Moreover, in his opinion, the reality of subjectival phenomena is totally evident (ibid, p. 149). However the subjectival reality covers only a small part of our total reality and of Being in general (ibid, p. 135).

[34] Similarly, Spiegelberg argues that the reality of non-subjectival phenomena of reality can never eliminate the possibility of an illusion or a mistake (Spiegelberg, op. cit., pp. 137), hence they are always doubtful. However, despite examinations, illusions "do occur and are bound to occur" and thus present non-subjectival phenomena as dubious. In his opinion, "one principle reason for such dubiousness consist in ultimate mutual inconsistencies between our various phenomena or reality" (ibid, p. 153), and only constructive synthesis of non-subjectival phenomena of reality might achieve certainty, though not complete. Yet he determined that "it is all we can reasonably expect, considering the nature of non-subjectival reality, our own predicament, and the nature of our cognition and understanding" (ibid, p. 163).

She contends that should the epistemologist begin the investigation with the sensory givenness, he or she will recognize its usefulness in the knowing of the external world (423), provided that the thing whose factual givenness is scrutinized is first taken purely in its sui generis essence (348). In any event, CM does not ground her realistic argument upon epistemological investigations, but on a phenomenological method that assumes the essential belonging of an essence to a specific phenomenal state of affairs (349).[35] Moreover, she contends that in order that something will argue for givenness, we must believe it has "existence" of its own – a confidence that in her opinion will transpire as justified. This means that what enables sensory givenness to be faithful to its assignment as mediating the external world is anchored in its fundamental bond to reality (423).[36]

CM clarifies that sensory appearance does not have the occupation of "uncovering" the "thing-in-itself" that lies there beneath, but "to bring the world in-itself to 'exposure'". The choice to focus on the outwardly revealing of the world permits her to "be satisfied with referring purely and for-itself to 'the manifest surface' (*sinnfällige Oberfläche*) that presents itself purely and as [existing] for-itself in the sensory appearance" (463), or alternatively as a self-standing entity (466). Here, the "manifest surface" does not designate only a contingent cut of a ready thing, but the exterior side of the matter in general that faces the "principally invisible" interior. This is unreachable by any possibly real cut from the given, for we are always walking upon the manifest surface (465, n. 1). The function of appearance brings about the material interiority of the body-thing (*Körperding*) to the presented appearance, and thus indicates the internal motion of the material being outwards (463). This self-presentation of the manifest surface is immediate, and is enabled only because in the sensory appearance, what presents itself describes at the same time the specifically constituting matter of the appearance. In any event, the sensory manifest surface is not only what is presented but also the existing in-itself (464). CM argues that "self-presentation" characterizes the sensory givenness so deeply that it distinguishes it from everything that "lacks a being-for-itself" and thus misses what might be presented externally (413). Simply, what appears as dependent in its

[35] The early phenomenologists understood Husserl's appeal "to return to the things themselves" as indifference towards epistemological questions. See: Ursula Avé-Lallemant, op. cit., p. 207. For the relations between phenomenology and epistemology and phenomenology, see: Spiegelberg, op. cit., pp. 130–131. Like CM, who characterized the epistemological approach as dogmatic (CM, "Zur Ontologie und Erscheinungslehre der realen Außenwelt", p. 347) and incapable of coping with its questions (Spiegelberg, op. cit., p. 351), Spiegelberg too criticized epistemology, which in its highly speculative accounts of how knowledge works omits its first and paramount obligation to be critical itself (ibid, p. 152).

[36] Like CM, Spiegelberg too emphasized the 'argument of reality' that is inherent in the real being. In this context, he designated as a "phenomenon of reality" the joining together of self-presenting of the phenomenal object and its arguing for being real. Therefore, reality's phenomenon is distinguished from all the "bare phenomena" that do not claim to be real. See: Spiegelberg, op. cit., p. 133. However, Heidegger did not pose this demand, but understood the phenomenon as showing itself. See: Heidegger, Martin Heidegger, Sein und Zeit (19th ed., Tübingen: Max Niemeyer, 2006), Chap. 27.

being cannot appear as what presents itself (413, n. 2).[37] Therefore, in spite of the fact that sensory appearance is not itself the entirety of the external world, addressing essence intuition to what is presented by this appearance by itself and out of itself, but not on what is above and beyond it, serves for CM as "a frame for the whole" of her study and a "guide to the order of the issue", that then might be given to differentiation in diverse directions of the sensory givenness (399).

Sensory givenness is composed of two fundamental phenomena of "feeling's givenness" (*Emfindungsgegebenheit*) and "appearance givenness" (*Erscheinungsgegebenheit*) or "manifest appearance" (*sinnfällige Erscheinung*) (hereafter: manifest appearance givenness). The discussion will now turn to the explication of these phenomena.

Feeling's Givenness

The Personal Touch

"Feeling's Givenness" designates the happening of "something's" direct touch on me (406). CM clarifies in this context that the term "feeling" (*Empfindung*) denotes a focus on the I's experience, and as such it is distinguished from whatever has objective orientation, such as "seeing" or "hearing" (460), that simultaneously indicates the occurring of something outside. The sensually given brings about "touching contact" (412) which enables a personal touch on the I (425) – a touch that is also designated as personal invasion, affection (474), pressure (518), and distance-less givenness. These are different modes of the real thing that enable the appearance of contents in the face of the I whose posture is quiet, passive, and relaxed of tensions coming from its internal being. This posture enables individuals to experience what naturally comes near as unmediated by their surrounding domain. For example: the wind that shakes me, the heat that encompasses me, the scent that envelops me, and everything that is simply there without one's needing to "get out of oneself" (404).

The grip of the real thing on me entails a certain amount of dependence on spirit that personally carries the object and signifies it as existing outside the I (441). This is indeed the way the external world reaches me personally (378), and the qual (often translated as "raw feel")[38] remains in the domain of the I (518). This belonging

[37] Spiegelberg explains that the very independence of the subject should not be considered as the essence of reality but as a "fundamental and essential result of reality" (excluding real acts of the subject that of course depend on him or her). See: Spiegelberg, op. cit., p. 132, n. 2.

[38] The world "Qualia" signifies the subjective content's experience of mental situations. The subjective aspect seems to resist any intra-subjective definition. Thomas Nagel characterizes "Qualia" as what "feels itself in a certain manner" (see: Thomas Nagel, "What is it like to be a bat?", in The Philosophical Review 83/4 (October 1974), pp. 435–450). Obviously, this characterization cannot be considered as ultimately valid, because it assumes that the content of the subjective experience

of qual to the I's domain has two consequences: the negative refers its lack of sensory recognizability, while the positive is embodied in its "appearance's stance" on the body that serves as the carrier of sensory givenness, because it is personally and in an unmediated manner pressed by the material content of the qual (525). On the face of it, the proximity to the I that is reflected in these consequences indicates its distancing and separating from the external world. While the first designates the distance from the world, the second locates the qual in a realm that is not the world but the body of the I. Yet CM harnesses these two for the grounding of the reality of feeling's givenness, specifically: the lack of involvement of the senses, which later transpires as showing the limited participation of one's spirit, leaves the arena of occurrence of one's feeling in the body as a realistic foundation. Indeed, as much as the personality of the touch is the thing that makes the phenomenon of feeling givenness what it is (450), from CM's perspective, what is more decisive in the very constitution of this phenomenon is the felt-thing (*Empfundene*), or the material givenness of the thing that is felt bodily (513). Although the felt-thing belongs to the "thing that is experienced sensorily" (530), for example, as rigidity or roughness (406), from the outset the givenness that brings about the feeling, by its very essence, "does not exist there for me". In this respect, the main part of this givenness is not anchored in its sensory appearance (455–456), but in the being that uncovers itself and thus acquires presence in the external world.

The Relation to Consciousness

CM determines that "**no direct sensory connection exists between feeling's being and consciousness in general**" (523), and despite being a sensory given "[there is] nothing in the feeling's being itself that is capable of leading it directly and personally vis-à-vis the spirit" in the same way as in the appearance of color and tone that directly and personally present themselves towards spirit. Indeed, spirit does not serve as a "carrier of the experience of sensory givenness", "it does not pick up the feeling's being and cannot **fill** itself with the sensory being of itself". CM characterizes as absurd the possibility of the involvement of senses in this context, since such an involvement would mean that the materially felt thing originated in consciousness (535), whereas indeed it grants itself as separated in principle from consciousness' sphere (441) that cannot personally encompass this kind of self-announcing being at all (520), and as transcendent to spirit. In her opinion, the transcendency of

has already been understood. Unlike CM, many philosophers deprived "Qualia" of reality. For example, see: Daniel Dennett, "Quining Qualia", in Anthony Marcel and Edoardo Bisiach (eds.), Consciousness in Contemporary Science (Oxford, Clarendon Press, 1993), pp. 42–77). Yet other philosophers as well as scientists regard the content of the subjective experience as undoubted. See: Ansgar Beckermann, Analytische Einführung in die Philosophie des Geistes (2nd ed., Berlin: De Gruyter, 2001). For further reading see: Clarence Irving Lewis, Mind and the World Order: Outline of a Theory of Knowledge (3rd ed., New York, Dover, 1991).

the felt-thing is reflected in the closedness characteristic of the real beings and in its capability to resist that places a barrier before the pure spiritual act. Hence, "the principle impossibility of 'catching' the qual and the 'phenomenal impossibility' of its presence 'inside' consciousness" (441–442).[39] CM explains that the separateness of qual from consciousness is so fundamental that no reference to spirit is enabled from qual, therefore the primordial characterization of the absorption of the qual in the I as "absolutely unspiritual" (526). Now the accurate meaning of the argument that the main part of sensory givenness does not depend on its sensual appearance transpires (456): the feeling's being in-itself and for-itself appears as sensorily given, but "**seeing itself is not sensory**" – as it could have been if consciousness had personally encompassed these experiences of sensory givenness (523). In other words, since consciousness does not bring about the qual, it cannot be subject to its influence, i.e., become sensorily recognized (525).

However, the "'consciousness-sphere' is not totally excluded" from feeling's givenness, but exists within it as "a residue of contact with an essential moment that belongs to it" (461). Firstly, the felt materiality, the subject's touch on the rigidity of the thing that generates the feeling, is accompanied by consciousness (427). Secondly, the experience of feeling (like that of the "manifest appearance givenness" that will be discussed below) achieves presentation (432), and thus poses itself in principle as an object for a possible I. At this point, CM vacillates between two opposed insights: one seeks to exclude consciousness from the realm of feeling, explaining that feeling's being in-itself and for-itself is incapable of presentation. The other necessitates the clarification of consciousness' part due to the understanding that consciousness is capable of pulling something into objective self-presentation thanks to its real touching of the I (452). These two insights join together in CM's effort to harness consciousness itself for the establishment of the independence of feeling's being towards consciousness and thus place it in the phenomenology of the external world.

CM describes in this context a "consciousness-beam" (*Bewußtseinstrahl*) that is directed to the feeling, that seems as if it generates or pulls outside for the first time the feeling as something that has been "experienced". She clarifies that we are not talking here about a phenomenon in which consciousness "keeps" its own objects inside itself (456), but about feeling's content that presents itself to consciousness from the outside (452). Additionally, the self-announcement of the felt-being, towards which consciousness-beam was directed, seems as "being towed to presentation", as a result of which it showed itself outwardly according to its innermost internality. Therefore, the particular involvement of consciousness in the presentation of the felt-being does not destroy the objective quality of its presentation (454). The characterization of consciousness' involvement in the material sensibility is portrayed as a kind of cunning typical of sensory givenness. In her words: "although the material felt-things lack a stance of appearance from themselves (i.e., in-themselves

[39] Spiegelberg presents the probe-resistance of objects to our will as an indication, sometimes even a strong one, of their reality, see: Spiegelberg, op. cit., p. 148.

and for-themselves they are not felt as manifest), nevertheless they achieve presentation outwards especially through the I if they have real touching-relation to it" – relations that are no more than presumption, though necessary, for feeling- experience (430). More importantly, CM assumes that feeling-experience has direct relations to things in the external world, and therefore this experience provides the actual foundation for the judging of the material sensibility of real things (432).[40] She concludes that the phenomenal state of affairs itself, and not consciousness, indicates the existence of the felt-being in the external world.[41] Otherwise, it would have made no sense to designate the materially felt-thing by the term "appearance" (433). This means that the main part in the rigidity felt by the senses is not within the influence on the I, but in serving as an indication of the reality of the materially felt-thing, hence rigidity should be understood as belonging to the object rather than to the subject. The trickery inherent in consciousness' involvement is attributed to its capability to blur the fact of the self-presentation of the felt-thing, so that mistakenly this presentation is assigned to consciousness and not to the felt-thing itself. Yet, CM determines that consciousness cannot disturb the very appearance of the felt-thing (514), since it grants itself to the observing spirit as autonomous in its being, and as such it makes itself known from its real position, outwardly or inside it. Moreover, in the felt-being "there is nothing" that somehow "presses" directly into "consciousness' domain" as for example the noise presses upon my hearing, or the blue sky my gaze (516).

Against this background we can understand the choice to characterize the felt-thing as "floating" and as existing "at the periphery" of the being of the I (441–442) – expressions that emphasize the steadfastness of the threshold that prevents the entrance of the felt-thing into consciousness' domain.[42] CM clarifies that we are not talking about a spatial or anatomic relation between what brings about the feeling's givenness and the I, but about a description of the way the subject experiences the touch of the felt-thing on him or her (443). She adds that here "**consciousness opens**

---

[40] CM clarifies that the novelty is not that such an approach might be misleading, but that the foundation of this experience is "'in truth' not in the belief that material sensibility has direct relations to the external world" (433), but in the actuality of these relations.

[41] Also at the present point, Heinemann's words recall those of CM. In his opinion, the primordial phenomenon of man is not that of consciousness but of appearance, namely entering into the appearance and changing within it (man lives in pictures before he knows that). In contrast, consciousness is an epiphenomenon, an ex-post-facto phenomenon, a reflexive act that exists only after appearances collapse. Therefore, he designates, exactly like CM, phenomenology as a "Doctrine of Appearance", see: Heinemann, op. cit., pp. 186–187.

[42] Spiegelberg characterized the peripheral field of our perception as "marginal openness", meaning that this field is never cut as sharply as its borders. However, he emphasized that peripherality does not designate non-reality. What we perceive at the periphery of field of perception are not only vague configurations, but mostly well-defined structures that are presented in decreased clarity. More importantly, we can still see via these modifications the phenomenon itself in its uninfluenced structure, rather than the structural openness of what is given in our perception's field. This implies that reality does not culminate at the borders of our perception, but continues beyond that. Openness teaches that the phenomena of reality stand on their own feet. See: Spiegelberg, op. cit., p. 147.

itself from its real-stance in the face of any feeling-being or towards the insides of it and thus bring it to its forum". Regarding the experience itself and the factual aspect, it is a feeling that is generated "from the outside inwards", meaning towards the I, although the feeling itself appears on the body (443–444). The fact that objective givenness' experience is "correlative by its being" to consciousness does not mean, of course, that beings as such that are accessible through this experience must also be "correlative by their being" to consciousness (458, n. 1). Obviously, consciousness does not serve here as a 'hatch' (*Gucklöchern*) through which something might really be seen as if "half inside and half outside", or alternatively half of it is on the body while the other half is inside consciousness. Not only that unlike the body, "consciousness does not have any real-margins (*Realbegrenzung*)", but also "there are no borders that are common to consciousness and the real-world". Obviously, when things or beings share no common borders they cannot have the same object (450). CM determines that it should be possible to argue without incongruity that something can somehow be "given" to me – even in a primitive sense – without my being "aware" of it. In such a case, consciousness is characterized as totally shadowy (459). In her opinion, the attempt to pull inside the feeling-being from its peripheral stance in the I into the center of consciousness is an essential contradiction. The reason for this is that from the viewpoint of feeling itself, feeling-being does not at all float "inside" or "above" consciousness' sphere, but is clearly strongly linked back into the "world" that is beyond consciousness or has a rooted stance inside it, even if this stance lies in the realm of the I (443).

The Relation to the Body

The experience of feeling can be determined only on specific stance of the living-body (*Leib*), and it occurs when the I, that is capable of experiencing it by itself, is "present" for itself in its own living-body. CM explains that real body (*Körp*) becomes a living-body capable of experience only when the I is somehow "inserted" into it. As combined, whole, and living, the I sustains bodily pressure by which it is able to undergo feeling-experience (532). So, when for example I wish to genuinely feel the solidity of something with my fingers, it inwardly presses my living-body into my fingertips. This means that in and for myself, I am always "inside" my fingertips, in which I can, more or less, be enfolded (534). In her opinion, the decisive moment of experience for genuine feeling's givenness is already located in every "entrance stance" to the living-body, and it is experienced as a factual oppression of the body (537) above what she designates as the "body periphery" or "from the outside inwards" (534). There commences and culminates the absorption of the felt-thing that appears on the living-body as material existence (529).

However, as much as the body plays a part in the sensory givenness, as a result of which this givenness appears as bodily or as such that occurs "at" the body, CM clarifies that the pure factual relation of this givenness towards the body does not describe unequivocally the scope of sensory givenness (401). Hence, in order to avoid a subjectivist and thus faulty understanding of the phenomenon of feeling's givenness, she restricts her initial determination regarding its being conditioned in real appearance in the I (430). Now, CM determines that the "touching contact" that the sensory given brings about for the I, denotes a "mode of comparison and not something genuine" (428), and that the main thing is that "there is no contradiction between the dependence on touch and the reality of the touched-thing", but "it is the real being of the external world that arrives as feeling's presentation". Therefore, "no touch-relation can be a real presumption, but every experience is totally posed at the same time from the I" (430, n. 2). In addition, "the touch-relation does not need more than to remain a presumption – though necessary – for the feeling-experience, and within itself it does not have an experienced-being" (*erlebter Bestand*) (430). Later on, we shall discuss the presentation as fundamental indication to real existence.

CM argues that the lived-bodily experience as such is not equivalent to the spiritual experience in which the objective receptive senses are contained (536). Therefore, the "carrying" of the experience of the I depends especially on its being a lived-body, and absolutely not thanks to its senses or its capability to objectively perceive objects by means of consciousness or a spiritual sense (533). Not only is the possibility of the feeling-experience to become actual, meaning to press the lived-body, independent of the creation of objective contact with consciousness, the being that appears as felt would not have arrived at sensual givenness in consciousness had it not been given as real. Moreover, in her opinion, it is not the feeling-givenness of the presented being on the living-body but its real being that illuminates the impossibility of sensory-objective contact with consciousness (537). So, the appearance of the felt-thing is achieved on the one hand by means of the splitting of the senses from the inclusive consciousness so that it is impossible to elaborate what is announced from inside by the senses, and on the other hand due to the openness of the living-body to the material felt-things (529). The experience of feeling is constituted then between two planes, the perceptive stance and that of the appearance-being; it appears where the rigid-thing (that enfolds within itself the phenomenal content of the experience) borders on the real lived-body. However, this location does not split the feeling-experience because the relation between its two constituting planes is real (530).

CM characterizes the real transcendence of the "sensory appearance" toward consciousness as a "dominant fact" whose meaning is as follows: when the body itself appears already as transcendent in principle to consciousness and thus irreducible to it, so also whatever is on the body – inside or outside – "arrives at appearance" from the aspect of feeling. And when the body "appears" as a being existentially rooted in itself, meaning as a real being in the rigorous sense, only then does whatever appears on it show itself as feeling-adhering and as what is connected to this real being (447). Yet, the transcendence and the embodied autonomy of the body in relation to the spirit does not signify a restriction of the spirit, but rather the essence of the spirit as "homeless", or alternatively as "deprived of its home in principle" (466). It transpires, then, that the body as a real foundation realizes the transcendent separation between feeling and consciousness. The personal distress, which is the content of the feeling, announces the involvement of the body within feeling's givenness. However, to the extent that this givenness is dissociated from consciousness, it touches the I, meaning what CM designates as the "I periphery" or alternatively "from the outside to here" (449) that restricts my living-body (530). Finally, unlike the senses that are portrayed as open, my living-body is depicted as closed (532).

The characterization of the phenomena of the spirit and my body as "real transcendent" is now accurately clarified: the body and the spirit cannot reach each other because they are spatially incommensurable, meaning that in terms of their expansion they belong to two different spheres (437). Surely, my body and my spirit actually appear together. Moreover, it is impossible to say about my body, to the extent that I feel it from the inside, that it presses me from a sphere that exists beyond my I, but inasmuch as it is experienced from the inside, it is also experienced as belonging to me in its entirety. Yet, one should distinguish the real transcendent against consciousness, that is part of the givenness of my body, from "everything that is beyond the I" that does not belong to it, including beings that sometimes personally press me from the outside inwards (447). In any event, there is no possibility for a genuine contact between my spirit and my body. As spirit and as consciousness, I am prevented from arriving at the real in the genuine sense, and in any case, I cannot have any influence upon it or capability to reshape it. Obviously, this transcendence is the reason for the impossibility of achieving a full and comprehensive concept about the real world. Indeed, CM argues that the objective perception itself already removes any moment that could become real (437).[43]

The traditional question of how the given changes itself from the outside into consciousness' content – a question that is usually diverted to the borders of our possibility to know and understand – is thus granted a clear answer: the felt-thing does not become a "content of consciousness", even if we understand this concept in the widest way (542). In spite of that, the genuine feeling-experience (in terms of itself) is beyond the inclusive dimension of possible conscious-experience, and according to CM, it is for this precise reason that it must be perceived; in her opinion, the felt-thing, to the extent that it is actually felt, does not deliver itself as known or clarified, but exactly as felt. It is impossible to achieve better accuracy in the presentation of the felt-thing, since givenness that is adhering to feeling is in an entirely sui generis mode of givenness, in order to understand which, one needs a much wider understanding of the I, such that exceeds the scope of her discussion in *Doctrine of Appearance*.[44]

[43] One of the arguments typical of the realistic approach in phenomenology deals with the difference in time between Being and being perceived. See: Moritz Geiger, Die Wirklichkeit der Wissenschaften und die Mataphysik (Bonn: F. Cohen, 1930), p. 170. Spiegelberg contended that "in principle, the situation is the same in all cases of sense-perception [...which] can never give what is present, but only what has just passed. And since the past no longer exists, we can never see the original object itself but only its 'trace' which means its cast or likeness", Spiegelberg, op. cit., pp. 156–157.

[44] CM exemplifies the questions with which such a future study must cope, meaning: What is the nature of the I that allows itself to be framed by the bodily-entity and thus be restricted by it? How do the relations between the I depict themselves phenomenally? (CM, "Zur Ontologie und Erscheinungslehre der realen Außenwelt", pp. 541–542).

The Giveness of Manifest Appearances

Manifest Appearance's Givenness vis-à-vis Feeling's Givenness

Manifest appearance's givenness concerns real being that arrives "here from a distance" that is rooted in the peculiar capability for self-presentation of the being that appears in it (430).[45] Despite both being included in sensory givenness, there are a few fundamental differences between "feeling's doctrine" and "manifest appearance's givenness": Firstly, in appearance's givenness something is determined through the senses (406), before which it stands out. In this context, CM describes the senses as establishing "real contact" that grants the appearance the character of self-presentation that adheres to it (425); or, alternatively, she typifies the senses' self-presentation as undergoing realization in the manifest appearance according to the boundaries that delimit the beings that appears in it (462). Secondly, the realization of the manifest appearance's givenness necessitates a distance from the I and the body – a distance that is bridged by the senses that are involved in the phenomenon (406). Whereas distance would have hindered the very existence of feeling's givenness that is apparent by touch and personal pressing of the living-body, the manifest appearance is depicted as a being that does not reach the I personally, but only as long as the distance in which it is essentially rooted and fixed is preserved (472). Moreover, it would never have been possible to reach the external world without the involved senses, thanks to which it is noticeable without abolishing the distance from the I. CM clarifies that the objective quality of the felt-thing imposes upon the I "an absolute self-restraint" from a direct act (473). She determines that the essential fact that hearing and seeing are far-senses (*Fernsinne*) necessitates referring to "the seeable and hearable as beings that are closed within themselves in any manifestation" and as "separated" from what has been sensed through them. Even when their real appearance-position is nearby me (for example, something is ringing near my ears), manifest phenomena are always experienced as a content that is kept at a distant position or as "closed for themselves" (473).[46] Finally, while feeling's experience is not dependent on the solidity and roughness of the bodies that bring it about having meaning for the I (406), the demand for distance regarding the manifest appearance enfolds at the same time the possibility of grasping or receiving them by means of consciousness. In her words: "a being is objectively given only when it is at a 'distant position' in which it is perceived by the I, yet remain separated from it and closed for itself". As we have seen, when approaching an object in order to represent it, consciousness must detach it from itself and position it against itself (470). Yet, unlike in the case of feeling's givenness, the understanding of the essence of the

[45] The discussion will refer to beings that appear in rigorous objectivity only, and not to what CM designates as the "loose givenness of manifest felt-being" (CM, "Zur Ontologie und Erscheinungslehre der realen Außenwelt", p. 504ff), which remains marginal in *Doctrine of Appearance*.

[46] See in the context: Husserl, Logische Untersuchun gen II, p. 254.

living-body-being of the I is totally irrelevant for the explication of the manifest appearance givenness that meets only the spiritual I's being (541).

The three fundamental characteristics of the manifest appearance givenness – the affinity to senses, the distance from the I, and the structural accessibility to consciousness – distinguish it from the phenomenon of feeling's givenness and thus fill with content the argument that the two constituents of sensory givenness are not located on the same plain (425).

Objectivity, Externalization, and Self-Announcement

The manifest appearance is created from an appropriate sensory qual, which establishes its essence and conditions its possibility of self-presentation as a ready whole (465). The qual does not denote a quality or feature that adheres to the manifest appearance, but rather the substance that establishes it (465). The sensory qual constitutes the type and mode of the "speak", of "externalization" (466) or the self-announcement, which means that the felt-thing grants itself from itself and at the same time remains by itself. CM describes the appearance of the manifest being as imposing itself upon the I that is there in its scope. She assumes that there is something in the manifest being that always attracts anew my gaze in an inescapable way. Whether or not this assumption is correct, she determines that no additional explanation is required. In her words: "it is totally obvious that existence (*Bestand*) in-itself and for-itself is totally independent of the being or the spirit to which it presents itself and stands against it as an appearance" (411–412).

The mode of appearance's presentation is possible because what constitutes the manifestation, meaning the qual that mediates the object, is itself composed of an objective material (471). Here the meaning of objectivity is only the following: mediation or externalization that is simultaneously being kept inside itself (468). The qual has a being that is surrounded by itself and thus grants it with a givenness-structure that is closed for itself (471), that has a specific appearance-place and appearance-shape (475).[47] Indeed, presentation, by its innermost essence, denotes a sensory-objective mode of givenness (506). Moreover, it is fundamental for the sensory qual that it is at the same time the mediator and the constructing material of the mediated. This means that the aspect that is used for mediation, i.e., that pulls the qual outside, is simultaneously being kept inside the thing itself. This double nature of the manifest appearance is rooted in that of the sensory qual as such (468).

[47] The objective closedness and shape can also describe a real moment that is not self-standing but needs to be filled inside another being in order to be able to appear concretely. This is the wide idea of objectivity. However, the narrower one, which according to CM is genuine, refers to a self-standing object. In other words, every object has an object-adhering being. But not everything that has such a being is purely for this reason an object in the narrow sense (CM, "Zur Ontologie und Erscheinungslehre der realen Außenwelt", pp. 475–476).

In fact, the being that shows itself in the manifest appearance does not need a means of presentation in order to be presented but, in CM's words: "**in the very being** that belongs to the manifest appearance as an entity (*Gebilde*) that is constructed from an announcing or appropriate material, rests **the mediation of itself**", or alternatively "by means of existing totally and only from an externalizing-material, **is itself** exclusively and entirely 'brought here and outside'" (466). The feature of presentation determines the nature of the color and the tone being realized within their very being that expresses itself independently of the presence or absence of an experiencing subject (412). When an appearance "arrives outside" by itself and according to its constituting materials, it simultaneously re-establishes the threshold of itself. Thus, the manifest appearance exists out of its self-presentation and at the same time it verifies itself by means of this self-presentation (467). Therefore, CM concludes that "the book of the real world" is opened within the manifest presentation (463).

Yet the function of the qual to carry itself towards me does not threaten to disrupt or abandon the objective framework that has been mediated. In CM's words: "no change takes place in the manifest surface of the body-thing" (468). She explains that the qual keeps its being "for itself" (471), inasmuch as obviously every real-being has being "for-itself" (472). Moreover, among all real beings the uniqueness of the "sensory given" results from its **purely out-of-itself** capability to be experienced, **noticeable**, and contained by spirit (473). In this context, CM characterizes the manifest appearance as "something that by its essence plays between the accepting consciousness and this real world that is rooted in itself and stands for itself".[48] However, this time she highlights something that is not obvious, that also this mere mediated real-being has a being that is closed in-itself, and that responds only through this closedness, and only by means of its connection to this objective form can it be experienced at all (472). Moreover, also the perception of the felt-thing is not accidental, but follows its own nature and its mode of appearance that she typifies as "appearance-adhering" (*erscheinungshafte*); meaning as appearance with a final objective shape in which it is experienced as self-standing. Hence the conclusion: manifest beings can be received adequately when they are experienced in an appearance-adhering concreteness of-themselves and for-themselves (478).

The Senses and the "Dead Spirit"

The appearance's disposition of the givenness of the manifest appearance does not include from the outset the 'being-in-itself' of the observing subject (378); meaning that the prominence (*sinnfälligkeit*) is a feature of the appearance and not a

[48] One should distinguish between the term "play" here that does not express a reduction from the reality of the appearance and the "play on the reality of objects" that is typical of the perceptive attitude that indicates the weakness of its reality dimension, CM, "Zur Ontologie und Erscheinungslehre der realen Außenwelt", p. 379.

description of an experience of the I. However, as discussed before, the manifest appearance comes into being for the I by means of the senses, in particular seeing and hearing, whose different modes of givenness and announcement present manifestly diverse sides of the material being (483). Although bodily ears and eyes must be there in order for a bodily individual to be able to hear and see (484), the manifest appearance is not dependent on "bodily" organs; especially as such they cannot serve as receptacles for objective experiences that stand in a real-transcendent relation to the I (480). CM's interest in these senses is due to there being essential and substantive "opponents" (*Widerparte*) for the absorption of manifest materials in general (484).[49] Seeing and hearing are visualizers as doors and windows through which the sensory given arrives "inside" the I (536), and characterized as "spiritual senses" or "spiritual organs" that have a specific accessibility to a certain objective mode of manifestation that is responsible for the presentation of the manifest appearance (484). So we experience from inside the eyes and the ears as the "body's gates" or as the "body's openness", through which the immediate contact between the spiritual-living I and the external and manifest world is positioned. Therefore, had we destroyed these bodily positions towards the external world, the body as such would have become closed for us (492). Seeing and hearing are described then as "filling" and "spreading" in the sense appropriate to them, unlike the "void, free, and accidental" stance of the felt-thing or that of the living spirit when it turns to another subject. This is why there is no place for an analogy between them and the manifest appearance (521–522). CM summarizes the genuine meaning of the term "manifest" (*Sinnfälligkeit*) as follows: the manifest is exactly what breaks in upon into these senses, meaning it is capable of specifically "delivering" itself to them (486).

The fundamental argument is, then, that the specific mode of announcing must "meet" the appropriate possibility of "perception" that is ontically rooted in the spiritual I, meaning that is not a result of a momentary influence of the living spirit (484). Moreover, the appearance of the manifest as a whole and as such, though not its very being, becomes possible through the inhibition of the spiritually living contact with the external world, namely avoiding a representative relation by consciousness. That is to say that the manifest is adequately absorbed by means of hearing and seeing without any considerable involvement of the spirit of the I (479). The same is true also in the opposite direction, i.e., when we break off the spiritually living contact with the external world, the manifest "remains" in its specific expanding and appearance-adhering position. Without the involvement of the spiritual I, the manifest appears as lacking the "depth", "rootedness", and "surroundedness"

[49] Spiegelberg's determinations, according to which "Ultimately, all these organs are themselves phenomena of reality and so are the causal links between them" illuminates the problem with which CM deals here as follows: "Is there a way back [...] from the retina via the cortex and the mental processes to the original object outside which supposedly started the whole chain of physical and physiological processes?". This problem "makes sense only on the assumption that the physical objects, as the 'stimuli' for our sense perception, our sense organs, and the physiological process within, are ascertained realities (...and) as long as it is possible to know some real objects themselves" (Spiegelberg, op. cit., pp. 150–151).

that could have been granted to it only a by means of spiritually living perception. These faces of the manifest resemble those that are mirrored by primitive observation. Nevertheless, CM argues that they uncover the right vision of the manifest appearance givenness (478–479).

However, if consciousness must remain open in the face of the appearance being of the external world in order to wholly apprehend it, where at all can exist such openness when the I does not actually participate in the spiritually living receiving of these beings? (479). Moreover, what should be the meaning of the manifest appearance on the part of the receiving I, when this meaning is understood in advance as not shaped by the "organ" that perceives it? CM regards these questions as essential for the realistic reasoning of the manifest appearance (480). Similar to feeling's givenness, here too she seems to harness consciousness in a careful, restricted, and accurate manner. Yet while the felt-thing is located at the "periphery of the I", meaning on its personally pressed body, the manifest appearance "makes itself noticeable simultaneously from-itself and outwardly by means of **personal pressing against consciousness' periphery**", where spirit is no longer directed by the senses to the external world (498). CM explains that precisely because the spirit is ahead of the manifest being, what remains to be received is what is being heard and seen (499).

The location of consciousness at the periphery is typified as a "spiritually dead acceptance" (*eine geistige tote Entgegennahme*) of manifest beings (490). This acceptance is merely the fixed reception of the thing that presents itself in its objective self-presentation. This appearance is experienced by the I as having no context, partly thanks to the appearance of the manifest being as an in-itself and for-itself fact (481). This context-less acceptance does not contradict the fact that the same I stands in the two experiences as carrier of givenness' experiences and as recipient of what appears in them. Yet since in each of them one experiences oneself from different places, they are considered as lacking context (482, n. 1). CM clarifies that indeed it is not the spirit itself that dies in the acceptance. Quite the opposite, the spirit is living and awake and conceives a world different from the externally manifest one, but the manifest appearance is what dies in it. In any event, even when the living spirit is at the periphery it is sufficient for the realization of the manifest appearance (490).

The argument that the recognition of the external world cannot be based on the living spirit is explained in that when it is directed to the manifest it does not achieve only what is being accepted by seeing and hearing, but in a much "improved" mode and with far-reaching implications; its achievement is not restricted to the fixed counter-acceptance of the manifest surface that is capable of presentation. As a result of the living envelopment (*Umfassung*) of the appearance by the spiritual gaze, the presented surface "becomes looser" and the latent content that underlies it comes to the fore. Now the manifest appearance is granted "depth" that enables it to appear before the pure observation as filled from inside. However, the content by which the manifest appearance is filled with does not belong anymore to the sensory material itself, and therefore it cannot be regarded as given (494). Simply, the internal is not manifest and thus is incapable of appearance. Elsewhere, CM adds that the living spirit changes the initial given as it inserts it into a unified, meaningful, and contextual world. Yet, the epistemological value of this achievement cannot be taken into consideration, since as a living spirit I can pose or grasp things, but not sensorily observe them as the observation basis of the sensory given lacks any spiritually living test. In CM's opinion there is an eternal gap, thus unbridgeable, between the spiritual achievement and the sensory given, meaning: as a living spirit one can only grasp and pose but not sensorily observe—which necessitates what is designated as "dead relations to givenness". Therefore, the consequences of the spiritual positioning cannot really be considered (488).

The question that is being evoked now regarding the condition of the living spirit in the "dead acceptance" is the following: What is the appropriate description of the spirit that has been shifted to the consciousness' periphery in favor of reception by the senses? (494). CM seems to portray certain accordance between the condition in which the manifest appearance is given and that of the I that is sensually opened towards it. Both are depicted in terms of restriction or prevention from expansion that is essential for the very coming into being of the appearance. The manifest appearance delivers itself from its closed-objective stance and from itself – a stance that is enabled when it is not conceived and grasped in a spiritually-living momentary of one's spirit. At the same time, the I is described here as subject to a condition of a "hold" (*fassen*) (495) that might enable (peripheral) conscious affinity towards the manifest appearance and at the same time remain separated from it. This duality is well reflected in the demand to carry out 'transcendence-performance' (*Transzendenzleistung*), meaning to constitute a relation toward an in-itself closed being while keeping a distance from it. The "openness" in particular is essential for the constitution of this relation by the living-spirit (474), which makes it possible for it to stand up against what is closed and sealed in-itself without violating the distance that lies between the I and the manifest appearance. Indeed the transcendence here is mutual—the spirit does not invade the manifest appearance, which itself remains located beyond it.

Against this background the accessibility of the 'dead spirit' to the manifest appearance can be understood. CM describes "a natural grasping-range" (*Natürlice Greifweite*) that enables spirit to conceive through the senses manifest beings as "dead" "situational-achieving" (*Zustandsleistung*) when the spiritually-living contact with the manifest and external world collapses (495). The capability of the senses to be equipped with a natural grasping-range, which facilitates the "dead acceptance", enables the senses to be regarded as the genuine and personal carriers of manifest beings (497); "natural" means here that there is no need for a special momentary act by spirit, but the involvement of the senses can suffice (495). Moreover, the living spirit can never rule an observation that has established itself as "dead acceptance". In spite of that, the sensory qual might be reached only if it encompasses itself with a kind of blindness to forms of perception, and when the senses "express" nothing from a perception of a unified world (477–478). The "capturing" of the manifest appearance itself and for-itself occurs "from itself", or alternatively from the natural grasping-range that replaces the momentary achievement of the living spirit (496).

However, this does not mean that the living spirit is completely eliminated. First and foremost, seeing and hearing are unthinkable at all without their actual insertion into the realm of consciousness, meaning: by the very existence of the senses there is a spiritual sphere, even if this is primitive by its constitution. Hence one should position the sensory sphere as a unique type of happenings that undergo realization in it, so that the spiritual sphere reaches as far as the possibility of these happenings does. Indeed, already the definition of the senses involved in the reception of the manifest appearance as "spiritual" hints at a certain affinity also in the vitality of spirit. Additionally, one should remember that not only are seeing and hearing considered "**open I-stances**", but also this openness is embodied in the possibility for objective experience—one that is actually realized in the self-presentation of the manifest appearance (506). Finally, just as the living spirit can withdraw from the senses and be located at consciousness' periphery, so too can this spirit return and be presented again inside seeing and hearing. Actually, when the living spirit is withdrawn from the manifest external world, it thus simultaneously withdraws also from the senses, and alternatively it is presented again inside when it is directed to the external world.

It transpires, then, that the spirit can have two modes of relations with the senses: it can let them remain peripheral as a gate to itself, and that enables what CM calls "dead acceptance" of manifest beings regarding which the senses serve as a sort of specifically accommodated reception places. Likewise, the spirit can itself enter inside the senses and thus be changed from "dead acceptance" to living spirit. To be sure, the senses thus continue to be able to serve as sensitive places for defined modes of presentation, but spirit "enters" them and "occupies" them from the inside, and thus becomes itself sensitive to certain modes of presentation (491)—although as mentioned before, it generates a transformation in the manifest given (488). In CM's words: transfer it from a formed condition into an unformed one. In any event, from her realistic point of view it is emphasized that "it is impossible that every being of external world will itself preform the movement back and forth or that by itself and for itself it can be one time 'formed' while in the other 'unformed'. It must be that the different hold will be attributed to a change in the givenness" (499) that already appears before the face of the I.

Now one can achieve a more accurate understanding of the spiritual element that is active in the appearance of the manifest beings: the adequate reception of something presenting itself manifestly does not occur by means of the senses, but "by the spirit inside which the senses enter" (497). Hence, "the natural grasping-range can be understood as a 'dead' residual achievement by a living-spirit, which remained at the periphery when the personal and living participation has been dropped in the 'dead' acceptance". The "impression", as CM put it, is that spirit equips the senses with the capability for receiving by means of its own natural grasping-range. This means that the spiritual senses can be placed only by means of the spirit that is capable of acting from within them and with their assistance (497); by this placing the senses are able to receive the manifest beings. CM summarizes "the only and important condition for the adequate counter-acceptance of the manifest appearance" as follows: "the spiritual senses must [...] be **carried out** in-themselves always in an actually living mode by the spirit as gates that are being posed personally" within the borders of the dead counter-acceptance (498). The demand from the I to carry out "transcendence-performance" transpires then as presuming the existence of the natural grasping-range together with a certain involvement of the spirit as such. In her words:

> We believe that there **is such a spiritual being that is a carrier that aids the situation of transcendence, which is natural to certain extent – in a sense there is no need for a special act or *Salto mortale*. Hence, the spiritual I does not live only by itself but also in a strange world.** It belongs to the nature of the spiritual being that an entire world can be contained within it. without it sometimes being asked for something that is beyond its belonging 'range' in order to guarantee an undisturbed spreading and development. [...spirit] can rest there without transcending its being superficially or by means of effort. (407–408)

The careful nurturing of the consciousness and of the I – meaning the withdrawing of the consciousness to the periphery, the opening through the senses, and the achieving of transcendence – are crucial to the experience of the manifest appearance. Hence one cannot simply take what seems more as a pretention of the manifest appearance to be the "beginning of givenness" (*Gegebenheitanfang*) of a being closed in-itself (494), since its very appearance involves the "dead acceptance" from a specific position of the I.

Epilogue: Immanence and Transcendence in the Phenomenology of the External World

How do the two composing phenomena of the "sensory givenness", i.e., "feeling's givenness" and the "manifest appearance givenness", relate to each other? Does the fundamental difference between the two eliminate the possibility of consolidating a united understanding of the external world? Seemingly, CM answers these questions directly as she argues that "the sphere of sensory givenness is divided into two layers that oppose each other" (425), and that her detailed discussion of them aims at "posing anew unequivocally the opposition between feeling's givenness and appearance givenness" (426). From time to time she refers to aspects common to the two phenomena, as for example in the following determination: what is for us valid as the specific designation of the genuine feeling's givenness is possible, though it does not always exist, also on the side of appealing's givenness (460); feeling's givenness can always be considered a provisory mode of the givenness of manifest beings (505); feelings are always already placed as the most extreme border case, that as such suffice for the pretensions of objective experiences' givenness (461), and that as sensually given beings, the felt-thing and the appearance share the real transcendence to consciousness' sphere (455). However, these few determinations do not lead to a substantial discussion in *Doctrine of Appearance* that is decisively devoted to a detailed explication that exposes the fundamental differences between the two phenomena without suggesting an overview of the extensive issues they involve, such as: the consciousness, the body, the spirit, materiality, etc. CM herself rightly referred to the need to complement the discussion in *Doctrine of Appearance*, and she actually deals with a few of the issues in her later writings.[50] In any event, the finding of meaning that binds together the two composing phenomena of the external world that occupies practically all of *Doctrine of Appearance* remains an assignment for the reader.

The following suggested interpretation observes CM's understanding of the phenomenon of the external world from the perspective of the relations between immanence and transcendence. The relevance of this perspective is apparent already in the general concept of the world that enfolds within itself immanent aspects as well as transcendent ones. Immanence lies in the very worldliness of the phenomenon of the external world, and contains within itself all the beings that establish the world from the inside, including the human consciousness and spirit. Transcendence is associated first and foremost with the nature of the appearance of the external world to the I and its consciousness. Additionally, transcendence is the depth of the reality of the world, because of binding together the very being of the world and its autonomy – in relation to the I, to consciousness, and to any other immanent aspect.[51] Indeed, the initial reference to the dimensions of transcendence and immanence appears already in CM's discussion.

Firstly, CM points to the transcendent aspect of the study of the external world. In her words: "obviously we think that no thesis of factual-science can turn into an essence-determination of an established principle" (355, n. 1).[52] Instead, she asks what actually underlies consciousness and exists out there in the "real world" or the "real external world"? (383). This question focuses on the beings themselves that compose the real world, while the transcendent denotes a hidden depth that dwells

[50] See: CM, "Zur Ontologie und Erscheinungslehre der realen Außenwelt", p. 373, pp. 395–396.

[51] Elsewhere I discussed in length the relation between immanence and transcendence, see: Ronny Miron, Karl Jaspers: From Selfhood to Being (Amsterdam/New York: Rodopi, 2012), p. 207.

[52] Similar to Husserl, who rejected what he designated in "Logical Investigations" as "positive sciences" that prefer facts to essences (Husserl, Logische Untersuchungen I, 81, 117; Husserl, Ideen zu einer reiner Phänomenologie und phänomenologischen Philosophie, I, §3), CM too binds the rejection of facts together with preference of essences. For further reading regarding Husserl, see: Jitendra Nath Mohanty, Phenomenology: Between Essentialism and Transcendental Philosophy (Evanston, Illinois: Northwestern University Press, 1977), pp. 3–5; Robert Sokolowski, Introduction to Phenomenology (Cambridge, UK: Cambridge University Press, 2000), pp. 9–10. For further reading, see Spiegelberg's distinction between "critical realism" and "critical phenomenological realism". The first "does not grant reality to any of the phenomena but tries to infer a real world different from the one presented to us", while overlooking "that a world beyond all appearances would of necessity remain inaccessible to us". This kind of realism "seems to be unable to cope adequately with the constant conflict among non-subjectival phenomena of reality". The second "does not transcend the phenomena of reality but either merely (...) strives for an integration of the phenomenal field by filling gaps left by a 'naïve' phenomenological realism that relies exclusively on isolated phenomena of reality", Spiegelberg, op. cit., pp. 165–166. (see also: Herbert Spiegelberg, "Three Types of the Given: The Encountered, the Search-found and the Striking", in Jitendra Nath Mohanty and Karl Schumann (eds.), Husserl Studies 1 (1984), pp. 69–78).

inside immanence and cannot enable exhausting the external world on the level of mere phenomenality. Moreover, the transcendent constituent shows a fundamental affinity to the essence that is characterized as transcendent to the appearance, or alternatively as its most profound immanent internality. Now, the superficiality becomes naturally clear of the two most frequent meanings of transcendence in the philosophical discourse that emphasize the external aspect in it, i.e., the ontological that identifies transcendence with everything beyond the world, and the epistemic that attributed to it everything that cannot become intelligible. Besides acknowledging the extending aspects of transcendence, CM regards the transcendent as a reality that is completely "inside" the world and hence it is principally accessible to consciousness, though it cannot be exhausted in our cogitations regarding transcendence.

At the same time, CM concentrates also on "what for us is immanent in the comprehensive phenomenon of the real external world" and on "natural consciousness" (378). Relevant here is the questioning of the quality of the phenomenal relations between "the sphere of the observable" and the real external world, or alternatively the elucidating of the affairs that connect the external world with the I. Thus, CM moves away from positivist and idealist approaches that carry out "external positing" (*Außenstellung*) of the external world while excluding the I from it and limiting it to the mere plane of phenomenality (383). Consequently, the phenomenon of the external world turns out to be a replication - direct or contradictory – of their idea of the I. CM shares with realistic approaches the ambition to establish the independence of the existent vis-à-vis consciousness. Yet unlike them, for her the explication of the I and even of consciousness is an indispensable part of the realistic reasoning of the external world. Her discussion exposes a circumspect and subtle effort to scrutinize the portion of consciousness and of the I in general in the reality of the external world. Phrases such as "consciousness-beam" and "dead spirit" express the search for minor measurements that will be sensitive enough to detect the rather limited presence of consciousness without turning it into a condition for appearance as such or violating the external world's transcendence and closedness in-itself before spirit of the external world (434).[53]

Undoubtedly, the distinct discussion about immanence and transcendence is critical to CM's approach, according to which givenness and reality are not identical (427). Moreover, it is important also for the more general reasoning of the independence of the external world against spirit without being thrown into an uncritical realism in which the I and consciousness are absent. However, in CM's philosophy of the external world the joining together also takes place of the immanent dimensions and the transcendent ones. This is noticeable in the indivisible cohesion between immanence, transcendence, and essence that arises from CM's

[53] Undoubtedly these phrases serve as founding stones in the investigation of the transcendental aspects that are involved in CM's realistic philosophy. Actually these appear to provide the essential infrastructure for CM's later idea of the subject, as she moved away from dealing with the phenomenology of the external world.

description of the expected understanding of the phenomenological investigation of the external world:

> [...] neither the "concealed" primordial phenomena nor the conceived ideas lie themselves in the "mere appearance's surface layer". Accordingly, the work that is demanded here can never be achieved by means of a passive gaze at the phenomenal differences that immediately arise from the actual appearance, even if this be descriptively captured to the last tiniest nuances. On the other hand, what must burst forth is that from this appearance's surface [...] are only the de facto essence-beings (*Wesensbeständen*) to which an accurate analysis of the appearance's surface currently leads as its own foundation; it does not have to lead to a totally transcendent related world of "resistance" (*Ahndungen*) or "metaphysical constructions". (353–354)

The expectation is then that the observation of the "surface" that is accessible to the senses will lead to "essence-beings"; meaning the innermost real kernel that establishes the thing whose mere "surface" is apparent in the appearance. There is no contradiction between her determination that this way "will not necessarily lead to a transcendent world" and what appears immediately afterwards, that "studies of essence in the sphere of actuality lead entirely to a transcendent realm" (355, n. 1).[54] CM expresses here an idea according to which the transcendent has an affinity to essence and no abyss separates it from the immanent understanding of the external world. Thus, she immediately rejects not only the approach that the transcendent is not included in the external world but exists beyond it – an approach that regularly reflects a metaphysical construction made by the human subject[55] – but also the world view that regards immanence and transcendence as two distinct authorities. In spite of that, the phenomenologist of the external world, as depicted by CM, unceasingly moves between two peripheries that are located on a boundary in which immanence and transcendence are tangent: the one is on the body's margin and brings about the feeling from the inside, while the other is on the margins of consciousness and directs one's observation toward what is outside there.

---

[54] In 'Realontologie', the first essay that appeared after *Doctrine of Appearance*, CM will clearly distinguish between the idea of transcendence upon which her realistic approach is based and the mistaken one. The first designates a "continuing maintaining" (*fortdaurende Erhaltung*) of the real thing in its real being that is established by-itself and in-itself. The second is characterized as fragile and suffering from possible dependence on immanence because of its rootedness in the human spirit, see: CM, Realontologie [1923], special print in Jahrbuch für Philosophie und phänomenologische Forschung VI (Halle: Max Niemeyer, 1924), §26, pp. 185–186.

[55] Throughout *Doctrine of Appearance*, CM stresses her rejection of the metaphysical approach as being unjustified (CM, "Zur Ontologie und Erscheinungslehre der realen Außenwelt", p. 346, p. 348) and unnecessary for her study (ibid., pp. 355–356). Later on, she will not reject metaphysics but argue that the presumption of the existence of the real world is an indispensable condition without which there is no metaphysics at all, see: CM, "Was ist Metaphysik?" [1931], in Schriften zur Philosophie, vol. 1(Munich: Kösel-Verlag, 1963), Vol. 1, p. 38. This argument is repeated also in: CM, "Bemerkungen über Metaphysik und ihre methodische Stelle" [1932], in Schriften zur Philosophie, (Munich: Kösel-Verlag, 1963), Vol. 1, pp. 49–88; CM, "Die fundamentale Bedeutung eines substantiallen Seinsbegriffs für eine theistische Metaphysik" [1931], in Schriften zur Philosophie, vol. 1 (Munich, Kösel-Verlag, 1963), Vol. 1, pp. 257–267.

Seemingly, the two fundamental phenomena of the external world are directly connected to the two questions that occupied CM's phenomenology of the external world. Dealing with the being that comprises the external world is associated with the manifest appearance that is at the same time what is real or "external", while the aspects that relate to the phenomenal relationship refer to feeling's givenness that enfolds within itself what is "inside" the world. Yet, as we have seen, the profound discussion of these two phenomena does not surrender to the banal division between outside and inside or between immanence and transcendence. Accordingly, CM does not adopt the logic that the I is more dominant in the experience of feeling compared to that of standing before appearance in which it seems as if the thing that appears is the central issue. In CM's opinion, "manifest appearance's givenness" and "feeling givenness" share the fundamental separateness from consciousness in an original way (451). Her discussion founds feeling's givenness on the independence of the felt-thing of the personal touch of the I.[56] Only after the realism of this phenomenon has been established does a minimalistic and reserved positive talk about the subjective dimensions that are involved in feeling's givenness commence. Thus, the aspect of consciousness is represented by the "consciousness beam", the I is withdrawn to the periphery and felt by a touch of the body's edge, and the senses seem to be totally eliminated. In spite of that, exactly in the explication of the appearance's givenness, one recognizes the effort to delineate the phenomenal affinities to the I whose representation here is much wider and includes the spirit, the senses, and consciousness in general.

The solution to what seems as a sort of "reversed logic" in the initial understanding of the two constituting phenomena of the external world lies in the different strategies that their explication demand. The affinity to the point of personal touch with the I in feeling's givenness requires a narrowing to a minimum, i.e., to "consciousness' beam", of the presence of consciousness. However, the distance that separates between the I and the manifest appearance necessitates the reversed action of uncovering and emphasizing the involvement of consciousness and spirit. What ruled the discussion of these two phenomena is not simply their different attitude to consciousness, but what is being shown through the observation of the way they present themselves in the face of the I (451–452), i.e., as a real bodily essence, as having spiritual senses and as a carrier of consciousness. The realistic understanding of the external world is thus not established upon a dogmatic ready-made principle; rather, being phenomenological, it is consolidated first from the observation of the appearance of the thing, yet it is not exhausted in it, but aims at its essence that dwells within its outskirts and enfolds inside it the profound reality that brings it about.

[56] Spiegelberg presents the non-dependence as an indication of the reality of phenomenal objects if not even of their total independence. See: Spiegelberg, op. cit., pp. 147–148.

Kant and the Starry Heavens or the Splendor and Misery of Speculative Rationalism

Rihards Kūlis

Abstract Kant stresses that scientific cognition can only materialize in conjunction with experience. Leaving the sphere of experience takes one into the world of fantasy. From the point of view of Kant's Copernican revolution in philosophy his description of the inhabitants of other planets in the *Universal Natural History and Theory of the Heavens* should also be viewed as unfounded wanderings of the mind. Such wanderings of the mind never occur in Kant's philosophy of mature years.

Kant's philosophical revolution strengthened the positions of scientific cognition by excluding metaphysical speculation. At the same time rejecting metaphysical claims for cognizing the supernatural and acknowledging man's inclination towards it, Kant imprints a sphere of freedom returning to the starry heavens above its mystique, its inscrutability that had actually been looted by speculative rationalism.

Kant is preoccupied with the problem of the objectivity of world cognition all through his lifetime and returns to it in his later-day treatise *Opus postumum*. In this manuscript Kant offers a special program of the metaphysics of nature that should ensure the objectivity of cognition in natural sciences. The program never saw its completion, yet it serves to demonstrate Kant's attempts at ensuring maximum objectivity to the world cognition process.

The history of science to this day gives positive appraisal of the origin of cosmos hypothesis, known as the Kant-Laplace hypothesis. The essence of the hypothesis is as follows: contrary to the ideas predominating in the eighteenth century as to the constancy of the universe an idea is voiced of the cosmos originating from initially chaotically scattered matter and its shaping itself in the further development. According to Kant, the formation of cosmos is a never-ending process.

R. Kūlis (✉)
Institute of Philosophy and Sociology, University of Latvia, Riga, Latvia
e-mail: rihards.kulis@gmail.com

A.-T. Tymieniecka (ed.), *Phenomenology of Space and Time: The Forces of the Cosmos and the Ontopoietic Genesis of Life: Book One*, Analecta Husserliana 116,
DOI 10.1007/978-3-319-02015-0_25,

Kant substantiates his theory of the origin of the Universe in his work *Universal Natural History and Theory of the Heavens or an Essay on the Constitution and the Mechanical Origin of the Entire Structure of the Universe Based on Newtonian Principles*, published in 1755. One cannot say that Kant's ideas were especially appreciated by his contemporaries. In 1796 the French astronomer Laplace published a hypothesis that is actually analogous to Kant's hypothesis. It is significant that Laplace had no idea of Kant's hypothesis.

Kant's hypothesis of the origin of the Universe is set forth in the first two parts of the work. These two parts are usually referred to by science historians describing Kant's contribution to the theory of the origin of the Universe. However, the treatise consists of three parts. The third, the last one, while speaking of Kant's cosmology theory remains in the shadows, as it were. Besides, the language giving an account of the philosopher's work more often than not draws the philosopher's conclusions nearer to contemporary scientific theories that kind of removes from the foreground the question that could arise in the reader's mind as to what "magic means" could have been at Kant's disposal allowing him to form a theory in the field of natural sciences – a theory that in its general outlines is not rejected in our time? The further theory of the formation of cosmos could be viewed as the continuation and specification of Kant's theory.

It is well known that Kant based his hypothesis on Newtonian physics, and it is well known that Kant was one of the most educated men not only in the humanities, but also in physics, astronomy, mathematics, medicine and geography. Could it all be enough for the formation of a magnificent *scientific* cosmological theory? By asking this we may finally come to raising *the question of the essence of scientific character and a scientific theory.* Writing his cosmological treatise Kant does not raise this kind of question. Much more significant, to his mind, is the question of how to combine the nebular hypothesis with the proof of the existence of God, the feasibility of which the philosopher is in no doubt at the time.

The problem of scientific character and the method of investigation becomes more acute as we read the work up to the end, not only the first two parts. However, for the sake of the experiment one could start reading the treatise from the third part forgetting for the moment the content of the first two parts. The third part is sure to have shocked any contemporary reader, especially if he did not know that the author of the text was the famous German philosopher Kant of whom, as is well known, it is said that he made a Copernican type revolution in Western philosophical thought. Maybe, it is this text that demonstrates the magnificence of the transformation?

What is it that Kant writes about in the third part? Neither more nor less than the inhabitants of other planets! Besides, the judgments and the argumentation claim to be no less scientific than those in the first parts. One must say that the third part brings to mind Voltaire's (Kant's contemporary) philosophical pamphlet *Micromegas* about an inhabitant of the planet Sirius – his stature being 24,000 times greater than man's stature – who travels round the Universe accompanied by a dwarf. The dwarf comes from Saturn and is only a thousand times bigger than man. In their travels they reach the Earth that seems rather deplorable to them. With the help of a microscope the Cosmos travelers manage to discover reasonable beings but are astounded by their stupidity. One might think that Voltaire's tract was a parody on Kant's theories, had it not appeared 3 years earlier than Kant' treatise. Kant had not read Voltaire's tract, just like Laplace, as has been mentioned, had had no idea of Kant's cosmology theory. It is so characteristic of the century. Just like, it must be mentioned, that speculations on the inhabitants of other planets and their characteristic features are characteristic of the eighteenth century. Probably it was a reaction to the collapse of the Ptolemaic geocentric (and at the same time anthropocentric) model of the Universe. The lack of anthropocentrism is compensated by sowing human beings in the now immeasurable Cosmos or, as it is done by Kant, forming a theory of knowledge that actually ascribes universal anthropomorphic features to any reasonable being – no matter where in the new Universe.

Following the traditions of the time Kant is convinced that the Universe is inhabited and based on the "scientific principles" of the time he indulges in speculation about the differences existing among the inhabitants of the Earth, Venus, Jupiter and Saturn. According to Kant, the Earth's inhabitants would dry up in the hot climate of Venus while those of Venus would become stiff and immobile getting to Earth. The inhabitants of Jupiter, according to Kant, should be made up of lighter and more gliding elements in comparison with the inhabitants of our planet so that the weak impact of the Sun could make them move.

Finally, the philosopher is so bold as to formulate even two "general laws".

1. "The material stuff of which the inhabitants of different planets, including even the animals and plants, are made must, in general, be of a lighter and finer type, and the elasticity of the fibres as well as the advantageous construction of their design must be more perfect in proportion to their distance away from the Sun".[1]
2. "That the excellence of thinking natures, the speed of their imaginations, the clarity and vivacity of their ideas, which come to them from external stimuli, together with the ability to combine ideas, and finally, too, the rapidity in actual performance, in short, the entire extent of their perfection, is governed by a particular rule according to which these characteristics will always be more excellent and more complete in proportion to the distance of their dwelling places from the Sun".[2]

Is it really written by the author of the famous origin of cosmos hypothesis, and how do these judgments go together with the ideas expressed in the first two parts?

Strange as it may seem, we may reasonably state that the two parts are not at variance with each other; they marvelously supplement each other by illustrating the method of scientific cognition and basic orientations of the time and highlighting the splendor and misery of speculative rationalism.

The author of *Universal Natural History and Theory of the Heavens* being the son of his time is sure both of the feasibility of giving "scientifically precise"

---

[1] Kant, I. *Universal Natural History and Theory of the Heavens or An Essay on the Constitution and the Mechanical Origin of the Entire Structure of the Universe Based on Newtonian Principles.* (Arlington, VA: Richer Resources Publications, 2008), p. 151.

[2] Ibid., p. 151.

description of the origins and history of the Universe, the inhabitants of other planets and offering convincing proof of the existence of God (although it should be added that Kant is not satisfied with the existing proof of the existence of God). The main thing is in finding and consistently implementing a method that would allow to cognize any sphere, and cognize it to the full. According to the representatives of speculative rationalism (this trend in Germany is most consequentially attested to by Christian Wolff), this possibility is provided for by logic and mathematics. The more we use mathematics in our investigations – whatever the object of research – the more precise will be the answers to the questions raised. At the initial stage of his philosophical activities, accepting the basic metaphysical orientations of speculative philosophy, Kant at the same time endeavors to base his investigations on the newest cognitions of natural science, Newtonian physics included. The result of the two trends produces a chimerical combination – a merger of cognition of natural sciences, logic-mathematical speculation and metaphysical fantasies.

Based on the mathematical method Kant undoubtedly achieves positive results in the explanation of the universal processes of nature; extensive knowledge of different spheres of natural sciences allows him to introduce into the vision of nature elements of historicity. It must be added that before Kant the Universe is thought of as created by God, ready and unchangeable. According to the German philosopher, God is the creator of the basic matter from which the world organization evolved according to the laws of mechanics. Henceforth, God does not interfere in the mechanics of the heavens and that is what enables the searching mind to penetrate into the secrets of the Universe. "Give me the material, and I will build a world out of it! That is, give me the material, and I will show you how a world is to come into being out of it".[3]

In the work under discussion Kant refers to Democritus and Epicurus. According to Kant, Epicurus groundlessly makes atoms change their rectilinear motion and collide with each other for no reason at all. Everything is given to chance. Kant offers a different theory in which matter would obey certain indispensable rules. Kant writes: "I see a beautiful and orderly totality developing quite naturally in its complete dissolution and scattering. This does not happen through accident or chance. By contrast, we see that natural characteristics necessarily bring this condition with them".[4] The philosopher is convinced that based on the universal laws of matter it would be possible to metaphysically show and substantiate the basic constructions of the edifice of the world. At the same time Kant is given to doubts whether just like previously one could say: "Give me the material, and I will show you how a caterpillar could have developed?"[5] The philosopher asks us not to wonder that he dares to state one could sooner discern the structure of all heavenly bodies in their motion, to learn of the origins of the current situation of the edifice of the world than to comprehend the emergence of one single stalk of grass or caterpillar.[6]

---

[3] Ibid., p. 17.

[4] Ibid., p. 15.

[5] Ibid., p. 18.

[6] Ibid., see p. 18.

It is interesting to note that already in the *Universal Natural History and Theory of the Heavens* an essential trend of thought is marked by the notion of *attraction* that makes one think of an especial liveliness of matter, a law-governed self-motion. In Kant's further works[7] the notion of *repulsion* is added, in the concept of the development of nature the theory of "physical monads" and "the centre of monadic power" occupy an ever more significant place testifying to an essential approximation to some of Leibniz's ideas. Although the theory, according to the philosopher, does not allow one to reach further than an understanding of the cosmic order, it embodies a peculiar poetical appeal, even pantheistic moods that address us from the pages of the *Universal Natural History and Theory of the Heavens.*

Kant tends rationally to reconstruct the process of the formation of the Universe. And his imagination conjures up a vision of the process that is a long way from the dry logic of mathematical formulae:

> We see at a glance wide seas of fire, raising their flames towards the heavens, frantic storms, whose fury doubles the intensity of the burning seas, while they themselves make the fiery seas overflow their banks, sometimes covering the higher regions of this world body, sometimes allowing them to sink back down within their borders. Burned out rocks extend their frightening peaks up above the flaming chasms, whose inundation or exposure by the seething fiery element causes the alternating appearance and disappearance of the sun spots. Thick vapours which suffocate the fire, lifted up by the power of the winds, make dark clouds, which in fiery downpours crash back down again and as burning streams flow from the heights of firm land of the sun into the flaming valleys, the cracking of the elements, the debris of burned up material and nature wrestling with destruction – these bring about, along with the most awful condition of their disorder, the beauty of the world and the benefits of its creatures.[8]

No matter to what extent Kant's cosmological conception reminded one of the teachings of Democritus, Epicurus and Leibniz, no matter how the mathematical method interlaced with poetical visions the rationalistic interpretation of the world combined with empirical knowledge gives fruit.

In Kant's time people knew of the existence of six planets – Mercury, Venus, Earth, Mars, Jupiter and Saturn. Kant advances a speculative supposition that beyond Saturn there are other as yet unknown planets. With time his suppositions proved to be right. Uranus was discovered during his lifetime, Neptune – in the nineteenth century and Pluto – in our time.

However, can logic and the mathematical method be regarded all powerful, can they be applied anywhere, in the investigation of any object? Let it be remarked that already in the *Universal Natural History and Theory of the Heavens* Kant does not venture to speak about the origin of life. In his opinion, mathematics is of no help here.

One could surmise that those were the attempts to consistently use the logic-mathematical method in the solution of any problem, the search for the proof of the

---

[7] See *Metaphysicae cum geometria junctae usus in philosophia naturali, cuius specimen I. continent monadologiam physicam (1756) 4 Metaphysische Anfangsgrunde der Naturwissenschaft.* (Riga, 1786).

[8] Kant, I. *Universal Natural History and Theory of the Heavens,* p. 121.

existence of God included, that made Kant re-evaluate the possibilities of rational speculations and the limits of their application. The entanglement of the problem is sharply delineated in the work *The Only Possible Argument in Support of the Existence of God* (1763). Kant is no longer satisfied only with considerations and proofs connected with an object or phenomenon – he raises the question of the specific character of veracity attributed to the object, i.e., actually the question raised is whether it is possible to speak of what is beyond this world in the same categories as of things of the physical world.

Writing the *Universal Natural History and Theory of the Heavens* Kant is also sure that he bases the work upon the material of experience, however when facts are missing he does not refuse to supplement the empirical experience and go beyond its framework making use of the synthesizing power of imagination and the conclusions of the intellect. It should be taken into account that hypothesis, speculation from the contemporary point of view as well not only can, but should go beyond the framework of the given experience, on the condition though that the hypothesis obeys the experiential content control. That is the borderline dividing the rational judgment, hypothesis from fantasy.

Kant increasingly tends to acknowledge that there are two forms of veracity: the veracity of natural necessity connected with the use of experience and logically mathematical method and the veracity pertaining to the sphere of metaphysics, to man's imagination and fantasy – as well as to the moral sphere. That marks the sphere of freedom. However, in doing scientific work man should not blunder into the realms of fantasy.

The further evolution of Kant's views proceeds under the influence of the epistemology of the English philosopher David Hume, especially his considerations on the causal relationship as a phenomenon of a psychological level. According to Hume, there are no causal relationships in nature, only processes, succession; causal relationships are a way one views things, a form of an attitude man uses interpreting phenomena in a given, inherited form.

In his conclusions Kant goes further than Hume. In his view there exists a whole line of a priori, i.e., independent of experience forms of world view and interpretation that man adjusts to the explanation of things and phenomena. Thus, the world turns into a man-made "scene of the world".

At one time under the influence of Neoplatonism (and not on the basis of empirical investigations) Copernicus came to realize that it was not the Sun that rotated round us, but we rotated round the Sun. Afterwards natural science confirmed Copernicus' Neoplatonic conclusion. Kant carries out a revolution in philosophy similar to the one perpetrated by Copernicus. He perceives that *in the process of cognition it is not the mind that adapts to things, but things adapt to the mind.*

The kind of understanding turns things into phenomena – occurrences – thus sharply raising the question of the feasibility of adequate cognition. *Is precise knowledge possible in the world of phenomena?* Prior to Kant the prevailing understanding in philosophy was that there was a consonance between the intellect and the world and the achievements of modern natural science seemed to confirm it, that Kant in no way wanted to question. According to Kant, there really exists an agreement between the mind and the world, but it exists in a very special way – not in the usual naïve kind of understanding, i.e. Kant believes that *the world science explains has already been arranged by the mind.* Man cognizes objective reality up to the level to which the reality is contained in the basic principles of the mind, and – one could add – to what degree those principles allow to cognize the world. Any cognition of the world takes place making use of categories inherent in the human mind. The specifically concrete character of scientific cognition derives from the mind embodied in man's world understanding, in his perception of things and phenomena. Kant convincingly proves that man's observation of the world is never neutral and free from previously given notions and judgments (in this respect contemporary science speaks of paradigms of scientific cognition). The world the man perceives and thinks about is formed as a result of world perception and judgment. Hence the question the scientist asks the world in principle arises actually from observation, from the existent horizon of understanding – judgment and notion. The question to a great degree embodies the answer. Thus, what we call the laws of natural processes are actually *the products of mutual interaction of the inner organization structures of the observer and the outside objects* and precisely for that reason the objects themselves are inscrutable because of the practical impossibility to separate the observer's inner organization structures from the object observed.

However, Kant stresses that scientific cognition can only materialize in conjunction with experience, i.e. fusing into one the sensual world, its perception and the grasping activity. Any attempt at leaving the sphere of experience takes one into the world of fantasy and dreams. From the point of view of Kant's Copernican revolution in philosophy his description of the inhabitants of other planets in the *Universal Natural History and Theory of the Heavens* should also be viewed as unfounded wanderings of the mind. Such wanderings of the mind never occur in Kant's philosophy of mature years.

Copernican Neoplatonic revolution at the time precluded man from occupying the central place in the Universe – the place he had ascribed to himself. Kant's philosophy robs us of the conviction that it is possible to unequivocally and truthfully cognize the Universe. This is the observation also made in many of the works by Anna-Teresa Tymieniecka who concludes that Kant's philosophy does not get as far as an objective viewing of the connection of cosmos, logos and the creative act – primarily being immersed in man's subjectivity. It is also stressed referring to Bertrand Russell by the contemporary German philosopher Wolfgang Welsch, in whose opinion "Kant has actually made an anti-Copernican revolution" or even "a Ptolemaic counterrevolution".[9]

We might only partly agree with Welsch. Really, while Copernican teaching decentralizes the world, Kant recentralizes it. Kant places man in the centre of the world that must be cognized, ascribing the primary role to the cognitive activity of man's mind that forms the scene of the world. By humanizing science Kant shakes the tradition that was at the basis of natural sciences as established by Descartes and

[9] Welsch, Wolfgang, *Mensch und Welt. Eine evolutionare Perspective der Philosophie.* (München: Verlag C. H. Beck, 2012), pp. 12, 13.

Newton. Kant's philosophical revolution undoubtedly strengthened the positions of scientific cognition by excluding metaphysical speculation (it is in this sense that Kant's contribution could be likened to the Copernican revolution). At the same time rejecting metaphysical claims for cognizing the supernatural and acknowledging man's insatiable inclination towards it, Kant imprints a sphere of freedom returning to the starry heavens above its mystique, its inscrutability that had actually been looted by speculative rationalism.

In Kant's opinion, the process of cognition can be endless, however, one should always bear in mind that the result of the process can never be characterized as absolute and final. That could only be achieved by a divine, not human intellect.

One must ask – as it has already been frequently done in the history of philosophy – hasn't Kant's theory of cognition been overpowered by supreme subjectivism? Hasn't there developed an insurmountable crack between the sensual and transcendental world, between nature and freedom? The questions seem to be troubling the philosopher himself. That is why the third critique – *Critique of the Power of Judgment* – tends to look for some intercession between the sensual world, the Cosmos being cognized by our understanding and the transcendental world of freedom. Such mediation, according to the philosopher, could be the principle of goal-directedness. However, the introduction of goal-directedness does not free the process of cognition from basic subjectivism. Kant is sure that our reflexive power of judgment should be capable of regarding nature in such a way "that in conformity to the law of its form it at least harmonizes with the possibility of the ends to be effectuated in it according to the laws of freedom".[10] Nature should be regarded in such a way as though it were in possession of a goal that is inaccessible to us, as though it had been energized by some transcendental substratum. It must be admitted that the principle of goal-directedness models our cognition process, formally it establishes unity between the sensual and the transcendental world, yet it hardly does anything to enhance either the cognizing activities or the objectivity of their results.

Kant is preoccupied with the problem of the objectivity of world cognition all through his lifetime. He returns to it in his later-day treatise *Opus postumum*. In this manuscript Kant offers a special program of the metaphysics of nature that should ensure the objectivity of cognition in natural sciences. Transcendental basic principles naturally retain their governing position while next to them is placed an elaborated system of physical notions. The go-between role, according to the philosopher, could be played by notions that could be derived from the physical notions with the help of specific transcendental idealization. It should be added that the program never saw its completion, yet it serves to demonstrate Kant's attempts at ensuring maximum objectivity to the world cognition process.

[10] Kant, I. *Critique of the Power of Judgment*, (Oxford: Oxford University press, 2007), p. 12.

Part VII

How to Approach Heideggerian Gods

Jani Vanhala

Abstract The question of Heideggerian gods is a multifaceted problem that has suffered from the lack of concrete methodological propositions. According to my interpretation this state of affairs is due to confusion concerning the possibilities that Heideggerian gods assumably can offer to philosophy. On the one hand Heideggerian gods seem to work in a similar way to gods of religions, namely like personal actors, but on the other hand their further explication, especially from the already familiar perspectives of religions, is usually dismissed even by Heidegger himself. In my paper I propose that this procedure of Heidegger's can be interpreted as a positive gesture that is supposed to function as a free, conceptually unlimited space to the further development of such a thematic. However, this means that the meaning of Heideggerian gods is not to be found by systematic interpretation of Heidegger's writings, but, instead, it is to be found from applied study that tries to reach these gods in their own appearance. The basic question concerning Heideggerian gods is, thus, to find access to the sphere of their effectual presence. Consequently, as a solution to this problem I present a model of communication, based on Heidegger's formulation of the discussion (*Gespräch*) of mortals and gods that can be found from our everyday practices and their intentional grounds.

Introduction

The theme of gods appeared in Heidegger's thought in the early 1930s. At first glance it may appear to be motivated mainly by historical perspectives. When we consider the beginning of mankind in the context of our historical self-consciousness we notice that gods are tightly connected to all the practices that have remained with

J. Vanhala (✉)
University of Helsinki, Helsinki, Finland
e-mail: jani.vanhala@helsinki.fi

A.-T. Tymieniecka (ed.), *Phenomenology of Space and Time: The Forces of the Cosmos and the Ontopoietic Genesis of Life: Book One*, Analecta Husserliana 116,
DOI 10.1007/978-3-319-02015-0_26,

us from the times of their appearance. This is to say that the question of gods has its justified ground in our past. It is fair enough to assume that if we learn to understand our ancestors' worldviews and ways of living, it might help us to understand better our present world and its structures. However, in his approach to gods Heidegger does not seem to be content only with these kinds of neutral perspectives. Instead, in addition to that, he seems to search for an interpretation of gods that could be valid even in our modern or postmodern world. This tendency naturally raises many questions. The most important of them is, I think, the question of *access* to the effective sphere of those gods. The urgency of this question is motivated by the widely shared experience of the *flight of gods* that is closely connected to the phenomenon of nihilism. This is to say, many of us are unable to encounter the presence of gods that would provide the experience of meaningfulness, and nevertheless we seem to be forced to search for some grounds for that experience, because without that our world is in danger of collapsing.

Depending on the needs and purposes of the study the question of gods can be divided into more specific sub-questions. Because my paper concentrates on contemporary possibilities of experiencing the presence of gods, I have chosen to pose the following questions: Where and how can we encounter gods and how do we know when we are dealing with gods? Thus, we have basically three questions to answer. The first two are methodologically motivated, while the last is epistemological. My proposed solution to these problems follows a line, in which the answers to the first two questions will render the need to ask the final question redundant. The supposed change in one's relation to the world affects the ways of posing and the need to pose epistemological questions. In this respect my proposal follows thematic paths concerning the *end of philosophy* outlined by Werner Marx in his book *Is There a Measure on Earth?*[1] The same kind of approach can also be found from more recent studies. For example, Hubert Dreyfus and Sean Dorrance Kelly have presented similar kinds of thoughts in their book *All Things Shining*, which is, in my opinion, the best presentation of this subject after Marx's classical study.[2]

My own contribution to this debate is the perspective of the *aesthetics of everyday life* and *aesthetics of existence* presented by means of Heidegger's notion of *discussion* (*Gespräch*) of mortals and gods that takes its factual form in communicative practices hidden in our being-in-the-world. Unlike Marx, who sees the locus of the changing of one's relation to the world in ethics, I would like to pay attention to such forms of experiencing aesthetic qualities that can be interpreted by means of discussion (Gespräch). They are usually hidden in our everyday practices and can be recognized in *moods* (*Stimmung*) that direct our intentionality. This is to say that mood (Stimmung) is not the object of intentional act but it is, rather, a mediating space through which we reach toward such intentional fulfilments that are not in our command. According to my hypothesis the nature of Heideggerian gods can be

[1] Werner Marx, *Is There a Measure on Earth? Foundations for a Nonmetaphysical Ethics*, trans. Thomas J. Nenon, Reginald Lilly (Chicago: Chicago University Press, 1987).

[2] Hubert Dreyfus, Sean Dorrance Kelly, *All Things Shining. Reading the Western Classics to Find Meaning in a Secular Age* (New York: Free Press, 2011).

approached most effectively precisely by these means. However, at the same time this means, unfortunately, that the more far-reaching ontological questions concerning the essence of these gods cannot be discussed meaningfully before we can gain enough understanding from the specific modes of communication in their actual happening.

Gods and Their Presence

Heidegger never gave any detailed and extensive interpretation of the ontological status of the gods that he talked about. However, many important features of these gods can be gathered from Heidegger's later writings that are usually concerned, some way or another, with art and aesthetics. Therefore it seems to me that the best way to enter this area of thought is to pay attention to their possible effects and appearances in the sphere of arts and aesthetics. In this respect the most important sources can be divided into two categories: the first consists of Heidegger's writings from the 1930s. The key texts in that era are *The Origin of the Work of Art* and *What is Metaphysics?* and essays concerning Hölderlin's poetry. The characteristic feature of these writings is their way of dealing with the question of the meaning of art combined with the question of the *origin* of human existence. The second category consists of Heidegger's interpretations of modern poetry, such as the poetry of Georg Trakl, Rainer-Marie Rilke and Stefan George. These writings are primarily written in the 1950s and their focus is, among other things, on the meaning of the prevailing *mood* (Stimmung) of the poem.

The combining of these themes at the level of concrete illustration is not an unproblematic task, but it can be made, at least in the form of a proposition, by following Heidegger's own descriptions. The examples on whose basis I will present my proposal are Heidegger's probably two best-known examples of the functioning of the works of art, namely his descriptions of a Greek temple and the shoes of a peasant depicted in Van Gogh's "A Pair of Shoes". Both examples can be found in the *Origin of the Work of Art*. The description of the Greek temple goes as follows.

> A building, a Greek temple, portrays nothing. It simply stands there in the middle of the rock-cleft valley. The building encloses the figure of the god, and in this concealment lets it stand out into the holy precinct through the open portico. By means of the temple, the god is present in the temple. This presence of the god is itself the extension and delimitation of the precinct as a holy precinct. The temple and its precinct, however, do not fade away into the indefinite. It is the temple-work that first fits together and at the same time gathers around itself the unity of those paths and relations in which birth and death, disaster and blessing, victory and disgrace, endurance and decline acquire the shape of destiny for human being.[3]

[3] Martin Heidegger, *The Origin of the Work of Art*, ed. David Farrell Krell, trans. Albert Hofstadter (London: Routledge, 1993), p. 167.

The description of the peasant's shoes is as follows:

> From the dark opening of the worn insides of the shoes the toilsome tread of the worker stares forth. [...] On the leather lie the dampness and richness of the soil. Under the soles stretches the loneliness of the field-path as evening falls. In the shoes vibrates the silent call of the earth, its quiet gift of the ripening grain and its unexplained self-refusal in the fallow desolation of the wintry field. This equipment is pervaded by uncomplaining worry as to the certainty of bread, the wordless joy of having once more withstood want, the trembling before the impending childbed and shivering at the surrounding menace of death.[4]

For reasons of space, I will go straight to the decisive difference between these descriptions, because through this difference we can explicate the point of *access* to the sphere of gods' effectual presence. Namely, as we can notice, the description of a Greek temple was based on the explication of the presence of god, whereas from the description of the peasant's shoes we cannot find any description of god. However, I propose that the latter description also contains a particular mode of the presence of god. The difference between these examples is, thus, in the ways they explicate the *effectual* presence of god. In the example of a Greek temple god's presence was given through the holiness that was primarily determined by a concealed precinct of the temple's interior, but that nevertheless was present in the whole existence of the people whose world that temple belonged to. In the case of the peasant's shoes we encounter quite a similar situation. The only difference seems to be that the god and the sphere of holiness is constituted differently. For the sake of its mode of constitution it must also be approached differently. The example of a Greek temple offers immediate and visible access to the sphere of god's effectual presence that appears in the experience of the holiness or sacredness of existence. The example of the peasant's shoes offers another kind of access. Namely, even the holiness that is the experienced appearance of god is not present in itself. It must be found from the *moods* (Stimmung) that are present in the example. The example of the peasant's shoes is, thus, based on the ways in which god is present in a camouflaged way or is *present in absence*, and the access that it offers to the presence of gods is constituted through these modes of being-in-relation with gods.

To say that god is present in a camouflaged way can be understood in a similar sense to the way that Heidegger interprets the following lines of Hölderlin's poem in his essay called *"As When On a Holiday..."*

> And hence the sons of earth now drink
> Heavenly fire without danger.[5]

According to Heidegger the god of wine, Bacchus, is to be encountered in grapes, fruits that are capable of mediating the holiness of god. That is, the "heavenly fire" that appears as an inspiration given by wine.[6] This model of explanation is, thus,

[4] ibid., p. 159.

[5] Martin Heidegger, *Elucidations of Hölderlin's Poetry*, trans. Keith Hoeller (New York: Humanity Books, 2000), p. 73.

[6] ibid., pp. 91–92.

based on epiphany. However, if god is considered to be present in absence, then the stress is on the remaining *remoteness* between man and god. Heidegger does not in fact use the expression *presence in absence* to describe our relation to gods, but, in my opinion, it can be applied to it. In this situation the relation of man and god does not appear in sudden states of aesthetic raptures but it prevails in certain *moods* (Stimmung) that structure man's relation to the world and his *final meanings*. These *moods* (Stimmung) are usually mixtures full of melancholy, yearning, silent tones of joy and gratefulness and, all in all, quite similar to those moods (Stimmung) that Heidegger described in the context of the peasant's shoes. The main difference to the epiphanic explanation is that in this kind of relation there is no experience of fulfilment in a sense of immediate encounter with god but instead the relation is constituted by forms of intentions that are forced to stay awake and wait their possible fulfilment. In practice we encounter both of these forms in our everyday life all the time. The decisive question is only whether we are able to recognize them.

On the general level it could be said that in the description of a Greek temple the effectual presence of god follows Hegel's way of understanding Greek gods. According to Hegel, Greek gods were entities that could be presented purely by artistic means. This is to say that their essence was purely sensual, at least concerning their features that were considered important to humans.[7] However, and this is an important notion, in this case the effectual presence of gods was not considered to be a thing that could be attained through thinking. Instead god's presence was considered to be an experiential thing that was given in an immediate, affective level. Unfortunately, Heidegger does not give many examples of this affection. To mention a couple, however, in *Parmenides* he says that gods could be described as the *attuning ones*, and in the essay "Language in the Poem" he mentions how Selanna, the moon goddess, makes all things lunar with her shining.[8]

That kind of attuning effect is an interesting topic for study, because it seems to contain meaning-bearing structures that cannot be found from purely bodily reactions, such as pain or pleasure. Instead the affection encountered in the context of gods could be described using terms associated with *attraction*. The modes of attraction in question share some features that are important concerning our possibility of approaching gods through phenomenology. First, because these forms of affection are experienced through attunement (*Befindlichkeit*), or more specifically as different moods (Stimmung), they also have intentional structure.[9] This enables us to approach them as meaning-bearing units and not just as subjective emotions. Second, because these affections are encountered through experiences, they also

---

[7] G.W.F. Hegel, *Introductory Lectures on Aesthetics*, ed. Michael Inwood, trans. Bernard Bosanquet (London: Penguin Books, 2004), pp. 11–12, 103–104.

[8] Martin Heidegger, *Parmenides*, trans. André Schuwer and Richard Rojcewicz (Bloomington: Indiana University Press), p.111. Heidegger, *On the Way to Language*, trans. Peter D. Hertz (San Francisco: Harper & Row, 1982), p. 169.

[9] Emmanuel Levinas, *Ethics and Infinity. Conversations with Philippe Nemo*, trans. Richard A. Cohen (Pittsburgh: Duquesne University Press, 1997), pp. 40–41, 119.

have an immediate effect on our actions, unlike judgements, which are made on the purely ideal level. Here we encounter a very important notion. Namely, through the experiential quality of affectivity our intentionality also has its practical side through which it is directed toward such fulfilments that cannot be reflected purely by means of reason. And, besides this, it always directs itself situationally in a sense that it considers its possibilities through given situations.

The mode of *practical intentionality* bears great significance when we consider the possibility of approaching Heidegger's gods. This is due to the fact that in practical intentionality our directedness towards possible fulfilments is constituted quite differently than in cases in which the fulfilment is considered to be a purely logical act immanent to consciousness. In practical intentionality we are all the time forced to encounter the existence of such otherness that disturbs our actions or rewards them. This appears in modalities of satisfaction and repulsion that we encounter as responses to our own actions. On the general level that leads us to an interaction with our environment that is not solely based on our attempts to rule things around us. Instead these dynamics are capable of *moving us* in a more adaptive relationship with our immediate environment.

Practical intentionality, understood in this sense, is relational in quite a different sense than intentionality that moves, for example, in the sphere of the *totality of useful things* presented by Heidegger in *Being and Time*.[10] The difference between the *totality of useful things* and practical intuition is based on their way of understanding human beings in relation to their own action. The *totality of useful things* and its constitutive relations such as *in order to* (*um-zu*) and *what-for* (*Wozu*) presents the situation of man from the viewpoint of his capability of directing his actions successfully. It could be said that it is a quite pragmatic approach that is based on horizontal relations. The relations of tools are understood mainly through the already-known possibilities they offer. Seen through the *totality of useful things* intentionality is, thus, tied closely to goal-oriented action. Compared to this, practical intentionality contains a peculiar form of reflectivity that separates it from goal-oriented action and points toward the *vertical dimension of being* whose modes of relationality are very different from those that can be encountered in the sphere of horizontality. For example, the mode of reflectivity inherent in practical intentionality elucidates well the difference between the horizontal and vertical dimensions of being. Namely, unlike reflection in the horizontal dimension, the reflection inherent in the vertical dimension does not appear *directly* as an act of the already-fixed self. The mode of reflection peculiar to the vertical dimension does not constitute itself through oppositions such as the difference between the reflecting self and the reflected contents. Instead it is willing to leave the definitions of such relations formally open and approach them through their processual self-opening that occurs through their structures of intentionality that are constantly in an adaptive condition. In this situation the direction of one's adaptivity is not constituted by relations that

[10] Martin Heidegger, *Being and Time*, trans. Joan Stambaugh (Albany: State University of New York Press), pp. 68–71.

are defined by goals that one has decided on oneself. Those kinds of goals would be decided by will. Instead the goal of the process is something that can be encountered only by taking part in the process of self-opening. Encountered this way, the goal appears to be a mediation in which fulfilment and the present situation are intertwined in a way that reminds one of Gabriel Marcel's definition of truth that he clarifies through an example of conversation: "[...] truth is at one and the same time that toward which the speakers are conscious of moving, and that which spurs them toward this goal."[11] As a result, the goal that can be encountered through reflection enabled by the vertical dimension serves as a medium of self-transformation that directs itself by means of experienced attraction or repulsion.

I claim that this kind of intentionality is very familiar to all of us, although we might not derive a theoretical model from it. Our natural sensibility, which does not make conceptual distinctions but flows through impressions, moods, tones, atmospheres etc., works in this way. Also Heidegger's notion of *meditation* (*Besinnung*) catches something essential from its nature.[12] I propose that the functioning of this kind of intentionality can be described by means of Heidegger's gods and our relation to them, although that kind of approach most probably cannot explain the nature of practical intentionality exhaustively. According to Vincent Vycinas, who follows in his interpretation the examples given by Walter F. Otto, Heidegger's gods can be understood by means of Greek gods and their effectual presence. Vycinas says that Greek gods can be understood as worlds in the sense of *logos*. This leads naturally to the opinion in which the meanings of encountered things such as cars, houses or flowers are interpreted through the world or *logos* to which they belong.

> In the worlds of two different gods, a thing is not the same in each because by reflecting a different essence of a god, it itself becomes different. Night in the world of Artemis and night in the world of Hermes are different phenomena because they disclose different worlds. Artemis' night is serene in its frightfulness and beauty, and Hermes' night is advantageous or disadvantageous cover in one's pursuits.[13]

Although Vycinas's explication is very clear, it leaves many practical problems. Namely, if we believe Heidegger's words according to which we live in times that are characterized by the absence of gods, how and where could we notice such effects of gods that Vycinas describes? Unfortunately, it is true that we very seldom encounter such strong emotions that we could honestly claim to be enchanting in a sense that a mood (Stimmung) attuned by god might be. This is the place where the approach enabled by the *aesthetics of everyday life* and *communication* (Gespräch) shows its usefulness.

---

[11] Gabriel Marcel, *The Mystery of Being: I. Reflection & Mystery*, trans. G.S. Fraser (London: The Harvill Press Ltd, 1950), p. x

[12] Heidegger, *On the Way to Language*, p. 1.

[13] Vincent Vycinas, *Heidegger's Earth and Gods. An Introduction to the Philosophy of Martin Heidegger* (The Hague: Martinus Nijhoff, 1969), p. 188.

Communication in Practice

Communication, in the sense that I use it in this paper, is a term that I have formed through Heidegger's interpretation of Hölderlin's poetry that he gives in his essay called "Hölderlin and the Essence of Poetry". In this essay Heidegger focuses on the following lines of Hölderlin's poem:

> Much has man experienced.
> Named many of the heavenly ones,
> Since we have been a conversation
> And able to hear one another.[14]

The word "conversation" is a translation from the German word "Gespräch", which means basically conversation or discussion in their everyday meaning. The translation is, thus, formally correct. However, if we try to think of the proper translation of "Gespräch" in the context of Heidegger's thought in general, there appear to be other options that are more convenient for comprehending how to approach Heideggerian gods. Concerning the effectual presence of a god the most interesting possibility can be found from Heidegger's pseudo-historical dialogue with a Japanese inquirer called *A Dialogue on Language*. There the word "dialogue" is a translation of the word "Gespräch". In this dialogue Heidegger mentions that in their discussion a "hidden drift" (*verborgene Zug*) prevails that leads their discussion and in some mysterious way ties all the topics together.[15] Heidegger also presents similar thoughts elsewhere. In the *Conversation on a Country Path* he talks about "the silent course of conversation that moves us".[16] This kind of vague expression suggests that being-in-conversation contains some kind of a *hidden vertical dimension* that works differently than understanding that moves on a horizontal level.[17] This vertical level must be termed *hidden* mainly because it is not transparent concerning its functions and constitution. In my opinion the functioning of this hidden vertical level can be described most accurately in terms of *existential communication*.

[14] Heidegger, *Elucidations of Hölderlin's Poetry*, p. 56.

[15] Heidegger, *On the Way to Language*, p. 30.

[16] Martin Heidegger, *Discourse on Thinking*, trans. John. M. Anderson and E. Hans Freund. (New York: Harper & Row, 1969), p. 70.

[17] By the concept of vertical I do not refer to such religious or neo-Platonic interpretations in which verticality inherently contains reference to some kind of absolute being or goodness that is the source of all meaningfulness. Instead in this paper the use of verticality should be understood in a phenomenological manner that stresses the difference of vertical and horizontal *givenness* and their different ways of enabling the experience of meaningfulness. The use of verticality in this article follows the definition of verticality given by Anthony J. Steinbock. "The spheres of experience and evidence that are more robust than just those of objects, I call vertical givenness [...]". Anthony J. Steinbock, *Phenomenology and Mystics: The Verticality of Religious Experience* (Bloomington: Indiana University Press, 2007), p. 1.

The decisive step in the proposition that I offer is based on the notion that the translation of the word “Zug” as “drift” can be understood from the viewpoint of existential communication. This is so, because in the heart of the discussion in which this *hidden drift* appears is the aesthetic concept of *grace* (*Anmut*), which is used in quite a peculiar way. For some unknown reason Heidegger decides to replace the concept of grace (Anmut) with the personified concept of “Huld”, which refers to the Three Graces, who are responsible for different kinds of artistic inspiration. This is to say that the *hidden drift* of discussion is based on the hidden presence of the goddess who appears as some kind of *attraction* that leads the discussion.

According to my interpretation this thought on “hidden drift” behind the discussion can be isolated from the sphere of dialogical thinking and can be applied to all our action. It can then be seen as a power that is present in our actions and that directs and motivates them in a way that surpasses the limits of goal-oriented action. It is not, moreover, a power that is effective on the level of discourse that bears a meaningful world as a referential complex, but it is, instead, a power that beckons us to reach existentially over the limits of the world as a referential complex. This is also the reason why I take it as a mode of *existential communication*. It appears experientially as an attraction that defines one’s situation as a meaningful whole and simultaneously points towards the limits of this given situation. This kind of contradiction implies that behind the horizontal surface of a situation a vertical dimension exists whose constitution might define the experienced meaningfulness.

Because my article concentrates only on concrete access to the effectual presence of Heideggerian gods, I must leave the questions of the constitution of this vertical dimension and the ontological status of Heideggerian gods for later papers. What can be said, however, is that the effectual presence of gods is based on a *mirroring* (*Spiegeln*) which forms the sphere of *intersubjective immanence*. Following this short definition I move on to the concrete appearances that this vertical dimension takes in our experiences.

One of the most interesting of these concrete appearances can be seen in those phenomena in which our actions and attitudes toward being reveal the fact that often we take being *as if* it were a personal actor. If we encounter a series of misfortunes we tend to get upset and bemoan our destiny. We might think that we do not *deserve* such a fate and therefore we feel justified calling for justice. Similar reactions can also be noticed in situations that are experienced as positive. A beautiful sunrise on a summer morning or the sight of a starry night may produce a mood (Stimmung) of *thankfulness* in which our understanding of what is valuable in life is suddenly radically changed. Such reactions would be absurd and intentionally inexplicable if we did not, in actuality, have such a relation to the world in which our actions are constituted *as if* they were responses to actions of other personal beings. According to my thesis, the effectual presence of gods can be seen most clearly in these kinds of situations. The reason that being is taken *as if* it were a personal actor has its basis in the functioning of gods that reflects the *invisible perspectives of other people.*[18] These invisible perspectives are constituted by socially determined attitudes, values, judgements, etc. of other people, but also by existential relations such as love, faithfulness, kindness, etc., that precede the previously mentioned socially determined relations.

All in all, this means that we take things that happen to us *as if* they were *gestures* towards us. We take them as given, like presents, and giving is an act that is usually regarded as a gesture from some person. Heidegger has described that kind of situation in his essay "The Thing", where he describes the complex meaning-structure of pouring wine into a jug.

> The gift of the pouring out is drink for mortals. It quenches their thirst. It refreshes their leisure. It enlivens their conviviality. But the jug's gift is at times also given for consecration. If the pouring is for consecration, then it does not still a thirst. It stills and elevates the celebration of the feast. [...] In the gift of the outpouring that is a libation, the divinities stay in their own way, they who receive back the gift of giving as the gift of donation.[19]

If we read Heidegger's example carefully, we may notice that the relation of mortals and gods is described through actions that are invisible and almost unrecognizable to us. It is said that "divinities stay in their own way" in the act of outpouring. This "own way" is essentially the same phenomenon as the example of the peasant's shoes. God is not present or encountered in sensual rapture, but is camouflaged in the actions and relations of things that bear references that reach beyond the given situation. The task of the aesthetics of everyday life is to make the divinities' "own way" visible through explicating those intentional acts that we mortals direct towards them in the act of communication. This also opens access to the effectual sphere of the gods. Namely, if we think that gods are the *attuning ones*, and that through our practical intentionality our acts are in communication with those gods, then we may assume that our communication with gods happens mainly in those acts in which we reach beyond the given situation. This, again, is to say that our practical intentionality mostly deals with something that is usually discussed by terms of *value* rather than terms of *facts*. However, in this case values are not considered to belong to the sphere of ideas or social conventions that can be approached through thinking, but instead they are something that we must recognize in their actual occurrence.

---

[18] Unfortunately there is not sufficient space to explicate this process here in detail. However, my explication is based on constructive reformulation of such notions of mirroring that Jean-Luc Marion presents in his book *God Without Being*. In his treatise Marion defines gods as *idols* that function as mirrors that freeze the gaze of man. My own interpretation of gods does not stress their character as idols but concentrates on their capabilities to make visible intersubjective structures of our reflective capacities. In general the problematic that arises from the basis of gods' reflective character is closely tied to the themes of our identities and our authenticity or inauthenticity. In short, this problematic is based on the fact that gods can reflect our own "essential" features, but also the features that we want to see as our essence because they are features that are desired by the other. If I, for example, desire a very expensive men's suit, most probably this does not tell so much about *my* desire but the desire of others. The moral implications of this problematic must be postponed to later papers.

[19] Martin Heidegger, *Poetry, Language, Thought*, trans. Albert Hofstadter (New York: Harper & Row, 2001), pp. 170–171.

Access to the effectual presence of gods is, thus, gained through reflection. But, as I said earlier, this form of reflection is a peculiar one. However, after the explication of the dynamics of communication the essence of this reflection can now be described in more detail. The basic feature of this form of reflectivity is its relational structure that differs considerably from the relation that prevails between subject and object. Namely, in subject-object relation the subject sets or defines itself through objects. In this situation reflection is a way to explicate or make visible the already full and independent self. However, in the case of reflection encountered in communication an individual does not try to determine its being but he leaves, instead, the whole process open and exposes himself to the transformation of his being and such possible fulfilments that he may not even be able to comprehend. In my view Heidegger's thoughts of *presentiments* (*Ahnung*) also refer to this kind of reflection.

Although this kind of teleological vision is characteristic of the Christian worldview, something similar can also be found from the Greeks' relation to their gods. Dreyfus and Kelly describe this well by giving an example from Homer's *Iliad*, an example that is barely understandable if we approach it from the moral conventions of our time. They mention a situation in which Helen, the wife of Menelaus, tells her story to Menelaus's court after the Trojan War. The Trojan War was, of course, brought about through Helen's infidelity when she fell in love with Paris, the Trojan prince. The amazing thing about her story is that after she had told it truthfully, Menelaus, the husband that she had betrayed, congratulates her for her story. The reason that makes the behaviour of Menelaus possible is, according to Dreyfus and Kelly, the Greeks' understanding of *excellence* (ἀρετή), which was quite different from what we usually understand by it. To the Greeks excellence was not a moral phenomenon, rather it was the capability of being in an appropriate relationship to that which is considered to be sacred. Helen's excellence, for which she is praised, is her way of responding to Aphrodite and her command.[20]

The decisive feature of this kind of interpretation of excellence seems to be that there are many different and even contradictory forms of excellence, not one universal form of perfection that harmonizes everything. Another important notion is that forms of excellence are given individually, they are not meant to be achieved by everyone. This is to say that a person's will to reach some particular form of excellence is not the decisive feature of an individual's quest. Instead one has to be responsive to the attraction that is directed towards oneself and follow it without any certainty of its possible goals. From the perspective of access to such phenomena this means that a person who seeks his way to excellence cannot direct his actions according to general knowledge or by calculative thinking. Instead he must approach the subject meditatively through his own actions. Not for the sake of gaining self-knowledge in the sense of self-enrichment or of cultivating one's own existence, but because the call that appears as attraction, the "hidden drift", is to be found from these actions.

---

[20] Dreyfus, Kelly, *All Things Shining*, p. 62.

The quest is, thus, to recognize this call in one's own actions. The question that follows is naturally where and how we can recognize the call. According to the model of communication it can be found from the features of our own actions that seem to be directed beyond the limits of a given situation. These features of our actions, thus, form a *surplus of meaningfulness* that cannot be derived from the meanings with which we deal in the sphere of goal-oriented action. These features of our actions also form the sphere of practical intentionality that is being directed toward such fulfilments that are not within our command, but that are nevertheless something that we can encounter experientially in the form of *fear*, *presentiments*, *promises*, *hints*, and so on. Their presence in the experience is usually a relatively unnoticeable phenomenon. Like in Heidegger's examples of pouring wine into a jug and a peasant's shoes. However, they can be recognized through *meditative thinking* (Besinnung) that directs its attention to the prevailing moods in an experience.

For a practical application we can refer to Charles Taylor's article "Heidegger on Language". Taylor talks about macho culture, which is an essential part of a biker's world. According to Taylor being macho is tied to certain gestures and expressions that enable its appearance.[21] I am quite convinced that Taylor's interpretation is correct, but I also think that those bikers who want to be macho do not want to be macho solely for cultural reasons. Surely they want to be part of a macho world because they appreciate it and respond to the call of its attraction, but still the final fulfilment that they seek through being part of the macho world is not something that is present in their factual situation. Instead the whole macho world with its gestures and expressions is directed towards something that is not present directly in the bikers' world. In this *objectless directedness* communication in all its forms of practical intentionality appears. The other participant in this communication is the god of bikers who, nevertheless, is formless and is often even unrecognized just like someone may be unnoticed or be hidden from us if we only pay attention to his or her explicit message. We may not notice the individual form of attraction that approaches us on the level of his or her personal style or charisma.

[21] Charles Taylor, "Heidegger on Language", in *A Companion to Heidegger*, ed. Hubert L. Dreyfus, Mark A. Wrathall (Malden: Blackwell Publishing Ltd., 2005), pp. 438–439.

Meaning in the Forthcoming Sciences of Life: From Nietzsche and Husserl to Embodiment and Biosemiotics

Ammar Zeifa

Abstract Towards the end of his life, Edmund Husserl became strangely interested by something like a new type of "*Science of Life*" or a "*Biology*", which could be, he said, a "universal science". Beside this, there was of course the new great topic of the World of Life, or, to say it like it is well-known, the "Life-World" (*Lebenswelt*). Even if Husserl was, for good reasons, very careful and prudent about "sciences" in general, and especially about the "Galilean mathematization of nature", which was making some "schizophrenia", some separation from life, and was pushing the "European humanity" to a treason of the deep inside of its life, it seems that these ultimate phenomenological ideas are really promising for us today, for our culture and future. But just before the founder of Phenomenology, Friedrich Nietzsche was always thinking and talking about a(n) (eternal) return to life, and, more precisely, about the *physiological roots* of so many things, and especially in our world, the world of human rationality. More recently, Francisco J. Varela, who was influenced by Maurice Merleau-Ponty, tried to inaugurate, after Hans Jonas and Jan Patocka, a new kind of "philosophical biology", which can be a radical research about "life in mind" and "mind in life". If the "embodied mind", and more generally, the "embodiment" became in our time something like a "mainstream", it is precisely because, as Nietzsche said, "Mind" and "Life" can never be separated from each other. What is even more important in that theory is that it could be the first step for a radical rethink of all that the philosophical tradition had called "Sensible" and "Intelligible", with their relations or interactions. It would also mean that "Matter" and "Form" can never be separated from each other. If we try like this to "listen" to the "life" of things, or to "see" them in this new way, this silent way, we will be, in some sense, more "living" and more "thinking". Some recent research in Biology, in "Biosemiotics", is talking nowadays about "Signs of Life and Life of Signs". This new "Biology of signification", with an evolutionary approach of a "natural history of intentionality", which could also be a radical theory of meaning, will be perhaps, some day, our greatest science and philosophy.

A. Zeifa (✉)
U.F.R. de Philosophie, Université de Paris I Panthéon, Sorbonne, Paris 75016, France
e-mail: azcnrs@gmail.com

A.-T. Tymieniecka (ed.), *Phenomenology of Space and Time: The Forces of the Cosmos and the Ontopoietic Genesis of Life: Book One*, Analecta Husserliana 116,
DOI 10.1007/978-3-319-02015-0_27,

L'œuvre de Husserl est marquée tout entière par des nouveautés successives, des questions qui surgissent, qui reprennent comme un héritage, et le porte encore plus loin, vers de nouveaux horizons. Aussi, le père de la phénoménologie, voulait-il même à la fin "tout reprendre à zéro".[1] Mais la question philosophique de la "vie", ou d'une "science de la vie", occupe quant à elle une place à part, et restera présente jusqu'à la fin, comme soubassement d'un problème, que Husserl appellera finalement "la crise de la science" ou même "la crise de l'humanité européenne".[2] Il apparaît clairement petit à petit que la "vie" de cette "humanité européenne" est la question la plus chère, le véritable souci du fondateur de la phénoménologie. C'est cette humanité-là qui souffre dans le monde moderne d'un rapport bizarre ou équivoque, presque schizophrénique, à sa propre vie, et même, pourrait-on dire, à la vie en général. Le tournant copernico-galiléen porte ses fruits, et imprime sa marque, celle d'une mathématisation de la nature, qui la rend exploitable industriellement à des fins utilitaires, et la rend de ce fait extrêmement vulnérable.[3] C'est pourtant, l'auteur de cette révolution, l'homme lui-même, qui se retrouve pris au piège de sa propre force, et comme victime de son propre narcissisme: il perd subrepticement sa place de maître et devient de plus en plus esclave de son propre projet. Ce qui sera repris par d'autres philosophes, tout au long du XXème siècle, trouve ici en Husserl un avocat ou un procureur d'une rare éloquence et d'une grande pertinence. Mais plus fondamentalement encore, l'apparition à la fin du thème du *Lebenswelt* parvient à reprendre toute la dynamique des recherches phénoménologiques, et à radicaliser cela même qu'il nomme le "retour aux choses mêmes". Où est donc

[1] Il s'agit d'une lettre tardive de Husserl dans laquelle il écrit: "Je ne savais pas que mourir fût aussi difficile. Toute ma vie, je me suis efforcé d'écarter toute frivolité. Et juste maintenant que je suis arrivé au terme de mon chemin, conscient de ma tâche et prêt à l'assumer, maintenant qu'avec les conférences de Vienne et de Prague […] j'ai jeté les bases d'un petit commencement – eh bien, je dois m'interrompre et laisser ma tâche inachevée. Juste maintenant, à la fin, maintenant que je suis un homme fini, je sais que je devrais tout reprendre à zéro". Manuscrit X, 1, 4 (cité d'après Claude Romano, "La tâche inachevée: la conceptualisation husserlienne de la *Lebenswelt* et ses limites", in Jean-Claude Gens (dir.) *La* Krisis *de Husserl. Approches contemporaines*. Revue *Le cercle herméneutique*. N°10, 2008).

[2] Nous parlons ici bien sûr des célèbres pages de la fameuse *Krisis: La crise des sciences européennes et la phénoménologie transcendantale* (traduction française par Gérard Granel, Paris, Gallimard, "Bibliothèque de Philosophie", 1976) ainsi que la célèbre conférence de Vienne et de Prague: *La crise de l'humanité européenne et la philosophie* (dans le même volume). Voir note 4.

[3] La problématique "écologique" a été en effet plus d'une fois reprise par les héritiers de Husserl, et on a parlé d' "éco-phénoménologie". Songeons au moins à l'œuvre de Hans Jonas (comme par exemple: *Une éthique de la nature*, Paris, Desclée de Brouwer, 2000). Voir surtout, plus récemment: Adam Christopher Konopka, *An Introduction to Husserl's Phenomenology of* Umwelt. *Reconsidering the* Natur/Geist *Distinction. Toward an Environmental Philosophy*. Ann Arbor, ProQuest, UMI Dissertation Publishing, 2011. Mais c'est Merleau-Ponty qui semble être la plus grande source d'inspiration dans ce domaine: Suzanne L. Cataldi, William S. Hamrick (ed.), *Merleau-Ponty and Environmental Philosophy. Dwelling on the Landscapes of Thought*, New York, New York State University Press, 2007.

passé le sens du "projet" philosophique millénaire de "l'humanité européenne"?[4] On a quelque peu perdu sa trace, on a rompu insidieusement avec lui, sans même le savoir: on a perdu justement nos liens aux "choses mêmes", au "monde de la vie".

Les sciences de la vie et la phénoménologie semblent vouloir depuis quelques temps entretenir des relations privilégiées. Depuis l'œuvre du fondateur, et jusqu'à nous, une certaine alliance s'approfondit jour après jour, et connaît aujourd'hui une véritable effervescence.[5] Il faut bien avouer tout d'abord que, dans la phase ultime de son œuvre, Husserl avait étrangement laissé entendre qu'un dialogue, entre la phénoménologie et une nouvelle forme de "science de la vie" ou même de "biologie", était non seulement souhaitable mais nécessaire; car cette science à venir, disait-il, pourrait même devenir un jour la "philosophie absolument universelle".[6] Cette intérêt inattendu pour ce qui apparaît de prime abord comme une "science de la nature" parmi d'autres, qui participe donc à l'hégémonie du "naturalisme", dénoncé par Husserl, n'est pourtant pas sans rappeler l'intérêt grandissant de Kant à la fin pour le "phénomène de la vie", dans son ultime tentative critique, dans ce qui allait donner la *Critique de la faculté de juger*.[7] C'est justement ce qui allait susciter au

---

[4] Cf. *La crise des sciences européennes et la phénoménologie transcendantale*, op. cit. Il s'agissait pour Husserl à la fois de développer "l'idée historico-philosophique" ou "le sens téléologique de l'humanité européenne" (Conférence, p. 347), et d'expliquer ou de montrer "la crise des sciences comme expression de la crise radicale de la vie de l'humanité européenne" (I, p. 7). S'il est question de la philosophie grecque et du platonisme, vu leur importance historique, il s'agit d'interroger principalement "la modernité philosophique": le questionnement porte sur "la façon générale *d'estimer* les sciences. Il ne vise pas leur scientificité, il vise ce que les sciences, ce que la science en général avait signifié ou peut signifier pour l'existence humaine" (I, 2, p. 10). Et, à ce sujet, Husserl tient surtout à nous dire une chose: "Dans la détresse de notre vie – c'est ce que nous entendons partout – cette science n'a rien à nous dire. Les questions qu'elle exclut par principe sont précisément les questions les plus brûlantes à notre époque malheureuse, pour une humanité abandonnée aux bouleversements du destin: ce sont les questions qui portent sur le sens ou l'absence de sens de toute cette existence humaine" (I, 2, p. 10). La "nouvelle tâche universelle" de la philosophie serait alors de montrer l'impensé radical de toutes ces sciences, en tant qu'il est le "monde-de-la-vie comme fondement de *sens* oublié de la science" (II, 9, p. 59) au point que "le problème du monde de la vie" devient le "problème philosophique universel" (III, 34, f.).

[5] Cf. par exemple *Sciences du vivant et phénoménologie de la vie. Noesis*, No. 14, 2008. Nous aborderons ensuite plusieurs contributions remarquables dans ce domaine.

[6] *La crise des sciences européennes*, op. cit. Appendice XXIII. Voir également l'article de Jean-Claude Gens, "La question en retour sur la vie et l'idée husserlienne de la biologie comme science universelle" in *Lectures de la* Krisis. *Approches contemporaines. Revue Le Cercle herméneutique*, No. 10, 1999.

[7] Nous faisons allusion évidemment à "la critique de la faculté de juger téléologique", qui, rappelons le, considère la vie du vivant comme un cas à part, impossible à expliquer véritablement par la physique mécaniste newtonienne. Le vivant est considéré comme étant "téléologique", poursuivant un but, et ce non pas d'un point de vue scientifique, en tant que nature, à partir d'un "jugement déterminant", mais seulement de notre point de vue à nous, subjectivement, comme "jugement réfléchissant". C'est ce qui cause encore aujourd'hui le plus grand embarras. Voir par exemple l'article remarquable de Francisco J. Varela et Andreas Weber: "Life after Kant: Natural Purposes and the Autopoietic Foundations of Biological Individuality", in *Phenomenology and the Cognitive Sciences*, 1, 2002, pp. 97–125. La question reste posée encore aujourd'hui.

moins à partir de Schelling et des "philosophies de la nature",[8] et jusqu'à nos jours, la plus grande fascination. Ce qui est moins connu, en revanche, c'est que l'un de ces admirateurs fut un jour aussi un certain Friedrich Nietzsche. Très jeune, en effet, il projetait même de rédiger une "dissertation doctorale" sur le thème de "La téléologie", et plus précisément encore sur "La téléologie depuis Kant". Cette tentative laissera des traces significatives, sans aboutir à un véritable ouvrage.[9] Mais le point commun qui rassemble autour de lui le mieux tous les protagonistes, de Kant jusqu'à nous, en passant par Nietzsche et Husserl, c'est la dénonciation d'un certain "mécanicisme", importé de la physique classique, et appliqué maladroitement à la "vie" ou au "vivant". Nous ne pouvons d'ailleurs que constater, au moins chez ces trois philosophes, que la question de la "téléologie" occupe pour le moins une place privilégiée.[10] C'est ce qui allait créer la plus belle tension dans ces œuvres, et fera surgir rien de moins que la question du sens, qui oscille ainsi entre deux extrêmes, qui se trouve tendue comme un arc, en attente d'une flèche, d'une finalité, et peut-être même… d'un but.[11] Quels seraient alors les ressorts cachés de cette histoire,

[8] Le romantisme et l'idéalisme allemand n'ont cessé de s'y intéresser, et Schelling, par exemple, avait écrit admirablement ce qui résume bien la situation: "La *Critique de la faculté de juger* est l'œuvre la plus profonde de Kant, celle qui aurait sans doute donné une autre orientation à toute sa philosophie si, au lieu de finir par elle, c'est par elle qu'il avait pu commencer" (*Contributions à l'histoire de la philosophie moderne*, SWX, 177).

[9] Ce texte de jeunesse (1868), "*Teleologie seit Kant*", n'a pu paraître dans l'édition de référence, *Nietzsche Werke, Kritische Gesamtausgabe,* établie par Giorgio Colli et Mazzino Montinari, au Tome I 4, qu'à la fin du XXème siècle, en 1999! Il a été d'abord traduit en anglais par les soins de la *North American Nietzsche Society* dans le volume 8 de sa collection *Nietzscheana*, et se trouve également dans le livre du même traducteur de la *NANS*, Paul Swift: *Becoming Nietzsche. Early Reflections on Democritus, Schopenhauer and Kant*, New York, Oxford, Lexington Books, 2005, pp. 95–105. Il n'a toujours pas été traduit, à notre connaissance, en français. Jean-luc Nancy y a consacré un article ["La thèse de Nietzsche sur la Téléologie", in *Nietzsche aujourd'hui*, Volume I, Paris, UGE, 1973], rare en son genre en français, mais qui, lui, a été traduit en anglais! ["Nietzsche's Thesis on Teleology", in *Looking after Nietzsche*. Albany, New York, New York State University Press, 1990, pp. 49–66]. On pourra lire aussi, avec intérêt, l'article de Paul Swift, "Nietzsche on Teleology and the Concept of Organic", in *International Studies in Philosophy*, vol. XXXI, No. 3, 1999; ainsi que le texte écrit par Christa Davis Acampora, "Between Mechanism and Teleology: Will to Power and Nietzsche's Gay "Science", in Gregory Moore, Thomas H. Brobjer (ed.) *Nietzsche and Science*, Ashgate, 2004. Et d'Alberto Toscano: "The Method of Nature, the Crisis of Critique. The Problem of Individuation in Nietzsche's 1867/1869 Notebooks", in *Pli*, 11, 2001, pp. 36–61.

[10] En effet, si l'on tient compte aussi du manuscrit 34 de Husserl, "*Universale Teleologie*", (manuscrit E III 5, *Husserliana* Tome XV, pp. 593–597, traduction française par Jocelyn Benoist, "Téléologie universelle", in *Philosophie* no. 10, Paris, Editions de Minuit, 1989, pp. 3–6) l'intérêt pour cette "téléologie" se trouve ainsi partagé par Kant, Nietzsche et Husserl. Mais n'est-ce pas ici finalement la question du sens, qui est sous-jacente, qui se cache ou se voile sous différents habillages, jusqu'à nos jours?.

[11] Nous nous permettons en réalité de reprendre à notre compte ce que dit Nietzsche à la fin de l'avant-propos de *Par-delà Bien et Mal*: "nous sentons encore en nous tout le péril de l'intelligence et toute la tension de son arc! Et peut-être aussi la flèche, la mission, qui sait? le but peut-être… Sils Maria, Haute-Engadine. Juin 1885." (traduction française de Henri Albert, Paris, Mercure de France, 1898, 1963).

alors qu'on parle désormais de "Biologie de la signification", ou de "*Biosemiotics*",[12] après avoir tant insisté sur le corps vécu, l'organisme ou la chair (*Leib, Body*), et la corporéité vivante ou l'incarnation (*Leiblichkeit, Embodiment*) de l'esprit et du sens?[13] Il y a là sans doute une idée ou un chemin à suivre, qui mériterait toute notre attention. Essayons d'apporter une synthèse, et allons jusqu'au bout des conséquences.

Parmi les interprètes contemporains de Husserl, il y en a heureusement, qui sont là pour nous rappeler aussi le caractère novateur de son œuvre. Outre Merleau-Ponty,[14] qui s'en est si brillamment inspiré, et qui ne manquait pas de le rappeler, certains commentateurs éminents tiennent beaucoup, à juste titre, depuis longtemps, à nous rappeler qu'il y a chez Husserl toutes les prémisses d'un renouveau de la phénoménologie elle-même, laquelle, peut-être, si l'on en reste au défenseur de la "pureté" de la logique, contre le psychologisme et le biologisme, n'aurait peut-être pas une telle présence encore aujourd'hui.[15] Dan Zahavi, par exemple, auteur d'une

---

[12] Même s'il y a très peu de textes, à notre connaissance, en français dans cette discipline, nous adoptons le terme "Biosémiotique" pour la désigner en français.

[13] Vaste sujet qui traverse toute la philosophie à notre époque, et représente, au moins depuis Nietzsche, un thème essentielle, non seulement chez Husserl, mais aussi chez Gabriel Marcel, Maurice Merleau-Ponty, Jean Paul Sartre... (cf. Richard Zaner, *The Problem of Embodiment. Some Contributions to the Phenomenology of the Body*, The Hague, Martinus Nijhoff, "Phaenomenoligica" 17, 1964, 1971; et plus spécifiquement, à propos de Sartre, *The Bodily Nature of Consciousness: Sartre and Contemporary Philososphy of Mind*, Ithaca, Cornell University Press, 1997) et plus récemment, chez des philosophes aussi différents que Michel Henry, Michel Foucault, Gilles Deleuze, ou Henry Maldiney. La problématique des *Stimmungen* chez Heidegger n'y est pas totalement étrangère, malgré l'oubli heideggérien de l'importance du corps, et de la dimension "pathique" de la vie. Tout cela rejoindra finalement et influencera, comme nous allons le voir par la suite, la problématique contemporaine de l'*Embodiment*, entre phénoménologie et sciences cognitives.

[14] Merleau-Ponty n'a cessé en effet avec une grande honnêteté de rappeler tout ce qu'il devait à ce qui se trouvait déjà, dans les manuscrits de Husserl qu'il a pu consulter à Louvain, avant leur publications, en disant, par exemple, "le terme est usuel dans les inédits" (*Phénoménologie de la perception*, Paris, Gallimard, 1945 pp. XIV) ou en reprenant souvent les mêmes expressions husserliennes en allemand, notamment dans les ultimes "notes de travail", de l'œuvre posthume *Le visible et l'invisible* (Paris, Gallimard, 1964).

[15] On ne peut en effet négliger le fait que la plupart des travaux de recherches sur Husserl aujourd'hui, après ceux de Merleau-Ponty, s'intéressent principalement à la "corporéité", à la "constitution corporel" ou "charnel" de l'*ego* lui-même, à la *Lebenswelt* (monde la vie), à l'importance décisive de l'affectivité, à la "synthèse passive", ou à l' "intentionnalité pulsionnelle", beaucoup plus qu'au "cartésianisme" husserlien du commencement, celui de la "logique pure", ou de l' "*ego* pur", et tout le domaine du "transcendantal". [Cf. en particulier: Nam-in Lee, *Edmund Husserls Phenomenologie der Instinkte*, Dordrecht, Kluwer Academic Publisher, 1993; Anne Montavon, *De la passivité dans la phénoménologie de Husserl*. Paris, PUF, 1999; A. Steinbock, *Home and Beyond: Generative Phenomenology after Husserl*, Evanston, IL, Northwestern University Press, 1995; Didie Franck, *Corps et chair: sur la phénoménologie de Husserl*, Paris, Editions de Minuit, 1981]. Merleau-Ponty a été peut-être le premier à avoir cherché à dépasser l'opposition entre le "transcendantal" et l' "empirique", en parlant notamment d'un "entre-deux"; ce qui va, comme nous allons le voir, attirer l'attention de plusieurs chercheurs, dont le biologiste philosophe Francisco J. Varela, et une bonne part des scientifiques et des philosophes, qui allaient travailler sur des thèmes comme le corps, le "phénomène de la vie", *l'Embodiment* et la *Biosemiotics*.

œuvre foisonnante, aux confins de la phénoménologie, de la philosophie de l'esprit (*Philosophy of Mind*) et même des neurosciences cognitives,[16] fut parmi ceux qui ont bien montré, il a déjà bien longtemps, que la phénoménologie de Husserl, loin d'être simplement une philosophie traditionnelle (cartésienne et kantienne) de la subjectivité, est déjà en elle-même un tournant: pour lui, il est clair que le fondateur avait déjà grandement pensé la relation fondamentale entre le corps et la (ou l'inter) subjectivité, qu'il inaugure par son œuvre la pensée de l' "*Embodiment*" du sujet transcendantal.[17] Cette façon de voir les choses a bien sûr été soulignée aussi, à plusieurs reprises, ailleurs.[18] Il semble en tout cas nécessaire aujourd'hui de dépasser la vision traditionnelle de la (première) phénoménologie husserlienne, celle des premières *Recherches logiques*, qui paraît finalement trop rigide ou trop *statique*, qui fait comme une fixation sur le thème de la conscience, et sur la "pureté" de la logique, alors que le même Husserl est aussi à l'origine d'une phénoménologie dynamique, ou pour être plus précis, d'une phénoménologie *génétique*, qui approfondit radicalement sa recherche des origines de l'esprit et du sens.[19] Cette phénoménologie nouvelle de la *genèse profonde* des actes intentionnels entraîne en effet avec elle un bouleversement, qui laisse apparaître de manière flagrante l'importance décisive du *corps* (*Leib*) dans le "monde de la vie" (*Lebenswelt*), la place prépondérante de *l'affectivité* dans la genèse, la dynamique, dans l' "originarité" même de la conscience et du sens, pour aboutir finalement à cette étrange "*intentionnalité*

[16]Voir notamment les travaux de Shaun Gallagher et de Dan Zahavi, par exemple: *The Phenomenological Mind: An Introduction to Philosophy of Mind and Cognitive Science*, London, Routledge, 2008, ceux de d'Evan Thompson, *Life and Mind: Biology, Phenomenology, and the Sciences of Mind*, Cambridge (Mass.), Harvard University Press, 2007, ou par example (ed.): *The Problem of Consciousness: New Essays in Phenomenological Philososphy of Mind*, Calgary (Alta.), Canadian Journal of Philosophy, Supplementary Volumes, 2003. Cf. aussi ce recueil textes: David Woodruff Smith, Amie L. Thomasson (ed.) *Phenomenology and Philosophy of Mind*, Oxford, Clarendon Press, 2005.

[17]Cf. Dan Zahavi, "Husserl's Phenomenology of Body", op. cit., p. 63.

[18]Voir notamment Taylor Carman, "The Body in Husserl and Merleau-Ponty", in *Philososphical Topics*, Vol. 27, No. 2, Fall 1999, pp. 205–226. Dans cet article, la différence et la continuité sont bien expliquées, surtout à propos de l'intentionnalité corporelle; James Dodd, *Idealism and Corporeity: An Essay on the Problem of the Body in Husserl's Phenomenology*, Dordrecht, Kluwer Academic Publishers, 1997; Natalie Depraz, *Lucidité du corps. De l'empirisme transcendantal en phénoménologie*, Dordrecht, Kluwer Academic Publisher, 2001.

[19]Au moins depuis Merleau-Ponty, on sait qu'un autre Husserl vient compléter, voir bouleverser le premier. Dans *Expérience et jugement*, dans les *Ideen II*, *La synthèse passive* ou dans la *Krisis*, Husserl approfondit les choses au point de découvrir ou dévoiler de nouvelles figures de son "savoir", qu'il n'avait pas vraiment traitées jusqu'alors. Ce qui allait engendrer le thème de la corporéité et de la chair, ainsi que celui de la genèse affective de la conscience. Et l'on verra ainsi apparaître une phénoménologie "génétique", qui viendra approfondir la première phénoménologie, "statique". Cf. par exemple, Bruce Bégout, Natalie Depraz, M. Mavridis et S. Nagaï, "Passivité et phénoménologie génétique" (L. Landgrebe, E. Holenstein, I. Yamaguchi, Nam-in Lee), in *Alter. Revue de phénoménologie.* No. 3, 1995, pp. 409–502; Alia Al-Saji, "The Site of Affect in Husserl's Phenomenology. Sensations and the Constitution of the Lived Body", in *Philosophy Today*, SPEP Publications 2000, Vol. 44, Chicago, DePaul University, pp. 51–59.

pulsionnelle",[20] qui *inverse* l'ordre les choses, en un sens quasiment nietzschéen, et renouvelle de fond en comble ou donne une nouvelle vie à la recherche phénoménologique.[21] Dans la mesure où il se retrouve ainsi à la recherche de ce qui est à l'arrière fond de la "conscience" ou de l' "esprit", de ce qui n'est encore que "vie", Husserl semble alors étrangement se rapprocher progressivement du dernier Kant, celui de la troisième *Critique*, de l'*Anthropologie* et de l'*Opus postumun,*[22] de Schopenhauer, le philosophe de la Volonté universelle,[23] mais surtout de Nietzsche, et de sa "volonté de puissance" *intentionnelle,*[24] sans parler de la psychanalyse, et de ses discours sur les "pulsions".[25] En allant d'une philosophie (ou d'une science) de

[20] Cf. *Universale Teleologie*, op. cit., traduction française, *Téléologie universelle*, op. cit.

[21] Brady Thomas Heiner, résume bien les choses, dans son introduction générale à un numéro spécial de *Continental Philosophy Review*, consacré justement à la corporéité, ou plus exactement à ce qui apparaît à l'heure actuelle comme "*Recorporealization of cognition*" dans la phénoménologie et les sciences cognitives: "How far Phenomenology has come from the methodoligical formalism and solipsism, the epistemological foundationalism and internalism, and the ontological Cartesianism of its initial phase. The RoC (Recorporealization of cognition), as a movement internal to phenomenology itself, destabilized this initial philosophical framework and – as we are only now beginning to fully appreciate – is expanding the horizons of phenomenological inquiry". ["Guest Editor's Introduction. The Recorporealization of Cognition in Phenomenology and Cognitive science", *Continental Philosophy Review*, 41, Springer, 2008, pp. 115–126, notamment p. 124 (pour la citation)]. Voir également l'article de Mary Jeanne Larrabee, "Husserl's Static and Genetic Phenomenology", *Man and World*, 9, 2, 1976.

[22] Il s'agit bien sûr de la *Critique de la faculté de juger*, notamment la "critique de la faculté de juger téléologique", de l'*Anthropologie du point de vue pragmatique* et de l'*Opus postumum.*

[23] Rappelons que c'est le fondement même de toute la philosophie d'Arthur Schopenhauer, qu'il développera principalement dans *Le monde comme volonté et comme représentation*, ouvrage dans lequel il considère que "l'essence la plus intime du monde" est quelque chose comme une "Volonté"; et il le découvre précisément dans l'expérience du corps, et grâce à elle. Mais cette "Volonté" est essentiellement "aveugle et irrationnelle"; le monde "intelligible", le monde "en soi", naguère, jadis "divin", ou au moins, plutôt bien, devient alors, de ce fait, quelque chose qui n'a plus rien de "bien", qui est vraiment "mauvais" et qui échappe à la raison: ce qui veut dire qu'il est totalement "absurde"! C'est la grande rupture avec toute la métaphysique traditionnelle, et l'un des commencements de la "mort de Dieu", mais aussi, à propos de l'homme, l'ouverture d'un chantier qui aboutira à "la découverte de l'inconscient". Schopenhauer a d'ailleurs été lu et par Freud et par Husserl. Rappelons également que, ironie de l'histoire, Franz Brentano, le père de la notion d' "intentionnalité", a été aussi le maître des deux. Ils ont décidément plus d'une chose en commun... (cf. aussi Jean-Claude Beaune (dir.) *Phénoménologie et Psychanalyse. Etrange relations*, Champ Vallon, 1998, et l'article très riche de Rudolf Bernet, "Inconscient et conscience: sur la nature de la pulsion, du désir, de la représentation et de l'affect", in Jean Greisch et Ghislaine Florival (dir.), *Création et évènement. Autour de Jean Ladrière*. Louvain-Paris, Editions Peeters, Editions de l'Institut supérieur de philosophie, 1996, pp. 145–164).

[24] Nous nous permettrons ici de renvoyer à nos précédents articles "Nietzsche and the Future of Phenomenology", in Tymieniecka A.-T. (ed.), *Transcendentalism Overturned. Analecta Husserliana CVIII.* Dordrecht, Springer, 2011, et "Le nihilisme et l'épuisement: Heidegger ou Nietzsche", in *Phénoménologies des sentiments corporels. Fatigue Lassitude Ennui.* Paris, Le Cercle Herméneutique, "Anthropologie", 2003.

[25] Quels sont donc les rapports entre la "pulsion" et l' "intentionnalité pulsionnelle"? Vaste question qui, un jour, sans doute, aura son temps. Voir cependant l'article remarquable de Bruce Bégout, "Pulsion et intention. Husserl et l'intentionnalité pulsionnelle", in J. -Ch. Goddard (ed.), *La Pulsion.*

l'esprit *(Geisteswissenschaft),* qui revendique d'abord son autonomie vis-à-vis des sciences de la nature *(Naturwissenschaften),* jusqu'à une philosophie (ou une science) de (ou du monde de) la vie, qui revendique sa proximité avec les sciences de la vie (*Lebenswissenschaften*), le fondateur de la phénoménologie nous laisse tantôt perplexes, tantôt enchantés par le caractère visionnaire de l'aboutissement ultime de son œuvre: mais c'est précisément ce que retiendra principalement la postérité au cour du XXème siècle, et c'est en tout cas ce qui hante manifestement la pensée actuelle.[26]

Le rapport à Kant tout d'abord. On ne peut passer sous silence une certaine ressemblance: elle est patente ou latente, mais elle est là quelque part, aux alentours, dans les parages de cette quête, qui commence par interroger l'esprit, pour aboutir finalement à une interrogation du "phénomène de la vie" en général, laquelle se révèle, dans l'ordre d'arrivée, comme dans l'ordre de "préséance", antérieure à l'esprit. Nous trouvons d'ailleurs l' "intentionnalité pulsionnelle" dans le fameux manuscrit sur la *Téléologie universelle*. De quoi s'agit-il? Il ne s'agit rien de moins que de cette question: "ne pouvons nous pas ou ne nous faut-il pas supposer une intentionnalité pulsionnelle universelle ?"[27] Et Husserl nous dit: "Cela nous conduirait à concevoir une téléologie universelle comme une intentionnalité (pulsionnelle) universelle".[28] Et qu'est-ce que cela englobe?

> Y incluse, l'infinité des monades pourvues d'*anima* (*animalisch*), animales (*tierisch*), préanimales, d'un autre côté montant jusqu'à l'homme, d'un autre encore des monades enfantines, préenfantines – dans la continuité du développement "ontogénétique" <et> phylogénétique.
>
> [...] la forme de la contexture générative, toutes les monades des degrés de monades, les animaux supérieurs et inférieurs, les plantes et leurs degrés inférieurs, et pour tous leurs développements ontogénétiques. Chaque monade essentiellement dans tel développement, toutes les monades essentiellement dans leurs développements génératifs.[29]

Paris, Vrin, 2006, ainsi que celui, tout aussi remarquable, de Rudolf Bernet, "Inconscient et conscience: sur la nature de la pulsion, du désir, de la représentation et de l'affect", op. cit.

[26] Il faut dire que Husserl a été marqué par l'opposition entre "Sciences de l'Esprit" (*Geisteswissenschaften*) et "Sciences de la Nature" (*Naturwissenschaften*), à une époque où Dilthey et les néo-kantiens voulaient absolument démontrer la spécificité et l'autonomie des "Sciences de l'Esprit". Mais il est clair que, même s'il tentera de surmonter cette dichotomie, il a d'abord fondé la "phénoménologie" par opposition ou comme réponse au "naturalisme" ambiant et dominant. C'était donc bien à l'origine comme une "science de l'esprit", fière de son savoir et de son indépendance, vis-à-vis des sciences de la nature. D'où le caractère intriguant de cette étrange "science de la vie" ou "biologie", qui apparaît en effet comme un intermédiaire entre les deux, étant à la fois "nature" et "esprit". Ce qui n'est pas à vrai dire une contradiction, mais ce qui, en réalité, allait attirer l'attention de la postérité, et se révéler prometteur au plus haut point, notamment par rapport au monde la vie (*Lebenswelt*). Un article de Peter Reynaert apporte une belle synthèse à ce sujet, avec des idées claires et distinctes: "Husserl's Phenomenology of the Animated Being, and the Critic of Naturalism". On peut le consulter sur Internet: http://heraclite.ens.fr/~roy/GDR/Animatedbeing.

[27] "Téléologie universelle", op. cit. p. 4.

[28] Ibid. p. 5.

[29] Idem.

On ne peut s'empêcher tout d'abord de penser à deux choses. D'abord, cette "téléologie universelle", qui est celle d'une "intentionnalité pulsionnelle", se trouve également chez les animaux et les plantes, et même dans les "degrés inférieurs" de celles-ci. On est alors en droit de se demander s'il ne s'agit pas ici finalement du vivant en général. N'y aurait-t-il pas alors une ressemblance avec Kant, avec l' "ultime" Kant, qui, pour le moins, s'est penché sur la question de la téléologie dans la vie des vivants? Si Husserl a bien hérité quelque chose de l'auteur de la *Critique de la faculté de juger*, c'est la rigueur des distinctions, celles qui séparent la philosophie de la science, celles qui autorisent ou n'autorisent pas telle ou telle "extrapolation" philosophique. Mais il est aussi de notre devoir d'attirer l'attention sur un enjeu qui nous semble d'une importance capitale. Car si l'on va jusqu'au bout de la logique des choses, c'est le "transcendantalisme" lui-même, aussi différent soit-il chez l'un et l'autre des deux philosophes, qui se trouve dès lors "contaminé". On peut continuer à faire comme si, comme s'il n'y avait rien, mais il peut y avoir ici indéniablement comme une "refonte" ou une "reforme" radicale du transcendantal, si l'on tient compte véritablement de *l'affectivité* chez Kant,[30] de *l'affectivité* et de *l'intentionnalité pulsionnelle* chez Husserl, pour comprendre la conscience ou l'esprit, ou pour repenser ce nous que appelons "subjectivité". L'*a priori* dans les deux cas se retrouve alors, en effet, bien plus du côté du "corps" que du...côté de "l'esprit",[31] qui, lui, se trouve à nouveau "fondée" sur quelque chose qui ne relève plus de lui, et se retrouve finalement comme à la surface, et non plus à l'origine des choses. Le fond caché de l'esprit est comme une "vie anonyme", comme le dit admirablement Merleau-Ponty, qui n'a pas encore d' "identité", ou d(e) "(morale) (d') état civil", comme le dirait Michel Foucault. Elle est encore étrangère à nos "catégories" morales et intellectuelles, et elle nous déconcerte parfois, parce qu'elle est plus profonde que nous, peut-être parce que nous l'avions perdu de vue, peut-être aussi par peur de la "déraison" ou de la "folie". Ce fond obscur, mais qui semble bien être pourtant à l'origine des "lumières" de la raison, et qui devient de nos jours l'objet d'un intérêt grandissant, est peut-être ce qu'il y a de plus prometteur, à la fois chez Kant, et chez le fondateur de la phénoménologie. N'est ce pas d'ailleurs au fond ce que Nietzsche a appelé le "dionysiaque"?[32]

L'importance de la "physiologie", et même pire, de la "biologie",[33] dans la philosophie de Nietzsche, est une question épineuse, qui embarrasse encore

---

[30]Voir notamment le livre de Jérôme de Gramont, *Kant et la question de l'affectivité. Lecture de la troisième critique.* Paris, Vrin, 1996, ainsi que: Eliane Escoubas et Laszlo Tengelyi (dir.) *Affect et affectivité dans la philosophie moderne et la phénoménologie*, Paris, L'Harmattan, 2008.

[31]Voir à ce sujet un livre suggestif de Karl Otto Appel: *L'*a priori *du corps dans le problème de la connaissance*, traduit par T. Simonelli, Paris, Cerf, 2005.

[32]Cf. nos précédents articles "Nietzsche and the Future of Phenomenology", op. cit., et "Le nihilisme et l'épuisement: Heidegger ou Nietzsche", op. cit.

[33]Cf. Le livre de Barbara Stiegler, *Nietzsche et la biologie,* Paris, PUF, Collection "Philosophes", 2001. Ce livre, malgré tout son sérieux et toute sa rigueur, n'a pas toujours été bien accueilli parmi les spécialistes de Nietzsche, et pour cause: la lecture "biologisante" est restée dans les esprits,

aujourd'hui les commentateurs, même les plus éminents.[34] Il y a comme une petite peur qui se profile à l'horizon, dès lors qu'il s'agit de prendre au sérieux ces "fragments posthumes", qui parlent en effet beaucoup de ce sujet, et qui, de surcroît, avaient d'abord eu le malheur d'être publiées, malencontreusement, sous le titre de *La volonté de puissance*.[35] Quand on découvre en plus, impliqués ou imbriqués, dans cette insupportable "physiologie", les concepts ou les thèmes de la "décadence", de la "dégénérescence" ou de la "dépression",[36] sans parler du "nihilisme", il n'y a plus alors qu'à s'en aller! Et ce pour une raison simple: ce qui fait peur à maints égards, à beaucoup de lecteurs, c'est souvent le sentiment que la philosophie elle-même risque ainsi, en dernière analyse, d'être *réduite* à la physiologie, même si c'est en passant par la psychologie.[37] Ne se disait-il pas tout le temps psychologue? N'est-ce pas la psychologie, et finalement la physiologie elle-même qui est considérée à la fin comme "la reine des sciences"?[38] Seulement les faits sont là:

---

profondément liée à l'histoire "noire" du livre *La volonté de puissance*, qui n'est pas vraiment un livre de Nietzsche.

[34] Voir l'article de Wolfgang Müller-Lauter "Décadence artistique et décadence physiologique. Les dernières critiques de Nietzsche contre Richard Wagner." in *Revue philosophique de la France et de l'étranger*, no. 3, 1998, pp. 275–292. Un certain malaise se dégage du texte à propos du "physiologique", du "pathologique", de la "décadence", de l' "épuisement" etc., à telle point que l'auteur semble vouloir trouver, comme il le dit lui-même, un "pendant au Nietzsche du physiologisme borné" (p. 288, note 1). Ce qui ne nous semble pas vraiment nécessaire, si l'on part du principe que ce "physiologique" est non seulement aussi "psychologique", mais il est précisément ce qui allait hanter le XXème siècle, qui s'est senti obligé de penser radicalement l'*Embodiment*, la "corporéité" de l'esprit, de la raison et de tant de choses, et même de parler, entre autres, d' "anthropologie physiologique" (F. J. Buytendijk, Viktor E. von Gebsattel ou Victor von Weiszäker). Tout cela n'était vraiment pas "borné"! Maurice Merleau-Ponty n'a cessé, de son côté, admirablement, de tenter de réunifier le "physiologique" et le "psychique", tout au long de son œuvre, de *La structure du comportement* jusqu'à *Le visible et l'invisible* (en particulier dans *La phénoménologie de la perception*, op. cit. "Première partie: le corps", pp. 80–232).

[35] Cf. le livre de Mazzino Montinari, *La volonté de puissance n'existe pas*, Paris, Éditions de l'Eclat, 1997.

[36] Rien que dans la *Généalogie de la morale*, nous pouvons observer qu'il n' y a pas moins de neuf occurrences de "dépression"; c'est dans la troisième dissertation sur les "idéaux ascétiques": pp. 155, 157, 160, 161, 162 (deux fois), 168, 170, 172.

[37] Parmi les plus grands lecteurs, Jean Granier, auteur d'une thèse mémorable et monumentale sur Nietzsche, dénonce par avance ceux qui seraient tentés, en prétendant suivre le philosophe, de réduire la philosophie à la psychologie, et la psychologie à la physiologie. Mais ce qu'il ne dit pas à vrai dire, c'est que Nietzsche a bien vu, avant le XXème siècle, qu'il ne s'agit pas là de réduire, mais de poser une question qui s'impose, si l'on cesse de croire aux "chimères incorporelles" et que l'on se rend compte, un peu à notre dépit, de l'importance de la "corporéité", dans une pensée profonde de tout ce qui est de l'ordre de l'esprit. C'est précisément ce qui constituera le paradigme de l'*Embodied Mind* et de *l'Embodied Meaning* à la fin du siècle dernier, sans parler de tout l'héritage phénoménologique, qui participe activement aujourd'hui, avec les sciences de la vie et les sciences de l'esprit (*embodied cognitives neurosciences*), à la montée en puissance des idées de ce nouveau paradigme. C'est ce que nous allons voir par la suite. {Il s'agit de l'œuvre de Jean Granier, *Le problème de la vérité dans la philosophie de Nietzsche*. Paris, Seuil, 1966}.

[38] Cf. O. P. C. *Fragments posthumes* Tome XIV 25 [1]. Nietzsche emploie en réalité l'expression suivante: "la grande politique veut que la physiologie soit la reine de toutes les autres questions".

Nietzsche n'a pas fait œuvre de physiologiste, il nous laisse un monument littéraire, philologique et philosophique qui intéresse encore aujourd'hui tant de *philosophes*, dans les plus prestigieuses instances académiques. C'est donc bien dans l'adversité, qui pimente parfois les choses de la vie, que son œuvre résiste encore, et pour cause. Le *réductionnisme*, qu'on pourrait appeler pour commencer "*matérialiste*", Nietzsche a largement eu l'occasion de le connaître, de son temps, de l'examiner avec beaucoup d'attention: ce n'est donc pas comme cela, que l'on pourrait "intimider" sa pensée ou son œuvre! On peut rappeler peut-être pour commencer qu'à un moment crucial de l'œuvre publiée de son vivant, il aborde curieusement la question dans *La généalogie de la morale*:

> Si [un homme] ne vient pas (à bout) d'une expérience vécue, cette indigestion n'est pas moins physiologique que l'autre – en fait elle n'est souvent qu'une suite de l'autre. Cette conception n'empêche aucunement, soit dit entre nous, de rester l'adversaire le plus intransigeant du matérialisme.[39]

Nous voyons bien que le risque est toujours là, et que Nietzsche le perçoit au point de le signaler et de l'écarter explicitement. Cependant, cette problématique ne cesse en réalité de s'enrichir, là où l'on croit qu'elle est une impasse: Nietzsche nous dit au fond ce que nous dira une bonne partie du XXème: c'est bien la vie, en tant que telle, au sens "physiologique", c'est-à dire au sens le plus large et général, qui est toujours prépondérante, notamment par rapport à ce qui est superficiel, autrement dit, "spirituel" ou "psychologique". Ainsi le corps est-il le soubassement de l' "esprit", il est sa face cachée, un peu ésotérique. Car le "physiologique" dont parle Nietzsche est déjà, par définition, "psychologique", mais il est surtout, vivement, au sens phénoménologique, "*prénoétique*", "*préréflexif*", "*antéprédicatif*", et son apparente "superficialité" cache jalousement sa profondeur, qui porte en sein, silencieusement, tous les "événements" qui ont laissé une trace, tout le passé, et même peut-être, en un sens, le futur qui risque d'avoir lieu un jour. Autrement dit, ce n'est pas du "naturalisme", critiqué par Husserl, dont-il s'agit ici chez Nietzsche.[40] Il s'agit bien au contraire de la "vie", de cet "entre-deux", qui bouscule le transcendantal et l'empirique, dont parlait si justement, après Husserl, avec lui ou contre lui, Maurice Merleau-Ponty.[41] Là, se trouve tout ce qui nous occupe encore aujourd'hui: l'âme, l'esprit, la conscience, le sujet et autres "choses" de ce genre, ne sont qu'autant d'ombres ou de fantômes, que pourchasse la pensée actuelle, mais pour découvrir enfin que c'est le corps qui constitue la véritable énigme, comme l'a bien

---

[39] *Généalogie de la morale*, troisième dissertation, 16, p. 154.

[40] Cf. le remarquable article de Keith Ansell Pearson, "Incorporation and Individuation: On Nietzsche's Use of Phenomenology of Life", in *Journal of the British Society of Phenomenology*, Vol. 36, No. 1, 2007, pp. 61–89 (notamment p. 62).

[41] C'est sans doute l'une des grandes tâches que se donne Merleau-Ponty, et qui sera reprise par l'un de ces plus brillants disciples: Francisco J. Varela. Il s'agit surtout pour Varela de reprendre les travaux de *La structure du comportement* et de *La phénoménologie de la perception*, mais nous pouvons dire que c'est une tendance générale dans l'œuvre de Merleau-Ponty, qui bouscule ainsi toutes les oppositions traditionnelles comme "le sujet et l'objet", "le transcendantal et l'empirique", l' "esprit et le corps" etc.

vu Nietzsche, il y a plus d'un siècle.[42] La "corporéité", cette "corporéité" de tant de choses de l'existence et de la nature, à commencer par celle de "l'âme", et qui rassemble ou réunit les hommes, et même les animaux et les végétaux, est devenu de nos jours le sujet et l'objet d'une quête, comme celle du sens, sans retour.

La problématique de l'*Embodiment* occupe aujourd'hui, et de plus en plus, sur la scène intellectuelle contemporaine, une place singulière, considérable: elle semble réconcilier et même rassembler autour d'elle la philosophie dite "continentale" et la philosophie "analytique". Cette question organise même une rencontre prometteuse entre la phénoménologie et la philosophie de l'esprit (*Philosophy of Mind*).[43] Contrairement aux idées reçues, on considère parfois à l'heure actuelle que cette réflexion remonte en réalité à Kant, et l'on voit dans le thème de la sensibilité, et surtout celui de l'affectivité, présent dans la troisième *Critique*, comme une preuve, ou une indication essentielle, qui nous montre que le criticisme kantien avait ouvert la voie pour une recherche de ce genre, en évoquant notamment quelque chose comme un "*Ideal*" ou un "*Transcendental Embodiment*", dans le cadre, plus général encore, de "*Kant's Theory of Sensibility*".[44] Même si l'on considère depuis longtemps que cette idée se trouvait déjà chez Husserl, Sartre ou Merleau-Ponty et d'autres phénoménologues,[45] le paradigme de l' "*Embodiment*", sous sa forme actuel,[46] apparaît en réalité dans les années quatre vingt, du siècle dernier. Lorsque

---

[42] Dans un article important, publié dans *The Body and the Self*, l'un des moments phares de l'histoire de la théorie de l'*Embodiment*, Marcel Kinsbourne tient à souligner à la fin que tout ce qu'il explique ici a déjà été dit en réalité par Nietzsche, il y a plus d'un siècle: "*I am body entirely, and nothing beside*". Cf. "Awareness of One's Own Body: An Attentional Theory of Its Nature, Development, and Brain Basis", in Bermudez J. L., Marcel A., Eilan N. (eds.), *The Body and the Self*, Cambridge (Mass.) – London, The MIT Press, 1995, pp. 205–223 (voir pp. 205 et 217–218: 11. "The Self as Emerging from Backround Body Sensation").

[43] Ce qui apparaît aujourd'hui dans de nombreux travaux, comme ceux, par exemple, de Dan Zahavi, de Shaun Gallagher ou de Hubert L. Dreyfus. Marc L. Johnson souligne par ailleurs le travail pionnier des grands fondateurs du pragmatisme américain: Charles Sanders Peirce (qui inspirera par ailleurs la *Biosemiotics*, comme nous allons le voir), William James et John Dewey. Voir également les projets de "naturalisation de la phénoménologie", et tous ceux qui se réclame de l'œuvre de Francisco J. Varela, ainsi que les travaux de Natalie Depraz.

[44] Cf. Angelica Nuzzo, *Ideal Embodiment. Kant's Theory of Sensibility*, Bloomington, Indiana University Press, 2008. Voir p. 200: "Kant establishes ontological and epistemological conditions that radically break with the modern paradigm of the mind/body dualism. His aim is to overcome such metaphysical dualism by proposing not only a new concept of rationality but also a new, broadly construed notion of human sensibility that includes *Anschauung, Empfindung, Gefühl, Affekt/Affektion, and Einbildungskraft.*"

[45] Cf. *The Problem of Embodiment. Some Contributions to the Phenomenology of the Body*, The Hague, Martinus Nijhoff, Phaenomenologica 17, 1964, 1971.

[46] Il faudrait distinguer en effet l'*Embodiment*, qui apparaît dans les neurosciences et les sciences cognitives depuis trois décennies, qui sera réexaminé et repris par des linguistes ou des philosophes comme Georges Lakkof et Marc L. Johnson, et *l'Embodiment* (*Leiblichkeit*) phénoménologique, qui est beaucoup plus ancien (cf. *The Problem of Embodiment: Some Contributions to the Phenomenology of the Body*, op. cit.) et qui n'a pas manqué d'influencer le "second" *Embodiment* à plusieurs reprises. On parlera alors du "retour" des sciences cognitives à la phénoménologie.

Mark L. Johnson publie, en 1987, *The Body in Mind. The Bodily Basis of Meaning, Imagination and Reason,*[47] cette idée n'est pas vraiment à la mode, ni dans son pays, ni dans le monde anglo-saxon en général.[48] Par ailleurs, les scientifiques et les philosophes, qui s'occupent de l'esprit ou/et du cerveau, ne sont pas encore attirés par cette réflexion ou cette méthode: comme le dira plus tard le célèbre neurobiologiste Antonio R. Damasio, dans *Descartes' Error: Emotion, Reason and the Human Brain*, l'idée dominante à l'époque, c'est que l'esprit et la raison n'ont pas grand-chose à voir avec les affects ou les émotions.[49] On peut donc dire que pendant longtemps, quand un scientifique comme Damasio, ou un philosophe comme Johnson, qui furent tous deux assurément des pionniers, parlait de ce genre de choses, il prêchait un peu dans le désert, et il n'y avait pas encore d'adhésion ou d'intérêt véritable pour l'*Embodiment*, pour la corporéité de l'esprit et du sens.[50] Il aura fallu attendre un concours de circonstances et d'heureux hasards, pour que les neurosciences cognitives soient petit à petit contaminées, et déclenchent, en faisant explicitement référence à la phénoménologie, un véritable mouvement qui se réclame de cette idée. Après Johnson, et avant Damasio, c'est surtout, la

---

[47] Marc L. Johnson, *The Bodily Basis of Meaning, Imagination and Reason*, Chicago, The University of Chicago Press, 1987.

[48] A cette date, en 1987, Johnson se trouve en effet encore assez seul, sans Varela, sans Damasio, et sans références continentales importantes, c'est-à-dire sans l'*embodiment* de la phénoménologie, et sans les "*embodied cognitive neurosciences*" (seulement quelques références à l'herméneutique de Hans Georg Gadamer, et une critique virulente de Gotlob Frege): il s'agit encore de quelques pionniers plus ou moins solitaires, dans la philosophie anglo-saxonne. Parmi eux, Eugene Gendlin, qui avait déjà publié, il y a un demi siècle: *Experiencing and the Creation of Meaning*, Evanston, Northwestern University Press, 1962.

[49] Cf. *L'Erreur de Descartes. La raison des émotions* Paris, Odile Jacob, 1994–2010, p. I–II. Damasio explique, dans sa préface à la nouvelle édition (2005), qu'à part quelques grandes exceptions notables comme Charles Darwin, William James ou Sigmund Freud, qui ont mis l'accent sur l'importance de l'émotion et des affects, "les sciences de l'esprit et du cerveau s'étant épanouies aux XXème siècle, leur intérêt s'est porté ailleurs et les spécialités que l'on regroupe aujourd'hui sous le terme vague de neuro-sciences ont plutôt tourné le dos aux recherches sur les émotions"; on peut même dire que les "exceptions n'ont fait qu'accentuer l'oubli dans lequel l'émotion, en tant que sujet de recherche, était tombée. Le béhaviorisme, la révolution cognitiviste et les neurosciences computationnelles n'ont pas atténué cet oubli [...] tel était, en gros, la situation lorsque l'*Erreur de Descartes* a été publié pour la première fois", à la fin du siècle dernier, en 1994.

[50] Pour se repérer et mettre en perspective l'histoire de ce courant, nous pouvons considérer que la montée en puissance du paradigme de l'*Embodiment* a eu lieu principalement selon un les étapes suivantes: (1) Francisco J. Varela, Evan Thompson, Eleonor Roch, *The Embodied Mind: Cognitive Science and Human Experience,* Cambridge (Mass.), The MIT Press, 1991. (2) Antonio R. Damasio, *Descartes' Error. Emotion, Reason, and the Human Brain*, New York, G. P. Putnam's Sons, 1994. (3) José Louis Bermudez, Antonio Marcel, and Noami Eilan (ed.), *The Body and the Self*, Cambridge (Mass.), The MIT Press, 1995. (4) Andy Clark, *Being There: Putting Brain, Body, and the World Together Again*, Cambridge (Mass.), The MIT Press, 1997. (5) Antonio R. Damasio, *The Feeling of What Happens: Body and Emotion in the Making of Consciousness*, New York, Harcourt Brace, 1999. (6) Shaun Gallagher, *How the Body Shapes the Mind*, Oxford, Oxford University Press, 2005.

publication en 1991, de *The Embodied Mind. Cognitive sciences and Human Experience,*[51] par Francisco J. Varela, Evan Thompson et Eleonor Rosch, aux presses du MIT, qui constitue un tournant décisif, et conduira à une vague de publications, qui ne cessent de se multiplier au fil du temps, jusqu'à nos jours.[52] On assistera ainsi à un tournant de la pensée contemporaine, qui s'inscrit en faux contre le "*linguistic turn*" du début du XXème siècle, pour se nommer finalement "*corporeal turn*".[53]

Parmi les initiateurs ou les précurseurs de cette "nouvelle vague", philosophique et scientifique, Marc L. Johnson occupe une place particulière, qui lui a donné, finalement, tout le recul nécessaire, pour penser globalement ce renouveau intellectuel. En effet, après avoir été l'un des pionniers, au moment de la publication de *The Body in the Mind. The Bodily Basis of Meaning, Imagination and Reason*, en 1987, il nous offre vingt ans plus tard, l'une des plus belles synthèses qui soient, avec *The Meaning of the Body: An Aesthetic of Human Understanding.*[54] On peut voir déjà dans le titre cet intérêt passionné, qui ne se démentira pas, pour la question du sens, ou, plus exactement, pour ce rapport énigmatique et révélateur entre le corps et le sens, puisque nous allons, au fond, du "*bodily basis of meaning*" jusqu'au "*meaning of the body*", expressions qui dessinent ainsi la grande voûte de ses interrogations scientifiques et philosophiques. Johnson fait partie, par ailleurs, des fondateurs de ce qu'on appelle la *Cognitive Linguistic*[55] ("linguistique cognitive"), qui rompt précisément avec la linguistique traditionnelle, et tout ce qui a fait la

---

[51] *The Embodied Mind: Cognitive science and Human Experience*, op. cit.; traduction française, *L'inscription corporelle de l'esprit. Sciences cognitives et expérience humaine*, Paris, Seuil, 1993.

[52] Pour une vision globale, vingt ans après la publication de *The Embodied Mind*, cf. *Enaction: Toward a new Paradigm in Cognitive sciences*. Cambridge (Mass.), The MIT Press, 2010.

[53] Cette expression est même devenu le titre d'un livre de John Tombomino, dans lequel il reprend les idées de Nietzsche sur l'importance du corps, dans la pensée et dans l'histoire, et propose d'en tirer les conséquences sur le plan sociétal et politique: *The Corporeal Turn. Passion, Necessity, Politics*, Lanham, Roman and Littlefield Publishers, 2002. Voir aussi à ce sujet le travail très riche de Lorenzo Altieri (sa thèse de doctorat): *Eidos et Pathos. Corporéité et signification entre phénoménologie et linguistique cognitive*, Bucarest, Zeta Books, 2009.

[54] *The Meaning of the Body. Aesthetics of Human understanding*, Chicago, The University of Chicago Press, 2007.

[55] La linguistique cognitive est un courrant important de la linguistique contemporaine qui émerge à la fin des années 1970, et qui vient contredire la linguistique "officielle", celle notamment de Noam Chomsky; mais ses auteurs critiquent en réalité toute la tradition formelle et analytique, celle qui, depuis Gotlob Frege ou Bertrand Russell se voue à la formalisation et l'analyse, et qui constitue le *Linguistic turn* au début du XXème siècle. Ce qu'elle oublie en revanche c'est l'importance cruciale du fondement "expérientiel", ou de *l'expérience vécue* dans la constitution du langage et du sens. L'un des livres fondateurs de cette tradition fut celui de Georges Lakoff et Mark L. Johnson: *Metaphors We Live By* (Chicago, The University of Chicago Press, 1980, traduction française, *Les métaphores dans la vie quotidienne*, Paris, Edition de Minuit, 1985), qui aboutira, vingt ans plus tard, à leur travail philosophique monumental: *Philosophy in the Flesh: The Embodied Mind and Its Challenge to the Western Thought*. New York, Basic Books, 1999. Notons la reprise, dans l'intitulé de l'ouvrage, de l'expression "*The Embodied Mind*", présente dans le titre du livre de Varela, en 1991. Ce qui, bien sûr, n'est pas dû au hasard, et montre l'affinité grandissante, qui apparaît ainsi, progressivement, entre tous ces auteurs.

fortune de la "philosophie analytique" et de la "philosophie du langage". C'est avec le linguiste Georges Lakoff, qu'il publie d'abord, en 1980, *The Metaphors we live By*,[56] dans lequel on peut voir déjà, indirectement, la présence du corps, mais surtout l'importance décisive de la "*métaphore*", dans notre usage du langage et notre maniement ordinaire du sens, dans la vie quotidienne la plus simple. Pour dire quelque chose en effet, en plus de (et y compris) nos paroles sur les objets, sur une chaise ou une table, nous faisons toujours *allusion* à des *expériences vécues* antérieures, pour nous *exprimer*, pour dire nos pensées ou nos sentiments. On opère donc toujours, à chaque fois, un appel à, ou un rappel de, pour renvoyer notre interlocuteur, à ce qu'il peut comprendre à partir d'une expérience commune, partagée. Mais cette expérience justement n'est pas une vérité "objective", une "vérité par adéquation" ou "par correspondance", ou une "représentation mentale" au sens traditionnelle, elle est toujours liée à des impressions et à des émotions, qui sont conservées, concentrées et véhiculées par et à travers des *métaphores*. Ce genre de formulations métaphoriques est absolument indispensable pour nous, pour *vivre* et *communiquer* entre nous, pour se faire *comprendre*. Il ne s'agit donc pas simplement de figures ou d'ornements rhétoriques, mais d'outils indispensables pour la vie quotidienne, qui occupent une place décisive et prépondérante. Ces "métaphores" sont inséparables "de leurs fondements expérientiels", et on peut dire que "c'est seulement au moyen de ce fondement que la métaphore peut servir d'instrument de compréhension".[57] Mais cette "compréhension" ou ce "fondement expérientiel" n'est rien d'autre précisément qu'une expérience vécue, "émotionnelle", et donc, en dernière analyse, corporelle. On voit bien alors comment le corps apparaît d'abord là où l'on ne s'y attendait pas encore, et prépare la venue de la question capitale du rapport entre le corps et le sens, qui allait se déployer sur plusieurs décennies, et entraîner, après Husserl, après Merleau-Ponty, un bouleversement presque obligé, de la pensée et de la recherche scientifique, à notre époque. Même si cela se prépare toujours silencieusement, comme la dérive des continents, à partir d'un certain seuil, le changement devient inévitable et souvent colossal.

Dans *The Body in Mind*, et surtout, plus récemment, dans *The Meaning of the Body*, les problèmes linguistiques cèdent quelque peu la place à une problématique plus radicale encore, qui tendra à montrer les origines ou les racines corporelles, sensori-motrices, de l'esprit, du langage et des concepts, même les plus abstraits, qu'ils soient ceux de la logique ou même des mathématiques.[58] La "linguistique

---

[56] Cf. *Metaphors We Live By*, op. cit., traduction française, *Les métaphores dans la vie quotidienne*, op. cit. Egalement leur article: "Why Cognitive Linguistics Requires Embodied Realism", in *Cognitive Linguistics*, 13, 3, 2002, pp. 245–263.

[57] *Les métaphores dans la vie quotidienne*, op. cit. p. 30.

[58] Voir surtout l'article de Vitorio Gallese et Georges Lakoff, qui a été souvent cité comme une référence: "The Brain Concepts: The Role of the Sensory-Motor System in Conceptual Knowledge", in *Cognitive Neuropsychology*, 22, 2005, pp. 455–479. (Lakoff est le linguiste, et Gallese, le neuroscientifique qui avait travaillé, entre autres, sur les "neurones miroir": [avec A. Goldman] "Mirror Neurons and the Simulation Theory of Mind-Reading", *Trends in Cognitive Science*, 2, 1998, pp. 493–501). Pour les mathématiques, on peut lire aussi avec intérêt l'œuvre de Lakoff avec le mathématicien Rafaël Nuñez, *Where Mathematics Comes From. How the Embodied*

cognitive" s'est ainsi approfondie et fondée sur la "nouvelle vague" des neurosciences cognitives, en rompant aussi bien avec la "philosophie analytique", issue en dernière analyse du projet de formalisation de Gottlob Frege,[59] qu'avec un certain cartésianisme de la pensée et du langage, et ses prolongements contemporains, comme la linguistique de Noam Chomsky,[60] et sa grammaire formelle, générative. Sans vouloir revenir aussi sur l'histoire de la cybernétique, qui n'est pas étrangère à ces problématiques, il importe de souligner cependant que les théoriciens de l'intelligence artificielle (AI) sont eux-mêmes revenus de leurs propres illusions, en découvrant l'importance radicale du corps, dans la vie et la constitution de l'esprit.[61] Marc L. Johnson a donc été un peu le témoin de l'avancée de ses idées, sans qu'il soit nécessairement lu par les protagonistes, et il en faisait partie, au hasard des lectures et des rencontres. Mais dans *The Meaning of the Body*, on découvre comme une synthèse, un état des lieux qui nous fait comprendre l'ampleur de l'enjeu et des progrès accomplis. Ce qu'il disait au départ avec Lakoff, il le redira presque trente ans plus tard: ce qui est vraiment "significatif", ce qui a un sens pour les hommes, ne se trouve nulle part dans tous ces traités "philosophiques", "analytiques" et "linguistiques"[62]; alors que le cœur de ses recherches à lui se situe justement dans

---

Mind Brings Mathematics into Being, New York, Basic Books, 2000, ainsi que leur article: "The Cognitive Foundations of Mathematics: The Role of Conceptual Metaphor", in *Handbook of Mathematical Cognition*, New York, Psychology Press, 2005, pp. 109–124; et les articles de Nuñez: "Do Real Numbers Really Move? Language, Thought, and Gesture: The Embodied Cognitive Foundations of Mathematics", in R. Hersh (ed.), *18 Unconventional Essays on the Nature of Mathematics*, New York, Springer, 2006, pp. 160–181; "Mathematicatics, the Ultimate Challenge to Embodiment: Truth and the Grounding of Axiomatic Systems", in Paco Calvo and Antoni Gomila (ed.), *Handbook of Cognitive Science: An Embodied Approach*, Elsevier, Academic Press, 2008.

[59] Il y avait déjà une grande critique de Frege, dans *Les métaphores dans la vie quotidienne*, op. cit. pp. 211–212.

[60] Cf. Noam Chomsky, *Language and Mind*, traduction française par Claude Bourgois, *Le langage et la pensée*, Paris, Editions Payot & Rivages, 2012; *Syntactic Structures*, traduction française par Michel Braudeau, *Structures syntaxiques*, Paris Editions du Seuil, 1979. Comme le dit Chomsky lui-même, il s'agit d'une "linguistique cartésienne".

[61] Cf. Rodney Brooks, "Intelligence Without Representation", in *Artificial Intelligence*, vol. 47, 1991, pp. 139–160. C'est précisément là que se trouve la fameuse "déception" de l'intelligence artificielle. Il s'est avéré, en réalité, qu'au fond, avant de partir en quête d'une "intelligence", on devait d'abord, auparavant, approfondir la problématique de la "vie artificielle", car, un esprit sans corps s'avérait de plus en plus impossible. Voir Francisco J. Varela, *Cognitive Science. A Cartography of Current Ideas*, traduction française par Pierre Lavoie, *Invitation aux sciences cognitives*, Paris, Editions du Seuil, "Sciences" 1996. Nous pouvons lire p. v que le problème était justement "la tendance de l'IA (ainsi que du reste des sciences cognitives) à l'abstraction, pour élaborer les perceptions et les capacités motrices". Mais "une tel abstraction ne peur saisir l'essence de l'intelligence cognitive, *qui ne réside que dans son intégration corporelle*".

[62] Cf. *Les métaphores dans la vie quotidienne*, op. cit. p. 7. "Les théories de la signification qui étaient dominantes dans la philosophie et la linguistique occidentales étaient inadéquates, et [...] le terme "signification", dans cette tradition, n'avait rien à voir avec ce qui, dans la vie des hommes, leur apparaissait comme significatif".

ces "*deep, visceral origins of meaning*",[63] à travers ce continent perdu, ce "*vast, submerged continent of non conscious thought and feeling that lie at the heart of our ability to make sense of our life*".[64] Pour cela, il se fonde précisément sur les "*recent developments in the new sciences of embodied mind*".[65]

Le cœur de l'argumentation de Johnson se trouve à la croisée des chemins entre ses théories philosophiques, qu'ils élaborent depuis longtemps, et des données scientifiques qui s'accumulent jour après jour, et qui viennent confirmer ses intuitions. Le point central et l'aboutissement de ses travaux restent au fond une interrogation fondamentale, et *une théorie du sens*, qui peut être désignée ainsi: "*the embodied theory of meaning*",[66] par opposition à ce qu'il nomme "*the objectivist theory of meaning*",[67] et qui a été, pratiquement jusqu'à Nietzsche,[68] la conception dominante dans la philosophie occidentale. Pour comprendre sa démarche, la première chose qu'il faut dire est précisément que *l'esprit, dans son fonctionnement, est enraciné dans l'activité du corps, plus exactement dans l'activité sensori-motrice et émotionnelle d'un corps, qui à son tour se trouve dans un environnement,*[69] ce dernier étant à la fois le milieu naturel et humain, avec ses dimensions écologique et intersubjective. Les avancées scientifiques

[63] *The Meaning of the Body. Aesthetics of Human Understanding*, op. cit. p. x.

[64] Ibid. p. xi.

[65] Idem.

[66] Ibid. p. 10. Voir aussi son très bel article sur Merleau-Ponty, dans lequel il développe également, à partir de l'œuvre du philosophe français, sa théorie du sens: "Merleau-Ponty's Embodied Semantics – From Immanent Meaning, to Gesture, to Language", in *EurAmerica*, Vol. 36, No. 1, March 2006, pp. 1–27.

[67] Ibid. p. 272. On peut dire que cette objection, contre "les théories objectives du sens", dans l'histoire de la philosophie occidentale, est une constante dans son œuvre, et ce depuis la publication de *Metaphors We Live By*, en 1980: la linguistique cognitive a été même une des premières "insurrections" contre l'oubli des racines "expérientielles" du sens. Elle annonçait déjà le *Corporeal Turn* ("tournant corporel") de la pensée contemporaine.

[68] Mais les allusions à Nietzsche restent malheureusement, extrêmement rares; ce qui montre que c'est un grand projet, encore à venir, vaste et prometteur, pour la pensée contemporaine, qui, sans faire exprès, sans le vouloir, lui donne raison, et de plus en plus. Elle le fait notamment face à Heidegger, qui malgré toute son importance, avait oublié le corps, et n'avait pas vraiment vu venir ce tournant majeur, ce qui allait être nommé "*Corporeal turn*". On peut lire tout de même—ce qui est encourageant—dans *The Meaning of the Body*, p. 105: "The history of Western philosophy, from the early Presocratics to the present day, reveals a succesion of attempts to identify and describe these universal, eternal norms. Wether they are believed to come from the mind of God, from Nature, or from Universal Reason, their function is supposed to be that of providing us with an always-fixed mark by which to navigate our way through the ever-changing, ever flowing waters of our temporal existence. But Nietzsche [...] and a host of subsequent thinkers have shown us that *life is change* and *existence is an ongoing process*. There is no eternal logic, no absolute form that could save us from grappling with change every moment of our lives. The logic we humans have is an embodied logic of inquiry, one that arises in experience and must be readjusted as situations change".

[69] Cf. *The Meaning of the Body. Aesthetics of Human Understanding*, op.cit. p. 11–12.

multiples dans ce domaine, celui des sciences de la vie et de l'esprit,[70] corrigent un égarement tenace et ancien, qui traitait le corps avec condescendance ou mépris, en prétendant ou souhaitant que la pensée soit une activité purement abstraite, déconnectée ou libérée du corps. Cette méconnaissance des choses se fonde sur des "*misconceptions*", qui peuvent être résumés ainsi:

> (1) the mind is disembodied, (2) thinking transcends feeling, (3) feelings are not part of meaning and knowledge, (4) aesthetics concerns matters of mere subjective taste, (5) the arts are a luxury (rather than being conditions of full human florishing)[71]

Cette façon de voir les choses est précisément ce que récuse l'auteur de *The Meaning of the Body*. Il est clair que ce que vise Johnson en priorité est cette théorie de l'esprit largement admise, surtout dans la "philosophie analytique de l'esprit et du langage".[72] Ce qui est étrange, en effet, avec cette tradition, comme avec la philosophie traditionnelle, c'est que l'esprit redevient quelque chose de désincarné, sans corps ni chair, sans émotions: on parle de logique, de langage, de pensée articulée ou standardisée, pour parvenir finalement à un traitement "informatique" des choses, qui n'a rien avoir avec l'homme en chair et en os. Les affects, les sentiments, l'art ou la beauté n'ont plus rien à voir, dans cette optique, avec le travail "sérieux", "austère" de la "philosophie", celle qui s'occupe du langage ordinaire, du langage formel, de l'analyse logique ou de l'analyse linguistique.[73] C'est cette même tradition, qui, ironie du sort, voulait et prétendait pouvoir surmonter le dualisme cartésien de l'âme et du corps, qui nous mène à un esprit "squelettique", désincarné, sans "âme" ni "états d'âme", calculateur comme un ordinateur, et qui n'a que faire de tous les tourments, de tous les bonheurs ou malheurs que connaît l'esprit humain. C'est sur les ruines ou les vestiges de ces erreurs et de ces égarements, célébrés au lieu d'être

[70] Il ne s'agit plus ici des *Geisteswissenschaften* de Wilhem Dilthey, mais bien de cette nouvelle vague des "*embodied cognitive sciences*". Cf. en particulier: *Handbook of Cognitive Science: An Embodied Approach*, op. cit. Johnson écrit par ailleurs, dans son livre, *The Meaning of the Body*, p. 1: "For at least the past three decades, scholars and researchers in many disciplines have piled up arguments and evidence for the embodiment of mind and meaning. [...] The best biology, psychology, cognitive neuroscience [...] available today teach us that our human forms of experience, consciousness, thought, and communication would not exist without our brains, operating as an organic part of our functioning bodies, which, in turn, are actively engaged with the specific kinds of physical, social, and cultural environments that humans dwell in".

[71] *The Meaning of the Body. Aesthetics of Human Understanding*, op. cit. p. xi.

[72] Ibid. p. 264: "analytic philosophy of mind and language".

[73] Ibid p. x: "much contemporary philosophy focuses exclusively on abstract conceptual and propositional structure, leaving us with a very superficial and eviscerated view of mind, thought, and language, These philosophers have developed elaborate conceptual schemes for indentifying the so-called cognitive, structural, and formal aspects of experience, thought, and language, but they lack adequate philosophical resources to plumb the depths of the qualitative feeling dimensions of experience and meaning". Et nous pouvons lire aussi page 9: "In the account of embodied meaning that I am developing [...] I am using the term *meaning* in a broader sense than is typical in mainstream Anglo-American philosophy of language and mind. I seek to recover most of the resources for meaning-making that are ignored in the writings of influential philosophers such as Quine, Searle, Davidson, Fodor, Rorty, and many others".

corrigés, par une majeure partie de la philosophie anglo-saxonne, et de la philosophie traditionnelle, que l'auteur poursuit inlassablement son chemin pour fonder une nouvelle approche du monde, de la vie et de l'homme dans son ensemble, et faire en sorte que la philosophie revienne sur terre, et reste près de la vie, dans sa réalité sensible, affective ou "pathique".

Parmi les scientifiques, Francisco J. Varela a été, de son côté, l'un des fondateurs, et l'une des figures majeures du paradigme de l'*Embodiment*, à notre époque. Il fut aussi brillamment le grand artisan d'un dialogue fructueux entre les neurosciences cognitives et la phénoménologie, et plus largement, entre les sciences de la vie et la philosophie.[74] S'intéressant d'abord, avec son maître et ami Humberto Maturana,[75] à la spécificité essentielle du vivant, de sa plus simple expression cellulaire jusqu'à sa forme anthropologique, il se retrouve ainsi, après Hans Jonas,[76] au cœur d'une "phénoménologie de la vie", qui tendra à comprendre finalement, et de façon radicale, les racines "vitales" et "corporelles", immémoriales de l'esprit. Sa "phénoménologie" à lui suivra cependant un chemin inversé: contrairement à Husserl, et à une certaine tradition "intellectualiste", comme le disait son inspirateur, Merleau-Ponty, son cheminement à lui sera surtout d'aller de la "vie" à l' "esprit", et non l'inverse, d'appréhender l' "esprit" dans sa forme la plus élémentaire, la "vie", jusqu'à son expression la plus complexe, l' "esprit". Cette originalité n'est pas sans rapport bien sûr avec le retour merleau-pontien au corps et à la perception, et sa réhabilitation ontologique du sentir et du sensible. Varela se réclame très explicitement de Merleau-Ponty,[77] et cherche à reprendre son projet philosophique à la lumière des sciences actuelles. Dans les deux cas, il y a indéniablement un retour à une "phénoménologie de la vie". Tout cela se fera bien au détriment de la tradition "objectiviste", "intellectualiste" et "représentationnelle", qui de Descartes à la "philosophie de

---

[74] Dès début de *The Embodied Mind,* il est question de l'œuvre et de la pensée de Maurice Merleau-Ponty, et c'est ce qui aboutira dans les dernières années, au projet de "naturalisation de la phénoménologie" ("The Naturalization of Phenomenology as the Transcendance of Nature", in *Alter. Revue de phénoménologie.* No. 5, Paris ENS, 1997, pp. 355–381), ainsi qu'à une "neurophénoménologie" (cf. "Neurophenomenology: A Methodoligical Remedy for the Hard Problem", in *Journal of Consciousness Studies*, 3, 1996, pp. 330–335). Voir aussi l'article très riche sur Kant, publié avec Andreas Weber, après la mort de Varela: "Life after Kant: Natural Purposes and the Autopoietic Foundations of Biological Individuality", op. cit.

[75] Varela avait commencé ses recherches en Biologie au Chili, avant son doctorat à Harvard, avec ce grand biologiste; et il retournera dans son pays pour le rejoindre, après avoir fini sa thèse. Il travailleront ensemble intensément, et avec beaucoup d'espoir, jusqu'au coup d'État de Pinochet, en 1973. A partir de cette date, il choisit le chemin de l'exil, et il ne reviendra au Chili qu'en 1985. Il s'installe finalement à Paris en 1986, et deviendra Directeur de recherches au CNRS en 1988. Il travaillera au CREA de l'Ecole polytechnique (fondé quelques années auparavant par Jean-Pierre Dupuy et Jean-Marie Domenach) et à l'Hôpital de la Salpêtrière jusqu'à sa mort en 2001.

[76] Varela s'est particulièrement intéressé à l'œuvre de Hans Jonas [notamment *The Phenomenon of Life*], qui apparaît encore dans l'ultime article sur Kant: "Life after Kant", op. cit.

[77] *L'inscription corporelle de l'esprit*, op. cit. p. 17: "Notre voyage au cours de ce livre peut être vu comme le prolongement moderne d'un programme de recherche fondé il y a une génération par Maurice Merleau-Ponty". Cf. aussi pp. 18, 19, 27, 28.

l'esprit" et aux sciences cognitives "computationnelles",[78] procédait à l'envers, et se retrouvait finalement dans une impasse, sur une voie sans avenir.[79]

Dans *Autopoiesis and Cognition: The Realization of the Living,*[80] écrit en collaboration avec Maturana, Varela tente de montrer dès le départ la nature de l'enjeu: il s'agit en effet de faire ressortir le caractère "cognitif" (ou "herméneutique" et "psychique") de la vie, dès sa plus forme la plus primitive, et sa capacité – qui pourrait même être considérée comme sa caractéristique principale ou sa définition, à entretenir une "relation", qui ressemble à un "dialogue" avec son environnent. C'est cette "compréhension" ou cette "interprétation" des choses et du monde qui doit tout d'abord attirer notre attention. Cette relation "égoïste" et "intéressée", par définition, qui fait qu'une cellule vivante profite de son environnement pour rester en vie, manifeste ainsi une capacité à "échanger", à "négocier" avec son milieu naturel, ou, autrement dit, parvient à "comprendre" ce qui l'entoure, et à "exprimer" finalement, activement, un certain "point de vue", correspondant à ses intérêts. Ces mots peuvent sembler n'être qu'une projection anthropomorphique inadéquate, inappropriée au vivant en général, en tant en que tel. La tâche que se donnent pourtant les deux grands biologistes, dans cet ouvrage, est de nous montrer qu'il n'en est rien, qu'une cellule se "comporte" réellement ainsi, selon une "habileté" ou une "intelligence" qui lui est propre. Précisons toute de suite, néanmoins, que le premier mot du titre de l'ouvrage, "autopoiesis" renvoyait d'emblée à la capacité du vivant à s'auto-(re)produire, en gardant, pour ainsi dire, sa "forme", tout en renouvelant sa "matière", grâce à son environnement, qui est censé lui être favorable, au moins sur ce point: lui permettre de se ressourcer, et préserver ainsi sa "vie" sous cette "forme". Mais se préserver veut dire préserver quelque chose, qui n'est autre que "soi", et ce point est particulièrement intéressant. En effet, non seulement il y a, comme par définition, préservation, dans l' "auto-production" (*autos:* soi, *poiein*: produire), mais il y a conservation d'une certaine "loi" ou d'une certaine "norme", qui constitue en quelque sorte "l'identité" de chaque vivant,[81] et exprime son "auto-nomie" (*autos:* soi, *nomos:* loi), sa vie sous sa loi propre. C'est là que se trouve les origines du "soi corporel", et chez les hommes, du "Soi" tout court, qui est encore et toujours,

[78] Il s'agit du premier grand courant des sciences cognitives qui allait aboutir, entre autres, à l'intelligence artificielle… et à ces déceptions. On considérait l'esprit, dans cette perspective, comme un opérateur formel, autrement dit, comme un ordinateur, qui fonctionne grâce à une certaine "logique", laquelle peut être modélisée, et donc reproduite, indépendamment du corps, dans une machine. Cf. *Invitation aux sciences cognitives*, op. cit. pp. 27–51 (notamment, p. 44–51: "L'enfant du cognitivisme: l'intelligence artificielle").

[79] Voir à ce sujet le livre de Jean-Pierre Dupuy, *Aux sources des sciences cognitives*, Paris, La Découverte, 1994; et son "éclairage historique", dans "Philosophie et sciences cognitives", in Jean Petitot et al. (ed.) *Naturaliser la Phénoménologie. Essais sur la phénoménologie et les sciences cognitives*, Paris, CNRS éditions, 2002, pp. 697–723.

[80] *Autopoiesis and Cognition: The Realization of the Living*, Dordrecht, Kluwer Academic Publisher, "Boston Studies in the Philosophy of Science" 42, 1980, traduction française, *Autonomie et connaissance: essai sur le vivant*, Paris, Seuil, "La couleur des idées, 1989.

[81] Cf. "Patterns of Life: Interwining Identity and Cognition", *Brain and Cognition*, 34, 1997, pp. 72–87: "An organism is fundamentally a process of constitution of an identity".

précisément à cause de ses origines, quelque chose de corporel. Il rassemble ainsi sous son joug les différents petits "soi" des cellules de son corps, qui ne sont plus vraiment de ce fait, de vrais "soi", mais, comme il dit, des "*selfless selves*", des "soi" dépossédés de leur soi, puisqu'ils sont au service de quelque chose d'autre qui les dépasse ou les transcende[82]: ils ne sont plus ni auto-poiétiques, ni auto-nomes, mais bien plutôt "allo-poiétiques" et hétéronomes. Alors que le vivant "maître de soi" se trouve fondé en quelque sorte sur sa propre loi ou sur sa propre constitution "morale" et "politique" ! Bien sûr nous sommes encore bien loin d'un *cogito*, d'un *ego*, ou d'un sujet transcendantal. Mais il y a toute même, ici, comme on peut l'imaginer, l'amorce de quelque chose, qui aura, au cours de l'évolution, des conséquences vertigineuses, pour le moins intéressantes, et non encore élucidées, suffisamment. Il faut bien dire qu'à l'échelle de l'homme, tout cela prendra une tournure on ne peut plus aiguë, et autrement plus tragique. Quoi qu'il en soi, malgré les blessures, les accidents de parcours, qui sont toujours là, on voit ainsi que l' "autopoeisis" et la "cognition" sont comme les deux piliers de l' "existence" du vivant, et expriment en réalité le même besoin et la même nécessité: se maintenir en vie, dans le "respect" de "soi", en (se) "comprenant" et en "dialoguant", dans le cadre d'un "échange" ou d'une "communication" avec son environnement.

Avec la publication de *The Embodied Mind*, Varela se lance alors dans une quête de quelque chose comme une "origine sans origine", ou un "fondement sans fondement". Le sujet principal du livre, cet "esprit incorporé", est d'abord à l'origine d'un renouveau véritable dans les neurosciences cognitives.[83] A partir d'une intuition profonde qui se réclame de Merleau-Ponty, Varela et ses amis parviennent à secouer les colonnes du temple cognitiviste, en montrant les limites flagrantes d'une conception abstraite, formelle et désincarnée de l'esprit. Loin du "computation alisme" et du "connexionnisme", qui s'enferment dans une conception de ce type, dont le modèle est l'ordinateur ou le réseau de neurones, et par-delà leurs oppositions,[84] les

---

[82] Cf. "Organism: Meshwork of Selfless Selves", in *Organism and the Origins of the Self*, Dordrecht, Kluwer Academic Publisher, 1991, pp. 79–107.

[83] Pour une vision rétrospective globale voir en particulier l'introduction de *Enaction. Toward a New Paradigm in Cognitive Science*, op. cit. pp. vii–xvii.

[84] Varela explique bien les choses, de façon claire et synthétique, dans son livre *Invitation aux sciences cognitives*, op. cit. Mais on peut dire pour résumer que le "computationnisme" constitue la version pure et dure du cognitivisme; il est l'héritier direct de la première cybernétique, celle des fondateurs, dans les années 1940, 1950, comme Warren MacCulloch, et qui compte notamment parmi ses adeptes, Herbert Simon, Noam Chomsky et Jerry Fodor. Pour cette école, l'esprit fonctionne comme un ordinateur, sur le modèle de la fameuse "machine" d'Alain Turing, qui fut précisément l'un des pères de la cybernétique. Le "connexionnisme", ou "émergentisme", viendra plus tard, et cherchera à s'éloigner de ce modèle, pour profiter des avancées des neurosciences, et des sciences de la complexité, et se fonder de préférence sur l'idée de réseau intelligent, celui qui peut se constituer entre des neurones connectés entre eux: ces neurones interconnectés peuvent ainsi faire émerger un système intelligent, grâce aux lois de l'auto-organisation. La cognition, l'intelligence, serait alors la propriété émergente d'un tout qui est plus que la somme de ces parties. Varela considère qu'il s'agit là de deux approches, qui sont non pas tant erronées, mais plutôt superficielles, et trop théoriques, et qui sont loin d'épuiser ce qu'est réellement l'esprit humain. Il s'agira pour lui de réinsérer ou réinscrire l'esprit non seulement dans son contexte biologique,

démonstrations de *The Embodied Mind* tendent à montrer l'intrication inexorable d'un corps et d'un esprit, pour leur existence mutuelle, en tant que telle. Après avoir montrer la "cognition dans la vie", ou la "vie cognitive" du vivant le plus simple, Varela parviendra ainsi à faire éclater les cadres du *Mind-Body problem* à l'échelle "supérieure", ou si l'on veux, comme le dirait Jean-Pierre Changeux, à l'échelle de "l'homme neuronal". Il n'y pas d'un côté un "esprit-cerveau" qui viendrait se greffer dans quelque chose comme un "corps". La "corporéité de l'esprit" ou son "incorporation" s'avère tellement incontournable, que la séparation entre deux entités distinctes perd tout son sens, parce qu'elle s'avère tout bonnement impossible, d'un point de vue scientifique. Il s'agit au contraire de revenir à *l'expérience vécue* d'un corps-esprit, qui n'a que faire de ses distinctions, puisqu'il vit de tout de façon ce qu'il vit, sans distinction. Ce qui veut dire que le caractère vécu et incarné de l'esprit doit devenir un fil conducteur, qui vient compléter de façon décisive l'investigation scientifique habituelle, "extérieure" (à la troisième personne), "expérimentale" ou "objective". C'est dire à la fois que le corps, avec ses capacités sensori-motrices et affectives, est le berceau de l'esprit, et que "l'expérience à la première personne"[85] est fondamentale et fondatrice d'une science, qui doit venir combler le déficit, dont souffre une démarche scientifique souvent bornée, par l'expérience "morte" de l'empirisme. Ce n'est pas d'un monde qui serait "pré donné", indépendamment du sujet, qu'il faut se soucier, mais de "la structure sensori-motrice du sujet (la manière dont le système nerveux relie les surfaces sensorielles et motrices)".[86] Car cette façon de "faire émerger un monde", qu'est justement une "perception" et une "cognition", "se fait au moyen d'un réseau, et de multiples sous réseaux sensori-moteurs interconnectés".[87] C'est dire aussi que la naissance du sens a lieu ici, et non pas dans les "hautes" sphères "sublimées" de

corporel ou charnel, mais aussi et surtout de revenir essentiellement à l'expérience vécue, en tant que telle, autrement dit, telle qu'elle est vécue par chacun, "à la première personne". D'où l'importance du corps, et surtout du "témoignage" de la personne. Ce qui peut avoir des applications thérapeutiques, qui ressemblent beaucoup aux idées de la "psychologie" et de la "psychiatrie phénoménologiques". [Varela y fait allusion, en évoquant Karl Jaspers et Ludwig Binswanger, au début de *L'inscription corporelle de l'esprit*, op. cit. p. 19, note 2]. Voir aussi notamment les travaux instructifs d'une des disciples de Varela, Claire Petitmengin: *L'expérience intuitive*, Paris L'Harmattan, 2001; "La neuro-phénoménologie: quels enjeux thérapeutiques ?", présenté à l'Université du bouddhisme, 13–14 novembre 2010, sur le thème: "La guérison: le fruit d'une interaction entre le corps et l'esprit ?"; "Un exemple de recherche neuro-phénoménologique: l'anticipation des crises d'épilepsies", in *Intellecta*, No. 40, 2005, pp. 63–89. Pour une liste plus complète, on peut consulter le site: http://claire.petitmengin.free.fr. Enfin, pour les conséquences ou l'impact des sciences cognitives et de leur conceptions, dans le domaine clinique, neurologique et psychiatrique: J. Vion-Dury, "Entre mécanisation et incarnation: réflexion sur les neurosciences fondamentales et cliniques", in *Revue de Neuropsychologie*, Vol. 7, No. 4, 2007, pp. 293–361.

[85] Varela accorde énormément d'importance à cette question cruciale. Cf. *The View from Within: First -person Methods in the Scientific Study of Consciousness*, Exeter, Imprint Academic, 1999.

[86] *L'inscription corporelle de l'esprit*, op. cit. p. 235.

[87] Voir aussi la belle présentation générale, de l'œuvre et de la pensée du biologiste, par Paul-Victor Duquaire: "Introduction à la pensée de Francisco J. Varela. A partir de *Autonomie et connaissance* et *L'inscription corporelle de l'esprit*", in *Les Cahiers de l'ATP*, juillet 2003, p.13.

l'esprit, puisque ce dernier s'enracine dans son existence même, dans sa possibilité ou son impossibilité, là ou se trouvent les premières perceptions ou cognitions, au niveau primordial de nos sensations et de nos perceptions, de nos capacités sensori-motrices. Il est donc bien évident que les liens originaires entre le "sens" d'un coté, et nos sens, indissociables de notre motricité, de l'autre, trouvent ici leur expression la plus éloquente. Ce qui nous explique pourquoi sans cette base, le sommet, l'esprit, avec sa conscience et son intentionnalité, ne peuvent ni fonctionner, ni avoir un sens, ni même exister.

Comme nous le rapporte son ami Evan Thompson, Varela commence à parler, à la fin des années 1980, au moment de la rédaction de *The Embodied Mind,* d' "*enactive approach*", mais pour exprimer en réalité ce qu'il appelait auparavant "*the hermeneutic approach*". Il voulait souligner par cela "*the affiliation of his ideas to the philosophical school of hermeneutics*".[88] Il est très instructif de voir ainsi que ce qui sera développé par la suite, dans maints articles et ouvrages, trouve son origine dans ce souci "herméneutique" ou, pourrait-on dire aussi, "(bio)sémiotique", comme nous allons le voir par la suite. L' "énaction", devenue elle-même paradigmatique dans les neurosciences cognitives,[89] vient accompagner ces théories, en soulignant le fait primordial qu'il y a "*interprétation*" dans l'activité du vivant, comme dans celle de l'esprit, et non pas "représentation", comme le croyait la philosophie traditionnelle et les sciences cognitives "computationnelles". L' "énaction" exprime ce lien essentiel entre *l'action* et *l'interprétation*, qui n'est d'ailleurs pas sans rapport avec le fait que la sensorialité et la motricité, à l'échelle cellulaire, animale et humaine, sont indissociables, et constituent ensemble la sensori-motricité fondamentale, sur laquelle se fonde la totalité indissociable du vivant, ou son "corps-esprit". Dans cet article, "*Sensorimotor Subjectivity and the Enactive Approach to Experience"*, Evan Thompson nous offre comme une explication rétrospective, qui nous assure justement que, dans cette perspective, l'esprit "*does not process information in the computationalist sense, but creates* meaning"; selon l'approche "énactive", "*the human mind is embodied in our entire organism and embedded in the world, and hence is not reducible to the structures inside the head"*. Il y a en réalité trois modes d'activités corporelles qui se trouvent impliquées, ou qui sont à la base de notre "vie mentale": "*our mental lives involve three permanent and intertwined modes of bodily activity – self-regulation, sensorimotor coupling, and intersubjective interaction*".[90] Si l' "auto-régulation" est là pour nous maintenir en vie, pour la régulation de la faim et de la soif, du sommeil et de la veille, le "*sensorimotor coupling with the world*" est cet échange permanent qui se fait avec le monde, ce travail continu d'exploration active, qui s'exprime dans l'émotion, la perception et l'action. L'intersubjectivité, quant à elle, constitue le monde de notre cognition et de nos interactions avec les "autres", avec tout ce qu'il comporte d'expérience affective avec soi et avec l'autre. Mais ce qu'il faut souligner en même temps, c'est que

[88] Evan Thompson, "Sensorimotor Subjectivity and the Enactive Approach to Experience", in *Phenomenology and the Cognitive Science*, 4(4), 2005, pp. 407–427, p. 21, note 1.

[89] Voir les textes très riches de *Enaction. Toward a New Paradigm in Cognitive Science*, op. cit.

[90] "Sensorimotor Subjectivity and the Enactive Approach to Experience", op. cit. p. 108.

le "*sense-making*" ou le "*meaning-making*" trouve précisément ces origines, dans cette auto- production-régulation, dans ce "dialogue" avec soi, avec les autres et avec le monde, dans cette dialectique fondamentale et originaire, bien avant l'émergence du langage articulé et de la pensée conceptuelle.

Comme nous l'avons vu, le thème de l'*Embodiment* cache au fond et révèle paradoxalement une problématique radicale quelque peu inattendue: la question du sens. C'est donc en tout et pour tout ce "*bodily basis of meaning*", entrevu de façon originale par Marc L. Johnson, dans les années quatre vingt du siècle dernier, qui exprime le mieux en réalité les tenants et les aboutissants de ce qui sera nommé finalement: "*the embodied theory of meaning*". Mais c'est justement au travers de cette "corporéité du sens" ou de ce "sens du corps", que nous pouvons aller encore plus loin, et découvrir un monde encore plus vaste, celui de la vie, et de la sémiotique qui s'intéresse au vivant. Grâce à un concours de circonstances et à la détermination de biologistes exigeants et résolus, qu'une "biologie sémiotique", ou "Biosémiotique" [*bios:* vie, *semion*: signe], voit le jour progressivement, et parvient ainsi à "sémiotiser" la biologie, pour aller ensuite jusqu'à "biologiser" la sémiotique. En effet, après une histoire longue et riche intellectuellement,[91] et grâce à une réunion quelque peu improvisée de plusieurs grands biologistes, qui avaient travaillé sur la question, chacun à sa façon, une nouvelle discipline a comme "atteint sa majorité" à Prague, en 2004,[92] en dépit de la diversité des approches, et malgré le doute et l'embarras qu'elle suscitait alors. Car, dès lors qu'il s'agit de parler de "signification" ou de "sens" dans les sciences, la démarche paraît suspecte, et est considérée comme une atteinte à l'intégrité du savoir scientifique. Malgré tout, chacun viendra à Prague avec toute sa science et son lot de pensées et d'idées philosophiques. Ainsi sera-t-il question tantôt de Martin Heidegger ou de Hans Georg Gadamer, tantôt de Charles Sanders Peirce, tantôt d'herméneutique, tantôt de sémiotique,[93] mais il y aura toujours au centre la réunification de cette approche de la biologie, qui débordait depuis quelques décennies sur ses propres découvertes, et inspirait tout le monde, sans faillir.

Mais qu'est-ce donc que cette "Biosémiotique" (*Biosemiotics*)? Si le naturaliste estonien Jacob von Uexküll est considéré comme un père fondateur, qui ne le savait pas encore, le mot "*Biosemiotics*" apparaît pour la première fois sous la plume de F.S. Rothschild, en 1962; et il fut employé dans la littérature sémiotique russe, par Yuri Stepanov, à partir de 1971. Mais il ne sera vraiment introduit, dans les travaux de recherches internationaux, que par le linguiste et sémioticien américain Thomas

[91] Voir l'article de Donald Favareau, "The Evolutionary History of Biosemiotics", in Marcello Barbieri (ed.), *Introduction to Biosemiotics. The New Biological Synthesis*. Dordrecht, Springer, 2008.

[92] Cf. Marcello Barbieri, "Biosemiotics: A New Understanding of Life. Review", *Naturwissenschaften*, 95, 2008, pp. 577–597. Voir p. 596.

[93] Anton Markos avait même préféré au départ le terme *Biohermeneutics*, parce qu'il voulait se rapprocher de la philosophie herméneutique, tandis que Marcello Barbieri proposait l'expression *semantic biology* ou *Biosemantics* en faisant référence à l'idée de "*science of biological semiosis*"; mais il parviendront finalement à un accord autour du terme *Biosemiotics*. Cf. "Biosemiotis: A New Understanding of Life. Review" op. cit. p. 596.

A. Sebeok, en 1986.[94] On peut définir cette science principalement de trois façons: (1) "L'étude des signes, de la communication et de l'information dans les organismes vivants".[95] (2) "La biologie qui considère et interprète les systèmes vivants comme des systèmes de signes".[96] (3) "L'étude scientifique de la *biosemiosis*",[97] c'est-à-dire l'étude du processus signifiant, de l'activité du signe biologique. Il s'agit en outre d'une "réunification moderne de la biologie", qui doit "se fonder sur la nature fondamentalement sémiotique de la vie".[98] On peut dire d'ailleurs qu'aux yeux de cette science le sens ou la signification est le caractère fondamental des systèmes vivants, et peut même être considéré comme une définition de la vie. De ce fait, la Biosémiotique peut être vue comme étant "à la racine et de la biologie et de la sémiotique (*as a root of both Biology and Semiotics*) plutôt qu'une branche de la sémiotique"[99]; ce qui est pour le moins important, pour comprendre la portée épistémique et philosophique du tournant amorcé par la biosémiotique, après celui de l'*Embodiment.*

Mais c'est peut-être Jesper Hoffmeyer, qui prédise à l'heure actuelle la société savante de *Biosemiotics,*[100] qui incarne probablement le mieux la synthèse la plus large et la plus philosophique, au sein de cette constellation. L'une des questions majeures qui le préoccupe est d'ailleurs la question de *l'intentionnalité.*[101] On voit bien, d'emblée, que l'enjeu est éminemment philosophique; et dans un souci pédagogique, pour retrouver les arrières fonds de cette problématique, on le voit revenir,

---

[94] Cf. l'article de Jesper Hoffmeyer "Biosemiotics", in P. Bouissac (ed.), *Encyclopedia of Semiotics*, New York, Oxford University Press, pp. 82–85.

[95] "The study of signs, of communication, and of information in living organism". In *Oxford Dictionary of Biochemistry and Molecular Biology*, Oxford, Oxford University Press, 1997, p. 72.

[96] "Biology that interprets living systems as sign systems". In C. Emmeche, K. Kull, F. Stjernfelt, *Reading Hoffmeyer. Rethinking Biology,* Tartu, Tartu University Press, 2002. p. 26.

[97] "The scientific studies of biosemiosis". in *Reading Hoffmeyer. Rethinking Biology*, op. cit. p. 9.

[98] Cf. Jesper Hoffmeyer "Biosemiotics: Towards a New Synthesis in Biology?", in *European Journal for Semiotics Studies*, 9(2), 1997, pp. 355–376.

[99] Cf. A. Sharov "From Cybernetics to Semiotics in Biology", in *Semiotica*, 120(3/4), 1998, pp. 403–419. Voir pp. 404–405.

[100] Il s'agit de la *International Society for Biosemiotics Studies*, qui publie sur son site Internet des textes d'un grand intérêt, et informe régulièrement sur ces activités et ses conférences internationales: www.biosemiotics.org. Et nous pouvons lire à la page d'accueil: "the interdisciplinary research project of biosemiotics is attempting to re-open the dialogue across the life sciences – as well as between the life sciences and the humanities – regarding what, precisely, such ineliminable terms as "meaning" and "significance" might refer to the context of living, complex adaptive systems. [...] Most fundamentally, the Society considers that one of its most important purposes is the promotion of a cross-disciplinary exchange of ideas between researchers who are actively studying any of the myriad forms of organismic sign use found throughout the natural and cultural world. *ISBS* thus welcomes the membership and collaboration of scholars from all relevant disciplines, including biology, philosophy, ethology, cognitive science, anthropology, and semiotics".

[101] "Evolutionary Intentionality", in Cf. E. Pessa, A. Montenanto and M.P. Penna (ed.), *Proceedings from the Third European Conference on Science Systems, Rom 1–4 Oct. 1996*, Rom, Edisioni Kappa, 1996, pp. 699–703. "The Natural History of Intentionality. A Biosemiotics Approach", in T. Shilhab et al. (ed.), *The Symbolic Species Evolved, Biosemiotics*, 6, Springer, 2012, pp. 97–116: www.springerlink.com

dans une admirable synthèse, non seulement à Husserl et à Brentano, mais aussi à Saint-Thomas d'Aquin et à la philosophie médiévale![102] Par ailleurs, l'une des figures principales auxquels il fait référence est le philosophe américain, père fondateur du pragmatisme, Charles Sanders Peirce. Mais il sera question surtout, à vrai dire, comme on pouvait s'y attendre, puisqu'il s'agit de biosémiotique, de la *sémiotique* de Peirce. En effet, Hoffmeyer et une bonne part du courant biosémiotique actuel choisiront, non pas le dualisme de Ferdinand de Saussure, celui du "signifiant" et du "signifié", mais la version peircienne "triadique" de *l'interprétation.*[103] Dans un article très récent intitulé "*The Natural History of Intentionality. A Biosemiotic Approach*", il parvient à rassembler une bonne part de son œuvre et à synthétiser les différentes problématiques, que recueille la biosémiotique et auxquelles elle est confrontée. Tout d'abord, il faut dire que l'interprétation, le fait d'interpréter ou "l'acte interprétatif" (*interpretative act*), dans le "réalisme sémiotique" (*semiotic realism*) de Peirce,[104] est la pierre angulaire sur laquelle se construit en majeure partie la théorie sémiotique appliquée du vivant. Mais il ne s'agit pas seulement à ses yeux, comme nous l'avions déjà dit, d'une recherche ou d'une théorie "appliquée", mais d'une problématique fondamentale qui engage toute la sémiotique, qui se trouve elle-même de ce fait refondée sur la base de l'acte d'interprétation propre au vivant. Autrement dit, si le fait d'interpréter a une réalité dans notre vie "consciente", c'est parce que la "vie", au seuil de son existence, est déjà, fondamentalement, un processus interprétatif, bien avant le langage et la pensée. Une interprétation plus profonde que celle de l'esprit traverse de part en part le vivant et la vie, bien avant que nous en soyons "conscients", dans le cadre des sciences du langage ou des sciences de l'esprit. Et son intentionnalité est bien plus originaire que tous nos actes intentionnels, nos pensées ou nos volontés, elle est comme une "cause originelle" qui se déploie et travaille en nous, comme dans un *projet herméneutique* d'écriture et de lecture, qui parvient jusqu'à devenir conscient de lui-même.

Le concept peircien de signe constitue en réalité une relation "triadique" entre un "*representamen*" (ou "*véhicule de signe*"), un "*objet*" et un "*interprétant*". Le plus important est de dire justement que "celui qui interprète", l' "interprétant", est d'abord une *activité*, qui peut être "consciente", "instinctive" ou, plus simplement et fondamentalement, "sensori-motrice".[105] Il ne s'agit donc pas nécessairement d'un "esprit" ou d'un "sujet conscient", mais bien plus radicalement d'*un processus vital d'interprétation*, c'est-à-dire de sensation, de perception, d'émotion, de "compréhension", plus ou moins grande, plus ou moins petite, à partir d'un certain

---

[102] "The Natural History of Intentionality", op.cit. pp. 97–105.

[103] Il faut dire cependant que ça n'a pas toujours été le cas: en effet, jusqu'aux années soixante la *Biosemiotics* était encore profondément divisée entre ceux qui avaient choisi le modèle de Saussure, avec son "signifiant" et son "signifié", et ceux qui défendaient la validité du système peircien du signe "triadique"; mais progressivement, et surtout à partir des années 1990, grâce notamment aux efforts de Thomas Sebeok, le modèle de Peirce est adopté, et devient la référence pour la communauté des "biosémioticiens". Cf. "Biosemiotics: A New Understanding of Life. Review" op. cit. pp. 494–595.

[104] "The Natural History of Intentionality", op.cit. p. 101.

[105] Ibid. p. 107.

"point de vue", celui d'une fourmi, d'un dinosaure, d'une amibe, d'Einstein ou de Proust! L'exemple que prend Hoffmeyer est d'ailleurs assez éloquent, et presque comique: fumer (véhicule de signe), feu (objet) et peur (interprétant). On peut voir ainsi comment le fait de fumer peut inspirer, par exemple, une certaine peur, en évoquant, d'une façon ou d'une autre, le risque d'incendie: "*smoke may act as sign (vehicle) that evokes a sense of fear by making us aware of the risk of burning*". Le signe selon Peirce, c'est cette relation "triadique" qui connecte un "véhicule de signe" (*sign vehicle*) (1) et un objet (2) à travers un "interprétant" dans un système réceptif (3): "*sign [is] a triadic relation connecting a sign vehicle with an object through the formation of an interpretant in a receptive system*".[106] Mais, de ce point de vue-là, une autre question nous guète et nous attend au tournant, celle qui a trait à l' "objet"; s'agit-il de l'objet au sens traditionnel, par opposition au sujet? Il importe donc de souligner qu'il s'agit aussi d'une "activité", ou d'un "phénomène", et non d'une chose qui se prétendrait "objective": elle est seulement ce que perçoit, au sens le plus élémentaire, chaque "forme de vie". Il s'agit en l'occurrence du feu, tel qu'un homme le perçoit, et non pas du feu comme le voit une mouche, ou tel qu'il peut être vu à travers un microscope électronique! Ce qui nous amène à dire qu'il s'agit à chaque fois d'un monde différent, pour une perception différente. Pour cela, la notion d'*Umwelt*, fondamentale chez Jacob von Uexküll, mais qui se trouve aussi admirablement chez Husserl,[107] permet à la biosémiotique de se fonder sur une distinction essentielle, qui montre à chaque fois, qu'il n'y a pas de monde "objectif", et de vérité "objective", de vérité "par correspondance" ou de "représentation mental" au sens traditionnel. Il s'agit toujours d'un "monde propre" au percevant, autrement dit de *perspective*, qui ne correspond à rien d'autre qu'au point de vue de celui qui sent, perçoit, mais aussi et surtout *interprète* ces "objets", qu'il a lui-même constitués, et compris à sa façon. Ce "perspectivisme" n'est pas sans rappeler Nietzsche et tout ce qui se tourne autour de la notion de "volonté de puissance", et

[106] Idem.

[107] Voir le même livre de Adam Christopher Konopka, *An Introduction to Husserl's Phenomenology of* Umwelt op. cit. Pour Jacob von Uexküll, c'est une notion centrale, puisqu'il ira jusqu'à fonder, en 1926, un institut, un centre de recherche à l'Université de Hambourg, qui sera nommé *Institut für Umweltforschung* (Institut pour la recherche sur *l'Umwelt* (le milieu, l'environnement)). Il s'agit en réalité d'une véritable exploration du monde (Welt) propre à chaque organisme vivant, c'est-à-dire l'environnement tel qu'il est *perçu*, *subjectivement*, par chaque être vivant. Marqué par l'œuvre de Kant, c'est en 1909, qu'il introduit cette notion, dans un ouvrage intitulé *Umwelt und Innenwelt der Tiere* (Berlin, Springer, 1909). On peut dire aussi que la physiologie sensorielle, qui sera l'une des bases de ses travaux, lui permettra justement d'étudier, entre autres, l' "espace subjectif" des animaux. Chaque "monde" sera ainsi considéré comme une "création" du système sensoriel ou perceptif et moteur de chaque animal. Dans chaque cas, nous aurons donc des "signes" différents, mais des signes quand même, qui viennent d'une "interprétation", à chaque fois différente, qui constitue les "liens" et le "dialogue" qui se créent entre un organisme et son environnement. Voir par exemple: Kull Kalevi, "Jacob von Uexküll. An Introduction", in *Semiotica*, 134 (1/4), 2001, pp. 1–59; Gudrun von Uexküll, *Jacob von Uexküll – Seine Welt und seine Umwelt*, Hamburg, Wegner, 1964; Thure von Uexküll, "The Sign Theory of Jacob von Uexküll", in M. Krampen, K. Oehler, R. Posner, T. A. Sebeok (ed.), *Classics of Semiotics*, New York, Plenum Press, 1987, pp. 147–179.

l'on voit qu'il y a là comme un "malin génie de l'herméneutique"[108] ou de la sémiotique, qui donne comme une portée ou un fondement "ontologique", ou plus exactement, comme chez Nietzsche, "anti-ontologique", au discours de la biosémiotique.

La thématique de l'intentionnalité que nous avions évoquée plus haut prépare ainsi son avènement impérieux. Hoffmeyer, qui revient à Husserl et à Brentano, pour rappeler le lien fondamental et exclusif, pour la future "phénoménologie", entre le mental et l'intentionnel,[109] veut parvenir plutôt à une "intentionnalité évolutive" (*evolutionary intentionality*), qui permet d'expliquer tout ce qui s'est passé au cours de l'évolution naturelle, sans avoir besoin pour cela de discontinuités ou de ruptures "surnaturelles". Il s'agira donc pour lui non seulement de parler d'intentionnalité "corporelle" ou "animale", mais d'aller encore plus loin:

> Rather than pursuing the question of animal intentionality (...) I shall address the question of intentionality as an even more general category of life, an evolutionary "aboutness" or evolutionary intentionality, i.e. the anticipatory *power* implicitly present in all [living] systems.[110]

Cette "puissance" dont il parle dans ce texte, "implicitement présente dans tous les systèmes vivants", pousse Hoffmeyer à rappeler avec bonheur que Merleau-Ponty considérait la "conscience" comme étant originairement un "je peux" et non pas un "je pense que".[111] Mais cette puissance a elle aussi une intentionnalité, qui est antérieure à celle de la conscience et qui en même temps la constitue, ou constitue sa base, et elle n'est pas sans rappeler, comme nous l'avions dit ailleurs, une certaine "volonté de puissance", avec son intentionnalité à elle, au niveau le plus élémentaire.[112] Ce qui veut dire, pour Nietzsche comme pour la biosémiotique, que l'intentionnalité de l'esprit humain n'est pas un "fantôme" venu de rien, mais a évolué, a émergé de quelque chose d'autre, et "devait être en germe"[113] dans quelque chose de plus général:

> Human intentionality has emerged as a peculiar corporeally individualised instantiation of a more general thirdness which is embedded as an irreductible element in the process of organismic evolution: evolutionary intentionality.[114]

Sans vouloir revenir aussi sur les "systèmes dissipatifs" de Ilya Prigogine, auxquels fait allusion l'auteur, et qui peuvent approfondir la question de l'ordre et du chaos,

---

108 Cf. Geneviève Hébert, "Nietzsche, Malin Génie de l'herméneutique ?", in Jean Greisch (dir.), *Comprendre et Interpréter: le paradigme herméneutique de la raison*, Paris, Beauchesne, "Philosophie" 15, 1993, pp. 311–341.

109 "The Natural History of Intentionality", op. cit. p. 103–104.

110 "Evolutionary Intentionality", op. cit. p. 701.

111 Idem.

112 Cf. nos articles: "Le nihilisme et l'épuisement: Heidegger et Nietzsche" op. cit.; "Nietzsche and the future of Phenomenology" op. cit.

113 Evolutionary Intentionality, op. cit. p. 701.

114 Ibid. p. 703.

même à l'échelle physico-chimique,[115] il est clair que Hoffmeyer veut parvenir à un approfondissement radical de cette problématique, en prenant en charge tout ce qui explique l'émergence naturelle, et non *ex nihilo*, d'un phénomène, celui de "l'intentionnalité humaine", dans un monde vu comme "matériel", et essentiellement non intentionnel.

Mais il apparaît surtout que la Biosémiotique porte finalement la problématique de *l'Embodiment* à sa plus grande radicalité ou généralité, et devient à la fois une "histoire naturelle de l'intentionnalité", une "histoire naturelle de la corporéité"[116] (*a natural history of embodiment*), et une "histoire naturelle du signe"[117] (*a natural history of sign*), reliées essentiellement à la question du sens. Cette "historicité" du sens, et sa dépendance, à chaque fois, de chaque organisme vivant, a mené la Biosémiotique à l'élaboration d'une véritable "théorie de la signification", qui rejoint l'intuition de son plus grand précurseur Jacob von Uexküll.[118] De quoi s'agit-il? Si la "corporéité de l'esprit" (*embodiment of mind*) nous a conduit finalement, comme nous l'avions dit, à reconnaître aussi la "corporéité de la signification"[119] (*embodiment of meaning*), et de là à découvrir le "sens du corps" (*meaning of body*), la Biosémiotique nous invite à venir à la rencontre de la "naissance du sens"[120] (*birth of meaning*), au raz de son existence, en élaborant une théorie plus générale encore, non seulement en partant du corps et de ces fondements sensori-moteurs, émotionnels et signifiants, mais plus radicalement, à partir de la vie elle-même, et de sa signification, dès sa plus simple expression naturelle. Ce qui est signifiant, sémiotique, c'est donc pour elle tout ce qui vit, et non ce qui raisonne ou calcule, comme l'ont cru les adeptes du "cognitivisme" ou du "computationnisme". Ils se trompaient à vrai dire doublement: il ne s'agit même pas exclusivement de l'esprit humain, ni même de son corps, mais de toute vie dans ce monde, qui se donne ou projette une signification, dans un monde, qui, en lui-même, objectivement, est

[115] Cf. Ilya Prigogine and Isabelle Stengers, *Order Out of Chaos. Man's New Dialogue with Nature*, Toronto, Bantam Books, 1984.

[116] "The Signifying Body. A Semiotic Concept of Embodiment", *Diagrammatology*, pp. 257–273.

[117] Ibid. p. 271.

[118] Cette idée se trouve essentiellement dans un texte de Uexküll considéré plus ou moins comme le texte fondateur de la Biosémiotique; il s'agit de: *Bedeutungslehre*. – BIOS. Abhandlungen zur theoretischen Biologie und ihrer Geschichte 10, Leipzig, Barth, 1940; traduction anglaise: "The Theory of Meaning", *Semiotica*, 42(1), 1982, pp. 25–82; traduction française "Théorie de la signification", in *Mondes animaux et Monde humain*, Hambourg, Gonthier, 1956–1965, pp. 83–155. Voir à ce propos: "A Natural Symphony? To What Extent is Uexküll's *Bedeutunsglehere* Actual for the Semiotics of our Time?", *Semiotica*, 134 (1/4), 2001, pp. 79–102; on peut consulter également un numéro spécial, de la même revue, consacré à cette même "Théorie" (l'éditeur de ce numéro est son fils): Thure von Uexküll (ed.), "Jacob von Uexküll's "The Theory of Meaning"", *Semiotica*, 42 (1), Special Issue, 1982.

[119] Voir aussi à ce sujet l'article de Roel Kerkhofs et Willem F. G. Haselager, "The Embodiment of Meaning", in *Manuscrito*, 29(2), 2006, pp. 753–764.

[120] "The Signifying Body. A Semiotic Concept of Embodiment" op. cit. p. 270.

insignifiant.[121] C'est cette relativité du *monde* et du *sens, généralisée à toutes les échelles des êtres vivants*, qui est peut-être l'idée plus la plus importante. Si la Biosémiotique rejette par avance tout vitalisme, comme tient à le rappeler Hoffmeyer lui-même,[122] elle parvient ainsi à définir le dénominateur commun de toutes les espèces, de tous les êtres vivants: le sens. Elle montre comment *la raison* ou *le langage de la vie* (*the* Logos *of the* Bios) est bel et bien là, avant nous, et il est, pour ainsi dire, déjà "parlant", bien avant l'apparition des formes linguistiques "supérieures", des formes scientifiques, culturelles ou religieuses, ou, comme le dit Ernest Cassirer, des "formes symboliques".[123] Dans cet étrange "retour à soi", auquel nous invite la Biosémiotique, une science vient nous rappeler nos origines oubliées et pourtant fondatrices, mais qui ont été si longtemps négligées, au profit d'un esprit, "déraciné", "exilé", ayant "rompu" ses liens avec ses sources naturelles, celles qui, pourtant, nourrissaient sa vie depuis toujours. L'interprétation originaire, le processus vital interprétatif, est donc non seulement déjà "intentionnelle", à sa manière, elle est à l'origine de l'intentionnalité de la conscience, à l'origine de la signification et du sens. Dans sa version la plus englobante, la Biosémiotique se voit ainsi comme une "sémiotique générale", et la traditionnelle sémiotique, qui étudie les systèmes humains des signes, est vue alors comme une partie spéciale: l'anthropo-sémiotique.

Ce qui apparaît clairement des thématiques de l'*Embodiment* et de la *Biosemiotics* est tout d'abord que c'est le corps, fondamentalement, qui porte et apporte, ou n'apporte pas, avec lui le sens: il a comme une *primauté* sur l'esprit, d'autant plus que tout ce que nous nommons "esprit", "raison", "langage", "concepts", et même "logique" et "mathématiques", est éminemment *embodied* ("incorporé"), dans son être comme dans son devenir. Le corps va ainsi jusqu'à "*modeler* l'esprit" (*shapes*

---

[121] C'est ce point précisément qui rapproche le plus la linguistique cognitive, l'*Embodiment* et la Biosémiotique, et leurs auteurs et leurs problématiques, de l'œuvre et de la pensée de Nietzsche... et ce n'est pas moins qu'une "révolution copernicienne", qui va plus bien loin que celle de Kant, et nous mène au "paradigme herméneutique de la raison", au "malin génie de l'herméneutique", après nous avoir dit, au fond, la "mort de Dieu" et celle du "monde intelligible", de la "vérité objective", du "monde objectif", de la "logique", de la "grammaire", et tout ce qui tourne autour depuis des siècles... et toutes ces "idées" n'étaient que d'admirables créations faites pour l'homme et par l'homme, pour défendre ses intérêts les plus profonds... Mais il faudrait pour cela, au moins, un autre article, si ce n'est beaucoup plus...

[122] "The Natural History of Intentionality", op. cit. p. 105: "Biosemiotics is not a new version of vitalism". (Il renvoie également à son article: "A Biosemiotic Approach to Health", in S. Cowley, et al. (ed.) *Signifying Bodies. Biosemiosis, Interactions, and Health*, Braga, The Faculty of Philosophy, Portuguese Catholic University).

[123] Voir le très bel article de Andreas Weber, consacré aux liens entre l'œuvre de Cassirer, sa "philosophie des formes symboliques", et les idées de l'ancêtre de la Biosémiotique, Jacob von Uexküll: "Mimesis and Metaphor: The Biosemiotic Generation of Meaning in Cassirer and Uexküll", in *Sign System Studies*, 33, ½, 2004, pp.297–307. Il dit cependant dans une formule saisissante, p. 300: "What Cassirer is missing in his theory of man as an "animal symbolicum" precisely is the animal. For this reason he is interested in Uexküll. And it is here where a further biosemiotic deepening can make Cassirer's already acclaimed philosophy still more important. We only have to follow the way he himself indicated: as Cassirer stresses, any critique of culture ultimately must be grounded on a critique of perception".

the mind),[124] et peut lui apporter bonheur ou malheur, avant même qu'il ne s'en rende compte, qu'il ne sache pourquoi et comment. La Biosémiotique, quant à elle, nous apprend finalement le respect ou, au moins, la prise en considération de la vie, en tant que telle: elle accentue ou radicalise l'approche de l'*Embodiment*, en donnant, non seulement au corps, mais aussi à la vie la plus simple, la capacité d'être sensible, créatrice et signifiante. Cette nouvelle science reprend l'héritage qui provient de la sémiotique, avec sa théorie du signe et de l'interprétation, mais pour enraciner ces activités humaines, dans le travail de la vie la plus simple. La vie elle-même est dès lors considérée non seulement comme la source originaire de l'interprétation et de la signification, mais aussi comme ce qui est définie par sa capacité à engendrer un "monde" doué d'un "sens", qui lui est propre, et qui sera son horizon, son champ de vie et d'action.

Mais cette source essentielle, qu'elle soit notre corps ou la vie en général, est fragile, sensible, et plutôt silencieuse: elle a besoin d'être au moins entendue, pour qu'elle puisse avoir réellement, dans notre existence, une présence véritable. Elle est aujourd'hui souvent broyée ou étouffée, d'autant plus qu'elle est bien moins "bavarde" ou "bruyante" que les discours du langage articulé, de la pensée discursive, ratiocinante; elle peut être aussi facilement dirigée dans la mauvaise direction, ce qui lui fait perdre à chaque fois toute sa saveur, et surtout, tout son sens! Et c'est, hélas, l'esprit, la conscience, la raison et le langage des hommes qui peuvent la conduire à sa perte, dans ce qu'elle a de plus précieux à dire. Ce qui risque de se produire, et se produit déjà, c'est une dégradation progressive, sournoise et inéluctable de la vie, de l'appareil sensori-moteur, vivant et signifiant, c'est-à-dire celui qui donne vie et sens, avant toute autre considération. Il est antérieur à tout, il est la condition de tout. Ce qui nous guète ou nous menace, il faut bien le dire, c'est donc l'apparition d'un homme vidé de son sens, de son humanité, de sa sensibilité, de toute sa substance, déraciné, désincarné, devenant comme un "esprit sans spiritualité", un peu comme un ordinateur, mauvais et maladroit. Un corps perdant sa "corporéité", une vie perdant sa "vitalité". Disons le encore: notre *sensibilité*, indissociable de notre *motricité*, est en danger, est le fondement même du corps et de tout ce que nous appelons "esprit"; et c'est elle surtout qui donne une signification au monde et aux choses, ou nous prive parfois de ce monde ou de ses choses. Prenons garde, de ne jamais oublier que si nous sommes *insensible* au *sens*, il sera, à coup sûr, insensible à nous. De nos jours, et de plus en plus, un certain "ascétisme mondialisé",[125] du "travail" commercial et industriel, une logique de la rentabilité avec sa curieuse "spiritualité" de la "productivité", "active" dans tous les domaines, s'imposent partout. Mais ils devront bien un jour céder la place à une nouvelle ère, non seulement plus

---

[124] Cf. Shaun Gallagher, *How the Body Shapes the Mind*, op. cit.

[125] C'est l'idée, ou l'une des plus grandes idées, qui se dégage principalement du fameux livre de Max Weber, *L'éthique protestante et l'esprit du capitalisme*. Mais nous avons été ravi de la voir aussi brillamment évoquée, à la fin d'un article consacré à Merleau-Ponty, l'*Embodiment*, l'écologie et... au capitalisme! Cf. John R. White, "Lived Body and Ecological Value Cognition", in Suzanne L. Cataldi and William S. Hamrick (ed.) *Merleau-Ponty and Environmental Philosophy*, op. cit. pp. 177–189 (voir surtout pp. 184–187: "Lived Body and Capitalist Ethos").

clémente ou respectueuse de la vie du corps, mais se *fondant* même sur le respect de sa sensibilité, en choisissant comme fil conducteur, ce *sens* primordial et essentiel de la "corporéité". Si nous voulons que nos vies aient une valeur quelconque, il est grand temps d'en tenir compte. A défaut, ce n'est qu'un simulacre de vie qui nous attend, et qui caractérisera sans doute toute existence humaine. Peut-on espérez, au moins, pour l'instant, que ce que Nietzsche nous dit depuis plus d'un siècle soit enfin saisi? "Par manque de repos, disait-il, notre civilisation court à une nouvelle barbarie".[126]

En somme, quelle que soit notre approche, on ne peut échapper au fait que l'homme plonge ses racines dans la vie d'un corps, dans la vie tout court, dans...son expression la plus simple, bien avant l'émergence de tout ce monde de la "conscience" ou de l' "esprit". Il y a comme une continuité essentielle entre la "vie" et l' "esprit", entre "l'esprit de la vie" et la "vie de l'esprit"; ils sont donc pour toujours indissociables, amoureux, inséparables. Mais l'homme d'aujourd'hui, et bien avant l'invention des ordinateurs ou la création des robots, réussit l'exploit incommensurable du grand écart entre l'esprit et le corps, et ce pour des raisons d'efficacité technique et économique. Un "esprit sans vie", "squelettique", une "vie sans esprit", absurde, insignifiante. Et nous voilà, à présent, dans une grande lassitude, une grande fatigue, un épuisement du corps et une crise de l'esprit, un appauvrissement de l'existence de l'homme, et de toute la vie sur Terre. Mais c'est dire aussi que l'homme est sans aucun doute, potentiellement, quelque chose d'autre, de tout autre, autre chose, en tout cas, que ce qu'il est aujourd'hui: il n'est plus que l'ombre de lui-même, un fantôme qui cherche sa route et même un "supplément d'âme", alors qu'il a perdu, en cours de route, et son corps et son âme. Et cette promesse mériterait vraiment qu'on s'y attarde, d'autant plus qu'à cause de notre état, nous n'en sommes qu'à peine conscient, et tout ce qu'on peut en dire, aujourd'hui, n'est que l'expression d'une idée vague ou fantomatique. Et ce n'est pas moins que le projet d'une civilisation radicalement autre, mais encore possible et peut-être à venir.

References

Acampora, C.D. 2004. Between mechanism and teleology: Will to power and Nietzsche's gay "science". In *Nietzsche and science*, ed. Gregory Moore and Thomas H. Brobjer. Aldershot: Ashgate.

Al-Saji, A. 2000. The site of affect in Husserl's phenomenology. Sensations and the constitution of the lived body. In *Philosophy today*, SPEP Publications, Vol. 44. Chicago: DePaul University.

Altieri, L. 2009. *Eidos et Pathos. Corporéité et signification. Entre phénoménologie et linguistique cognitive*. Bucharest: Zeta Books.

Barbieri, Marcello (ed.). 2008a. *Introduction to biosemiotics. The new biological synthesis*. Dordrecht: Springer.

[126] *Humain, trop humain* I, Chapitre V, "Caractères de haute et basse civilisation", § 285, « L'inquiétude moderne ».

Barbieri, Marcello. 2008b. Biosemiotis: A new understanding of life. Review. *Naturwissenschaften* 95: 577–597.
Beaune, J.-C. (ed.). 1998. *Phénoménologie et Psychanalyse. Etrange relations*. Seyssel: Champ Vallon.
Bégout, B. 2006. Pulsion et intention. Husserl et l'intentionnalité pulsionnelle. In *La Pulsion*, ed. J.-Ch. Goddard. Paris: Vrin
Bégout, B., N. Depraz, M. Mavridis, and S. et Nagaï, 1995. Passivité et phénoménologie génétique. In *Alter. Revue de phénoménologie*, ed. Landgrebe L., Holenstein E., Yamaguchi I., and Lee N.-i, no. 3, 409–502. Fontenay-aux-Roses: Ecole normale supérieure.
Bermudez, J.L., A. Marcel, and N. Eilan (eds.). 1995. *The body and the self*. Cambridge, MA: The MIT Press.
Bernet, R. 1996. Inconscient et conscience: sur la nature de la pulsion, du désir, de la représentation et de l'affect. In *Création et évènement. Autour de Jean Ladrière*, Editions de l'Institut supérieur de philosophie, ed. J. Greisch and G. Florival, 145–164. Louvain-Paris: Editions Peeters.
Bouissac, P. (ed.). 1998. *Encyclopedia of semiotics*. New York: Oxford University Press.
Brooks, R. 1991. Intelligence without representation. *Artificial Intelligence* 47: 139–160.
Brown, K. 2006. *Nietzsche and embodiment*. Albany: State University of New York Press.
Carman, T. 1999. The body in Husserl and Merleau-Ponty. *Philososphical Topics* 27(2): 205–226.
Cataldi, S.L., and W.S. Hamrick (eds.). 2007. *Merleau-Ponty and environmental philosophy. Dwelling on the landscapes of thought*. New York: New York State University Press.
Chomsky, N. 2012. *Language and mind* (traduction française par Claude Bourgois), *Le langage et la pensée*. Paris: Editions Payot & Rivages.
Chomsky, N. 1979. *Syntactic structures* (traduction française par Michel Braudeau), *Structures syntaxiques*. Paris: Editions du Seuil.
Clark, A. 1997. *Being there: Putting brain, body, and the world together again*. Cambridge, MA: The MIT Press.
Damasio, A. 1994a. *Descartes' error. Emotion, reason, and the human brain*. New York: G. P. Putnam's Sons.
Damasio, A. 1994–2010. *L'Erreur de Descartes. La raison des émotions*. Paris: Odile Jacob.
Damasio, A. 1999a. *The feeling of what happens: Body and emotion in the making of consciousness*. New York: Harcourt Brace.
Damasio, A. 1999–2002. *Le sentiment même de soi*. Paris: Odile Jacob.
Depraz, N. 2001. *Lucidité du corps. De l'empirisme transcendantal en phénoménologie*. Dordrecht: Kluwer.
Dodd, J. 1997. *Idealism and corporeity: An essay on the problem of the body in Husserl's phenomenology*. Dordrecht: Kluwer.
Dupuy, J.-P. 1994. *Aux sources des sciences cognitives*. Paris: La Découverte.
Duquaire P.-V. 2003. Introduction à la pensée de Francisco J. Varela. A partir de *Autonomie et connaissance* et *L'inscription corporelle de l'esprit*. In *Les Cahiers de l'ATP*, juillet.
Emmeche, C., and J. Hoffmeyer. 1991. From language to nature: The semiotic metaphor in biology. *Semiotica* 84(1/2): 1–42.
Emmeche, A., K. Kull, and F. Stjernfelt. 2002. *Reading Hoffmeyer. Rethinking biology*. Tartu: Tartu University Press.
Escoubas, E., and L. Tengelyi. 2008. *Affect et affectivité dans la philosophie moderne et la phénoménologie*. Paris: L'Harmattan.
Favareau, D. 2008. The evolutionary history of biosemiotics. In *Introduction to biosemiotics. The new biological synthesis*, ed. M. Barbieri. Dordrecht: Springer.
Franck, D. 1981. *Corps et chair: sur la phénoménologie de Husserl*. Paris: Editions de Minuit.
Gallagher, S. 1995. Body schema and intentionality. In *The body and the self*, ed. J.L. Bermudez, A. Marcel, and N. Eilan. Cambridge, MA: The MIT Press.
Gallagher, S. 2005. *How the body shapes the mind*. Oxford: Oxford University Press.
Gallagher, S., and Zahavi D. Dan. 2008. *The phenomenological mind: An introduction to philosophy of mind and cognitive science*. London: Routledge.

Gallese, V., and G. Lakoff. 2005. The brain concepts: The role of the sensory-motor system in conceptual knowledge. *Cognitive Neuropsychology* 22: 455–479.

Gendlin, E. 1962. *Experiencing and the creation of meaning*. Evanston: Northwestern University Press.

Gens, J.-C. (dir.). 2008. *La* Krisis *de Husserl. Approches contemporaines.* Revue *Le Cercle Herméneutique*, numéro 10, premier semestre.

Gens, J.-C. (dir.). 2008. La question en retour sur la vie et l'idée husserlienne de la biologie comme science universelle. In *La* Krisis *de Husserl. Approches contemporaines.* Revue *Le Cercle Herméneutique*, numéro 10, premier semestre.

Goldstein, K. 1934. *Der Aufbau der Organismus.* Den Haag: Martinus Nijhoff. 1951. *La Structure de l'organisme.* Bibliothèque de philosophie (traduction française par le Dr E. Burckhardt et J. Kuntz). Paris: Gallimard.

Gramont, J.de. 1996. *Kant et la question de l'affectivité. Lecture de la troisième critique*. Paris: Vrin.

Hébert, G. 1993. Nietzsche, Malin Génie de l'herméneutique? In *Comprendre et Interpréter: le paradigme herméneutique de la raison*, Philosophie, vol. 15, ed. Greisch Jean, 311–341. Paris: Beauchesne.

Heiner B. T. 2008. Guest editor's introduction. The recorporealization of cognition in phenomenology and cognitive science. *Continental Philosophy Review* 41, Springer 115–126.

Hoffmeyer, J. 1997. Semiotic emergence. *Revue de la pensée d'aujourd'hui*, 25–27(6): 105–117.

Hoffmeyer, J. 1995. The semiotic body-mind. In *Essays in honor of Thomas A. Sebeok. Cruzeiro Semiotico, 22–25*, ed. N. Tasca, 367–383. Porto: Almeida Foundation.

Hoffmeyer, J. 1996a. *Signs of meaning in the universe*. Bloomington: Indiana University Press.

Hoffmeyer, J. 1997. Biosemiotics: Towards a new synthesis in biology? *European Journal for Semiotics Studies* 9(2): 355–376.

Hoffmeyer, J. 1998. Biosemiotics. In *Encyclopedia of semiotics*, ed. P. Bouissac, 82–85. New York: Oxford University Press.

Hoffmeyer, J. 1996. Evolutionary intentionality. In *Proceedings from the Third European Conference on Science Systems, Rom 1–4 Oct. 1996*, eds. Pessa, E., A. Montenanto., and M.P. Penna, 699–703. Rom: Edisioni Kappa.

Hoffmeyer, J. 2012. The natural history of intentionality. A biosemiotics approach. In *The symbolic species evolved.* Biosemiotics 6, ed. T. Shilhab et al., 97–116. Dordrecht: Springer.

Hoffmeyer, J. 2013. A biosemiotic approach to health. In *Signifying bodies. Biosemiosis, interaction, and health,* ed. S. Cowley et al.. Braga: The Faculty of Philosophy, Portuguese Catholic University.

Husserl, E. 1976. *La crise des sciences européennes et la phénoménologie transcendantale.* Bibliothèque de Philosophie (traduction française par Gérard Granel). Paris: Gallimard.

Husserl, E. 1989. *Universale Teleologie*, manuscrit E III 5, No. 34, *Husserliana* Tome XV, 593–597 (traduction française par Jocelyn Benoist) Téléologie universelle. In *Philosophie* no. 10, 3–6. Paris: Editions de Minuit.

Hyder, D., and H.-J. Rheinberger (eds.). 2009. *Science and the life-world. Essays on Husserl's crisis of European sciences.* Stanford: Stanford University Press.

Jacob, F. 1970. *La logique du vivant. Une histoire de l'hérédité. Préface de M. Foucault,* Bibliothèque des sciences humaines. Paris: Gallimard.

Janicaud, D. (ed.). 1995. *L'Intentionnalité en question: entre phénoménologie et recherches cognitives.* Paris: Vrin.

Janicaud, D. (dir.). 2008. *Sciences du vivant et phénoménologie de la vie. NOESIS* No. 14. [Revue philosophique du Centre de Recherche d'Histoire des Idées]. Nice: Revue Noésis.

Johnson, M.L. 1981. Metaphor in the philosophical tradition. In *Philosophical perspectives on metaphor*, ed. M. Johnson. Minneapolis: University of Minnesota Press.

Johnson, M.L. 1987. *The body in the mind: The bodily basis of meaning, imagination and reason.* Chicago: University of Chicago Press.

Johnson, M.L. 1993. *Moral imagination: Implications of cognitive science for ethics.* Chicago: University of Chicago Press.

Johnson, M.L. 1997–1998. Embodied musical meaning. *Theory and Practice*, 22–23: 95–102.
Johnson, M.L. 2006. Merleau-Ponty's embodied semantics – from immanent meaning, to gesture, to language. *EurAmerica* 36(1): 1–27.
Johnson, M.L. 2007. *The meaning of the body. Aesthetics of human understanding*. Chicago/London: University of Chicago Press.
Jonas, H. 1966. *The phenomenon of life. Toward a philosophical biology.* New York: Harper & Row. 2001. *Le phénomène de la vie. Vers une biologie philosophique* (traduction française par D. Lories). Bruxelles: De Boeck Université.
Jonas, H. 2000. *Une éthique de la nature*. Paris: Desclée de Brouwer.
Kalevi, K. 2001. Jacob von Uexküll. An introduction. *Semiotica* 134(1/4): 1–59.
Kerkhofs, R., F.G. Willem, and W.F.G. Haselager. 2006. The embodiment of meaning. *Manuscrito* 29(2): 753–764.
Kinsbourne, M. 1995. Awareness of one's own body: A attentional theory of its nature, development, and brain basis. In *The body and the self*, ed. J.L. Bermudez, A. Marcel, and N. Eilan, 205–223. Cambridge, MA: The MIT Press.
Konopka, A.C. 2011. An introduction to Husserl's phenomenology of Umwelt. Reconsidering the Natur/Geist distinction. Toward an environmental philosophy. *Ann Arbor, ProQuest*, UMI Dissertation Publishing.
Lakoff, G., and J.L. Johnson. 1980. *Metaphor we live by*. Chicago: Chicago University Press.
Lakoff, G., and J.L. Johnson. 1999. *Philosophy in the flesh: The embodied mind and its challenge to western thought*. New York: Basic Books.
Lakoff, G., and J.L. Johnson. 2002. Why cognitive linguistics requires embodied realism. *Cognitive Linguistics* 13(3): 245–263.
Lakoff, G., and R. Nuñez. 2000. *Where mathematics comes from: How the embodied mind brings mathematics into being*. New York: Basic Books.
Lakoff, G., and R. Nuñez. 2005. The cognitive foundations of mathematics: The role of conceptual metaphor. In *Handbook of mathematical cognition*, 109–124. New York: Psychology Press.
Larrabee, M.J. 1976. Husserl's static and genetic phenomenology. *Man and World* 9(2): 163.
Lee, N.-i. 1993. *Edmund Husserls Phänomenologie der Instinkte*. Dordrecht/Boston/London: Kluwer.
Maturana, H., and F.J. Varela. 1980. *Autopoiesis and cognition: The realization of the living*. Boston studies in the philosophy of science 42. Dordrecht: Kluwer. 1989. *Autonomie et connaissance: essai sur le vivant.* La couleur des idées (traduction française). Paris: Seuil.
Maturana, H., and F.J. Varela. 1987. *The tree of knowledge: A new look at the biological roots of human understanding*. Boston: Shambhala/New Science Library.
Merleau-Ponty, M. 1945. *Phénoménologie de la perception*, Bibliothèque des Idées. Paris: Gallimard.
Merleau-Ponty, M. 1964. *Le visible et l'invisible*, Bibliothèque des Idées. Paris: Gallimard.
Monod, J. 1970. *Le hasard et la nécessité. Essai sur la philosophie naturelle de la biologie moderne*. Paris: Editions du Seuil.
Montavon, A. 1999. *De la passivité dans la phénoménologie de Husserl*. Paris: PUF.
Montinari, M. 1997. *La volonté de puissance n'existe pas*. Paris: Éditions de l'Eclat.
Moore, G., and T.H. Brobjer (eds.). 2004. *Nietzsche and science*. Aldershot: Ashgate.
Moravia, S. 2002. Contre la phénoménologie: réflexions sur le point de vue des sciences cognitives. In *Le Médecin Philosophe aux prises avec la maladie mentale,* Etudes de Lettres 2–3, ed. R. Celis., H. Mesot. Lausanne: Le Cercle Herméneutique, Collection "Phéno".
Müller-Lauter, W. 1998. Décadence artistique et décadence physiologique. Les dernières critiques de Nietzsche contre Richard Wagner. *Revue philosophique de la France et de l'étranger* 188: 275–292.
Nancy, J.-L. 1973. La thèse de Nietzsche sur la Téléologie. In *Nietzsche aujourd'hui*, vol. I. Paris: UGE.
Nuñez, R. 2006. Do real numbers really move? Language, thought, and gesture: The embodied cognitive foundations of mathematics. In *18 Unconventional essays on the nature of mathematics*, ed. R. Hersh, 160–181. New York: Springer.

Nuzzo, A. 2008. *Ideal embodiment. Kant's theory of sensibility*. Bloomington: Indiana University Press.
Pearson, A.P. 2007. Incorporation and individuation: On Nietzsche's use of phenomenology for life. *Journal of the British Society for Phenomenology* 38(1): 61–89.
Petitmengin, C. 2005. *L'expérience intuitive*, Paris L'Harmattan, 2001; La neuro-phénoménologie: quels enjeux thérapeutiques?, présenté à l'Université du bouddhisme, 13–14 Novembre 2010, sur le thème: La guérison: le fruit d'une interaction entre le corps et l'esprit?; Un exemple de recherche neuro-phénoménologique: l'anticipation des crises d'épilepsies. In *Intellecta*, No. 40, 63–89. http://claire.petitmengin.free.fr
Petitot, J., F.J. Varela, B. Pachoud, and J.-M. Roy (eds.). 1999. *Naturalizing phenomenology. Issues in contemporary phenomenology and cognitive sciences*. Stanford: Stanford University Press.
Petitot, J., F.J. Varela, B. Pachoud, and J.-M. Roy (eds.). 2002. *Naturaliser la phénoménologie. Essais sur la phénoménologie contemporaine et les sciences cognitives*. Paris: Editions du CNRS.
Prigogine, I., and I. Stengers. 1984. *Order out of chaos. Man's new dialogue with nature*. Toronto: Bantam Books.
Reynaert, P. Husserl's phenomenology of the animated being, and the critic of naturalism. http://heraclite.ens.fr/~roy/GDR/Animatedbeing
Rodrigo, P. 2009. *L'intentionnalité créatrice. Problèmes de phénoménologie et d'esthétique*. Paris: Vrin.
Schauvrtz, P.S. 1993. The status of Nietzsche's theory of the will to power in the light of contemporary philosophy of science. *Internationnal Studies in Philosophy* 25: 85–92.
Smith, D.W., and A.L. Thomasson (eds.). 2005. *Phenomenology and philosophy of mind*. Oxford: Clarendon Press.
Steinbock, A. 1995. *Home and beyond: Generative phenomenology after Husserl*. Evanston: Northwestern University Press.
Stewart, J., O. Gapenne, and E.A. Di Paolo (eds.). 2010. *Enaction. Toward a new paradigm for cognitive science*. Cambridge, MA: The MIT Press.
Stiegler, B. 2002. *Nietzsche et la biologie*, Philosophies. Paris: PUF.
Stiegler, B. 2006. *Nietzsche et la critique de la chair. Dionysos, Ariane, le Christ*, Epiméthée. Paris: PUF.
Swift, P. 1999. Nietzsche on teleology and the concept of the organic. *International Studies in Philosophy* XXXI(3): 29–41.
Swift, P. 2005. *Becoming Nietzsche. Early reflections on Democritus, Schopenhauer and Kant*. New York/Oxford: Lexington Books.
Thomson, E. ed. 2003. The problem of consciousness: New essays in phenomenological philososphy of mind, Calgary (Alta.). *Canadian Journal of Philosophy*, Supplementary Volumes.
Thomson, E. 2005. Sensorimotor subjectivity and the enactive approach to experience. *Phenomenology and the Cognitive Science* 4(4): 407–427.
Thomson, E. 2007. *Mind in life: Biology, phenomenology, and the sciences of mind*. Cambridge. MA: Harvard University Press.
Tombomino, J. 2002. *The corporeal turn. Passion, necessity, politics*. Lanham: New York, Oxford, Rowman and Littlefield Publishers.
Toscano, A. 2001. The method of nature, the crisis of critique. The problem of individuation in Nietzsche's 1867/1869 notebooks. *Pli* 11: 36–61.
Uexküll J. von. 1909. *Umwelt und Innenwelt der Tiere*. Berlin: Springer.
Uexküll J. von. 1940. *Bedeutungslehre* – BIOS. Abhandlungen zur theoretischen Biologie und ihrer Geschichte 10. Leipzig: Barth.
Uexküll, J. von. 1956–1965. Théorie de la signification. In *Mondes animaux et Monde humain*, 83–155. Hambourg: Gonthier.
Uexküll G. von. 1964. *Jacob von Uexküll – Seine Welt und seine Umwelt*. Hamburg: Wegner.
Uexküll, Th. von., Thure, von Uexküll ed. 1982. Jacob von Uexküll's "the theory of meaning". *Semiotica*, 42(1) Special Issue.

Uexküll Th. von, Thure von Uexküll. 1987. The Sign Theory of Jacob von Uexküll. In *Classics of semiotics,* ed. M. Krampen, K. Oehler, R. Posner, and T. A. Sebeok, 147–179. New York: Plenum Press.
Varela, F. J. 1979. *Principles of biological autonomy.* Amsterdam: North Holland.
Varela F.J. 1991. Organism: Meshwork of selfless selves. In *Organism and the Origins of the Self,* 79–107. Dordrecht: Kluwer Academic Publisher.
Varela F.J., E. Thompson, and E. Rosch. 1991. *The embodied mind : Cognitive science and human experience.* Cambridge, MA: The MIT Press.
Varela, F. J. 1996. Neurophenomenology: A methodological remedy for the hard problem. *Journal of Conscienceness Studies* 3: 330–350.
Varela, F.J. 1996. *Cognitive science. A cartography of current ideas* (traduction française par Pierre Lavoie). *Invitation aux sciences cognitives.* Sciences. Paris: Editions du Seuil.
Varela, F.J. 1997. The naturalization of phenomenology as the transcendance of nature. In *Alter. Revue de phénoménologie.* 5: 355–388.
Varela, F.J. 1997. Patterns of life: Interwining identity and cognition. *Brain and Cognition* 34: 72–87.
Varela, F.J. 1999. *The view from within: First–person methods in the scientific study of consciousness.* Exeter: Imprint Academic.
Varela, F.J., and A. Weber. 2002. Life after Kant: Natural purposes and the autopoietic foundations of biological individuality. *Phenomenology and the Cognitive Sciences* 1: 97–125.
Vion-Dury, J. 2007. Entre mécanisation et incarnation: réflexion sur les neurosciences fondamentales et cliniques. *Revue de Neuropsychologie* 7(4): 293–361.
Weber, A. 2004. Mimesis and metaphor: The biosemiotic generation of meaning in Cassirer and Uexküll. *Sign System Studies* 33(1/2): 297–307.
Zahavi, D. 1994. Husserl's phenomenology of the body. *Etudes phénoménologiques* 19: 63–84.
Zaner, R. 1964, 1971. *The problem of embodiment. Some contributions to the phenomenology of the body.* Phaenomenoligica 17. The Hague: Martinus Nijhoff.
Zeifa, A. 2003. Le nihilisme et l'épuisement: Heidegger ou Nietzsche. Du nihilisme comme état psychopathologique ou de l'impossibilité du nihilisme ontologique, et ses conséquences médico-philosophiques. In *Phénoménologies des sentiments corporels. Fatigue Lassitude Ennui,* Anthropologie. Paris: Le Cercle Herméneutique.
Zeifa, A. 2009. Merleau-Ponty and the eternal return to the life-world. Beyond existentialism and phenomenology. In *Phenomenology and existentialism in the twentieth century,* Book 2, Analecta Husserliana, vol. CIV, ed. A.-T. Tymieniecka. Dordrecht: Springer.
Zeifa, A. 2011. Nietzsche and the future of phenomenology. Der Wille zur Macht and the criticism of modern transcendentalism. In *Transcendentalism overturned,* Analecta Husserliana, vol. CVIII, ed. A.-T. Tymieniecka. Dordrecht/Heidelberg/London/New York: Springer.
Zeifa, A. 2013. Nietzsche and Merleau-Ponty: The sense of the earth and the earth of sense. In *Phenomenology and the human positioning in the cosmos. The life-world, nature, earth. Book two,* Analecta Husserliana, vol. CXIV, ed. A.T. Tymieniecka. Dordrecht/Heidelberg/London/New York: Springer.

Motion in Crisis: Why the Analytic Principles of Thought Destroy Motion and Life in the Cosmos

Ion Soteropoulos

Abstract By motion in crisis we mean the interruption of motion—that motion is logically impossible or unintelligible because it defies our analytic principles of thought. Motion in crisis originates therefore with the Greek philosopher Zeno (fifth century BCE) who attacked the thesis that there is motion in the universe by deriving contradictory consequences from it. Because life is something kinetic—as a matter of fact life is the power of self-motion—Zeno's attack of motion is essentialy an attack of life within the cosmos. We will show in this paper why our analytic principles of thought are incapable of comprehending motion and life, which means that if we live and move, it is precisely because we do not function according to the analytic principles of our Euclidean individual senses that underlie empirical science.

Courtesy SetarehKorkchi

I. Soteropoulos (✉)
Independent Research Philosopher and Founding Member of the Apeiron Centre, Apeiron Centre, Paris, France
e-mail: ion.2@orange.fr

A.-T. Tymieniecka (ed.), *Phenomenology of Space and Time: The Forces of the Cosmos and the Ontopoietic Genesis of Life: Book One*, Analecta Husserliana 116,
DOI 10.1007/978-3-319-02015-0_28,

The Analytic Principles of Thought Declare Motion From *a* to *b* Impossible

According to the reports of Plato and Aristotle the Greek philosopher Zeno, who was a disciple of Parmenides of Elea, wrote an astonishing treatise in which he attacked the thesis that there is motion in the world by deriving from it contradictions. To obtain contradictions from motion he used the analytic principles of thought, which are abstractions of our Euclidean individual experience. Among the analytic arguments Zeno designed to show that motion is impossible, we will concentrate our attention on the most famous, the *argument of the dichotomy.*

In the eighth book of *Physics*, Aristotle presents two formulations of Zeno's argument of dichotomy.[1] The first, which is the most familiar, states: If we are to traverse the Euclidean distance *ab* of 1 unit, we must first traverse half the distance, or $1/2^1$; and then half of what remains, or $1/2^2$; and again half of what remains, or $1/2^3$; and so on ad infinitum. Thus, an infinite series of halves ($a_1 = 1/2$; $a_2 = 3/4$; $a_3 = 7/8; \ldots$) must be traversed successively if we are to reach the limit $b = 1$. But if an infinite series is a series without limit, how is it possible to traverse an unlimited series and reach the limit $b = 1$? Don't we have here a contradiction between the unlimited series and the imperative requirement to reach the limit that destroys motion along the unit distance *ab*?

Aristotle attempted to solve the contradiction of unlimited/limited by assuming that the unit distance *ab* is *actually* limited and *potentially* unlimited.[2] Thus, when we traverse the distance *ab*, we *actually* and by *essence* traverse with one step or with a finite number of steps the whole distance *ab* and only *potentially* and by *accident* traverse with an infinite number of steps the unlimited series of halves ($0 + 1/2^1 + 1/2^2 + 1/2^3 + \ldots$) (Fig. 1). In this way, motion from $a = 0$ to $b = 1$ is saved at the cost of excluding the cumbersome infinite by assigning to it a potential or accidental ontological status. This solution is incomplete, as it solves the problem of motion under only the particular condition that the infinite is *uniquely* potential and not actual.

We would like, however, to present a universal solution to the problem of motion in which motion takes place under all conditions independent of whether the infinite series of halves is potential or actual. What happens, in fact, if the above particular condition changes and the infinite series of halves is an actual infinite? Can we traverse *actually* an unlimited series and reach *actually* its limit? At this point, our initial contradiction of unlimited/limit reappears in a vicious circle.

---

[1] Aristotle, *Physique* VIII, trans. Henri Carteron (Paris: Les Belles Lettres, 1986), 263 a4 – 11.

[2] Ibid., 263 b3–9.

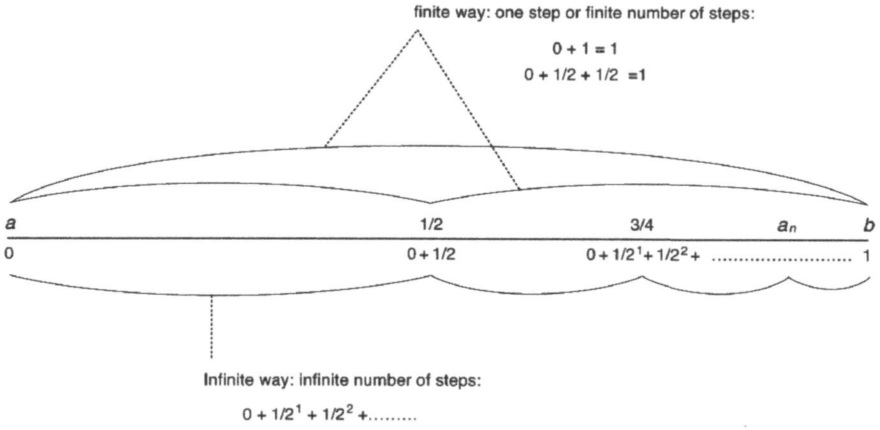

Fig. 1 There are two ways to move along the Euclidean unit distance *ab*: (*a*) the finite way, which consists of one step, for example, 0+1 or a finite number of steps—say, 0+½+½—and (*b*) the infinite way, which consists of an infinite number of steps: for example, $0+½^1+½^2+\ldots$. Because we are finite bodies with finite energy and speed, we perform actually one step or a finite number of steps and only potentially an infinite number of steps. Therefore, we access infinity only potentially and indirectly. Is it possible in the future to access infinity actually and directly?

The Analytic Principle of Inequality and Temporal Order Destroys Continuous Motion

Some mathematicians such as Whitehead, are convinced that Zeno's thesis of the impossibility of reaching the limit in 1s by passing through the unlimited series of halves is based upon a mathematical fallacy that can be resolved with simple arithmetic.[3] To see whether simple arithmetic can lead us to the limit in one instant by passing through the unlimited series of parts, let us try to *actually* pass this unlimited series by adding successively its halves. In the first $1/2^1$ instant, we compute the partial sum $a_1=0+1/2=1/2^1$; in the next $1/2^2$ instant, we compute the partial sum $a_2=0+1/2^1+1/2^2=3/4$; in the next $1/2^3$ instant, we compute the partial sum $a_3=0+1/2^1+1/2^2+1/2^3=7/8$; and so on forever.

What have we obtained from this successive addition of an infinite number of halves? The answer is: merely an infinite sequence of partial sums: $a_1=1/2; a_2=3/4; a_3=7/8; \ldots a_n=2^n-1/2^n$; and so on. The limit and total sum *b* is nowhere. To put it another way, *b* is somewhere but is inaccessible to the indefinitely approaching

[3] See A.N. Whitehead, *Process and Reality* (New York: The Free Press, 1978), Part II, Chap. II The Extensive Continuum, 69.

a_n. Indeed, no matter how close to its limit the converging sequence of partial sums a_n is, a_n *must* always be infinitesimally before and less than its limit and total sum b if the analytic principle of inequality and temporal order is to be satisfied and the Euclidean principle that the whole is greater than any of its proper parts is to be confirmed: $a_n < b$.[4] This means that simple arithmetic employing temporal order can in no way realize the passage from the unlimited series a_n to the limit b. In fact, the best thing we can do with simple arithmetic is to compute up to the partial sum a_n leaving then the total sum b incomputable. It follows that insofar as we perceive through the finite cognitive faculties of our individual sensibility and analytic understanding the multiplicity of things, say a_n and b, as contradictory points existing successively on a Euclidean line, motion between consecutive points, such as the computation of the total sum b by the successive addition of its parts a_n, is an impossible task.

Analytic Convergence and the Apparent Passage to the Limit

Because inequality and succession between the contradictory points a_n and b of the Euclidean line destroy continuous motion from a_n to b, the main question posed by the modern analytic theory of infinite convergent series is how to realize the passage to the limit despite succession, and without refuting the analytic principles of organization of our Euclidean observable world. This is accomplished by reducing inequality and succession into something infinitely small but not minimum. Indeed, any reduction of their inequality and succession into zero means that unequal things are equal, thereby violating the analytic principle of inequality and temporal order that governs the Euclidean observable world. The reduction of the inequality between a_n and b into something infinitely small enables us to count it as if it is zero, and hence *as* if it is an equality, without committing an appreciable error. This apparent equality between a_n and b enables us in turn to replace the variable a_n by its equal and constant b, and therefore to effect the passage to the limit b.

Thus, since the time of Cauchy (nineteenth century), the modern Aristotelian mathematician defines motion from the unlimited series a_n to the limit b by the following formula:

$$\lim_{n \to \infty} a_n = b,$$

where n is a variable, finite number that increases without limit, and ∞ is a constant, infinite number taken as the inaccessible maximum limit of the indefinitely increasing n. This formula gives an analytic definition of the infinite convergent series. The series a_n converges (tends) to its maximum limit b without ever reaching

[4] The analytic principle of inequality and temporal order states that different things are unequal and consecutive. Thus: If $a \neq b$, then either $a < b$ or $b < a$.

b if, and only if, the infinitely small difference $b - a_n = 1/n$ converges (tends) to its maximum limit $1/\infty = 0$ without ever reaching 0 as n of a_n, by passing through the infinite succession of finite numbers, increases indefinitely without ever reaching its maximum limit. We call this unlimited series a_n deprived of its maximum limit b *indefinite* or *potential infinite* (Aristotle). It is a Euclidean *relative infinite* verifying analytic principles of organization, such as the principle of contradiction, which stipulates that nothing unlimited is limited.

The infinitely small inequality between the consecutive terms $a_n < b$ enables us to behave *as if* there is an equality and continuous motion between them. In reality, however, there is a persistent discontinuous inequality that, no matter how small, destroys continuous motion. To put it another way, this infinitely small inequality is simultaneously an infinite inequality generating an infinite separation and distance between a_n and b. Indeed, insofar as the difference $\infty - n$ is always equal to ∞, the distance between n and ∞ is always ∞: that is, infinite or maximum. No matter how great n is, n is as far off from ∞ as the least finite number; hence, n is always the least part of ∞.

Let us, establish a one-to-one correspondence between the infinite sequence of partial sums a_n whose limit is b and the unlimited sequence of integers n whose limit is ∞:

1	2	n	∞
a_1	a_2	a_n	b

If we have $\infty - n = \infty$ and $b - a_n = 1/n$, and if by virtue of the assumed one-to-one correspondence we have

$$n = a_n \text{ and } \infty = b,$$

then, if we replace in the difference $\infty - n$, ∞ by b, and n by a_n, we obtain $b - a_n = \infty$.

We have then:

$$b - a_n = 1/n \text{ and } b - a_n = \infty,$$

which shows that the infinitely small difference and distance between a_n and b is infinite. The ancient problem of motion is posed again: How can we traverse the infinite distance that has no limit and attain the limit b?

To sum up, no matter how much the increasing a_n approaches b, a_n is both at an infinitely small distance from b and at an infinite distance from b similar to the least finite number a_0 of the series of finite numbers a_n. Not only is the indefinitely approaching a_n incapable of reaching its end and last number b at point b, but it is also incapable of moving beyond itself insofar as we regard it as similar to the least finite number a_0 at point a. Continuous motion along the Euclidean unit distance ab is impossible because the indefinitely varying a_n can neither begin nor end its motion. Such is the paradox of motion in which the impossibilities to begin and to end motion are different aspects of impossible motion—of the impossible change of a given state of motion in the immobile Euclidean analytic space.

The analytic theory of the infinite convergent series has failed to deliver the promised solution to the problem of motion because it developed according to our finite mode of perceiving the physical world's multiplicity of things successively in conformity with the analytic principle of inequality and temporal order. Because inequality and succession between different things, for instance $a_n < b$, interrupt continuous motion, the aim of the analytic theory of the infinite convergent series is to allow the discontinuous inequality to become infinitely small so that it can be counted *as if* it is a continuous equality without committing an appreciable error. This attempt has proved unsuccessful, however, as the desired infinitely small separation and distance between a_n and b can in no way neutralize the *actually* infinite separation and distance between n and ∞ and therefore the actually infinite separation between a_n and b. The apparent equality based on the infinitely small difference and inequality between a_n and b disappears to reveal the naked truth: the existence of an actually infinite gulf between a_n and b that no analytic theory of the infinite convergent series can bridge.

What the analytic theory of the infinite convergent series accomplishes is not the solution of the problem of motion; it is merely its ephemeral *regulation* by dissimulating the impossibility of solving it by way of analytic means. In fact, because our finite retinal cells at rest detect uniquely the finite part of light travelling empty space at the finite speed c, we are incapable of perceiving the multiplicity of things $a_n \neq b$ simultaneously as a unitary whole; instead we perceive them successively as contradictory and isolated individuals deprived of unity and motion, and separated by an infinite distance. This shows us that our finite analytic experience, far from being the solution to the crisis of motion, is its very origin!

It follows that solving the problem of motion and therefore of life having the immanent property of motion resides outside the analytic principles of our Euclidean time-conditioned observable world generated by our finite particular senses and their corresponding cognitive faculties of particular sensibility and analytic understanding and inside their negations. I leave to the imaginative reader the freedom to discover them.

"Heraclitus/Nietzsche/Heidegger in Πόλεμς" - "τὰ δὲ Πάντα οἰακίζει Κεραυνός" – Heraclitus

Kimiyo Murata-Soraci

Abstract This paper will attempt to rethink a common ground of kinship and love among myriad lives in a thoughtful remembrance of Nietzsche's conception of nature in *The Birth of Tragedy.* By drawing special attention to an anonymous play of forces and "a fraternal union" of Apollo and Dionysus in the text, we will highlight the shared traits of nothing, irreducibility of the contraries, and *Auseinandersetzung* in the thoughts of Heraclitus, Nietzsche, and Heidegger.

Nihilism is a faceless nemesis we encounter in daily lives. Like anxiety, it silently captures our entire being. Unlike anxiety which enables us to become in touch with an abyss-like ground of human existence and predicament, nihilism shrouds an abyss (*Abgrund*) of life by sticking our face to an illusory surface of life's mere appearances and casts our mindset in the reprehensive mood of "a loss of meaning for life,"[1] and dejection over human impotence in the (re-) creation of world. As nihilism retains traditional axioms of transcendence/immanence and self/the world, it tends to block us from encountering the essence of nothing and the intrinsic interexistence of all lives of things.

In this paper, I should like to retrieve Nietzsche's (1844–1900) thought-experiment of nothing that is "the nature/the Dionysian," in *The Birth of Tragedy* (1872) to highlight his overcoming of nihilism. I will attempt to interweave into our reading Heraclitus' (521–487 B.C.) and Heidegger's thoughts of nature (*Φύσις*)-a nonreflexive process of becoming and *alētheia*, respectively. I recall that in Western philosophy Heraclitus, Nietzsche, and Heidegger share in common (1) love for nothing, (2) eye and ear to a simultaneity of coming and going of finite life, (3) emphasis on an

[1] Charles E. Scott, *The Lives of Things* (Bloomington: Indiana University Press, 2002), p. 185.

K. Murata-Soraci (✉)
Tama University, Fujisawa City, Japan
e-mail: soracik@gmail.com

A.-T. Tymieniecka (ed.), *Phenomenology of Space and Time: The Forces of the Cosmos and the Ontopoietic Genesis of Life: Book One*, Analecta Husserliana 116,
DOI 10.1007/978-3-319-02015-0_29,

irreducibility of the opposition, and (4) wonder at the finite lives' "divine power" in their fiery, self-wrenching emergence with loss and with no purpose.[2]

I recall also their common emphasis on a hearing attunement to the nonpresence of life just as the Heraclitean *Fragments* opens aloud with such a call (Fr. B1).[3] Heraclitus incites his fellowmen to remember the nonpresence of, in, and with all lives' appearances both in their present and absent forms (Fr. B50, B114). To Heraclitus, his people look "uncomprehending" (*ἀξίνετοι, axynetoi*, Fr. B1) of the *Logos*; thereby, they appear out of joint. He wants to help his fellowmen become aware of their being at war with both themselves and the logos in their turning gaze away from the nondeterminate roots of life. Heraclitus wants to help them refigure what it means to bear out a common body of language and memory and how to live genuinely with one another in mindfulness of the self-enactment of life.

If Heraclitus the Obscure stands before Heidegger as "the thinker of *Alētheia,*" Heraclitus appears to Nietzsche as the predecessor of "a Dionysian philosophy" and the teacher of the eternal recurrence.[4] To them both, Heraclitus retains amazingly an awareness of the *Wesen* of the nothing and helps their self-overcoming, through dialogue and genealogy, of a nihilistic history of Being and a nihilistic Socratism and its cultural heritage. I shall now turn to Nietzsche's proto-genealogical work *The Birth of Tragedy* and re-collect the nature of nothing and of a friendship between the Apollonian and the Dionysian forces, and highlight their kinship to Heidegger's understanding of nature (*Φύσις*) and love in his reading of the Heraclitean fragment B16 in "*Alētheia*." I so doing, I wish to pay heed to a Nietzschean way of like-making art which is other to the Platonic and the Aristotelian *mimesis*.

In *The Birth of Tragedy*, Nietzsche assumes nature to be the *genesis* of all lives (*BT*, 38).[5] By nature, he means a reservoir of natural potencies which affect things' organic forms, dispositions, and their range of sensibilities.[6] Unlike the figure of transcendent aseity or fixed substance, Nietzsche's conception of nature is essentially fluid; like the Sōtō Zen Master Dogen's (1200–1253) view of Vast Emptiness (Kokū),[7]

---

[2] I draw attention to Heidegger's explanation of the Greek word Zao with its intensification with *za* in His "*Aletheia*" essay. See Martin Heidegger, *Early Greek Thinking,* trans. David Farrell Krell and Frank A. Capuzzi (San Francisco: Harper & Row Publishers, 1984), p. 116; henceforth, *EGT.*

[3] Fr. B1: "Although this account holds forever, men ever fail to comprehend, both before hearing it and once they have heard. Although all things come to pass in accordance with this account, men are like the untried when they try such words and works as I set forth, distinguishing each according to its nature and telling how it is. But other men are oblivious of what they do awake, just as they are forgetful of what they do asleep." Charles H. Kahn, *The Art and Thought of Heraclitus* (Cambridge: Cambridge University, 1979), p. 62. Fragments cited in this paper are from Diels-Kranz' edition.

[4] Friedrich Nietzsche, *On the Genealogy of Morals, Ecce Homo,* trans. Walter Kaufmann (NY: Vintage Books Random House, 1969), pp. 273–274.

[5] Friedrich Nietzsche, *The Birth of Tragedy and the Case of Wagner*, trans. Walter Kaufmann (NY: Vintage Books, 1967); henceforth *BT*. "...as artistic energies which burst forth from nature herself, without the mediation of the human artist–energies in which nature's art impulses are satisfied in the most immediate and direct way–", p. 38.

[6] Ibid. 1, p. 122.

[7] See my paper "*Wonder of Emptiness*," in *Art, Literature, and Passions of the Skies,* ed. Anna-Teresa Tymieniecka, *Analecta Husserliana, Vol. CXII* (Chicago: Springer, 2012).

nature purely enacts "itself" simultaneously with self-presencing of life, pulsates and infiltrates thoroughly throughout each thing's finite trajectory.[8] Nature occupies no place apart from finitude of life. Just as the Apollonian and the Dionysian artists bear out immediately their respective qualities of creative impulses thus play a role of mere "medium" for the form-creating forces of nature to which their art-forms imitate (*BT,* 52), all living things embody primordial imprints of forces-at-play and unconsciously mirror a silent play of forces. The theatric interplay takes place beneath human conscious and unconscious mentation thus it is other to our memory (anamnesis);[9] yet, constituting a complex of life and saturating all tissues of lived-experience, it is most intimate to our memory in memory's loss of it.

Thus, from an unheard perspectivism of the utterly "unreal" *genesis* and "unreal" appearances of myriad lives, Nietzsche experiments to re-collect the birth and death of Greek tragedy. His retrieval does not aim at a production of objective knowledge about tragedy and its bygone cultural world. Rather, through a combined strategy of his disciplined research and scholarly interpretation of gathered documents with an element of the "fable," he envisions giving an account of an origin from which the current spiritual make-up of his existence and his culture comes to brook. Nietzsche attempts to bring forth a birth of new being beyond a loop of subjectivity and a protocol of "the theoretical man." (*BT,* 94)

According to Nietzsche, the death of tragedy was brought about by a dominant draft of force called "logical Socratism" (*BT,* 89).[10] Socratism stands for a torrent of life-energies which seek to release before a presence of meaning and tends to react repulsively in the face of purely nonteleological coming-and-going (*genesisphthoran*) of life to the extent to redress the *letheic* facets of life (*BT,* 87). Thus, in seeking a site/sight of clarity and wholeness, this complex of forces bears inevitably an *agon* with a natural movement of life both within and beyond itself and seeks to feed its own empowerment by maintaining a constant war against its matrix and by turning gaze and ear away from an abyssal source (*Lethe*) of life (*BT,* 89).[11]

The likes of Euripides and Socrates incarnate the influx of the life-energies; especially in the latter's personality, it blossoms forth into the paradigm of "the theoretical man" (*BT,* 94–95).[12] Socrates's excessive demand of clarity in truth and his demand of accuracy in speaking and thinking through an art of argumentation attests

[8] Friedrich Nietzsche, *The Will to Power*, ed. and trans. Walter Kaufmann (NY: Random House, 1967). Section 1067: "...a play of forces and waves of forces, at the same time one and many...a sea of forces flowing and rushing together ...with an ebb and a flood of its forms."

[9] See Heidegger's discussion of intricate connections between logos, memory, and forgetfulness in his *Parmenides*. Martin Heidegger, *Parmenides*, trans. André Schuwer and Richard Rojcewicz (Bloomington: Indiana University Press, 1992), p. 185.

[10] Nietzsche, *BT*, 5, pp. 88–89. "...the enormous driving-wheel of logical Socratism is in motion, as it were, behind Socrates, and that it must be viewed through Socrates as though a shadow."

[11] Ibid. 5, p. 89. "To this eye was denied the pleasure of gazing into the Dionysian abysses."

[12] Ibid. 5, p. 95. "...a profound illusion that first saw the light of the world in the person of Socrates: the unshakable faith that thought, using the thread of causality, can penetrate the deepest abysses of being, and that thought is capable not only of knowing being but even of correcting it. This sublime metaphysical illusion accompanies science as an instinct and leads science again and again to its limits at which it must turn into art–which is really the aim of this mechanism."

to a violent truncation of the force from its Apollonian stem (*BT,* 91).[13] As Socrates embodies more of the same complex of beyond-/meaning-bound life energies than his friend Euripides (*BT,* 73, 86, 87), we can say that nature supplies not only a figurative account of qualitative and quantitative differences of force in an individual but gives also an account of kinship among individuals even across a genre, just as a friendship between Euripides (art) and Socrates (philosophy) is the case at point.

All living things wear thus masks (*personae*) of unpresentable forces at play which beat at the heart of each being. Things' singularities, their likeness and difference, and their sociality with the other are drawn figuratively by nothing but a chance-bound play of forces which unrolls beyond a control of human will and awareness even exceeding a monstrous power of Socrates. Thus, such a fabulous *genesis* with its nonreflexive movement (*Wesen*) of forces is younger and older than any "ontopoietic" sources of life and the formations of world relations produced in the Socratic heritage since the "logical Socratism" itself billows out of the Apollonian creative force (*BT,*94). Although a violent sweep of the "Socratic," metaphysically- attached forces brings about catastrophic demises of tragedy and myth as well as a "degeneration and transformation" of the Greeks' personality and sensibility (*BT,* 137), Socratism and its cultural heritage are essentially unfixed thus open to a radical change and a new birth. We shall now turn our gaze and ear to the *Wesen* of tragedy and heed a birth of forces to which we, along with Nietzsche, hold a long-standing relation in the fall (*lethe*) of our memory.

According to Nietzsche, tragedy refined into an art form an archaic experience of the Greeks expressed in the folk songs wherein an earlier vestige of the contact between Apollo and Dionysus is found. Tragedy preserves, in a double *mimetic* structure and with a form of the two gods' "fraternal union" (*BT,* 130), the initial inception of the visual and the sonorous arts in the folk songs. Thus, being composed of the opposites, tragedy concerns essentially a matter of how to gather a common understanding of seeing (the gaze) and hearing (the ear) into their primordial oneness and to heal our blindness and deafness pertaining to the nature of life. Through a form of the union, tragedy allows the audience to stand outside (*ecstasis*) of their self-identities, daily perceptions of life, and social norms and to originate and release their lives once again in consonance with a natural movement.

Apollo and Dionysus are personae of nature; their opposition dissimulates a contrary double movement (*Wesen*) of self-rising-and-falling of nature so as to evoke in the audience an unrepresentable play of the forces of nature. Thus, the two gods are not identical with the primary nature, but together they reenact, from afar, not only the tragedy's loss but also human forgetfulness of nature in an artifice of radiant, dream-like appearances.[14] As we shall see, both Apollo and Dionysus carry out equally the contradictory double movement of nature. Apollo does not take part only in a trajectory of self-opening nor does Dionysus simply do so in self-concealing as though they constitute a oneness by sharing responsibility for an alter dimension of nature. Let us now take a look at a form of their friendship.

---

[13] Nietzsche, *BT,* 5, p. 91. "Here philosophic thought overgrows art and compels it to cling close to the trunk of dialectic. The Apollinian tendency has withdrawn into the cocoon of logical schematism."

[14] Charles E. Scott, *The Time of Memory* (Albany: SUNY Press, 1999), p. 113.

True to his epithet ("*Der Scheinende*"), Apollo's creative power rests in that of "shining." (*BT*, 35) His power is **with** life's self-presencing towards light. And yet, releasing life into motley sparks of imagery, Apollo shrouds neither an impenetrable backdrop of life's appearance nor an indelible lightness-nonsensical, evanescent, and fragmentary-of its occurrence. He lets life appear "as a mere appearance of mere appearance" (*BT*, 45).[15] In letting life come to light, Apollo lets each life draw forth, in its own terms, a space of an individual as though he allows each life to wrench a self-distance from concealment (*lethe*). The ignited space of a self appears essentially groundless, indeterminate, unstable and destined to fade away like fragile touches of reveries or nocturnal dreams. By the same token, the individuated space of a self appears fabulous because it hosts a bare presence of all possibilities regardless of their incompatibility and entertains impartially "all compossibility"[16] belonging to each life for future repetition. Each life stands out thusly to light in an obscure and yet beautiful outlook of being-in self **without** ownness.

Thus, as "the reflection of eternal contradiction, the father of things" (*BT*, 45), Apollo merely gives every self-showing life a bearing of suffering **with** joy. Here, he evokes a figure of "*pharmako(s-)n*."[17] He makes the concealed nature (*Lethe*) to stand outside (*ecstasis*) of itself in the inside (lit space) of each emergent life by doubling the fold into an innermost passage and abyssal space of a self as Apollo co-bears a wakeful trajectory of each life's coming to appear out of nothing. Apollo lets the concealed arise **with** what is shown; conversely, he lets what surfaces seen as nothing more and less than what is shown. From the outermost to the innermost of the concealed, "the rising of what is [barely] concealed"[18] scintillates immediately, as Apollo bears out the spaced distance of self-differen(ce-)tiation of life. Thus, his titanic power of shining involves necessarily as much power of resistance (*polemos*) against an inevitable threat of self-concealment so as to save each life **with** the concealed.

Dionysus reigns in the titanic power of "rapture" (*BT*, 59).[19] The Greeks used to experience a complete self-oblivion in the archaic Dionysian festivals; in tragedy, chorus reenacts it by plucking the audience's eardrum through which a "collective

[15] Nietzsche, *BT*, 5, p. 45. "...we shall then have to look upon the dream as a mere appearance of mere appearance, hence as a still higher appeasement of the primordial desire for mere appearance. And that is why the innermost heart of nature feels that ineffable joy in the naïve artist and the naïve work of art, which is likewise only "mere appearance of mere appearance."

[16] Jean-Luc Nancy, *The Fall of Sleep*, trans. Charlotte Mandell (NY: Fordham University Press, 2009), p. 7.

[17] John Sallis, *Crossings* (Chicago: The University of Chicago Press, 1991), p. 51. Sallis designates the word "*Pharmakon*" exclusively to Dionysus and the Dionysian duality in this remarkable book; however, in my view, it applies equally to Apollo and the Apollinian.

[18] Jacques Derrida, "*Heidegger's Ear: Philopolemology (Geschlecht IV)*" in *Reading Heidegger*, ed. John Sallis (Bloomington: Indiana University Press, 1993), p. 192.

[19] Nietzsche, *BT*, 5, p. 59. "For the rapture of the Dionysian state with its annihilation of the ordinary bounds and limits of existence contains, while it lasts, a lethargic element in which all personal experiences of the past become immersed. This chasm of oblivion separates the worlds of everyday reality and of Dionysian reality. But as soon as this everyday reality re-enters consciousness, it is experienced as such, with nausea: an ascetic, will-negating mood is the fruit of these states."

release" of "the Dionysian music in particular excited awe and terror" (*BT,* 40)[20] comes to reverberate from afar. Nietzsche tells us that the Dionysian dithyramb artist used to undergo heightened "struggles for utterance" in his ecstatic loss of "all his symbolic faculties" and found in music an immediate expression of his union with nature (*BT,* 49).[21]

We note in passing that the Dionysian artist's way of giving voice to his ecstatic union with nature is akin to a way of Apollo's "shining" of lives in which individuals, really, mirror them in a hollow of self-identity/propriety. The artist's expression in music, albeit immediate, comes to appear with a loss in his memory of his origination in the primary nature; thus, even the Dionysian music, too, comes to pass in and with the loss of nature so that it can't be identified with nature, nor can the artist be identified as an authoritative source of creator. The artist merely highlights the individual's out-of-mind experience of the dismemberment-and-remembrance of all determinations and measures by which individual and communal identities are configured and set in order.

Thus, tragic chorus intones a range of tonality, pitch, and rhythm far beyond a moderate range of harmony which sounds pleasant to the ear of civilized individuals, and carries adrift the audience's normal range of attunements without their knowledge of what is taking place. Like a coming to pass of a sleep, chorus wondrously draws them to fall into the concealed (*Lethe*) heart of nature, to originate therein with myriad other imageless lives, and therewith to embody once again the lost dimension in both individual and collective memory. In a shared experience of the dismemberment-and-recollection of common memory, Dionysus gives suffering and joy at once.

Dionysus carries both lethe and shining. His gift of joy is released as nothing which is immediately borne in people's remembrance of the shared voices since both nature and their primordial union with nature will never come to pass as an object for recognition; thus, the Dionysian music appears truly a "re-echoing"[22] of the cosmic rhythm and dance of self-showing earthly life with lives out of nothing/nature to which Dionysus releases the artist, tragic chorus, the audience, and all other present and absent participants sharing a historico-cultural context of the theatric performance. The Dionysian joy remains forever closed off from a sighting of articulation and identification; and yet, at every time of "repetition" through tragic chorus, Dionysus enables the audience to totally unfasten their memory of who they are and to redeem how they can be by intertwining it with the immemorial roots of common life. To ignite a fire (πῦρ) of immemorial past in people and to enable them

[20] Nietzsche, *BT,* 5, pp. 46–47. "...the ecstatic sound of the Dionysian festival; how in these strains all of nature's excess in pleasure, grief, and knowledge became audible, even in piercing shrieks;......The individual, with all his restraint and proportion, succumbed to the self-oblivion of the Dionysian states, forgetting the precepts of Apollo. Excess revealed itself as truth. Contraction, the bliss born of pain, spoke out from the very heart of nature."

[21] Ibid. 5, p. 50. "...The Dionysian musician is, without any images, himself pure primordial pain and its primordial re-echoing."

[22] Ibid. 21.

to stand in and out of their historical and non-historical dimensions of life, Dionysus must maintain his titanic power of resistance to all forces of clarity and their traces of figuration and formation. In this light, he too dons a mask of "*pharmakos(-n)*."

So then, Apollo and Dionysus appear equally in "being-outside-oneself" (*Selbstvergessenheit*) in their union. In their relation to both self and the other, each god bears a twofold of shining and concealing and guards the opposition unresolved by virtue of his full attendance to the twofold and his total expenditure of energies therein. To be sure, their unyielding resistance to a constant threat of collapse streams from nature, nonvoluntarily, without a reason; thus, their resistance is other to human will and temperaments. Their amazing power of carefree resistance attests that each god is fully and already playing out the spiritual fires of freedom and necessity without seeking them beyond oneself.[23] Thus, neither human intentionality nor the Will of transcendent God gathers each god's com-position of opening-and-concealing. Apollo and Dionysus are purely devoid of self-identity; they appear truly "unreal."

At the same time, the distention of god's resistance spawns a just and free space of the "in-between" for their sociality.[24] Their loving bond is fastened by their "monstrous" power of resistance to a constant threat of life's de-com-position which, in turn, unfastens their desire for self-possession. Their loving kinship is purely contracted by a double power of self-postponement with respect to a nonpresence and a presence of singular life. Their self-renunciation is stainless of cultural norms and moral codes and shaded by the opposite traits of unbreakable hardness and deep tenderness. Their affiliation appears unrealistically light and loose rather than "lawful" and fixed; thus, their union remains in-finitely airy and hollow, while each god offers to the other a gift of nothing but a bare life ("*genesisphthoran*"). Being grounded and welded by nothing, they let the other be in a full strife of forces and traverse jointly the "unreal" distance across the nonhistorical and historical dimensions of nature.[25] Their joint-venture is fragile and their friendship is barely kept intact insofar as their spiritual fire of self-resistance-and-self-abandonment flickers reciprocally.

Regarding the matter (i.e., gift) of nothing which releases all lives ecstatically to the world of their being-in and gathers a salutary rapport of singular lives, Nietzsche's conception of the fraternal union evokes Heraclitus' "ever-unsettling fire"(Fr. B30) and reminds us of Heidegger's interpretations of the Heraclitean logos as "*ὁμολογεῖν*" – gathering in "*Logos*" essay. Nietzsche's gaze into the love of fraternal union brings to mind Heidegger's translation of the Heraclitean "*φιλεῖ*" intrinsic to *φύσις* (Fr. B 123) into "the inherently reciprocal favoring" in "*Aletheia*" essay.[26]

---

[23] Charles E. Scott, *Living with Indifference* (Bloomington: Indiana University Press, 2007), p. 122.

[24] I have in mind here a mimetic likeness to the conceptions of "dif-ference" and "Auseinandersetzung" in Heidegger's thought.

[25] Ibid. 1, p. 62.

[26] Ibid. 2, p. 114.

According to Heidegger, the Heraclitean logos/legein indexes the originary spring of language in and beyond the two forms of "φωνή" and "σημαντική."[27] Just as Nietzsche experiments an imperceptible play of forces as the genesis of all forms, Heidegger assumes that the Speaking of language wells up and down from an abyssal origin of language to all forms of man's shared voices expressed in either speech or writing, or in either private or public, and flows the fro and to the same dimensions of life's ecstatic entry into the world and its whiling therein; in effect, gathering force of logos is borne with man in an indelible mark of forgetfulness (*ἐπιλανθάνομαι*; *EGT,* 109) whose mode of passing happens so naturally to man.

This middle-voiced pattern of self-oblivion and its anonymous force of gathering is akin to man's experience of "the fall of sleep" that Nancy has remarked (*ouvre*) in a title of his book (*oeuvre*). Thus, like Nietzsche's primary nature, Heidegger's view of *φύσις* recollected through Heraclitus' fragments stands in and out of the historical bodies of shared memory about life, truths, norms, and moral virtues and values, and calls man to correspond (*Entsprechen*) to the speaking of language (*das Sprechen der Sprache*) so as to re-member an intrinsic bond of friendship which is prior to any forms of *φιλία* formed on a base of man, his desire, and claim for a transcendent presence of meaning for life.

For Heidegger and certainly for Nietzsche, Heraclitus the Obscure takes a seat in an immemorial origin as well as a future of Western Philosophy and the Socratic Culture. Heraclitus does not occupy a supreme place of "the paradigm philosopher;" he does occupy, however, a daimonic place of "the seers"[28] or "the sages" (*οφὸς ἀνήρ*) who bear an etymological linkage of "sage" with "*sapío*"[29] and retain the "keenest taste" and ear for "the echoes of the world symphony."[30] His cosmology of becoming lights a signal fire of "the archetypes of philosophical thought"[31] which are disposed to articulate the knowledge, "first, on its own abyss"[32] and as such, which are to come in wonder at the gathering of nothing which is, albeit untotalizable, incredibly mightier than nihilism.

Thus, in the end, tragic chorus voices no meaning. In lieu of a substantial meaning or a technical solution to daily problems, tragedy gives the audience a dawn of the eternally same day ("*dies illa*")[33] wherein they co-exist with others in a common loss of self-identity and in an awareness of an irreducible necessity of the opposites

---

[27] Ibid. 2, p. 77.

[28] Ibid. 18, p. 191.

[29] Friedrich Nietzsche, *Philosophy in the Tragic Age of the Greeks*, trans. Marianne Cowan (Chicago: Gateway, 1962), p. 43.

[30] Ibid. 29, p. 44.

[31] Ibid. 29, p. 31.

[32] Jean-Luc Nancy, *The Creation of the world or Globalization*, trans. François Raffoul and David Pettigrew (Albany: SUNY Press, 2007), p. 89.

[33] Ibid. 32, p. 71.

in and for life.[34] Nietzsche calls "metaphysical comfort"[35] such an ecstatic gift of historicity that tragedy offers to people. In remembering the death of tragedy in *The Birth of Tragedy*, he has unexpectedly found a dawn of his own historicity, a possible way-out from nihilism and an immense opening for his genealogy and the will to power in the originary spring of nature.

[34] Nietzsche, *BT*, 5, p. 52. "-for it is only as an aesthetic phenomenon that existence and the world are eternally justified-."

[35] Ibid. 5, p. 104. "...a metaphysical comfort tears us momentarily from the bustle of the changing figures. We are really for a brief moment primordial being itself, feeling its raging desire for existence and joy in existence; the struggle, the pain, the destruction of phenomena, now appear necessary to us, in view of the excess of countless forms of existence which force and push one another into life, in view of the exuberant fertility of the universal will. We are pierced by the maddening sting of these pains just when we have become, as it were, one with the infinite primordial joy in existence, and when we anticipate, in Dionysian ecstasy, the indestructibility and eternity of this joy. In spite of fear and pity, we are the happy living beings, not as individuals, but as the one living being, with whose creative joy we are united."

Lebenswelt and Operational Methodology in the Philosophical and Epistemological Reflections of Hugo Dingler

Dario Sacchi

Abstract Hugo Dingler (1881–1954), a German philosopher and scientist who was taught by Husserl, among others, devised a singular and original form of operationalism based on phenomenology which while being very different from the perspective of P. W. Bridgman and American pragmatism, was no less interesting than either of them. According to Dingler, finalistically orientated behaviour is the ground on which we build all theoretical constructions and represents their criteria of validity: the actions that constitute scientific theory are not essentially different from those that make up our everyday lives, and like these must be declined *methodically* and determined *univocally*. Technically characterised "doing" is not the arbitrary result of human artifice but coincides with that *order* which emerges from the same context of the vital procedures that Husserl refers to in *The Crisis of European Sciences*.

In § 27 of the first volume of *Ideen zu einer reinen Phänomenologie und phänomenologischen Philosophie,*[1] Husserl observes that the world present in natural attitude is not made up of things but of things with a determined value for me, of things with a practical significance, things to be used, and this concept of the world as it appears to man in natural attitude is also developed in detail by Scheler and, in particular, by Heidegger in *Sein und Zeit*.

Since its very beginning, phenomenological philosophy has shown itself to be aware of the impossibility of identifying a "pure" or "neutral" original moment of perception on which the practical-manipulative operations of the subject can

[1] Originally published in a single volume and usually entitled *Ideen I* (Halle a.d.S.: Niemeyer 1913). The whole work is currently available in three volumes, the last two of which were published posthumously, and are edited by W. Biemel (Husserliana III, IV, V, The Hague: M. Nijhoff, 1950–1953). English translation: *Ideas pertaining to a pure phenomenology and to a phenomenological philosophy*, 3 vols., The Hague: M. Nijhoff, 1980–1982.

D. Sacchi (✉)
Università Cattolica del Sacro Cuore, via Giuditta Sidoli 2, 20129 Milan, Italy
e-mail: dario.sacchi@unicatt.it

A.-T. Tymieniecka (ed.), *Phenomenology of Space and Time: The Forces of the Cosmos and the Ontopoietic Genesis of Life: Book One*, Analecta Husserliana 116,
DOI 10.1007/978-3-319-02015-0_30,

subsequently be grafted, because there cannot be perceptions or apprehensions of the true nature of something that have not always been included in an emotional and dynamic context. Intentional consciousness undoubtedly creates the conditions for establishing the phenomena, but we should not forget that it is always the consciousness of a *living* subject, one that possesses instincts, desires, needs and self-movement abilities, together with the ability to transform-assimilate environmental elements for the purposes of its own growth. The sensory-motor conditioning of perception implies that the emotional, volitional and operational context where the living subject inserts the perceived things is the condition why the ideal sense of the things originally appears, and every object from our experience can only be perceived as a single, substantial structure on the ground of an operational interest in the object itself.

Because of all these reasons, it would be particularly interesting to go back and re-read the original, singular type of operationalism on a phenomenological basis that was elaborated in the first half of the last century by Hugo Dingler (1881–1954),[2] a German philosopher and scientist who was taught by, among others, Husserl and is considered to be the founder of "methodical constructivism" that in more recent times is regarded to be best expressed in the works of Paul Lorenzen and the Erlangen School. According to Dingler's operationalism – a perspective that is not merely epistemological but also authentically philosophical and undoubtedly more organic and profound than both the subsequent operationalism of Bridgman and American pragmatism – finalistic oriented behaviour is the ground on which all theoretical constructions are built and it represents their criterion of validity: the actions that form the basis of scientific theories do not essentially differ from those that form our daily lives, and as such they must be declined *methodically* and determined *univocally*. Technically characterised "doing" is not the arbitrary result of human artifice but coincides with that *order* which emerges from the same context of the vital procedures that Husserl refers to in *Die Krisis der europäischen Wissenschaften*.

In this essay we will attempt to highlight some of the most significant moments in this thinker's itinerary that emerge both from his last, posthumously published book, which he finished the year before he died, and from the fundamental and voluminous treatise that he dedicated to *Die Methode der Physik* in 1938.

Every form of philosophy that wants to give itself a systematic structure is confronted with the question: where to begin and with what? "All of life, the whole world and the entire mental sphere seem to be at our disposition for such a choice.

---

[2] He graduated in mathematics, physics and astronomy after having been a student of F. Klein, D. Hilbert and H. Minkowski. Taught at the University of Munich and at Technische Hochschule in Darmstadt. He was accused of being pro-Jewish by the Nazi regime and was forced to leave teaching. His most important books are: *Grundlinien einer Kritik und exakten Theorie der Wissenschaften*, Munich: Ackermann, 1907; *Der Zusammenbruch der Wissenschaft und der Primat der Philosophie*, Munich: Reinhardt, 1926; *Das Experiment. Sein Wesen und seine Geschichte*, Munich: Reinhardt, 1928; *Metaphysik als Wissenschaft vom Letzten*, Munich: Reinhardt, 1929; *Philosophie der Logik und Arithmetik*, Munich: Reinhardt, 1931; *Geschichte der Naturphilosophie*, Berlin: Junker und Dunnhaupt, 1932; *Die Grundlagen der Geometrie*, Stuttgart: Enke, 1933; *Das Handeln im Sinne des Hochsten Zieles*, Munich: Reinhardt, 1935; *Die Methode der Physik*, Munich: Reinhardt, 1938; *Grundriss der methodischen Philosophie*, Fussen: C. F. Winter'sche Verlagshandlung, 1949; *Die Ergreifung der Wirklichkeit*, Munich: Eidos Verlag, 1955.

Where is the beginning of Arianna's thread from which we can start in order to unravel the tangle?".[3] Where can we find immediate assertions whose certainty is absolute and, at the same time, not limited to logic but extended to correspondence with reality? And since concepts represent the essential ingredient in every assertion, the question can also be formulated thus: where can we find absolutely univocal concepts?[4] According to Dingler, there is certainly a group of assertions that satisfy the requisites referred to above, which are the indications, or directives, for action (*Handlungsanweisungen*).[5] If, for example, I say: to make scrambled eggs you put a bit of butter in a pan, pour in the eggs and stir until they begin to set and become soft, these are indications for carrying out an action, and from the logical point of view it is a definition of the concept of scrambled eggs. Such an assertion is not just valid for an individual reality but also for an unlimited number of cases where the action can be carried out and is carried out, so that we are faced with authentic universal propositions that are valid in reality. According to current opinion – which is not valid and which we will look at later – such propositions are obtained through induction.

But what do we mean by the *starting point* of thought and of every mental construction? We are obviously dealing with a different question from the one we looked at before, which concerned the beginning as meaning the *ground* of knowledge. Based on this second, no less important meaning the beginning is characterised by the demand to be free from any conscious mental addition that we can bring to its content, an immediate that is no longer epistemic, as in the previous case, but phenomenological. It is a somewhat self-evident fixed point which, according to Dingler (whose position on this was very different from Husserl's), can be deliberately adopted by anyone, at any time, without effort, and what we have in this fixed point is the "intact" ("untouched", "virgin", "non-manipulated" – *Unberuhrte* in German). In the Intact the world is not yet "perceived" through the sense organs (a conception that involves a causal mental construct), it is simply *there*. Things, movements, other people, other thoughts (and always in a way that is placed before any conceptual delineation and clarification) *are there*, simply and immediately. The Intact is therefore the true immediate world. Deliberate, conceptual and causal mental construction that begins here then leads to the theory of sense organs and their function, and to the resulting consequences.[6] Compared to the Intact, this theory is therefore something secondary, a scientific construction. Now, when the final theory is falsely transformed into a starting point, as sensationalism did and still does, a solipsistic and sensationalistic image of the world is created, the impossibility of which was in truth always known but cannot become wholly visible until a way is found to obtain the fixed point of the Intact.

But what role must be assigned to "unconscious inferences"? This in effect is the most suitable point for understanding the difference between Dingler and Husserl,

[3] *Die Ergreifung der Wirklichkeit*, cit., p. 39.

[4] Cf. ivi, p. 47.

[5] Also cf. *Grundriss der methodischen Philosophie*, cit., p. 32.

[6] Cf. *Die Ergreifung*..., cit., pp. 24 on.

and in our opinion it is a point in favour of Dingler. It could be thought that these inferences should be seen as additions, so that whatever is concluded on the basis of these must be left aside in order to obtain the Intact, but a more careful study shows that this would in reality be wrong. This very notion of unconscious inference, since it is unconscious, can only be the content of an hypothesis and certainly not of immediate data, and as such it already represents an addition to the Intact. In itself it is an element of the scientific explanation and therefore does not belong to the Intact. The only things that need to be deducted from the Intact are the conscious additions, therefore you can't expect the results of the unconscious inference to be removed in order to arrive at the Intact. An example of this is the question of the opposite parts of objects.[7] In the theoretical model of physiology and sensationalism I can only see the side of the thing that is facing me, since there is nothing in this model that can help me to see the other side. But in the Intact I experience objects as "multisided". The theory explains it by referring to inferences that are triggered unconsciously as a result of previous experiences, suppositions and so forth. These unconscious inferences can subsequently easily reveal themselves to be false, and many illusions and errors are based on them. Nevertheless our philosopher strongly emphasises how the Intact gives the impression of multi-sidedness.

Moreover it is said that consciousness and thought are the product of a material, bodily organ, the brain.[8] This is already a real theory because "product" means causality, and the fact that the brain produces a thought is a causal construction that can only take place in the mental elaboration sphere and requires a rigorous basis, so that it can never form the starting point of scientific thought, as we are led to believe here. That it is indeed like this can be seen by the fact that this causal linking of the central nervous system to thought is rather late if you consider, as Dingler rightly pointed out, that even for Aristotle the brain was an organ in charge of cooling the blood while it was the heart that was responsible for mental activities.

The ideal methodology must therefore be capable of ensuring that I can start at any time to reconstruct physics from its beginning (*the principle of reconstructing physics whenever we want*) and always with the same results.[9] It must provide an exact, systematic and explicit justification of all the steps and principles applied according to it and it clearly cannot make use of any proposition that is immediately admitted as valid, not even propositions that constitute the apparently evident foundations of current official opinion, for example the principle of induction or the conviction that only mathematical hypotheses and their proof through experience are important. Since all methodology refers to activities that must be carried out (naturally only intentional, conscious activities), the beginning of a methodological system must include *a declaration of the aim of the activity* on which all the activities of methodology itself must be concentrically directed.[10] This aim of a group of conscious activities naturally must be subject to *a choice on the basis of a free voluntary*

[7] Ivi, p. 31.

[8] p. 147.

[9] Cf. *Die Methode der Physik*, cit., pp. 103 on.

[10] Ivi, pp. 110 on.

decision, and in our case the choice is made saying: *the aim of our methodology is to arrive at the conscious, conceptual and manual dominion of reality in the widest ambit that is currently possible and in a univocal way*. Dingler adds that the expression "systematic methodology" means the exact continuous reproductibility of a scientific result, and he emphasises that a result achieved using systematic methodology is *indisputable* since it can be reproduced with complete certainty at any time. When we enter methodology we enter, at the same time, the reign of the *will*, and in point of fact all methodology is the doctrine of intentional *activities*. In the light of this it would appear to be contradictory to maintain that there could be unassailable and absolute methods since neither the idea of methods as activities decided by free will nor the free choice of aims would appear to be consistent with the idea of unassailability, which in itself contains the idea of inevitability and, consequently, of an absolute motive. We can immediately resolve this paradox if we look more closely at the circumstances that dominate voluntary intentional activities. In the univocal methodical system totally free choice or, if we want to call it by another term, free will has its place *only in a unique point*, which can only be found in the very first starting point of the system where the first decision is taken to identify the aim with the achievement of a univocal and systematic science. All the rest is no longer subject to free choice but is linked to this first decision about the aim and is a necessary consequence of it. From this we can see how methodology that comes from free voluntary choice can still lead to absolute methods. Whoever takes this first decision to identify the aim remains tied to it, must subscribe to the same univocal methods and, consequently, to the same results. Naturally this systematic methodology can only produce univocal results if it is constructed with univocal rules. These types of developments must be supported by *a form of axiomatics*, with axiomatics here meaning something that is fundamentally different from formal axiomatics as it is used nowadays in mathematics, where it occupies the place it deserves. Here we must speak of effective or real activities, and with this we must in the first place speak about sense or content (*inhaltlich*). There is no room here for propositions whose significance or content are put aside, as happens in mathematics. This naturally means that the hypothesis that all logic must be logistic is wrong, even though it is supported by some exponents of mathematic logic or logistic.[11] Logistic is a system of formal calculation that, like all things similar, has no intrinsic significance whatsoever. When you want to interpret this formal system as logic, you have to add a content or semantic meaning to it, and this addition on its own is naturally not of a logistic nature. If this addition is univocal, which all logicians admit is evident because otherwise all deduced logical calculations would have no sense, it requires that first, before any logistic, the concepts necessary to provide a content for logistic are univocal. But only the rules that Dingler demonstrates in his own methodology can in his opinion guarantee this univocity. Before logistic there is real logic that consists in propositions that are imperative and which logistic need if its claims are to have any sense, and the somewhat wide-spread belief that a description can only be "exact" if it uses logistic is wrong. The instrument that more than any other

[11] pp. 134 on.

ensures the exactness of our methodology is *pure thought*, or thinking based on ordered activities that are univocally, clearly defined, which is thinking that can also be called "planning thought" and which can never be completely replaced by logistic because it is practically part of preparing logistic. The description of some of its parts that have already been sufficiently formalised by means of symbolic signs and their combination rules can only be planned and carried out by using planning thought itself. Logistic is not able to provide anything that hasn't already been provided by pure thought, while logistic can never entirely cover the field of pure thought.

Methodology is made up of a series of steps that must always possess a *first element* if it is to be really carried out, while in order to put into action this first element there must already exist something that makes this action possible. In other words, at the beginning we must already be in possession of specific "capacities", otherwise we cannot "do" or begin anything, and it is these capacities that in general we use to start science.[12] For example, the capacity to carry out intentional activities as part of a specific pre-established plan; in the case in question, the capacity to be aware of an aim, to continue to want it, to use our minds and focus them on the aim, and also to use our hands. In addition, the capacity to think in a very generalised way (initially pre-logical), to use daily language in a sensible manner, both oral and written, to give words a specific meaning and to maintain this meaning during a description or explanation, etc. No thinking or axiomatics or calculation can provide us with these capacities or guarantee them for us. *We must simply have them*, otherwise every beginning becomes impossible. If we want to obtain them through axiomatics or a calculation, we must use them to construct the axiomatics or to use the calculation.

In addition it should be remembered that science is the ultimate aim of a series of very complex aims. The procedures illustrated in the methodology explained by the Munich philosopher should be the intermediate aims in the series of aims directed towards this ultimate aim.[13] When man is driven by an aim, he needs to have a supreme, absolute aim that supports all his activities so that these are suitable and not contradictory, and in this way this aim is his "general supreme aim". Since we are talking about the thing that should drive all his activities, without any exception whatsoever, the doctrine involved can only be *ethics*. The aim of reaching science is an intermediate aim of the supreme ethical aim. The determination of this ultimate ethical aim is for everybody an act of his own free will. Dingler emphasises that the supreme ethical aim cannot be proved, since the only way to prove an ultimate aim would be the logical deduction of it from a higher aim, something that on the basis of the same definition of the ultimate ethical aim cannot exist. The ultimate ethical aim can therefore only be the object of free choice. In *Das Handeln im Sinne des Hochsten Zieles*, Dingler proposes to determine what influence the formal and therefore the objective and the "absolute" have on ethics, which is why the book was sub-titled *Absolute Ethics*, and he reaches the conclusion that only a

[12] pp. 113 on.

[13] pp. 118.

teleological ethic could be methodically consequent and feasible. He argues that the ultimate basis of ethics can never be formal or absolute but always subject to a personal choice that in the final analysis comes from pre-rational innate vital forces, which means that only the construction erected on that basis and not ethics as a whole can contain decisive and therefore absolute formal elements. Naturally in various cases we can see that the ultimate aim and the intermediate aim do not have a direct subsumable relationship. The fact that the intermediate aim effectively helps us to reach the ultimate aim of the series is above all a *real relationship* that can either be empirically established by trial and error or suggested by a previous knowledge of the effects of the intermediate aim, and it is clear that here the question may arise as to why the intermediate aim helps us to reach the aim of all the series (a question that leads us into the environs of the much larger problem of causality, which we will look at later). For the time being it is worth noting, from the purely methodological point of view, that what was previously said about the series of aims made no reference to the quality of the external world but merely highlighted some fundamental capacities whose use depends on our will and are constantly at our disposition. There were also defining conceptual assessments and statements about the activities which are to be carried out and can always be carried out thanks to fundamental capacities. Nevertheless here we were dealing with those types of statements that possess, as has been referred to since the very beginning, security and immediate certainty. It is therefore not surprising that at this point Dingler says quite clearly that: "The ideal methodology of physics can only consist in the *precise indication of the conceptual and manual activities that make it possible to arrive at the complete foundations of a univocal domination of nature, starting from ground zero of physics*, that is from untouched nature and from a complete ignorance of physics. *All the rest is imposed on us*, once this aim has been accepted, whose justification certainly cannot be questioned".[14] Only when methodology is constructed in such a way as to start again at any given time with physics, obtaining the same results, can it be complete and convincing.

Fundamental capacities also include some *immediate knowledge of relations,*[15] which is already present in all immediate sensations that because of their univocity do not require any theoretic conceptual constructions. This type only includes knowledge of *equality, resemblance and diversity* that under no circumstances could be replaced by others. If diversity, resemblance or equality were not immediately apparent (only in a qualitative form, naturally), no mental construction could lead us to them. Lastly, everything that is usually designated as "relationships" relies on these fundamental relationships and when applied can only be defined by means of them. They only belong to the starting point of systematic methodology, but in that position they cannot be eliminated. On the contrary the assertion that *the world has been presented to us as a collection of sensory perceptions* does not belong to the starting point of systematic methodology, as we saw before. Because of its content, this assertion already represents a theoretical construction and

[14] p. 124. Italics in text.

[15] Cf. pp. 128 on.

therefore cannot belong to the starting point for all theoretical science, which coincides with the starting point of systematic methodology. The immediate feeling is not yet clearly set out in the various sensory fields, nor can it be said that it contains a distinct division between internal and external experience, but rather it runs or moves along as a result of what is essentially continuity and compactness. The internal-external division and the division into various sensory fields represent causal theoretical constructions, which need a good deal of scientific support and deduction in order to be vigorously supported, and in some points this has not yet happened (Dingler offers the example of the conflict between Kantian and empiricist philosophers in deciding if geometric space and then, above all, time are "internal" or "external" to the subject). Methodology does not need such an interpretative model, apart from the fact that it is logically defective because it contains a circle. *Methodology works in a world that is immediately real*, which it teaches to dominate systematically. Its results are valid for this real world, teaching us how to work it scientifically according to our definition of science, and not for a world constructed behind the sensationalistic screen, as is claimed nowadays not only in physics but in philosophy too (the so-called critical realism). "Also in this case methodology, instead of pale metaphysical constructions and arbitrary conceptions, provides clear, useful circumstances of the immediate and fresh world where we all live".[16] Our philosopher designates the building he strives to construct as the *univocal-methodical system*, while designating the *visual angle of the pre-systematic situation* as the logical situation in which everyone finds themselves before beginning the construction. As we have already seen, the only point that remains arbitrary in this project is the beginning, where free will decides to establish a univocal science (we could say that the only arbitrary point in this project is the project itself as such). All the rest must come, in an obligatory, unequivocal way, from this will.

Since the following point can often be misunderstood, it is worthwhile explaining it clearly: it can never be the responsibility of physical methodology to explain or work out the concepts we use in our everyday language, and its only responsibility is to obtain concepts that are strictly univocal and therefore exact that make it possible to dominate reality.[17] The two spheres must be kept clearly apart if we don't want to take wrong directions, without any way of turning back. The difference between pre-methodical and methodical concepts lies in the fact that pre-methodical concepts are only qualitative and only apply to certain immediate spheres of everyday life, while it's possible that they clash at some point, that they're not sufficient in all cases and they're not always univocal. They therefore can never represent anything that is absolutely certain and, above all, they do not provide a systematic explanation of anything at all. However, from this it is easy to draw the wrong conclusions and say that pre-methodical concepts are to some extent evidence of the "presence" of a clear, purified concept and consequently of something real, almost as if they were only temporary attempts, and therefore still imprecise and maybe even contradictory, to formulate something *real, something which is*

[16] p. 130.

[17] Cf. p. 140 on.

self-consistent and would then be discovered and analysed with science. For example, if we want to deduce from the existence of spatial expressions in everyday language the existence of a unitary entity by name "space", without contradictions, that pre-systematic man would find in reality and would strive to express in words, we want to draw a conclusion which is totally unjustified and not univocal, because there are other ways of interpreting the relationship, and in addition contains metaphysical elements that have nothing to do with the question (such as the objective metaphysical existence of this entity by name "space", which is outside of the mind but at the same time not real). Consequently, the creation of a similar entity is not necessary and therefore falls under the strokes of Occam's famous razor (*entia non sunt multiplicanda praeter necessitatem*), and in fact it isn't even necessary in other ways since it doesn't offer anything that is practically useful and it is theoretically unsustainable. It is worth looking at what Dingler says about space, because this is a point that is particularly suitable for helping us to understand how he interprets the construction of the physical world. Above all he maintains that "exact space is defined by using exact geometry",[18] and he justifies this idea by saying that geometric science is the only thing that makes it possible to establish what is or what is not spatial. In the first place it is necessary to distinguish between experience that only comes from spatial relationships, which means from spatial *constants*, and experience that does not come from these. If a body changes its form and size, this can be a pure spatial experience when the changes are exclusively due to my movements with regard to the body, and it cannot be a spatial experience if the body itself changes its form. Thus it can quickly be seen that the decisive criteria is whether or not the body behaves like a "rigid body": if it does, then the change is merely spatial, and if it doesn't then it is not. *But the concept of a rigid body is a strictly geometric concept.* Besides, all changes where we don't know if they are real or perspective are an example of this, and to decide we need to employ the rules of perspective, which if developed strictly are a part of geometry. However, if we obtain exact space using exact geometry, how do we obtain exact geometry? "We take away all "depth" from the surfaces, lines and points and we keep only their concatenation (…). It is precisely because we are looking for ideal forms, free from the highest possible number of specific qualities, so that we can define ideal forms by using only the application of the immediate experience of diversity, that we take away these elements from them. We are able to find these forms because we can form the idea of an infinitely thin continuous surface, after we were first made aware of them after their discovery by an unknown Greek. Here we are faced with the need to eliminate all the casual individual material elements until only the purely formal ones remain. In this case *it isn't that we arrive at the idea through the abstract, but that we arrive at the abstract through the idea*, since abstracts are always *guided* abstracts, and you can only have this guide with the direction of an idea. The ideas of points constructed in this way are indistinguishable and therefore identical. The same is not true for surface and line ideas. From this point of view, the body idea corresponds to the surface idea (closed) from which all the material characteristics have been

[18] p. 143.

taken away. The surfaces and lines can also be of different forms. These concepts therefore do not yet belong to the univocal system we are looking for. In order to construct elements that belong to it, it is necessary to try to obtain at least one form among the possible surfaces and lines that is univocal and explainable through determinations such as to make the individual examples indistinguishable, identical. We don't have many ways of establishing these determinations. The application of any empirical criterion can be excluded immediately because the univocal surface looked for must be an *idea*. This is the only way we have of constructing an ideal, univocal doctrine with spatial forms, while the empirical is never univocal in the strictest sense".[19] With immediate experience we are able to understand if there are any differences or not, and so we can determine a surface with a specific form by establishing that its two sides must be indistinguishable, both as a whole and in each point. In this way you can say that this surface is effectively univocal, since all such surfaces can be superimposed on each other. Cutting two of these surfaces would produce a univocal line, while cutting two of these lines would produce a univocal point. Thanks to these determinations that generate univocity, starting from the general idea of surfaces and lines, we can obtain the *plan* and the *line*, although these elements are still not sufficient to univocally determine any figure whatsoever of space. In general terms, we cannot univocally establish and determine the non-deformability of a figure with these elements, and this can be expressed by saying that a rigid body is not yet determined univocally by using them. When a rigid body is univocally determined, the resulting whole can finally be called "geometry", and this is univocally determined. If we conserve the plan, the line and the point and yet do away with even just the determinations that concern the rigid body, the length of the segments will no longer be defined univocally.[20] It could be possible, with the application of means that are certainly not yet present in the pre-systematic stage, to establish another determination of the length of the segments that is different from the rigid body determination (as long as it doesn't clash with the determination of the plan, line and point), and this would also produce univocal figures, but they would be deformed in relation to those from the previous geometry. These geometries are called *non-Euclidean*, while the previous one is called "Euclidean". However, in order to establish a non-Euclidean definition of length, it is necessary to use more complicated determinations. Only Euclidean geometry comes from simple experience of differences, and it is not possible to determine a different geometry only through the application of our fundamental capacity.

In this way it can be shown that between Euclidean geometry and non-Euclidean geometries there isn't just a gradual difference but, from the methodical and epistemological point of view, an essential difference and an unbridgeable fracture. The geometry outlined above in its methodical operations is the one that is utilized as soon as you carry out additional physics research, and we designate it *conventional*

---

[19] pp. 145–147. The italics are ours to highlight a statement that we think is very true but also extremely Platonic, although without being inconsistent with Dinglerian thinking (we come back to this at the end of the present essay).

[20] Cf. pp. 149 on.

geometry. It is necessary when constructing exact equipment, particularly measuring instruments, and it is also used for the logical-mathematical structure of physics problems (for example, in geometrical optics). First of all it is noteworthy that for this aim we need a *univocal* geometry, a geometry that should be determined in a univocal way up to and including its last properties. The determination must therefore be ideal, so that it does not contain, for example, any elements and constants that have to be established empirically. Conventional geometry must be able to univocally guide the manipulations necessary for preparing the measuring equipment and instruments (for example, the geometric instruments). In *Grundlagen der Geometrie*, Dingler carries out a detailed analysis of how the determinations of this geometry are necessary for all of these construction operations, and how *all* its independent determinations (its so-called axioms) are also necessary. Since the determinations guide these construction operations they must have an *ideal* nature, because they must direct that activity on behalf of intellect, and because every precise geometric instrument represents only one passing moment in an unlimited process of ever increasing exactness, just one idea (indeed an infinite idea....) can guide this process. On the other hand, as our philosopher says, nobody has yet managed to indicate the ideal criteria for segment equality that can lead to non-Euclidean geometry.[21]

Thought has often been given to the necessity to determine by measuring the nature of geometry that is applied to our space. The precision of geometrical measurements, carried out by using those measuring instruments, does, however, depend entirely on the precision of these instruments, that is on the precision the manufacturer has employed in the criteria applied when constructing the instruments. These criteria correspond to the rules of conventional geometry, so that it can be said that the precision of the geometric measurements depends entirely on the precision that is employed in applying the rules of conventional geometry when constructing the measuring instruments. Therefore it is never possible to obtain independent decisions concerning geometry by using these measurements. Thus we find ourselves in the interesting situation that even though theoretical geometry does not reflect any properties of nature, despite this it is determined univocally. This, as our thinker notes to his great satisfaction,[22] *is not an arbitrary convention*, as Poincaré thought, *but we are led to it in a univocal way* simply by the circumstances and consequences of the methodical-practical procedure and of acting. The possibility of explaining in this way (the right way) the nature of geometry has never figured in the many attempts made, because we have always thought that there is complete freedom of choice in geometry, since it was discovered that it wasn't determined by nature or innate forms. In such a situation univocity would appear to be unthinkable, but in point of fact systematic methodology demonstrates that this possibility actually exists.

With the realisation of the ideas contained in the basics of physics we are already involved in an unlimited, manual process directed by ideas, and this process

[21] Cf. p. 157.

[22] p. 161.

generates the science of physics and, as Dingler observes, is possibly the most incredible instrument the human mind has ever developed. The first infinite idea that we come across is the *plan*. By studying the realisation we can study the dominant lines of the realisation of infinite ideas. The accurate realisation of plans takes place in precision-engineering factories when constructing precision measuring instruments.[23] In these activities there is the continuous need for accurate plans based on various levels of precision, and it is necessary above all to construct a plan-basis of maximum precision. This is produced by rotating alternatively against each other three steel discs, A, B and C, that have previously been roughly smoothed, so that each one adheres completely to the other ones. It should be noted that it wouldn't be sufficient to use only two discs because they could produce a spherical surface. All the plans put into practice refer to these fundamental plans. Therefore the logical part of geometry must be implemented at the same rate as the "manual" part, during the construction and practical realisation of the geometric forms.

Many people give in to the temptation to regard a concept as independent from the activities that were involved in making it, once it has become intuitive or has obtained a mathematical expression. As a result of this, "pragmatic circles" are created, which are logically vicious circles that do, however, involve the praxis. We forget, if we stay with the example referred to, that the measuring instruments depend pragmatically on the geometric principles used to construct them, and we end up expecting that the latter can, for their part, be pragmatically dependent on the measurements taken by the instruments.

"Just as when we want to turn on a light we have to press the switch or when we want to play a particular note on a piano we have to press the right key, in everyday reality we also look for the 'switch' or the 'key' that produces the desired effects. The desire to command our natural surroundings involves *finding a system of 'keys'* that produces all the changes we are looking for and produces them in a *univocal way*",[24] which is where the desire to find reciprocal dependence among the phenomena originates. And here we see the reappearance of the same relationship between the pre-systematic phase and the univocal system, a relationship that we always find. In the pre-systematic phase, everyday life, we already come across an extraordinary number of dependencies of this type. Our entire life, working with everyday objects, and also our relations with people like us and all the other living beings are full of them, although these dependencies share a lack of security and absolute univocity. As everyone knows, in these cases sometimes we don't attain the desired effect or it has a different form, and we must always be aware of this uncertainty. We can therefore understand the effort that is put into finding methods that are suitable for discovering, or at least preparing the highest number of *univocal* dependencies available to us at any one time. This is actually the job of rigorous science, of the univocal-methodical system. But even moving from the dependencies that we come across every day to "experimental" dependencies the uncertainty remains, or at least it does until the univocal system has exact rules that can ensure univocal

---

[23] Cf. p. 163 on.

[24] p. 189.

dependence. Amongst all this uncertainty we can catch a glimpse of an *experience-element*, which makes up the basis of *all* the dependencies that we subsequently possess by means of the univocal-methodical system. We often immediately notice or sense that if we move a body A in a certain way, this always produces a phenomenon B which is not produced if we don't move A, and we call this experience a *dependence-experience*. However, this has produced a lot of uncertainty, such as: is B the direct effect of A or is there an intermediary element between the two and if so what is it? For this and other reasons the dependence experience is not similar to a "natural law", and to promote it to the level of a "natural law" would obviously require many other things. However, here we want to deal with immediate experiences. The dependence-experiences which we have seen belong to immediate feelings from the pre-systematic point of view, like the immediate experience of diversity. Nature only really offers these two pre-scientific elements.[25] The first of these is the immediate experience-material for the variable sphere, and so is the second for the static one. All the rest are performed by methodology, with the aim of constituting science. In actual fact both diversity-experience and dependence-experience are always individual experiences, while it is necessary to have universal assertions in order to constitute science. In the case of dependencies this means that you must try to make them repeatable at will, and for this you need to study the *conditions*, since everything depends on their constancy and, in the final analysis, on the possibility of keeping general natural circumstances constant. "The univocal system intervenes here creating figures and shapes, as (...) already happens in logic, arithmetic and geometry. Now our business is to prepare these univocal, reproducible forms also *with reference to the things that change*, because it is clear that constant static forms cannot do this by themselves. Only the univocal forms of the univocal system make it possible to fix in reality more secure constancies by creating them within the limits we are capable of nowadays. Whatever combinations of these univocal forms are produced, in these combinations, which are known as *apparatuses*, everything that refers to these forms is well known, fixed and constant. Only the application of the univocal form of the system makes it possible to keep conditions, and consequently dependencies, effectively constant. *General dependencies that are reproducible, namely that belong to general assertions ('laws'), therefore depend in every way and for everything on the univocal system.* It is only when, by using the system, it becomes possible to make them reproducible that they become general dependencies or *causalities*, and the *general principle of causality* is only valid for them, with the result that *this principle is therefore an exclusive product of the system.* It is not included in any way in 'nature', but is a consequence of our methodical way of proceeding".[26] Nature as such does not contain universal laws, but we are the ones who construct universal nature laws. It is only our conscious desire to dominate the world both mentally and manually that produces these laws. The world appears to us directly and immediately in a form (the Intact) that represents the material to which we can bring order in compliance with the above-mentioned aspiration of

[25] Cf. p. 193.

[26] pp. 95 on.

ours. The Intact is the task we are faced with and it is always inexhaustible from the point of view of any rational construction. The rational is always a creation of our intellect that can never state anything that is ontologically valid and can only serve as a practical aid to dominating reality.[27]

The principle of sufficient reason thus becomes a *purely methodical principle*. If the fundamental element of dependence-experience has to be constructed in the univocal system, we have to create ideal univocal forms relative to the reciprocal dependencies. In order to obtain clear relationships in this area that are exactly distinguishable and determinable and that have purely ideal and logical conceptual and constructive forms, it is necessary to make agree a specific property of these forms with every specific effect ascribed to them. If, for example, given the determinations of a form already indicated, in some cases we find a new effect without any particular determination added to the form, there would be no possibility of establishing when the effect is due to the form and when it is not, and the decision would be undetermined and therefore *not univocal*. From this we see that the principle of sufficient reason is simply *a consequence of the principle of univocity*, which is the principle that is present at the beginning of all our development and guides it. So the principle of sufficient reason can be formulated in this way: *if a difference in the effect is noted, then an exactly corresponding difference must also be found in the thing that determines the effect.*[28]

Dingler said that of the reviewers of his text on the methods of physics, those with a theoretical leaning (physics and similar, but also the majority of philosophers) passed over the exactness-process and only considered the theoretical, namely logical-formal relationships, while on the contrary almost all the reviewers from the world of technology displayed a clear understanding of the exactness-process, whose theoretical significance however they were often unable to understand because of their short familiarity with philosophical thinking. It was also quite common for the theoretical reviewers to be unable to understand the systematic construction of a methodology, since they were only able to think of the possibility of a systematic order in the logical-formal area. In all the parts not handled with logical symbolism, the thought of rigorous logicality in forming concepts and drawing conclusions was increasingly disappearing, while formerly it had supported every form of serious science. The intimate reason for all of this is an unconditional but wholly unjustified adherence to a basic empiricism which is believed in an uncritical way, almost like a form of religion. It is also a kind of escape from our own responsibilities, because (a) all rigorous thinking is restricted to the apparently safe field of the rules of mathematical calculus, so that we can rid ourselves of the need to deal with it, and (b) "nature" is regarded as ordering and deciding by itself, so that we don't have to bother about method and order. It has been forgotten that absolute rigour also exists in non-mathematical, non-calculating thinking. On the other hand, technics and *modern technicians* have still not been able to find a bridge that allows us to insert technical thinking and working into the general development

[27] Cf. *Die Ergreifung des Wirklichen*, cit., pp. 196 on.

[28] Cf. *Die Methode der Physik*, cit., pp. 201 on.

of Western thinking, demonstrating how they are *the immediate and intimate bearers of the great spiritual tradition of ideal thinking*. Here there is a big deficiency in the training of our technicians, which takes away from them the possibility they are entitled to of operating in the conceptual area.[29] By "system" modern theoreticism can only mean a logical-formal, hypothetic-deductive system, an axiomatic system with the addition of its logical consequences, and those who have grown up in theoreticism cannot get away from this field, no matter how hard they try. Here we are dealing with something that is fundamentally different, an *activity* system, which is a well-established succession of intentional activities carried out for a particular aim, an *activity plan*. You could actually say that a theoretician would fly into a panic if he came into contact with the manual basis of something that is physical, so much so that he would usually shy away from it and find refuge in his beloved field of things that can be formalised, although in the long run this is never possible. The tradition of exact logical thinking *with content*, which was found by the Greeks and has supported all forms of rigorous science for 2,000 years, has been lost by those people who nowadays are only expert in symbolic calculations.

Theoreticists "*don't grasp that realm that exists in the middle, that extraordinary and decisive sphere that exists between untouched nature and thinking, the intellect*, the sphere where the hands, guided by the spirit, work effectively, model the real material and prepare and transform the things from reality so that they have determined properties and provoke univocally determined effects. They miss the convergent *process* of *realisation*, which is incredibly complicated and collectively extended throughout that part of humanity that is the custodian of a culture and divided into millions of small everyday operations, destined to prepare and connect in an increasingly perfect way objects that are ideally determined, a process that we have identified in the 'exactness-process'. This divorce of the conceptual from real events, this extraneousness from the world that are found in theoreticism are probably only possible thanks to the division of labour that took place in the nineteenth century. Experimental physicists were induced to ignore the constructive work of hands that alone can carry out their measuring experiments, because *more and more frequently they receive the ready-made measuring instruments from the factory*. As a consequence of this the hands that deal with the *realm in the middle* and apply the products to quantitative research are very different from these. The former is in the hands of the engineers and workers in precision-engineering factories, who have no contact whatsoever with the theoretical and methodical problems of physics, while the latter is the job of physics professors, who are actually in contact with certain theoretical problems but often ignore the manual realisation of the fundamental forms. That brings us to the attitude of modern theoreticians, who do indeed know something about the activities of physics professors who carry out their measurements but know nothing about the realm in the middle, which is entrusted to other hands".[30] And since that realm in the middle is unknown to them, they believe they can propose unconditionally and without limits their theoretical

---

[29] Cf. pp. 411 on.

[30] pp. 422–424. Italics in text.

formulations, which are simply the numerical results of their measurements and are considered in themselves, without any thought for their coming from the realm in the middle that influences them.

Naturally, our philosopher observes, a person can also behave in such a way as to *want no* order and consequentiality in thinking and in methodology – in short, in science. So the researcher must decide if he wants univocalness or not. In the former case, Dingler is quite sure, there is nothing but the univocal-methodical system. In fact if the empiricists and the theoreticists (which nowadays are the same thing) have not any difficulties that prevent them from conserving and pursuing physics, it is because unknown to them Dingler's codified methodology is applied in the areas where they operate, and consequently they never find themselves in the position of having to effectively start physics *from the beginning*, from untouched nature, according to *their* principles. But in the theoretical field as well they continually use the so-called classic patrimony, and it is only on the margins of this that they carry out their "revolutions". They are guaranteed this patrimony by solid exactness, even though they are not able to justify it with their methods. It is a curious thing that *empiricism has proved to be a long way from reality, while only the system that comes from the idea can properly comprehend what is real in physics.*[31]

It must not be denied that man's desire (to be ethically grounded) to be as far as possible the master of his own environment, dominating from the manual and conceptual point of view the natural circumstances with an ever-increasing exactness, has been present for a long time and has always been conscious, and if in the past it has operated in a more instinctive way, nowadays we have become more and more conscious of it. The univocal-methodical system is still completely under the control of the conscience and *it is assumed with a full conscious will,* so that every instinctive movement is abandoned and replaced by intentional acts only, based on clear aims, and only those who discard the final aim can raise objections. Nevertheless, according to Dingler the will of all men strongly and spontaneously tends towards this aim that they will never abandon, since men have a very strong sense of what is helpful and what is not helpful for preserving their community.[32]

Our philosopher then observes that theoreticism is necessarily disjointed if it refuses to explain some deviations from a fundamental law, postulating the presence of a disruptive element (which nowadays we would say: by means of an *ad hoc* hypothesis). It is already forced in thousands of cases to apply this procedure and cannot provide any exact criteria for the cases in which it must or must not be applied: in this way it is faced with arbitrariness and inconsequence if in some cases he refuses to apply it. In everyday life we continuously work with unobservable elements, and if we only wanted to concern ourselves with immediate sensory perceptions, we would never again make head nor tail of our existence. E. Mach set out this programme for physics and attempted to implement it in its phenomenology, although neither he nor his students were able to do so. It was believed that such a programme would entail being particularly "close to reality", but critical research,

[31] Cf. pp. 426 on.

[32] p. 435.

as we have already seen, shows that true reality is ingenuous experience of the things that make up everyday life, and that the limitation to the sensationalistic screen of the colour-spots etc., which the phenomenology aims at, represents only a poor arbitrary abstraction. This, however, is not real but is a *theoretical construction*. Philosophy has produced entire libraries in an attempt to establish what is real, without success. How can it start from such an uncertain and artificially unilateral abstraction, when in addition any conclusion lies completely outside its field? Luckily philosophy has no need to build on those unclear philosophical speculations. It can be demonstrated that without assuming theoretical elements which are already complex and rely themselves on physical considerations it is not possible to identify a clear primary division between what is purely phenomenological and what is "added" to experience. So this division cannot constitute the basis of physics, and this is enough for the "philosophical" aspect of the matter. Unobservable elements that are larger in number and different in principle from the usual everyday life are not introduced into the univocal-methodical system. No physics can do without this, and in point of fact everything that is not part of direct temporary perception is an "unobservable element", as is everything that is "behind" observed objects and everything that is inside non-transparent bodies. Therefore the need to eliminate these elements represents an arbitrary cutting operation and can never be strictly defined. The only need that must be imposed is that non-perceivable temporary elements must be thought of in such a way as to become *observable* in such cases. This need is only satisfied by the univocal-methodical system, whose constructions are always so planned that when there is the possibility of a direct perception it turns out to be of the type that you experience directly and is therefore "observable".

Summarising the Dinglerian perspective in the way it appears from these analyses, we can say that every physical result obtained from quantitative physics comes from the following three stages: (a) the manufacture of measuring equipment and the preparations for the experiment; (b) the dressing and execution of the measuring experiment, which provides the measurement numbers; (c) the theoretical and mathematical explanation and treatment of the result of the experiment. Unfortunately, nowadays these three stages are generally found in different hands: (a) takes place in specialist factories; (b) is carried out by "experimental physicists"; (c) is carried out by "theoretical physicists". In this way (a) provides (b) with ready-made equipment and (b) provides (c) with the results of the experiment in the form of measurement-numbers. Naturally (b) is not concerned with what happens in (a), i.e. in exactly how the equipment is constructed and which decisions are included in it; (c) is even less interested in what happens in (a), and is only interested in the "experiment scheme" in (b). Modern physics' "philosophy" places no value on the fact that "at least half" of the definitive measurement results are determined by (a), that the forms of (a) "fit in with the measurement results" and that therefore (a) includes the second source of the measurement results together with the measured process. This philosophy is therefore false, so it is necessary to review the fundamental philosophical axiom of all modern physics, which is expressed in different forms: "Everything from experiments", Everything from experience". If by experiments and experience we mean an *exclusively passive behaviour* with regard to nature, then both of the conceptions are false, and nowadays physics *tacitly* means these two concepts just in a passive sense. This conception contains an objective error, as should be clear to anyone who has studied even just once how the equipment is manufactured. If it is now unlikely that the concept of experience is not meant in the passive sense, it is not necessary that this applies to the concept of experiment. Ever since experiments became conscious, which was more or less around the beginning of the seventeenth century, this concept has always referred to *total progression* that leads from untouched nature to ordered knowledge. Yet this total progression also includes the activities used to manufacture the equipment and the measuring instruments. From this we can see that the slogan "everything from experiments", which has become so popular with modern physicists, can be safely repeated only if we are always conscious that the concept of experiment can never be meant in a purely passive sense but also refers to the *active and formative* part, which is really essential, of research into nature. On the other hand, the expression "everything from experience" should be completely banned from the science of nature because it is false in principle.

Dingler states that, as we have already seen in part, mathematics is no longer an auxiliary instrument that is used to render in a symbolic form the thoughts imposed in another way in order to reap the consequences with its automatized and therefore unfailing method, but is *increasingly becoming the object itself.* Real mental research, in physics, now *no longer comes from the pre-mathematics field* and mathematics doesn't just lead to the mathematical consequences of this research, but *the research takes place in the mathematical formulas themselves.*[33] We've seen that the foundations of classic mechanics are developed in the form of Newtonian equations and Galilean transformations. If we carry out a study based on detailed knowledge of the history of the effort put in over a period of 2,000 years to produce Newtonian mechanics, if we are aware of the difficult conceptual problems that had to be solved over this long period, then we will be aware of the enormous amount of formative work that was necessary just to arrive at the intuitions from which Newton successfully began when building his theories. All of this is completely ignored and disappears in the mathematizing conceptions. If we were only dealing with preliminary work that could be forgotten once the aim of the work had been achieved, we could quite easily forget it, but this is not the case. This preliminary work also contains, certainly not in an obvious way, the real future, the intrinsic validity of the Newtonian foundations, or at least it contains many circumstances that can help to see the profound validity of these foundations. Theoreticists now preclude this aspect of the situation, remaining true to the legend that these foundations were obtained simply through experience, and that even where this didn't happen it could have happened. At this point there is decisive proof that such an empirical origin is not possible. The heart of this proof lies in the fact that all natural forms, from which it is believed that these foundations can be drawn, represent extremely specific cases. If we liberate ourselves of unconscious prejudices, which consist in only considering these

[33] Cf. pp. 455 on.

particular cases, we can immediately recognise that to the unprejudiced eye nature *offers* not just these but *innumerable other* cases. For example, for those who don't know about universal gravitation, nature doesn't just offer bodies that fall but also those that rise and move in all directions. Therefore how could such people determine "from experience" that all bodies fall? *Only if they had decided to classify as "falling"* all bodies could they also explain the movement of bodies that rise as a distorted case, and obtain a unitary, methodical treatment for all bodies. However, this doesn't come from experience but from *a mental decision* and from an interpretation of what is immediately experimented. Mathematicism not only precludes all of this profound, far-reaching knowledge but also the simple position of the problem, when it advances its legend "everything from experience". As soon as I fell into this legend, the entire problem of the type of validity and evidence of the foundations was excluded. Newtonian equations and Galilean transformations for me simply represent an "economic compendium" of the measurement-results obtained up until now, so that I can certainly decide to replace them with other formulas if these only have to correspond to the measurement-results.

A big danger with this mathematicized approach lies in the possibility that purely mathematical properties of a similar equation-system are regarded as physical facts without carrying out a detailed study. But *no formula of physical theory is an immediate expression of nature*: each one of them can only be obtained after a series of methodical activities which are manually carried out on the basis of conceptual ideal forms, therefore it is obtained either on the basis of our elementary ideal forms or, if measurements are involved, through the use of measuring instruments which have been constructed on the basis of these ideal forms. In conclusion, each formula possesses a methodical layer between itself and untouched nature, a layer without which it would not be possible, and these formulas, when the construction, order and foundations of physics are involved, can only be applied in a correct, reasonable way on and with this methodical basis.

The three traditional solutions to the problem of universals or, to say it in the modern manner, laws of nature (the Platonic or realistic-exaggerated, the Aristotelian or realistic-moderate and the nominalistic), hold that the reason that determines the universal laws of nature lies in nature itself as its extreme background. The fourth and appropriate solution, which holds that such reason consists in the ideas that we ourselves produce when trying to achieve certain goals and that are implemented in reality by means of our instrumental equipment, has only been considered in more recent times. So we recognise that one of the essential points of the modern natural philosophy, which is the question of the real essence and type of existence of the laws of nature themselves, is none other than the exact reproduction of the medieval problem of universals. The explanations proposed for this show how nowadays we are still concerned with the heart of the question. The fourth and most appropriate solution must distinguish: (1) Those laws of nature (if in general we want to call them so) that constitute the fundamental forms and according to which we construct our measuring equipment (the forms that can always be freely constructed with the hands); and the other combinations of these forms. These themselves are not assertions concerning nature, but they form the preliminary conditions which generally make it possible to make the assertions in an exact manner. This group includes Euclidean geometry laws and Newton-Huygens laws of mechanics. These "laws of nature" are therefore ideas that are finalistically formed in us, which we force ourselves to carry out in a better way in reality. With this equipment we then approach natural phenomena in order to manually manipulate them according to these fundamental forms and measure them. (2) This produces a second group of laws of nature, known as "empirical" laws. Using the "theory", we'll try to connect these as closely as possible with the fundamental forms so that they are increasingly more structurable and controllable by the forms that can be manually manufactured. There is a tendency to construct them as pure combinations of the fundamental forms and as a consequence to transfer them into the first type of laws of nature. In addition, these empirically measured laws are not found as such in nature, since they are only chosen and clearly defined in nature through the equipment created by us. (3) A third type of laws of nature is represented by the constancies and uniformities that we can experience in reality. These are not absolute in a rational sense, and we must always bear in mind that they can be revised. They are never universals, they are constantly *singularia*, but they are found in natural datum itself and do not need to be identified by any equipment or be measured.

We hope that this look at the Dinglerian doctrine concerning the nature of scientific knowledge has in some way succeeded in its aim, which is to outline the image of a singular and original philosophical personality who was capable of uniting, in a novel and in many ways fascinating combination, classic aspirations to the ultimate foundations of knowledge and to absolute epistemic rigour with a rare, lively sensitivity for the type of rationality that unfolds inside the technical-operational dimension of existence. Just as Bergson acutely said that Kant had subjectified Platonism, we can say that Dingler, whose inspiration was authentically based on Plato, made it pragmatic.

The Permanent Creativity of Self

Stefano Polenta

Abstract Within the area of psychoanalysis and epistemology of complexity, a conception of the subject intended as originally creative, active, and constantly expanding has been spread. This idea has been ignored for too long, because of the prejudices of a reductionist science that could not conceive how subjects would always be considered as agent, intentional, and able to self-generate their own world of meaning. Creativity is also the principle which allows the creation of value, because each person brings a "differential" with respect to the reality deriving from its own irreducible subjective look.

Creativity, however, if it does not want to degenerate into a self-reverie, has to be intended as a principle of inclusion, since it cannot be separated from the world, from the "object"; creativity demands to take reality into account and to include it in the construction of the world of meaning of the subject, to operate "synthesis" increasingly mature and complex between subject and object. In the epistemology of complexity this concept is expressed by the assertion that individuals are simultaneously self-referential and hetero-referential. This also origins an ethic implication, because man is obliged to "take charge" of reality (from his subjective point of view), reaching higher and higher levels of integration between subject and object, rebuilding both the sense of the whole humanity (C.G. Jung, E. Fromm, L. Sander) and the meaning of the whole universe (T.S. Eliot, W. Bion, A.N. Whitehead).

Introduction

Allow me to comment on the topic of this conference, from which I took inspiration for my contribution. When my colleague Daniela Verducci talks about the topic of the *62nd International Congress of Phenomenology – The forces of the cosmos and*

S. Polenta (✉)
University of Macerata, Via Einstein, Pollenza, MC 62010, Italy
e-mail: polenta@unimc.it

A.-T. Tymieniecka (ed.), *Phenomenology of Space and Time: The Forces of the Cosmos and the Ontopoietic Genesis of Life: Book One*, Analecta Husserliana 116,
DOI 10.1007/978-3-319-02015-0_31,

the ontopoietic genesis of life – in my opinion it seemed very "explosive". My attention was primarily drawn from the use of the term "forces of the cosmos" rather than "laws of the cosmos". This terminology appeared to me particularly appropriate because it refers to a conception of the universe understood not as a mere mechanism governed by laws, but rather as something "alive" and able to generate the new, to be creative.

The idea of a creativity of the universe was formulated with great force in the early decades of the last century by philosophers such as Alfred North Whitehead and Henri Bergson and it has reappeared recently in the so-called theory of complexity. This latter has tried to encode a perspective of thought that matured during the twentieth century and is one of the most important intellectual achievements of this century. This perspective is well summarized in the title of a famous 1972 paper by the physicist and Nobel laureate Philip W. Anderson: *More is different*. Here, Anderson argues that

> at each level of complexity entirely new properties appear. At each stage [of the science] entirely new laws, concepts, and generalizations are necessary, requiring inspiration and creativity to just as great a degree as in the previous one. Psychology is not applied biology, nor is biology applied chemistry.

From the perspective of the complexity the phenomena are not reducible to the properties of the parts of which they are composed (*reductionism*) because they develop self-organizing processes, which generate "emergent properties". Another way to express this concept is to say that "the whole is more than the sum of its parts". It follows that breaking the phenomena in their component parts, without taking into account their overall functioning, prevents the capture of their essence. In that there is a phenomenological attitude. With this, the theories of complexity are not a remake of "vitalism". On the contrary, the appearance of emergent properties is seen as an expression of "normal" dynamics of the systems.

Which consequences can we draw from this perspective about the self and the subjectivity?

First, the Self is not to be understood as a mere aggregate of parts, but as an integrated system result of a self-organization. How can it be understood that "self-organization" that "pushes" the self to be structured?

> How the principle of wholeness, or coherence, operates remains one of the mysteries of the life process, which, for the most part we resist confronting – or of which we remain totally unaware – taking it for granted, without thinking (Sander 2002: 17).

Donald Winnicott wrote beautiful pages about the topic of creativity of the Self, highlighting how this ability is not something "technical", but the more profound essence of being subjects; this essence is not a "substance", but the innate spontaneity of the subject, his way of self-organizing external and internal experience.

Secondly, creativity can also be seen as the process that led man to emerge from the nature and constitute himself as autonomous and separate individuality. This separation, however, does not exclude man from nature, from the psychobiological dimension that exists prior to the individual one. It may be thought that the self has a dual affiliation: on the one side, the self is part of the biological and natural world; on the other side, there is a separate individuality as the result of processes of emergence. Man is apparently detached from nature, but in fact continues to be part of it: "Man is in the nature, he is the nature itself that thinks" (Pirandello 1910). Although man has an agency, the "forces of the cosmos" continue to act in him. In achieving his "life project", the subject organizes and shapes the nature of which he remains a part. This has ethical implications, because man is required to account for the profound logic of the reality of which he is part, to set free his potentiality, so that every human fulfillment is a sort of living hologram of all that exists. This process of "expansion" of the self to account for the mystery of his own soul and, at the same time, the mystery of all that exists cannot have an end. Heraclitus said: "You will not find the boundaries of the soul, as far as going forward, so much deeper is its reason" (fragment 45). The creativity of the self must be understood as a permanent process.

To develop what is said above, the contribution will be divided into two parts:

1. First, it will be demonstrated that the self is active, with an agency, and that this agency is a primary quality of the Self. It is the result of a process of "emergence" from a biopsychic size of which the Self remains part. This "emergence" of the self is a process, a continuous reorganization of the experience in a subjective manner. As such, also, the Self has a bivalent nature: it exists both as an individual and part of the whole. For this reason, the self is never completely thematizable, because it cannot objectify the all-pervading background of which it is part;
2. In the second part some ethical and pedagogical implications of this perspective will be drawn, recalling the thought of the philosopher Alfred North Whitehead and of some psychoanalysts who investigated what can be defined as the "mystical" side of psychoanalysis (which examines not only the emergence from the individual biological dimension, but also its remaining a part of it).

At the Origins of Creativity of the Self

The creativity of the self can be understood almost as a synonym for "being a subject". It is necessary to clarify this statement. There is "a real artistry in the person" (Thomson 1997: 6). Everyone has its own subjective "style". It is appropriate to speak of "style" because each person has a peculiar way to "touch" things. Think of an artist that we like. It is not "what" he says, but "how" he says it. This "how", however, is not only a technical aspect: it represents his way to relate to reality. We are well aware when, for some time, we lose this precious subjective dimension, which is to possess a subjective view on our world, our "creativity" that introduces a "differential" with respect to the reality. When it is not present, we feel like a mere "thing among things". Sartre well described the "nausea" that we feel in these situations.

> Now I know: I exist – the world exists – and I know that the world exists. That's all. It makes no difference to me. It is strange that everything makes so little difference to me: it frightens me. Ever since the day I wanted to play duck and drakes. I was going to throw that pebble, I looked, and then it all began. I felt that it *existed*. Then after there were other Nauseas, time to time objects start existing in your hand (Sartre 1938, tr. engl. 2007: 122–123).

Subjective style is clearly visible and recognizable already in very young babies. The *Infant Research* also shows how infants are capable of interacting as real subjects. Colwyn Trevarthen is one of the most careful of these researchers. He says:

> One of the difficulties in working with babies is that they have their own minds. Sometimes, when they don't do certain things, is because they do not want to do them, and when they want to do something, they make every effort (Trevarthen 1993, tr. it. 1998: 149).

For Donald W. Winnicott creativity is central. He thinks that "at the centre of every person there is an incommunicado element, and this is sacred and most worthy of preservation" (Winnicott 1963: 187) He calls it "true self". It belongs to the creative and spontaneous gesture of the subject. Winnicott's true self is "being before doing". An individual feels himself if he keeps in touch with his own creativity. Leading a creative life means "not being constantly killed or destroyed by being complacent to [...] a world that treats the individual with violence, being able to see everything in ever new ways" (Winnicott 1970, tr. it. 1990). In contrast with the true self is the false self, which was developed on the complacent basis. In fact, without the true self there is a mere reaction to circumstances.

The perspective of complexity of which we are reminded allows us to provide an additional theoretical framework for the insights of Winnicott.

The biologist Stuart Kauffman calls *agency* to be an agent of the subject. The term "agency" very well gives the idea that living creatures can be "active" by virtue of the processes of self-organization. The agency exists naturally in the universe like the moving particles. It is not a mere illusion.

> A couple in love walking along the banks of the Seine are, in real fact, a couple in love walking along the banks of the Seine, not mere particles in motion. [...] Emergence is therefore a major part of the new scientific worldview. Emergence says that, while no laws of physics are violated, life in the biosphere, the evolution of the biosphere, the fullness of our human historicity, and our practical everyday worlds are also real, are not reducible to physics nor explicable from it, and are central to our lives (Kauffman 2008, p. x).

The idea that individuals have an irreducible agency is very important on the pedagogical side. Today, in fact, it rather seems to have a "usable" concept of subjectivity and to give a less weak status to that *unicum* which each of us is, without falling into the dualism between mind and matter that is currently one of the central philosophical problems (Searle 2004). Speaking of self or subjectivity, many scholars fear to reintroduce a *mysterious central agency* (Allport 1955), or something very "close to the idea of soul" (cf. Jervis 2011: 59). But how is it possible to define an educational project with no anchor to the "singularity of the person" (Frabboni and Pinto Minerva 2003: 82) or to its "own reasons" (Laporta 2001: 28)?

The perspective that there is a creativity in the universe, of which the human agency is an expression, returns in many of those who adopt the complex thought. Although it may seem not so "scientific", this perspective aims at integrating the idea of reality that the classic science gave us. This latter thought of the universe as a mechanism or as a "big clock". Later, when with the discovery of thermodynamics and disorder it became evident that this model was inadequate, we began to see disorder everywhere. To counter the tendency to disorder (entropy) we can, however, force the nature to work in an orderly way, using a machine. In this way, we started thinking – as noted by W. Köhler (1947, tr. it., 1989: 75) – that nature imitates the operation of the machines we built. Where the machines break, chaos returns to reign supreme. For the theories of complexity, however, there are constructive processes that occur naturally in the universe. Prigogine observes in this connection that the universe produced not only chaos and dispersion, but "these beautiful flowers arranged in a vase by my wife" (Prigogine 1996, tr. It. 1997: 54). With this statement he wants to convey the idea that complex structures are the result of the normal dynamics of the forces of the cosmos. This "creative" dimension of the universe has been interpreted as a sign of a "new second law of thermodynamics" (Waldrop 1992), yet to be discovered, that – contrary to the tendency to entropy and disorder provided by the second principle of thermodynamics – would attest the existence of constructive and integrative forces in the reality. The laws of physics are not understood by complexity theorists as the logical structure of the universe: rather they have a "local" validity and follow, rather than precede, the radical creativity with which phenomena occur. Also Kauffman believes that the history of the universe is animated by a profound creativity that can generate the new. Also the human agency is an expression of this creativity.

> If no natural law suffices to describe the evolution of the biosphere, of technological evolution, of human history, what replaces it? In its place is a wondrous radical creativity without a supernatural Creator. Look out your window at the life teeming about you. All that has been going on is that the sun has been shining on the earth for some 5 billion years. Life is about 3.8 billion years old. The vast tangled bank of life, as Darwin phrased it, arose all on its own. This web of life, the most complex system we know of in the universe, breaks no law of physics, yet is partially lawless, ceaselessly creative. [...] I believe we can reinvent the sacred. We can invent a global ethic, in a shared space, safe to all of us, with one view of God as the natural creativity in the universe (Kauffman 2008, pp. xi–xiii).

Seeing creativity as a driving force of physical/natural processes including the Self, has spread in various disciplines. For example, the psychoanalyst Louis Sander long since has adopted an epistemology of complexity and he tried, within his activity, to find "the basic principles governing life as an ongoing creative process" (Sander 2008: xiii). He writes:

> We think [at the consciousness] as an outcome of the evolutionary process and not in terms of human consciousness as being with the "direction" of a process of change over longer spans of time in the history of time and evolution – an overarching creative process moving us toward a global ecology that we are within but, on the level of the individual, as yet do not conceive. The evolution of human consciousness will reach a new level of both complexity and unity, and a new domain of creativity once we comprehend the inclusiveness and overarching direction of the creative process we are within and the principles that govern it (Sander 2008, p. xv).

In the above passage, Sander uses the term "process" several times. The transition from a way of thinking in terms of "structures" to a way of thinking in terms of "processes" is part of the paradigm shift that the theories of complexity are seeking to develop. The agency provides a systemic organization in living organisms, without which they could not develop their own "internal models" to interact with reality and could not survive; but this organization is constantly changing, reconfiguring on the basis of the interactions with others and with the environment. Sander uses the very effective term of "living systems". So, the integrity of the living system is not so rigid but it is constantly modified by interactions with the environment. Therefore, there is an integrity which is maintained despite the continuous changes. Indeed, integrity is maintained not "despite", but "thanks" to these continuous changes because the lack of change would lead to stagnation and death. When we talk about topics such as the Self, the identity or characteristics of a person, it is then important to understand these concepts in terms of *process* and not as a rigid structure. If we start with the self as unchangeable structure, as a monad, or as a static essence, it becomes difficult to consider the process of living. In fact when we refer to terms such as self, subject, identity, we should use more appropriate terms such as "individual process", "subjectivation process". "Life is a continuous flow that we seek to arrest, to fix in certain stable forms inside and outside of us" said Luigi Pirandello. It is more useful to understand the self as a process and, from this, to highlight the persistence of structural stabilities; yet this should never be seen as static, but as the result of an imbalance between moments of tension and moments of balance, harmony and disharmony, as John Dewey noted. The moments of harmony are celebrated as precious because they stand out against a background of disharmony, they are moments of nirvana in the chaos of the world, because when nirvana is absolute there is only stiffening and decadence. But where there is neither integration nor balance, there is only chaos and dispersion. The intimate precariousness of the moments of harmony does not undo them, but makes them more real and precious, because every moment is different from the previous one. It is not merely different: every moment contains the previous one and exceeds it, having accepted the new moment and converging in the next one. Every moment is lost and found in the following one. Bergson had given to this stream the name of "duration". Both moments are needed in their dialectical tension: stasis and change, structure and anti-structure, order and chaos, life and form. This is a paradox of life that is not easy to think.

I think I'm moving closer to the central theme: the permanent creativity of self.

Even in Winnicott there are many paradoxes of this kind. In his opinion, creativity of the Self grows interacting with reality, but the Self "realized" is not the nucleus of the creative self. For Winnicott the self is never completely objectified. For example, he says that he could not conceive of an artist who comes to the conclusion of his work because in every artist there is "an inherent dilemma, which belongs to the coexistence of two trends, the urgent need to communicate and the still more urgent need not to be found" (Winnicott 1963: 185). Even Friedrich Nietzsche often had this feeling: "Everyone is the furthest thing for himself" he said in *The Gay Science* (Book IV, Aphorism 335).

Initially the True Self does not contact reality. Without "object" the creativity of the Self cannot assume any consistency. One can imagine that the baby has creative impulses or hallucinations without an "object". If the mother is able, with empathy, to provide the object that the child is hallucinating and …voilà! The object magically appears. The child is thus able to contact reality without any effort, because it is the mother who provides "objects" to his hallucinations. A beautiful omnipotence! For Winnicott, a little creative omnipotence remains in our acts of adult life. For example, an artist has a blank canvas in front of him. Suddenly he takes the brush and throws a bit of color onto the canvas.[1] The painting can begin. As the child learns to interact with reality he "binds" his creativity to it. Creativity, then, starts addressing precise aspects of reality, which it takes into account. The result is a process of growth and education of the true creative self. Winnicott calls this "negotiation" between true self and external reality "transitional area". This area initially regards the "transitional objects" that some children take (the teddy bear, Linus' blanket etc.), but then it becomes the area where people live as healthy adults, is the area of cultural, religious, political, artistic interests, etc. In the transitional area there is an ongoing "transition" from the perspective of the inalienable true self and equally incontrovertible existence of a reality. Just because both poles are irreducible, this area contains a paradox. In fact, for Winnicott, as it says, the creativity of the true self is an infinite potentiality of meaning, he can never be completely realized, discovered, communicated, saturated in "object".

> I suggest that in health there is a core to the personality that corresponds to the true self [...]; I suggest that this core never communicates with the world of perceived objects, and that the individual person knows that it must never be communicated with or be influenced by external reality. This is my main point, the point of thought which is the centre of an intellectual world [...] Although healthy persons communicate and enjoy communicating, the other fact is equally true, that each individual is an isolate, permanently non-communicating, permanently unknown, in fact unfound (Winnicott 1963, p. 187).

Also Mauro Laeng thinks that there is a core of personality, which he calls "I", which is eternally unknowable:

> The I thinks the Me, and there is a distinction between the two: the first is the transcendental all-embracing horizon, which cannot be bypassed, the second is always a particular empirical content, linked to the identification of *hic et nunc* [here and now] in space and time and to present being and to previous becoming (Laeng 2004, p. 57).

Since creativity of the self can never be fully "captured" by its concrete realization, it is always declined to the present, as a project to be always carried out and never definitively achieved. That's why the process of formation of the Self can never have an end.

With regard to the "transcendental all-embracing horizon" mentioned by Laeng, in the following part it shall be pointed out – with input from some philosophers, psychoanalysts and poets—that such a non-thematizable dimension of the self can be understood as our rootedness in the universe. The Self is an agency glimmer which emerges from the bottom of the shadowy life to make his own individuation path, but which has to go back to that life, in an endless circularity.

Permanent Creativity of the Self Between Part and Whole

The Self has a dual affiliation: it is simultaneously a separate individual and a part of a whole; it has a "personal" agency and it is a part of the creativity of reality.

[1]This example is in Strokes (1965).

Talking about a "creativity of reality" may seem a risky assumption. This intuition, however, is present – in a more or less explicit way – in different scientific and cultural perspectives of thought. One of the more mature formulations of this concept is found in Alfred North Whitehead, for whom reality is a cosmic creative process which increases its value thanks to the "generation of individual actual entities" (Whitehead 1929: 75). In this perspective, the permanent creativity of the self derives from the creativity of reality, that generates individual entities to reach an "intensification".

Even psychoanalysis has placed at the center of its reflections the theme of the emergence of man from a natural dimension that pre-exists him, confronting it with the "paradox that arises between mind and matter organic", quoting C. Trevarthen.

C. G. Jung and E. Fromm interpret this paradox in a non-reductionist way, asserting that being creative is the specific "task" of man, who must develop the activity of the nature to a specifically "human" level.

We have seen that in Winnicott there is a specific focus on the theme of creativity as a core of the True Self. For Winnicott creativity is permanent and can never be definitively "saturated" in the "object". The True Self is a "becoming" rather than a "being," a process rather than a structure.

From such considerations some ethical implications derive.

For some authors such as Hans Loewald and, above all, Wilfred R. Bion it can be argued about a "mystic" development of psychoanalytic thought.

In Bion, in particular, man has a paradoxical relationship with the truth, because it belongs to a dimension – reality – that man cannot fully express and realize. Man must be conscious of the radical partiality of his achievements, because this is the result, we could say, not using Bion words, of an agency that is separate from the creativity of reality, though it is an expression of this latter. From here it derives the consciousness "that all thought as we normally know it, that is as an attribute of the human being, is false, and the connected problem is the degree and nature of the falsehood" (Bion 1976, tr. It. 1981: 29) and a radical profession of "humility", according to T.S. Eliot in *East Coker* (1940): "The only wisdom we can hope to acquire/Is the wisdom of humility: humility is endless". Man must be careful not to interpret the achievements of his creativity as a possession, from which "perversion" would arise (Bion). Creativity is a "being at the service of", a being available to listen, develop and "realize" (Pirandello 1897) reality, a widening the boundaries of one's own thinking to include and integrate all aspects of reality, without any exception.

Alfred North Whitehead

Whitehead – who anticipated many typical themes of complex thought – believes that reality should be thought of in terms of relation between parts and whole. This is, in fact, a "philosophy of organism" for which the reality consists of a cosmic creative process that realizes itself through a "generation of individual actual entities" (Whitehead 1929: 75). These entities apparently abandon the universal network to which they belong, since they have "self-creativity" and "subjective goal". But this abandoning is only apparent, because the individual is still part of the universe and maintains a substantial "solidarity with the universe" (ibid., 56). So, for Whitehead, "every actual entity is present in every other actual entity" (ibid., 50). It follows that every action has a double character, because it allows creatures to achieve their goals and influences the whole universe: "everything that in any sense exists has two sides, namely its individual self and its signification in the universe. Also either of these aspects is a factor in the other" (Whitehead 1938: 11). Whitehead called "concrescence" this co-evolving of parts-everything. Also the individual entities, as part of the cosmic creative process, are to be considered as a creative process. Whitehead believes that separating the two concepts is a serious mistake (Whitehead 1938). In this conception, he is near to three great philosophers who emphasized the procedural and dynamic nature of the self and its making in relation to reality: James, Bergson and Dewey. The ultimate goal of the creative process – of which God is the "eternal primordial character" – is "intensification". From Whitehead's reflections an important ethical implication derives: as part of the process, every subjective realization is also a realization of the cosmic creative process of which it is part.

Creativity and Mysticism in Psychoanalysis

Psychoanalists Sandor Ferenczi, Ignacio Matte Blanco, Erich Fromm, Carl Gustav Jung, Hans Loewald, Wilfred Bion, Louis Sander – who have a large speculative imagination – focused on the study of the subjectivation process by which the human emerges from the pre-existent natural and bio-psychical dimension. Thinking of this emergency, without separating with a "metaphysical abyss" (Searle 2004), the biological dimension from the mental one has been the focus of psychoanalytic thought. For psychoanalysis, man is a "human animal" and man's thinking is also a nature thinking in him. "Man is in nature, is nature itself that thinks" (Pirandello 1910). Sigmund Freud believed that what allows the emergence of the human is the taboo of incest. This is relived by each in the Oedipus complex and it prevents the full satisfaction of pulsions. Man, says Freud, preferred the security that comes from a civilized life to the full satisfaction of the unconscious and its pulsions. The "Civilization and Its Discontents" is unavoidable. Ferenczi, Matte Blanco, Fromm, Jung, Bion, and Loewald have started thinking about the unconscious not only as a place where low pulsions lie, the legacy of our animality, but also as an original size of membership by which man is off to fulfill his path of humanization, but he must return eternally, in an endless circularity. They discovered the "creative" power of unconscious, his being an original dimension with which man must stay in contact to make his own subjectivation process. This is what might be called the "mystique" of psychoanalytic thought, with an observation that, of course, psychoanalysis as a scientist and psychoanalysis as a mystic must think together and not against one another (Carere-Comes 2003). Thus, in psychoanalysis there is the thematization of the relation between parts and whole, the emergence of the individual from one dimension of which he continues to be a part. That's why the authors mention that the individual is conceived as an "individuation process" in constant growing, where the relationship with the unconscious is central.

For Jung, the man is a "hero" that separates from the "Great Mother", which is nature, to do his "individuation process". He becomes conscious of himself and of his separateness with dismay and, simultaneously, comes into contact with a large array of meanings that is the collective unconscious; this is the memory of the process that man has made to emerge from nature and to develop a mind, and it contains the kind of thinking and feeling of our ancestors, their way of experiencing life and the world, men and gods. It is an expression of continuity between mind and nature, the long line of derivation that led from the animal to humans. The unconscious contains such useful elements to support and nurture the individuation process.

Loewald (1960) gives us a particularly evocative metaphor. He speculates that, originally, all our mental acts were part of a "primal density", similar to the concentration of matter before the Big Bang. Subsequently, the matter has expanded, separated, organized. But the primal density remains as a place full of meaning, of integration of the parties. It is necessary, says Loewald, that the development and the articulation of our psyche maintains a connection with the primal density, otherwise the content of our minds loses lifeblood and impoverishes, lacking passion and imagination, what Winnicott calls "escape into mental health".

Also for Bion the subject constructs himself with a specific function of personality, which he calls *alpha function*. It allows to extract emotions and thoughts (essential for the development of personality) from a "protomental" system, which represents the biopsychical-animal size, common to all men and animals. Gradually, Bion began to emphasize the subjectivation principle as working with "protomental material" assumed for "private purposes" (Bion 1970, tr. it 1973: 20). In this way, the subject can have his own mind and live his life. Therefore, the subject always exerts a "force" (the term is not Bionian) to include the protomental within some categories (psychic "containers") useful to the subject. It follows that, for Bion, in each subjective transformation there is an unavoidable portion of untruth: "falsity is the characteristic of thought within an individual, or of thought within a container. It follows that all thought as we know it normally, that is as an attribute of human beings, is false, and the problem is the degree and nature of falsity" (Bion 1976, tr. it. 1981: 29). At the same time, Bion begins to consider the protomental not only as a primordial dimension, but also as a higher dimension, the place of truth, which he denoted with the letter "O". O is both a starting point and an ending point, source and destination. Creativity comes from O. Bion also uses these ideas on the psychotherapeutic field: it is not the analyst or the patient who "does" something or has an aim; it is the "O" that is common to both allowing the evolution of the session. Any real change is a "transformation in O". As for Jung, for Bion the individuation process is constructive if man gets into a dialogue with a pre-existent truth – as the artist who gets to listen to the work that he intends to do. If someone believes that reality belongs to him, "envy and possessiveness arise" (Bion 1970, tr. it. 1973: p, 143).

A peculiarity of these authors, as you may have noticed, is to highlight the integrative-thaumaturgic-creative power of unconscious. The "man-hero" who engages in his individual process perceives the unconscious as the original dimension of belonging and meaning. So he may experience moments in which he returns to "lazily exist in the unconscious" (Neumann 1949, tr. it. 1978: 35), moments of "rest of the hero", which allow an experience of integration and relaxation. Even sexuality can be understood in this way. In *Thalassa* (1924) Ferenczi suggests that coitus may be the return to an original state of rest. Even D.H. Lawrence was attracted by the idea that "unconscious-nature-silence" could be a source of true health. Such a concept, but much simplified, is also present in many "new age" philosophies and could be summarized in the formula "let you heal from your unconscious". But the creativity of the unconscious may act if the person, as Jung well showed, is secure in his individual process, but not for who has abdicated from it, thus becoming only the occasion for regression and confusion.

It should be remembered also that Freud speculated, with the great genius that characterizes him, on these issues. In *Beyond the Pleasure Principle* (1920) he assumes that the pulsions do not obey the "pleasure principle", as he had thought until then, but they are the expression of a tendency to restore the inorganic state that existed before the living one. Freud does not admit the existence of internal mechanisms of development of organisms because the organic evolution are the consequence only of "external factors of disturbance and deviance". However, towards the end of the work, Freud recalls the *Symposium* of Plato, from which one can draw the hypothesis of the "need to restore a previous state". This may be the result of a deeper need to rediscover the original unity of matter that was lost and must be found again. Freud will continue to be attracted to this theme so much so, as to search a contact with Romain Rolland, a pacifist who was awarded the Nobel Prize in literature. In one of his letters to Freud, Rolland spoke of the religious feeling of communion with life and with others which he calls "oceanic feeling". Freud, in his answer, admits that the "oceanic feeling" gives him no peace. He will mention it in *Civilization and Its Discontents* (1929), however, interpreting it as a mere remnant of primary narcissism in which ego felt in unity with the environment.

Fromm explores the ethical side of this perspective, highlighting how the individual process implies to extend our consciousness and to make it more inclusive. This requires a comparison with all aspects of the human, according to the famous saying of Publio Terenzio Afro: "I am a man: nothing that is human is alien to me". Fromm believes, quoting Marx, that man will be "humanized" only if, meanwhile, he will be "naturalized".

> In any culture, man has all potential, he is at the same time, the dawn of man, the animal of sacrifice, the cannibal, the idolater, and a he is capable of reason, love and justice. But then the content of the unconscious is neither good nor evil, neither rational nor irrational: it's all these things together. [...] The non-consciousness is the universal man, the entire man, rooted in the cosmos; it represents, at the same time, the vegetative, animal and spiritual part; it finally represents the past until the dawn of human existence and the future until the day when man becomes fully human, and in which nature is humanized inasmuch as man in turn will be "naturalized" (Fromm 1960, tr. it. 1968, p. 113).

The integration of opposites and the openness to the creativity of the unconscious that Fromm recommended could occur only if man is rooted in his singularity. Without his singularity, he returns to be simply a "collective man". Let's go back to the paradox parts/whole, network/emergence. Creativity can only flourish because there is a subjective assumption of the power of the unconscious.

Conclusions

It is possible to draw two conclusions from the perspectives analyzed

1. The first is that individuality is unsurpassed. We cannot ignore the "real" individuality. Whitehead underlined the "insistent particularity of things", which he called "stubborn fact". Only individuals are real. Individuals cannot be "overcome" in the name of an abstract ideology. The Person is the central moment of every authentic pedagogical reflection.

   A corollary of this position is that, pedagogically, any true understanding and creation of a common plan – as in democratic life – is a result of the interaction "among" persons who have irreconcilable points of view but are able to change and communicate. As Dewey showed (1916), every authentic communication is the result of an incessant confrontation among different yet interconnected "minds". Edgar Morin uses the term "dialogic" in this sense (1999, tr. it. 2000: 99). There isn't an abdication from one's own perspective because it is the tension created among the different points of view that produces ever new imbalances and new syntheses.[2] For Louis Sander there is a continuing dialectics between being "distinct from" and "being-with" others. Within this dialectics, there are "moments of meeting" between individuals which, however, are still different. This acknowledgment of being distinct causes an "expansion" of conscious experience (Tronick 1998).
2. Second, if individuals can be understood as an expression of the cosmic creative process to which they belong, then every individual is the "condensation" of all reality in a point of view. Each individual is a "living archive" of all the steps that nature made to reach him as an individual with agency and subjectivity. But then every individual is like a hologram which reflects the entire cosmic creative process. "Each point of view reflects the spatiotemporal world" (Whitehead). We can extend this dialectics to all things which are, at the same time, themselves and part of the cosmic web, "Everything is always everywhere at any time" (Whitehead). William Blake says:

   > To see a world in a grain of sand/And a heaven in a wild flower/Hold infinity in the palm of your hand/And eternity in an hour (*The Auguries of Innocence*, William Blake).

---

[2] This, of course, does not mean that everything is interpretable in terms of process "bottom up" and that is not essential, in democracy as communication, including the processes governed from above.

A corollary of this position is that creativity is like a coin with two faces. On the one side there is the subjective impulse, being the agent of the self in its own sense, on the other side there is the individuality as an expression of a cosmic creative process to which it belongs. Creativity cannot be just a "private matter". "If the thinker believes to be essential to the thought that he has thought, envy and possessiveness arise" (Bion 1970, tr. it. 1973: 143). The result is a paradox, very similar to the dilemma that faces the artist, who puts his creativity at the service of his work. Luigi Pareyson writes that "the artist fails if he does not do the will of the work" (Pareyson 1966: 21) – even though, of course, "he himself institutes this will" (*idem*), with his creativity. He defines the artistic process as "formativity" because the subjectivity is at the service of the objectivity. The true art speaks of the world, despite being created by an individual. Who gets in adoration of his subjective products transforms them into fetishes. Nor, on the other hand, creativity can achieve a full realization, the "total object" because it is part of the cosmic creative process. "Life does not end. It cannot conclude. If tomorrow ends, it's over" (Pirandello 1926).

This perspective is expressed in a very deep way in *East Coker* (1940) by T.S. Eliot, where creativity is perceived as a continuous widening the gaze so as to welcome and give reasons for the complexity of the world and respect the "insistent particularity of things" (Whitehead), including the petrel and the dolphin. For this to happen it needs to be done in the dark ("Be still, and let the dark come upon you"; "In my beginning is my end"; "You must go through the way in which you are not/ And what you do not know is the only thing you know") so that all aspects of reality can find a mental space to be greeted. The only authentic creation springs from having put oneself at the service, in having a vacuum ("in my end is my beginning").

We must be still and still moving
Into another intensity
For a further union, a deeper communion
Through the dark cold and the empty desolation,
The wave cry, the wind cry, the vast waters
Of the petrel and the porpoise. In my end is my beginning.

References

Allport, G.W. 1955. Is the concept of self necessary? In *Becoming: Basic considerations of a psychology of personality*, 36–40. New Haven: Yale University Press.

Anderson, P.W. 1972. More is different. *Science* 177: 393–396.

Arena, L.V. 1989. *Comprensione e creatività. La filosofia di whitehead.* Milan: Franco Angeli.

Bion, W.R. 1970. *Attention and Interpretation. A Scientific Approach to Insight in Psycho-Analysis and Groups.* London: Tavistock Publications. 1973. *Attenzione e interpretazione* (tr. it.). Rome: Armando.

Bion, W.R. 1976. *Group and Organization Studies.* Colchester: University Associates. 1981. *Il cambiamento catastrofico. La Griglia, Caesura Seminari Brasiliani, Intervista* (tr. it.). Turin: Loesher.

Carere-Comes, T. 2003. La terapia è naturalmente dialogico-dialettica. In *Il futuro della psicoterapia fra integrità e integrazione*, ed. G.G. Alberti and T. Carere-Comes. Milan: Franco Angeli.
De Toni, A.F., and E. Bernardi. 2009. *Il pianeta degli agenti*. Turin: UTET.
Dewey, J. 1916. *Democracy and education*. New York: Macmillan. 1974. *Democrazia ed educazione* (tr. it.). Firenze: La Nuova Italia, Firenze.
Eliot, T.S. 1940. East coker. In *Opere 1939–1962*. Milan: Bompiani, 2003.
Frabboni, F., and F. Pinto Minerva. 2003. *Introduzione alla pedagogia generale*. Bari: Laterza.
Freud, S. 1920. "Jenseits des Lustprinzips", *Internationaler Psychoanalytischer Verlag*. Leipzig: Wien und Zürich (tr. it.) "Al di là del principio di piacere". In *Opere*, Vol. 9, 189-249. Turin: Bollati Boringhieri.
Freud, S. 1929. "Das Unbehagen in der Kultur", *Internationaler Psychoanalytischer Verlag* (Leipzig-Vienna-Zurich), GW, XIV, 419-506 (tr. it.) "Il disagio della civiltà". In *Opere* Vol. 10, 555-630. Turin: Bollati Boringhieri.
Fromm, E. 1960. Zen Buddhism and psychoanalysis. In *Zen Buddhism and psychoanalysis* ed. Fromm, E., D.T. Suzuki, and E. De Martino. New York: Harper & Brothers. 1968. *Psicoanalisi e Buddhismo Zen*, 83–147 (tr. it.). Rome: Astrolabio.
Jervis, G. 2011. *Il mito dell'interiorità*. Turin: Bollati Boringhieri.
Kauffman, S. 2008. *Reinventing the sacred: A new view of science, reason, and religion*. New York: Basic Books.
Köhler, W. 1947. *Gestalt Psychology*. New York: Livering. 1989. *La psicologia della Gestalt* (tr. it.). Milan: Feltrinelli.
Laeng, M. 2004. *I gradini dell'ascesa*. Brescia: La Scuola.
Laporta, R. 2001. *Avviamento alla pedagogia*. Rome: Carocci.
Loewald, H. 1960. On therapeutic action of psychoanalysis. *International Journal of Psycho-Analysis* 41: 16–33.
Maturana, H., and F. Varela. 1980. *Autopoiesis and cognition. The realization of the living*. Boston: Reidel.
Morin, E. 1999. *Le tête bien faite*. Paris: Seuil. 2000. *La testa ben fatta* (tr. it.). Milan: Raffaello Cortina.
Neumann, E. 1949. *Ursprungsgeschichte des Bewusstseins*. Zürich: Rascher Verlag. 1978. *Storia delle origini della coscienza* (tr. it.). Rome: Astrolabio-Ubaldini.
Pareyson, L. 1966. La contemplazione della forma. In *Conversazioni di estetica*, ed. L. Pareyson, 20–24. Milan: Mursia.
Pirandello, L. 1897. Sincerità e arte, 7.3. *Il Marzocco*. Firenze.
Pirandello, L. 1910. *Leviamoci questo pensiero*. In *Novelle per un anno*, Vol. III. Milan: Mondadori, 1990, I Meridiani.
Pirandello, L. 1926. Uno nessuno e centomila. In *Tutti i romanzi*, Vol. II. Milan: Mondadori, 1973, I Meridiani.
Prigogine, I. 1996. *La fin des certitudes. Temps, chaos et les lois de la nature*. Paris: Odile Jacob. 1997. *La fine delle certezze* (tr. it.). Turin: Bollati Boringhieri.
Sander, L. 2002. Thinking differently. Principles of process in living systems and the specificity of being known. *Psychoanalytic Dialogues* 12(1): 11–42.
Sander, L. 2008. *Living systems, evolving consciousness, and the emerging person: A selection of papers from the life work of Louis Sander*. New York: Analytic Press.
Sartre, J. P. 1938. *Nausea*. New York: New Directions Books, 2007.
Searle, J.R. 2004. *Mind. A brief introduction*. Oxford: Oxford University Press.
Strokes, A. 1965. *The invitation in art*. London: Tavistock Publications.
Thalassa, S. Ferenczi. 1924. Versuch einer Genitaltheorie. Leipzig/Vienna: Internationaler Psychoanalytischer Verlag. 1993. *Thalassa. Saggio sulla teoria della genitalità* (tr. it.). Milan: Cortina Raffaello.
Thomson, M. 1989. *On art and therapy. An exploration*. London: Virago, (2nd ed., London: Free Association Books, 1997).
Trevarthen, C. 1993. Playing into reality: Conversation with the infant communicator. In *Winnicott studies*, 7, 67-84. Spring (tr. it.). 1998. In *Empatia e biologia*, 145-165. Milan: Raffaello Cortina.

Tronick, E.Z. 1998. Dyadically expanded states of consciousness and the process of therapeutic change. *Infant Mental Health Journal* 19(3): 290–299.

Waldrop, M.M. 1992. *Complexity. The emerging science at the edge of order and chaos*. New York: Simon Schuster.

Whitehead, A. N. 1929. *Process and reality*. New York: Macmillan, corrected edition, 1978.

Whitehead, A.N. 1938. *Modes of thought*. New York: Macmillan.

Winnicott, D.W. 1963. Communicating and not communicating leading to a study of certain opposites. In *The maturational processes and the facilitating environment*, ed. D.W. Winnicott. London: The Hogarth Press, 1972

Winnicott, D.W. 1970. Living creatively. In *Home is where we start from – Essays by a psychoanalyst*, ed. D.W. Winnicott. London: Norton, 1986 (tr. it.) "Vivere creativamente". In *Dal luogo delle origini*, ed. D.W. Winnicott. Milan: Raffaello Cortina, 1990.

Part VIII

Cognition and Emotion: From Dichotomy to Ambiguity

Claus Halberg and Simen Andersen Øyen

Abstract In this article, we pose the question of the relation between cognition or reason and emotion or affectivity as it emerges in contemporary debates in psychology, particularly in the debate between Robert Zajonc and Richard Lazarus. Basing themselves on experiments involving priming, Zajonc and Lazarus have offered contrasting accounts of the roles of cognition and emotion respectively in the processing and evaluating of information. On this issue, Zajonc has typically privileged emotion, whereas Lazarus insists on the efficacy of cognition. Despite their disagreement, however, we claim that both scholars make the same basic philosophical assumption of an in-principle dichotomous relation between the two orders. In order to bring out the limits of such a conceptual schema, we turn to Merleau-Ponty's discussion of the patient Schneider in *Phenomenology of Perception*. We argue that his analysis of the sexual function – or rather its absence – in Schneider makes salient how the subject's cognitive and affective capacities more generally are inseparably intertwined through their concrete and lived exercise.

The relationship between cognition and emotion, and the correlative relation between mind and body, has persisted as a constant concern in the history of Western thought from its inception up to present-day debates in experimental psychology and other fields. In Plato, the emotions (correlatively, the body) provided only knowledge of the sensory world, while the powers of rational thought afforded entry to the realm of the unconditionally true and eternal being of formal reality. To judge from certain of Plato's statements, the sensuous and affective potential of the flesh enslaved the rational powers of the mind and prevented access to the supremely and eternally true, good and beautiful realm of ideas. The denigration of the flesh in many Christian

C. Halberg (✉) • S.A. Øyen
Centre for the Study of the Sciences and the Humanities, University of Bergen, Bergen, Norway
e-mail: Claus.halberg@skok.uib.no; Simen.oyen@svt.uib.no

A.-T. Tymieniecka (ed.), *Phenomenology of Space and Time: The Forces of the Cosmos and the Ontopoietic Genesis of Life: Book One*, Analecta Husserliana 116,
DOI 10.1007/978-3-319-02015-0_32,

traditions attests to a basic inspiration from this basic dualistic scheme of the human found in Plato. The division between body and mind was reinforced in René Descartes through his distinction between *res extensa* (that which has extension) and *res cogitans* (that which thinks), and it was carried forward in Kant's wholesale exclusion of all forms of inclination from the definition of the moral agent. Hegel's solution was to suppose that the rational and the real come back to the same thing and that the factor of negativity and *Zerissenheit* in the element of life can be consumed without remainder in the element of work and discourse. But, as Georges Bataille profoundly showed, this was effectively to leave the persisting non-rational element of life in the black night outside work and discourse, and the dualistic situation of rational cognition and irrational emotion was not done with.

These distinctions are still to be found in various disciplines today. In social science, especially in decision theory and different versions of sociological utilitarianism (for example, rational man theory), it is assumed that people freely participate in social interaction only if the rewards outweigh the costs, or benefits are greater than the disadvantages. It is assumed that the subject is a rational agent. The subject is constituted as a purely rational creature, and as a rational and benefit-maximizing actor, driven by utilitarian affects and self-interest. Even views of the subject developed as reactions to this approach to human action can be seen to operate with a similar reductive division. Habermas' concept of communicative rationality places emphasis on language as all-encompassing for human interaction. Habermas connects rationality to communicative competence, and with this, the human way of being is linguistically mediated, thereby derived from, and preconceived by language. The philosophical problem here is obvious: a concept of the rational actor can be fruitful for a methodological reductionism and appropriate, for example, in explaining exchanges in limited economic spheres. However, the concept is problematic when it receives an ontological status, constituting the subject per se. The dualistic schema to which one has reduced subjectivity and intentionality receives the status of factual being, and thus leaves untheorized the complex interlockings of all of the subject's capacities, bodily and spiritual, emotional and rational, habitual and spontaneous in his or her concrete cognitive and active life.

The named examples are not theoretical exceptions, neither are they unique for quantitative and utilitarian positions in social science. Also theoretical disciplines which focus on problem complexes such as political processes or democratic development, and related topics tend to construct a similar reductive understanding of the subject. In this context, the contract theory approach to democracy is illustrative. John Rawls, for example, with his hypothetical-rights thought experiment of the "original position", operates with an understanding of the subject in which the individual is rational and seeks to maximize benefit. In the following, we will explicate these theoretical and ontological problems as they appear in the psychology of perception. Taking our point of departure in Merleau-Ponty's existential phenomenology, we bring to light the subjective character of the body and, correspondingly, the embodied character of the subject as a blind spot in this field of knowledge. In schematic terms, perception and understanding in the cognitive approach are seen as the processing of information in which the analogy to the computer is central, while in the biological approach comprehension is reduced to neuro-physiological and genetic causality. These perspectives in psychology are not just trends, but also expressions of a more comprehensive paradigm for how the question of cognition and emotion, body and mind, are comprehended. Merleau-Ponty's existential phenomenological analysis of the phenomenon of affectivity, especially as it is revealed through sexuality, offers a problematizing of the conceptual dichotomy which is quietly accepted as common ground in psychological interpretations, being mutually exclusive only for a superficial consideration. It also points the way to a more concrete and comprehensive approach to phenomena.

Cognition and Emotion in Evidence-Based Psychology

The study of psychology has, as is the case also with medicine, been strongly influenced by a natural scientific view of the body – at least in periods – in which the separation of the human into cognition and emotion, mind and body is taken as points of departure. The Zajonc–Lazarus debate illustrates this. Emotions are usually defined in psychology through three relations: arousal, behavior and affect (Hilgard 1993). Affect covers a wide spectrum of experience such as emotions, attitudes and general appraisals of a situation or similar event. Emotions refer more often to short, but intense experiences. The assumption we make of a situation has meaning for our emotional experience in that situation. Such assumptions can become a part of our evaluation of the situation, and the resulting understanding of the situation. It was this relationship Schachter and Singer (Hilgard 1993) attempted to illustrate in their famous experiment, in which the assumption was that emotions are a result of an interaction between physiological activation (arousal) and the cognitive interpretation (appraisal) of this arousal. In this approach, emotions are a function of both physiological arousal and cognitive factors where the individual searches the immediate environment for emotionally relevant and available cues to label and interpret physiological arousal and affectivity.

The core of the debate between Robert Zajonc and Richard Lazarus revolved around the degree to which the emotional experience demands cognitive evaluation. The core of this dispute is the question as to whether a stimulus must be processed cognitively for the manifestation of an emotional response to it. Zajonc (2004) argues for the possibility that an emotional evaluation of stimuli can be obtained independently of cognitive processes, even though emotion and cognition normally appear simultaneously. Emotions appear primary, on the understanding that affective processing can be quicker than cognitive processing, and that the first emotional processing is different than the later cognitive. He has found empirical evidence for this in various experiments with priming. In this context, priming can be understood as different types of stimuli which are presented subliminally (Zajonc 1980), therefore impressions that are not conscious, appearing under the threshold of consciousness, but which nonetheless affect an individual's behavior. It is an automated process in which a stimulus word or picture activates various components in one's long-term memory. For example, the stimuli in one study were either smiling faces or angry faces, presented for 4 ms, and followed by Chinese ideographs or signs. The study subjects were then asked to evaluate the different ideographs according to which they liked best. They chose the typical objects they previously had been primed for. The study subjects developed, in other words, preferences. This was evaluated against a control group.[1] The preferences for the Chinese ideographs were equally great even when faces were presented once before, and with a red filter over them. This is a condition of exposure that does not allow for any form of recognizable cognitive processing (Zajonc 2004). This evidence gives support to the view that there has occurred a positive affective reaction to a certain stimulus, without cognitive processing. The phenomenon is termed "mere exposure effect".

Lazarus argues, for his part, that some cognitive processing is an essential precondition for an emotional response to stimuli. He approaches the question with support from various experiments showing that cognitive evaluations occur parallel, or prior to, emotional states, but that these cognitive processes are not necessarily conscious. Lazarus, with colleagues (Speisman et al. 1964), performed several tests to show the importance of cognitive judgment in connection to emotions. One experiment presented a group of subjects with an angst-inducing film, containing, for example, work accidents, surgeries, etc. Cognitive judgment was manipulated by varying the sound level of the films and comparing the stress experienced with a control group exposed to a film with no sound. The test conditions were further manipulated by the denial of content in the films, or intellectualized. Test subjects were told that the workers in the accident were actors, or were told to watch the film as if they were anthropologists and the film documenting an ancient rite from a faraway culture. Various psycho-physiological measurements were taken of the subjects while they saw the films. Lazarus, et al., found out that when the events of a film were intellectualized or denied as real, there was a substantial reduction in the stress response. Lazarus argues, on the basis of this experiment, and similar ones, that there occurs a cognitive evaluation parallel or prior to emotional states, but that these cognitive processes are not necessarily conscious. Lazarus (Smith and Lazarus 1993) would call this a primary evaluation of the surrounding environment and stimulus situation in which the world is evaluated pre-reflexively as positive, stressful or irrelevant for an individual's well-being. This primary evaluation is part of a series of evaluative components, and different emotional states can be distinguished on the basis of these components.

To sum up, Zajonc (1980) argues for the primacy of affect because he maintains he can document affective evaluations without the appearance of conscious cognitive recognition. Lazarus (1982) would contend, on the contrary, that also subliminal and automated processes of consciousness can contain cognitive components, and that these components substantially determine the outlook of the subject's experience.

[1] In a similar study, in which the priming stimuli were feminine or masculine faces, when the study subjects were asked to evaluate the Chinese ideographs according to a feminine/masculine dimension, the results were opposite (Murphy and Zajonc 1993).

Both positions have been criticized on methodological grounds. Zajonc's studies have been criticized for not being relevant to actually lived and factually emotional states, that they are artificially construed in an experimental setting (Leventhal and Scherer 1987). But they also raise more fundamental questions connected to our definitions of consciousness and cognition. Lazarus is, on his side, criticized from several positions. The psychologists Brian Parkinson and Anthony Manstead (1992) argue that Lazarus' approach represents a limited and context-dependent view of emotion. The individual is passively confronted with threatening stimuli while being held separated from the complexity of the life-world. In a case, the positions in this debate generally represent a pronouncedly dichotomous view of the relation between cognition and emotion. It is assumed that affect occurs in the body as a passive object for causal processes to externally work upon, and that cognition is a second-order process that either autonomously elaborates the affective input or else emerges as a second level of meaning on account of the efficacy of this input. In thus according primacy to either one or the other, their basic definition as self-contained and determinate categories is perpetuated. It is further supposed that psychological processes occur through representations or allow of the domination of objectivizing functions, or bear the characteristics of such functions, whether it is thought of as emotional or cognitive. But their critics also partially accept a similar distinction, in which the theoretical objections are directed more at methodological problems, for example, an experiment's value for application to the life-world.

It is possible to indirectly reveal with examples from pathology how emotions and cognition are intertwined. This also has the benefit of short-circuiting the methodological issues pertaining to the Zajonc-Lazarus debate. Consider, for example, the case of patients with bilateral amygdala damage. This condition involves a reduced ability to recognize the emotional meaning of facial expressions, and indicates a failure to perceive emotional aspects of stimuli and situations. In an article entitled "Cognition and Emotion", Yiend and Mackintosh (2005) describe a case in which a person was involved in a car crash together with his wife. He received a lasting injury to his head. The wife was also injured and sent to the hospital, but later recovered. This man refuses to believe that his wife survived the crash. He thinks she is an impostor. When this man was tested for SC-change while looking at pictures of his wife, emotional arousal was absent. SC-changes signal an emotional response that normally occurs when one sees an emotionally charged expression in another's face, or the face of someone known. Without this response there is no emotional resonance to the experience. Apparently, this man lacks this interpretation process, lacking an emotional feedback which could prove a close relation to another person. Although the physiognomic traits of the person confronting him were objectively identical in his own eyes to those of his wife's, he nevertheless assumed she must have been someone who closely resembles his wife or an impostor.

Yiend and Mackintosh's approach to the puzzling condition of this patient is to suppose that, in the normal subject, cognitive and emotional meaning are processed in separate but strictly parallel systems that supplement and complement each other so as to compose a comprehensive situation for the subject. From this point of view, they further suggest that the reason why the patient in question has lost his ability to recognize his wife as truly his own wife despite the cognitive facility of his perceptual capacity being intact, is that his injury has directly struck the system where emotional meaning is processed (Yiend and Mackintosh 2005). By interpreting the case in this way, they perpetuate the same tendency to dissociate cognition and emotion for methodological purposes as we have already seen in Zajonc and Lazarus, although they admittedly offer a more complex picture. Whereas Zajonc assumes emotional responses to stimuli to be autonomous, Lazarus claims that emotional responses are cognitively filtered. On the strength of a pathological case Yiend and Mackintosh propose that cognitive and emotional processing work in tandem in the normal subject, but nevertheless consider them to be two ideally isolable and determinate phenomena or systems.

Yiend and Mackintosh' case evokes, however, a similar and also classical case in the history of neuro-pathology, namely, the case of WWI veteran Johann Schneider, whose cortical lesion had, among numerous other effects, the effect of literally abolishing his sexual function. It was on the basis of this pathological material that Merleau-Ponty developed a phenomenological existential analysis of the phenomenon of affectivity, an analysis that decisively questioned the dichotomous schema that we have seen is still prevalent in today's experimental psychology. Given this prevalence, it seems to us that Merleau-Ponty's phenomenological insights into the phenomenon of affectivity still retain their currency as a way to problematize the basic conceptual and ontological postulates subtending contemporary research in the psychology of cognition and emotion. To this we now turn.

Cognition and Emotion in Merleau-Ponty's Phenomenological Account of Sexuality

In Merleau-Ponty's early major work, *Phenomenology of Perception* ([1945] 2012), he offered an elaborate critique of both positivist or empiricist and intellectualist excesses in the science of psychology in his time. Although these labels are of course too coarse and obsolete to correspond to anything actually occurring in the contemporary scientific world, the terms of Merleau-Ponty's critique of these positions as he perceived them seem to us still to be relevant as a diagnosis of the current situation in experimental psychology. Focusing in particular on the phenomenon of perception, he pointed out how the psychologies and epistemologies of his time tended, on the one hand, to reduce perception either to a fortuitous outcome of gratuitous events in the objective world and in the body considered as a mechanism. On such an account, the subject of perception would be reduced to its capacity for being sensuously affected by the external world. Furthermore, on the same account, the coming together of disparate sensory givens into unified objects and ideas had to be accounted for in terms of the operation of a "mental chemistry" that could be established on empirical grounds just like any other natural fact. On the other hand, the intellectualist alternative as reconstructed by Merleau-Ponty envisioned the subject of perception as a thinking agency that would perceive things by synthesizing heterogeneous sensory material according to laws, forms or ideas that it represents to itself. In the first account, the external world is given as determinate, and the subject of perception and its perceptions are determinable effects produced within it; on the second account, the internal world of the subject is given as determinate through its representations, and the external world and its objects are the determinable products of the subject's synthesizing and representational activity.

According to Merleau-Ponty, these two seemingly contrary accounts of perception and subjectivity share a basic fundamental prejudice which he calls the prejudice of "determinate being" (Merleau-Ponty 2012: 51, n. 60). The one-sided privileging of either the order of causality, affectivity and passivity or else of the order of reason, cognition and activity are borne, according to Merleau-Ponty, of a prior ontological decision in favor of being as determinate or as ideally determinable. This prejudice persists even as the intellectualist line of thought takes on board the positivist conception of sensation as the exigency that perception must have a given "stuff" to work upon. Reason and causation, cognition and affection, action and passion remain sealed upon themselves as ontologically determinate categories. The two sides of the resulting dichotomous picture of perception may relate to each other in the mode of logical exclusion (as in positivism) or in the mode of complementarity (as in intellectualism), but in either case both sides of the dichotomy are conceptualized as essentially determinate in themselves. In light of our above diagnosis of current trends in experimental psychology, the general picture that Merleau-Ponty drew in 1945 of the psychological and philosophical accounts of perception of his time would seem to have retained its currency.

It is the ontology of determinate being, as it is tacitly assumed by epistemological and psychological positions that are only summarily opposed, that Merleau-Ponty purports to undermine as he moves through the world of corporeally lived phenomena. These phenomena, Merleau-Ponty contends, will force us to recognize "the indeterminate as a positive phenomenon" (2012: 7) and to recognize at the heart of human existence a "principle of indetermination". What makes for such a principle of indetermination is the peculiar relation between two dimensions of human existence that Merleau-Ponty repeatedly refers to as the "personal" and the "anonymous" respectively. It is in the articulation between these two dimensions that we will be able to determine on a phenomenologically more sound basis the relation between the cognitive and the emotional registers of human experience. Before coming more specifically to his analysis of Schneider's curtailed sexual function, then, a few general remarks must be made regarding Merleau-Ponty's distinction between the personal and the anonymous.

According to Merleau-Ponty the personal self of the reflective and volitional "I" is dependent on and sustains itself by the grace of an anonymous life of bodily, social and cultural existence which for its part retains an autonomy vis-à-vis the personal. And yet, this autonomy of the anonymous life of the "one" (*on*) is never absolute, it never achieves the character and closure of a self-contained substance (as traditionally understood) that I may inspect and analyze as if under a microscope. The anonymous "one" withholds itself from total appropriation by the personal "I" and exceeds it precisely because the anonymous can never be said to be *utterly* anonymous or general, just as the personal "I" is never utterly personal or individual. Ultimately, from a phenomenological point of view, the anonymous "one" surrounds the personal "I" as a horizon or margin of "almost impersonal" existence, which it may try to approach or from which it may try to stand back in order to make it into an object, but with which it can never totally coincide or fuse, nor yet encompass with either gaze or symbol:

> [M]y life is made up of rhythms that do not have their *reason* in what I have chosen to be, but rather have their *condition* in the banal milieu that surrounds me. A margin of *almost* impersonal existence thus appears around our personal existence, which, so to speak, is taken for granted, and to which I entrust the care of keeping me alive (Merleau-Ponty 2012: 86).

The horizontal or marginal character with which the anonymous is presented to the personal self is experienced, Merleau-Ponty notes, as in a dream or a sort of stupor into which we are immersed when, for example, "we truly live at the level of sensation" (2012: 223). Breaking off from our daily practical concerns, laying our bodies down, closing our eyes, stopping up our ears, trying to close ourselves up in some bodily affect, we may vaguely sense the constant pulsing and buzzing of our bodily senses that persists beneath our waking, practical and active life, "just as the constant hum of a large city serves as the background for everything we do there" (2012: 343). Moreover, we may add, in the oneiric landscapes we visit during sleep, or in certain movies (such as Alain Resnais' *Last Year in Marienbad* and many of David Lynch's films), we may be offered an inkling of the anonymous temporality or "natural time" secreted by our bodily existence, with its characteristic dismemberment of chronology into tableaus of cyclical, folded and stratified simultaneities. The anonymous existence of our body, our language and the social field in which we find ourselves escapes us in a way not unlike the way in which the dream of last night, once we wake up, starts to slip between our fingers.

If the anonymous "one" is a margin, then the personal "I" is so as well. The anonymous and the personal, the general and the individual are, for Merleau-Ponty, limit-concepts; they do not denote entities or substances that are externally related, but rather two orientations or tendencies of one single current of existence. The anonymous life of the body, society, language and culture in general constitutes, at the core of the personal self, "that internal weakness that forever prevents us from achieving the density of an absolute individual" (2012: 452). Conversely, personal or individual existence is virtually present in the anonymous as an outline that is continually and cyclically sketched, undone and re-sketched: "My organism is not like some inert thing, it itself sketches out the movement of existence. It can even happen that, when I am in danger, my human situation erases my biological one and that my body completely merges with action" (2012: 86). The personal and the anonymous cannot be distributed between, for example, the psychic and the physiological as if between two orders of reality or two causal orders; rather the psychic and the physiological are provisional expressions of a single and continuous to-and-fro movement of existence in its personal and anonymous phases respectively. And between these two phases, only an "imperceptible shift" or "insensible turn":

> [T]here is no single movement in a living body that is an absolute accident with regard to psychical intentions and no single psychical act that has not found at least its germ or its general outline in physiological dispositions. (...) [T]hrough an imperceptible shift (*un tournant insensible*), an organic process opens up into a human behavior, an instinctive act turns back upon itself and becomes an emotion, or, inversely, a human act becomes dormant and is continued absentmindedly as a reflex (Merleau-Ponty 2012: 90).

This notion of an "imperceptible shift" between different registers of experience and levels of being seems to entail a vastly different and, phenomenologically speaking, far more salient conceptuality than the dichotomous distribution of experience into separate and determinate registers still perpetuated in contemporary experimental psychology. As it happens, the notion of this imperceptible shift is developed most intensely by Merleau-Ponty in his discussion of the phenomenon of affectivity, as emblematized by the sexual function. "It is impossible", Merleau-Ponty suggests, "to identify the contribution of sexual motivation and the contribution of other motivations for a given decision or action, and it is impossible to characterize a decision or an action as 'sexual' or as 'nonsexual'" (2012: 172). His discussion of this "principle of indetermination" will aid us in our confrontation with the situation we have diagnosed in the field of contemporary experimental psychology.

To begin with, it is clear for Merleau-Ponty that sexuality is not an autonomous cycle or reflex mechanism seated in the sexual apparatus, as if – as is so often assumed even today – it were a program designed by who-knows-whom to make individuals procreate in the service of the species. If it were, then one should expect a cerebral lesion, such as the one suffered by the patient Schneider, that curtails the subject's power of representation (visual or otherwise) to enhance the subject's sexual impulses, since many of the subject's representations work to inhibit and censure these impulses. As we know, Schneider's injury was accompanied not by an intensified sexual function, but rather a practically abolished sexual function. On the other hand, Merleau-Ponty notes, one cannot base the sexual function purely on a power of representation or imagination. The deficiency of Schneider's power of visual representation does not support the hypothesis that the sexual function is based on such representations, because one would then have to explain as well why the tactile stimulations obtained during foreplay and intercourse do not appeal to him sexually, since one cannot maintain that, for the normal subject, there is no tactile representation or significance of sexual acts. A hypothesis of a generalized loss of representations, whether visual or tactile, has no explicatory force when one confronts it with the concrete and particular figure assumed by the abolishment of sexual dynamics in Schneider's life, such as the infrequency of his nocturnal emissions, which are, moreover, unaccompanied by dreams (2012: 157–158).

What Schneider's case brings to light, according to Merleau-Ponty, is that sexuality is internally linked with thinking, feeling and acting human existence in its entirety and synchronized with it, and that, accordingly, we must see Schneider's impotence in light of the altered affective and existential modality of his whole personality: "Just as he is generally no longer within an affective or an ideological situation, Schneider can no longer place himself in a sexual situation (…). The world is affectively neutral" (2012: 159–160). The whole difficulty lies, then, in accounting for that insensible turn by which a particular perception, fantasy, memory or traumatic encounter inserts itself into and modulates the circuit of sexual energy and by which, conversely, sexual energy "emanates like an odor or sound from the bodily region that it occupies most specifically" (2012: 172) in order to imbue our personal existence with its particular degree and style of vitality and fecundity. Crucially, however, while one thus submits the sexual function to the accidents, articulations and particular style of one's personal existence, Merleau-Ponty nevertheless finds it incumbent to avoid submerging sexuality in existence and reduce it to a "mere reflection of existence" or an "epiphenomenon" (2012: 162). To avoid submerging sexuality in existence is to recognize that "life is particularized into separate currents". To recognize that life is particularized into separate currents is, for Merleau-Ponty, to be open both to the fact that a vigorous and inspired public, political and activist life need not be accompanied by an equally dynamic sexual life and vice versa, and to the fact that neurotics, more often than not, vigorously seek expression for their relational troubles through sexual symptoms: "sexuality is not simply a sign, but in fact a privileged one" (2012: 162).

In an attempt to take hold of this ambiguous situation philosophically, Merleau-Ponty proposes that the relation between the personal and the anonymous, such as it is brought to light through the phenomenon of sexuality, is one of "reciprocal expression". On the one hand, personal existence is an expression of the anonymous life our senses, our motricity, our sexual system and in general our body to the extent that it "takes up" and "gathers" in a particular manner or style these inchoate outlines and sketches of a genuine presence in the world. This is just a matter of recognizing that a "lived experience" is rooted in, although of course not determined by, the more primordial and inchoate operation of "living" (*vivre*). In short, we must

> eat and breathe prior to perceiving and reaching a relational life…be directed toward colors and lights through vision, toward sounds through hearing, and toward the other person's body through sexuality, prior to reaching the life of human relations. Thus vision, hearing, sexuality, and the body are not merely points of passage, instruments, or manifestations of personal existence. Personal existence takes them up and gathers in them their given and anonymous existence (Merleau-Ponty 2012: 162).

On the other hand, Merleau-Ponty suggests, sexuality may express existence and even occupy a privileged position at the core of our being because it is "like a passive experience, given to everyone and always available, of the human condition in its most general moments of autonomy and dependence" (2012: 170). Sexuality is one of many loci where the personal and the anonymous are converted one into the other, while at the same time, in a sense, it sums up or contracts them all into an acutely and intensely felt crystallization of the lot of being both autonomous and dependent: the lot of being delivered over to and at the mercy of the elemental adversity and unpredictability of pre-human nature while at the same time being able to feel it, take a view of it, paint it and write of it in books; the lot of having to borrow oneself from (m)others and from language and the constant risk of abandonment, desertion, dispersion and displacement that this entails, while by the same token being given something that may be understood, expressed and re-fashioned, a situation and a field of possible individual and collective action and re-invention. Sexuality, for Merleau-Ponty, is the acute and yet inchoate awareness that "[n]o one is fully saved, and no one is fully lost" (2012: 174).

Conclusion

As we have seen, a phenomenological approach to the problematic of cognition and emotion entails a problematization of the dichotomous schema to which they are submitted in the operationalizations of experimental psychology. It suggests that emotions and cognition are mutually dependent, and that a temporal priority is irrelevant. Thus, the dichotomy becomes meaningless and the debate becomes merely a question of definitions. From a phenomenological point of view, consciousness does not contain purely cognitive structures. It already contains an emotional processing level. Consciousness is a holistic activity comprising cognitive and emotional fields inseparably and indeterminately. On the basis of Merleau-Ponty's notion of a "principle of indetermination" or the "insensible turn" between the personal and the anonymous, the cognitive and the affective, we may no longer assume that psychological processes use representations of any kind as a conduit, nor submit themselves to objectivizing functions. The human is not only a purely rational creature, acting from principles to consequences or from means to goals, but neither are we subject to a blind mechanical causality. The subject is woven within a life-world praxis in which processes of consciousness are synergistically coordinated, the decentering and separation of which appears above all as a pathological condition.

References

Hilgard, E. 1993. *Hilgard's introduction to psychology*. Fort Worth: Harcourt Brace College Publishers.

Lazarus, R. 1982. Thoughts on the relations between emotion and cognition. *American Psychologist* 37: 1019–1024.

Leventhal, H., and K. Scherer. 1987. *The relationship to emotion to cognition: A functional approach to a semantic controversy*. Wisconsin: Lawrence Erlbaum Associates Limited.

Merleau-Ponty, Maurice. 2012. *Phenomenology of perception* (trans: Donald A. Landes.). London: Routledge.

Murphy, S.T., and R. Zajonc. 1993. Affect, cognition, and awareness: Affective priming with optimal and suboptimal stimulus exposures. *Journal of Personality and Social Psychology* 64: 723–739.

Parkinson, B., and A.S.R. Manstead. 1992. Appraisel as a cause of emotion. *Review of Personality and Social Psychology* 13: 122–149.
Smith, C.A., and R. Lazarus. 1993. Appraisal components, core relational themes, and the emotions. *Cognition and Emotion* 7: 233–269.
Speisman, J.C., R. Lazarus, A. Mordkoff, and L.A. Davison. 1964. The experimental reduction of stress based on ego-defense theory. *Journal of Abnormal and Social Psychology* 68: 367–380.
Yiend, J., and S. Mackintosh. 2005. Cognition and emotion. In *Cognitive Psychology*, ed. N. Braisby and G. Angus. Oxford: Oxford University Press.
Zajonc, R. 1980. Feeling and thinking: Preferences need no inferences. *American Psychologist* 35: 151–175.
Zajonc, R. 2004. *Feeling and thinking – Closing the debate over the independence of affect*. Paris: Cambridge University Press.

The Meeting of Man with Man

Leszek Pyra

Abstract The phenomenon of a meeting in reference to men seems to be something normal, on the one hand, but also something peculiar, on the other. During a meeting one has an opportunity to know the other in his/her different aspects. In this context the views of Ingarden, Husserl, Buber, Tymieniecka and Kępiński have been compared. Man's cognitive autonomy, a phenomenon appearing on the basis of human freedom, has been discussed and evaluated in this context. One of the most characteristic traits of a meeting is mutuality (reciprocity). Thanks to it, meeting is possible, but at the same time reciprocity does not assume absolute symmetry. Emotion is a starting point for every meeting, without emotions no meeting is possible. A meeting is a process and its nature reshapes, sometimes to a great extent. For example, friendship may turn into love after some time, or change into hatred. Each participant of a meeting both knows the other and is known by the other. Each participant has some self-knowledge and tries to project it onto the other. But self-knowledge is not always adequate, often the partner of dialogue can better evaluate certain aspects of a given man's personality than a self-knowing man himself. One can be known through the verbal sphere and through the non-verbal sphere (e. g. by the look of one's eyes, body language). In this context different ways of knowing the other within the non-verbal sphere have been presented and discussed. The category of a meeting seems to be a very interesting phenomenon, about which we still do not know enough.

The reflection concerning the meeting and knowing the other man reaches to ancient times, definitely. It is especially well seen in the case of Socratic dialogues led in order to reach the truth about man's soul and human moral obligations. Axiology

L. Pyra (✉)
Pedagogical University, Cracow, Poland
e-mail: Lepyra@op.pl

A.-T. Tymieniecka (ed.), *Phenomenology of Space and Time: The Forces of the Cosmos and the Ontopoietic Genesis of Life: Book One*, Analecta Husserliana 116,
DOI 10.1007/978-3-319-02015-0_33,

and anthropology were present in the dialogues presented by Plato, although no such names existed in ancient times. Throughout the ages such issues did not evoke any greater interest, but then in the twentieth century the situation changed. The cognition of the other man was conceived naturally as knowing the other man's "inner states", "psychical states", etc. The other was usually conceived as a certain material-spiritual whole. The following considerations contribute to the meeting and cognition of the other with whom one engages oneself in some closer contact, in which there appears a mutual influence and co-creation of some kind. One can get access to the other when getting some information about him, mainly verbal information, connected with the indirect cognition of the other. But very important, in my opinion, is especially the direct cognition of the other, not necessarily connected with words, but rather with body-language, acting, etc.

Husserl claims that the cognition of the other man is based on empathizing (*Einfuhlung*). This is the most adequate way of knowing the other. On the lowest level the other appears as a certain material object. But there is no possibility to reach the others' body directly. Empathizing means that I treat the other as Alter-ego. Husserl writes: "... Ego and Alter-ego are always and inevitably given in a primordial joint in pairs" (Husserl 1974: 142). Thanks to the constitution of the feeling body of the other it becomes present as a "physical-psychical" object, which obviously means that we begin knowing the other from its body. But there is more than this. The cognition of the other reveals, according to Husserl, some further layer which is of a personal character. Only in personal attitude the other appears as a personal being. To know the other is to "experience him understandingly". The partners of dialogue exert a motivating power on each other. Husserl writes: "Bonds of understanding are shaped in this manner: an answer follows the question, theoretical, evaluative requirements are followed by the practical ones, as one imposes on the other (...), agreement or non-agreement, or a sort of counter-proposition" (Husserl 1974: 372). This understanding experience requires obviously the mediation of a body. In this context Husserl notices that human body is "wholly a body saturated by a soul. Every movement of a body is full of soul, coming to and going away, standing and sitting, running and dancing, etc. The same with every human achievement, every product" (Husserl 1974: 338). Empathizing means primarily getting insight into motives driving the other. A person can be understood only when I understand by what he is driven, what motives stimulate a persons' actions. The action of empathizing is described by Husserl as follows: "I shift myself into the other subject: through empathizing I catch what, how strongly, with what power motivates it (...). I perceive this motivation when I shift myself in his situation, into his level of education, into his development in his youth, etc., and in this shifting into depth I must gain solidarity with the motivations, I feel not only as if I entered his thinking, feeling, acting, but I must follow him in this, his motives become quasi-motives (...) I am with him in his temptations, in his false conclusions, and in this 'together' there is an inner co-feeling of inner motivating stimuli" (Husserl 1974: 384–5). Of course it is often difficult to imagine someone's "level of education" or "development in youth". We do not have direct access to such things without some additional prior knowledge of the other. The knowing person must have solidarity with the motives of the other, but this does not mean that he wholly accepts these motives, it rather means that they become quasi-motives of a knowing ego.

The problem of the cognition of the other was for Husserl very important in context of his transcendental philosophy. The philosopher was absolutely certain that only one's own consciousness may be given originally and absolutely, whereas in case of the consciousness of the other one can even assume its nonexistence. He pays attention how in empathizing the other man is being constituted, how and in what order particular layers of a man in the process of knowing him are being discovered. He pays particular attention to the problem of the body of the other. The constitution of the body of the other was a sine qua non condition to the constitution of higher layers of man, to the constitution of psychical layer and ultimately spiritual layer, the soul. In this process the attitude to the other should be changed from naturalistic to personalistic. Of course in this context a question may appear whether an access to a higher level is possible without some lower layer? Perhaps Husserl was too dogmatic assuming that the three layers are accessible only in one order of the cognition, namely from the lower (bodily) to the higher (psychical) and to the highest (spiritual). According to Ingarden, Husserl did not adequately solve the problem of the cognition of the other. Therefore the philosopher from Cracow writes: "(...) the whole conception of knowing the alien I and its experiences seems to be highly unsatisfactory and one should somehow start from the beginning" (Ingarden 1963: 500). One should notice however that there are many interesting remarks and points in Husserl's theory concerning knowing the other. In this context it should be remembered that Husserl pays attention to the fact that the body of the knowing person is also important as far as the process of knowing the other is concerned. It enables the analogization, which means the shifting of some sense given in experiencing one's own body unto the body of a person we try to know. But such category certainly needs some further elaboration.

The meeting of man with man is certainly a very important phenomenon. According to W. A. Luijpen, authentic encounter is characterized by: "a certain kind of the participation in the existence of the other, for whom I care" (Luijpen 1972: 255–6). According to the author the presence of the participants of a meeting means in fact their engagement. There is mutuality involved in the meeting, but it does not mean that on both sides there is equal mutuality, and in addition to this one must notice that the partners of a meeting do not necessarily have the same intentions. But mutuality as such is a sine qua non condition of a true relation. On different stages of a meeting mutuality changes, once it is stronger, once weaker. According to Józef Tischner encounters are truly events. He writes: "meeting, we feel: we are looking for another, new sphere of being. Everything must be begun from anew. Previous gestures and previous words must acquire new sense. What's more: my so far mode of existence becomes problematic" (Tischner 1978: 75). The basis of a meeting is a kind of an emotion, usually. According to Tischner the meeting may be based on positive feelings, as for example sympathy, friendship, love, but on negative feelings also, such as: hate, anger, envy. Such bases definitely shape the atmosphere of a meeting. When the meeting lasts it may change its character, it may change for example from love to hate. This fact is presented, analyzed and deeply elaborated by Stephen A. Mitchell, a well-known American psychoanalyst, the founder and chief editor of *Psychoanalytic Dialogues*, writing on passion (love). According to Mitchell: "Fact that aggression is connected with love leads to the feeling of guilt (…) close to romantic love there is also aggression" (Mitchell 2008: 194). Therefore it may happen that positive feelings may be substituted for negative ones. Changes in the basis of a meeting are turning moments, they stimulate its development but quite often weaken it or even contribute to its end. Of course it is obvious that the meeting is a phenomenon of a temporal character. There are so-called "open meetings", in which the time of their end is not known (the participants sometimes expect that they will last forever), but there are also meetings defined in time, their nature is such that the participants know when the meeting will end.

In order to enlighten some problems concerning the category of a meeting let me refer to Martin Buber's philosophy of dialogue. As it is well known, Buber assumes that dialogue may be led within some spheres, namely: nature, spiritual beings, man and the Eternal You. From the point of view of actual consideration the second sphere of dialogue, the dialogue with man is most important. Writes Buber: "Here the relation is manifest and enters language. We can give and receive the You" (Buber 1970: 57). A man sends a call to dialogue by means of a speech, using one of the languages. This is the essential kind of dialogue. Only in the dialogue with man one can achieve full mutuality; only in such cases does a call equal an answer in regard to quality, therefore in regard to the quality of beings engaged in dialogue and in regard to means of communication (human language). It is worthy of remembering that Buber is sometimes called "personalist" by some theoreticians. They are perhaps right in the sense that Buber often stresses the fact that the effort connected with achieving and sustaining the relation I-You confirms the development of personality, the process of a person being created.

The main category of Buber's theory is the relation I-You. It stands in opposition to the relation I-It, which is a separation and which means treating the other only instrumentally. In order to achieve a grown-up personality a man should take a relation I-You as often as possible.

The construction of the relation I-You looks, according to Buber, as follows. The word "You" means man's attitude towards a partner in dialogue characterized by respect due to a partner's existential value. It is characteristic that the notion under discussion means not only an object of relation, but something that appears between partners, and this appears to be the bilateral acceptance of the partner's value. When the other component of the relation is considered, namely the "I", it seems to be internally well-ordered and it is ready at every moment to lead a dialogue with a being from which a call comes, it can be the named "person".

The Buberian philosophy of dialogue is heavily loaded with values of different kinds. The relation I-You is characterized by the following features: presence, unmediating, exclusiveness, mutuality, impermanence, responsibility. The feature of presence means readiness of man for the meeting, readiness to accept a call and answer it. Writes Buber: "What is essential is lived in the present" (Buber 1970: 63–4). The feature of being unmediated means that nothing appears between men engaged in dialogue. I think that it can be interpreted in two ways; cognitively, as the lack of any prior assumptions, but also valuationally. In the first case there is nothing conceptual between I and You. In the second case it means that all additional aims, however worthy they might be, should be excluded from the relation, because the only aim of the dialogue is the dialogue itself.

The feature of exclusiveness means that only two men can participate in a given dialogue at a given moment, because man's possibilities of initiating and sustaining a given dialogue are limited. It means that true concentration upon a given partner of dialogue excludes the possibility of being involved in dialogue with other partners at the same time.

The feature of mutuality means interaction between the two men involved in dialogue. Buber writes: "Relation is reciprocity. My You acts on me as I act on it. Our students teach us, our works form us" (Buber 1970: 67). Before mutuality appears, we have a situation in which one man sends an invitation to dialogue, and the other man answers it. In the Afterword to "I and Thou", published in 1957, the thinker answers the question concerning the character of mutuality given to man as response to a given call. He states that full mutuality is not possible in inter-human relations. In addition to this he remarks that there are some I-You relations in which a priori there appears no full mutuality. This refers to the following relations: educator-pupil, psychotherapist-patient, clergyman-worshipper. As we know, Buber was very much interested in pedagogy and so he explains what he means referring to the pedagogical context. In case of the relation educator-pupil, the educator, in order to bring out what is best in his pupil, must participate in the meeting by also looking at it from the pupil's point of view, practicing the kind of relation which embraces the whole dialogical situation. The same refers to the other relations mentioned above.

The feature of impermanence means that the I-You relation does not last forever, that it has a tendency to become the I-It relation. According to Buber constant effort is needed to sustain the relation I-You, because it has a tendency to disintegrate.

I am inclined to distinguish two kinds of responsibility in Buber's philosophy. The first one is explicitly expressed by the author himself and means the responsibility of the I for the You. The second one means an obligation imposed upon man to get involved in the I-You relation as often as possible in order to develop one's personality.

There appears the question whether the dialogue happens on the basis of positive feelings only or is also possible in regard to the negative ones. Buber does not give an explicit answer in this respect, but the conclusion can be drawn, on the basis of some of his formulations, that he allows the possibility of meetings based on negative feelings. Writes Buber: "Yet whoever hates directly is closer to a relation than those who are without love and hate" (Buber 1970: 68). No further comments are needed as far as this problem is concerned, I think.

The problem of changes appearing in men engaged in a meeting is quite interesting. Buber notices that during a given meeting man's potentiality is strengthened. This means that man is both given reciprocity and the meaning of his own life "here and now" becomes clear to him. Buber repeats again and again that every meeting constitutes a fact, individualized and unique to the highest degree. Therefore the author writes: "The meaning we receive can be put to proof in action only by each person in the uniqueness of his being and in the uniqueness of his life (Buber 1970: 59)."

Buber claims that "no prescription can lead us to the encounter" (Buber 1970: 59). I cannot agree with his opinion without objections. Because even if no method can guarantee a concrete encounter I think that a man engaging himself in true encounters acquires some knowledge how encounters proceed, how to stimulate their development, how to sustain encounters. Some self-knowledge and the knowledge of other men is needed in this respect. Generally, Buber is of the opinion that the most fundamental vocation of man sustains a realization of man's humanism by means of different relations of the I-You type.

Continuing the discussion of the problem of cognition of the other one should notice that a meeting has a tendency to last thanks to the fact that there appears the knowing of the other. Such knowing changes in each particular phase of a meeting, each partner of a meeting is the knowing person and the known. Knowing the other is at the same time being known by the other. It does not mean that one talks only of purely epistemological cognition. Both participants of a given meeting allow to know each other to a certain degree, they open up, so to say. One's own being is manifested not only directly, by acting and reacting mainly, but also by verbal activity. This activity is defined differently in different languages (recall Buber's opinion on this!), and of course it is dependent on one's knowledge of one's language. There is a trivial example: more educated men usually have richer language than those who do not read much, therefore they are often able to define and express their feelings more adequately. But it is sometimes difficult to know one's own feelings. They are often so fragile and difficult to express. Full self-knowledge is usually very difficult to achieve. The notion of dialogue appears very often in Anna-Teresa Tymieniecka's philosophy. She writes for example: "Both a philosopher, a poet and scientist should lead a dialogue…" (Tymieniecka 1988: 145). Let us give voice to the interpreter and translator of Tymienieckan works into Polish, Małgorzata Wiertlewska, who writes: "A leading thread upon labyrinth of themes appearing on the border line of a body, psyche and spirit that co-create creative human condition is self-individualization of a being in the process of becoming" (Wiertlewska 2011: 397). It should be noticed that Tymieniecka refers to the problem of dialogue somewhat similarly as Husserl does, stressing the existence of three layers of personality, but the difference is that she is much more interested in the phenomenon of life, whereas Husserl seems to be much more interested in the world of physical objects.

Self-knowledge allows effectiveness of an insight into the other's feelings, but not always. The verbal expression of one's feelings may be more or less adequate. Sometimes self-knowledge gives most adequate contents of the soul but to describe it properly in words seems almost impossible. The indirect insight into the other is conditioned certainly by the understanding of the other's utterances which in turn is also better or worse. Sometimes it may be quite inadequate. In verbal behaviors besides language layer one can distinguish beyond-language layer. Human voice, different in different men, enables different realizations of the same verbal utterance. Also the same man can supply different realizations of the same sentence, depending on the mood he is in. A man thus in a very individual way manifests his own being. This manifestation includes different spheres of human being: emotional, intellectual, volitional, etc.; this is indirect manifestation of one's being. As we see there are two layers of verbal behavior. The first one is pure verbal uttering, and this means indirect knowing of the other. The second one is connected with the way something is being uttered and it means at the same time the direct knowing of the other. The other way of knowing the other, the direct manifestation of oneself, is through gestures, looks, grimaces (generally: body language). Man manifests himself spontaneously when there is no restraining oneself, when he for example is overwhelmed by joy or anger, etc. In such contexts human being manifests itself directly and sometimes pretty strongly.

The meeting as such is a sequence of different continuous actions. It has its phases, its dynamics, and also its turning points. Presence as such reaches to the past, whereas future is based upon presence. Each action is performed upon what has already been done but at the same time it throws some light upon what is going to be in the future. Only the partner of dialogue is able to understand the sequence of actions, their mutual dependence. The sequence of actions refers to the behavior of both participants of the meeting. The action of a given participant refers not only to his own behavior in the past but also to the behavior of the partner of dialogue. This mutual co-influence is described by Ingarden like this: "One should not forget that my stimulation is emotional co-stimulation in contact with other man and it reveals to the other when I am close to him, and that my manner of behavior, my way of co-stimulation is perceived by the other person and causes in her correlate co-stimulation, and there is a play of different feelings and emotions, different desires and revulsion, etc., and in result men co-living with each other know more and more about each other in each phase of living through and they stimulate each other somehow" (Ingarden 1981: 100). It seems that the effectiveness of knowing the other depends on self-knowledge, the better self-knowledge the greater the chance that we know the other better, when in turn we know the other better we know ourselves better. Self-knowledge definitely seems to be intertwined with knowing the other. This knowing includes both bodily existence of the other, as well as a psycho-spiritual dimension of one's personality. It refers to the other's feelings, experiences, psychical states, one's personal aspirations and desires, tendencies, life plans, dreams, etc. Some of them are well seen, some are not quite clear. Sometimes it seems that one knows a lot about the other, but not a full understanding of the other is possible. It seems that human being as such is unpredictable. After all humans are not machines. It also seems that in man there is a certain contradiction which makes it impossible to know the other fully.

Concluding, one can say that during the meeting of the other man there appear some phenomena worthy to be discussed. One can agree with Ingarden that Husserl's remarks concerning the cognition of the other man are vague and inaccurate. Of some importance is the distinction between direct and indirect cognition of man. Some controversies appear because man's cognitive faculties in man engaged in dialogue are somehow distorted because both partners are active in cognition when engaged in true relation understood as Martin Buber understands it. In this context the well-known Polish psychiatrist, Antoni Kępiński, writes about a mask using the following formulations: "Mask plays a role of social mirror. One cannot see oneself as one perceives somebody outside oneself. One only sees the reflection of oneself in the eyes of surrounding environment" (Kępiński 2002: 83). Kępiński as a psychiatrist seems to know very well how to lead the dialogue with other man, although it is a specific case of a patient. He insists on stable, lasting dialogue with the other, and repeats again and again that it is impossible to know the other man fully (Kępiński 2002: 9). The cognition of psychical states of the other precedes the cognition of the objective world. A baby begins the cognition of its mother's emotional sphere. The same underlines Buber writing about the primordial I-You relation (*das eingeborene Du*) (Buber 1970: 9). Kępiński insists that the psychiatrist must keep some distance from one's own emotional states and from the patient's emotional states. He must create the character of an ideal observer who, without emotional involvement into the relation keeps an eye on the patient's reaction and the psychiatrist's emotional reactions. Of interest is Kępiński's remark that what has not been said during dialogue is often more important than what has been said (Kępiński 2002: 35–6). The direct cognition of the other reaches deeper than the indirect cognition of man, verbal cognition in this case. Also the axiological dimension of the dialogue is of importance. Particular attention is paid to them by the author of *I and Thou*, who analyzes the characteristic traits of the relation I-You and stresses the importance of ethical dimension of an encounter much more than Husserl. The Polish philosopher of dialogue, Józef Tischner, also underlines the importance of ethical categories when writing on encounter. Anna-Teresa Tymieniecka does not consider ethics as a primary object of her study but nevertheless her theory is loaded with ethical considerations, especially when the relation to Everything-What-Is-Alive is considered.

References

Buber, M. 1970. *I and Thou. A new* (trans: Kaufmann, W.). New York.

Husserl, E. 1974. *Idee II* (trans: Gierulanka, D.). Warszawa: PWN.

Ingarden, R. 1963. Z badań nad filozofią współczesną. In *O Formalnej i transcendentalnej logice*. Warszawa: PWN.

Ingarden, R. 1981. *Wykłady i dyskusje z estetyki*. Warszawa: PWN.

Kępiński, A. 2002. *Poznanie chorego*. Kraków: Wydawnictwo Literackie.

Luijpen, W.A. 1972. *Fenomenologia egzystencjalna* (trans: Chwedeńczuk, B.). Warszawa: Instytut Wydawniczy PAX.

Mitchell, S.A. 2008. *Czy miłość trwa wiecznie?* (trans: Boniszewska, A.). Warszawa: G+J.

Tischner, J. 1978. *Fenomenologia spotkania*, *Analecta Cracoviensis*, Vol. X. Znak: Cracow.

Tymieniecka, A-T. 1988. *Logos and life*: *Creative experience and the critique of reason*, book I of treatise on the introduction to the phenomenology of life and the human condition, Analecta Husserliana, Vol. XXIV. Kluver. Kluver Academic Publisher: Printed in the Netherlands.

Wiertlewska, M. 2011. Nota o Autorce. In *Anna Teresa Tymieniecka*, *Życie w pełni logos* (trans: Wiertlewska, M.). Poznań: Wydawnictwo Poznańskie.

Humour, an Enlightening and Restorative Force of the Inner *Cosmos*: A Phenomenological Approach

Tereza-Brindusa Palade

Abstract Among all the potentialities of the soul, one of the less researched upon, even within the complex field of phenomenology, is *humour*. Could the sense of humour play a significant role in restoring the inner order or harmony (the *cosmos*) of the soul, and could it be seen as one of the powers mobilized by the self in order to reach a more thorough knowledge of the world? I will try to provide an answer to this question by applying Edith Stein's phenomenology of the structure of the soul to the experience of receiving a humourous account or a witty explanation.

The paper starts with a reconstruction of Stein's triadic structure of the unity of the soul from her *Endliches und ewiges Sein*, which consists of *essence*, *potency* and *life of the soul*. Then, it focuses on the experience of receiving an enlightening humourous account or explanation that helps the *I* to reorder the inner system of beliefs and meanings, or the inner *cosmos*. However, if this enlightenment is indeed produced, due to the activity of the intelligible powers of the soul, as well as to the implicit capacity of the receiver to share some cultural meanings with the source or author of the humourous account, there will be, arguably, also an increase in the vitality of the receiver. In Stein's terms, a restoration is produced within the life of the soul whenever it receives something of a spiritual nature which gives life, by bringing about cheerfulness, or pure wonder or joy. This restorative potential of humour will be related to the intelligible character of a humourous account, by using Aquinas' correlations between an act of understanding (*intellectus*), will (*voluntas*), delight (*delectatio*) and joy (*gaudium*), and will be further explained also by the paradoxical nature of wittiness.

Finally, I will argue that this restorative force of humour is especially welcome whenever the soul, given its fragility, passes through an even slightly confusing or painful event, which diminishes its vitality. Yet, in order to receive the restorative

---

T.-B. Palade (✉)
National University for Political Studies and Public Administration, Bucharest, Romania
e-mail: btpalade@gmail.com

A.-T. Tymieniecka (ed.), *Phenomenology of Space and Time: The Forces of the Cosmos and the Ontopoietic Genesis of Life: Book One*, Analecta Husserliana 116,
DOI 10.1007/978-3-319-02015-0_34,

effect of humour, one should foster a willing cognitive disposition towards the outer world, which allows a new understanding, a reordering of the inner *cosmos*, and hence a new impulse for the life of the soul.

Although the philosophical interest for humour has been vivid throughout the whole history of Western thought since Plato and Aristotle, humour is often seen nowadays as a marginal or even insignificant topic of philosophical debate. Despite the fact that, unlike plain laughter, humour involves a cognitive internal move, there is a scarce theoretical concern for its mechanisms, which are often relegated to the status of mere psychological or emotional phenomena. It seems that the more psychological theories of humour have somewhat eclipsed the philosophical approaches to humour, such as the incongruity theory, the superiority theory, or the play theory.

In what follows I will attempt to provide a phenomenological account of humour, which relies upon the "structure of the soul" elaborated by Edith Stein, the Scholastic connection between understanding, delight, and joy, and the classic incongruity theory.

The Triadic Structure of the Soul and Its Open *Cosmos*

In her philosophical opus *Finite and Eternal Being*, Edith Stein revisits the classic Trinitarian structure of the human soul elaborated by Augustine of Hippo in his *De trinitate*. Augustine had posited two triads of the inward men composed of soul-love-knowledge and memory-intellect-will. Stein distinguishes a similar triad, which is seen as a dynamic structure of the unity of the soul consisting of *essence*, *potency* and *life of the soul*. This triadic structure of the soul seems to be discovered and depicted through Stein's original endeavour to combine Husserlian phenomenology with Scholasticism.

The Steinian notion of *essence* expresses the singularity of the soul and its capacity to form the *potency* and the *life of the soul*. Within the *essence*, Stein distinguishes a *sense*, which is the form of the end, according to which the soul reaches its essential determination, and a *potency*, which is the power of the soul to become what it ought to become, and is developed within the *life of the soul*. There is, accordingly, an intricate and purposeful relationship between these three dimensions of the human soul. Yet, they are not to be understood as "dividing" the soul, since the soul always remains a unitary whole.

If the *potency* is developed within the *life of the soul*, the latter is, in its turn, explained according to its *potency* to become what it ought to become. But this development of the *life of the soul* by its *potency* could hardly take place only within the limits of the inward life of the human soul, and thus it also depends on the impressions received by the soul from the external objects. These impressions are first collected, and then elaborated inwardly by the *I*. According to Stein, the

I receives, elaborates, and responds throughout the multiple powers of the soul which derive from one *potency*.[1]

The multiplied potencies of the soul follow different routes and play different functions, some being open towards the world of objects and receiving knowledge, others conserving and elaborating the received knowledge inside the soul, and some others providing answers by the appetite and will. Although this complex process of conscious interaction with the world is based on the capacity to receive knowledge from the universe outside the soul and to elaborate it inwardly, we may however remark that the conservative power of the soul maintains also an inner harmony which could be seen as a *cosmos* of the soul. Hence, the impressions received from the outer world are often elaborated by the *I* so as to be somewhat adjusted to the inner *cosmos*. On closer inspection, this *cosmos* of the soul is not, however, a static structure, since it has to integrate meanings and knowledge coming from the intentional contact with the external world, which are able to decrease or increase the *life of the soul* through the mediation of *sense*. So, the term *cosmos* is used here for inner harmony, rather than for an ordered system with no inputs and outputs whatsoever.

The structural "openness" of the soul to the external world emphasized by Stein is of course to be read primarily in the universal sense in which the soul is exposed to *all kinds* of impressions from the world of objects that are gathered through the power of knowledge. Nonetheless, I will confine here to a more particular reading of this account of the movements of the soul and of the adjustment to the inner *cosmos* made by the *I*, namely the reception of a humourous account that is not only entertaining for the reader or the listener, but is also to some extent enlightening, even if not in a strict logical sense, since humour is often received by way of paradox. Then, I will focus on the way in which the *I* elaborates such an account and provides the inner *cosmos* with a restorative force that increases the life of the soul.

Receiving an Enlightening Humourous Account

In order to react to a joke or a humourous account, one should first understand its meaning. If a joke is told in a foreign language or by using certain expressions unknown to the listener, the most likely effect would be that he or she will not grasp its significance and, therefore, would not be able to react to its content.[2] But grasping the literal meaning of a joke is not sufficient for understanding it and providing an answer. Another prerequisite is the capacity of the receiver to share some cultural

[1] E. Stein, *Finite and Eternal Being: An Attempt at an Ascent at the Meaning of Being*, trans. Kurt F. Reinhardt (Washington, DC: Institute of Carmelite Studies Publications, 2002). I refer to the Italian translation, *Essere finito ed essere eterno. Per una elevazione al senso dell'essere*, trans. Luciana Vigone and introd. Angela Ales Bello (Rome: Città Nuova Editrice, 1988), p. 449.

[2] S. Critchley, *On Humour*, (London: Routledge, 2002), p. 4.

or contextual meanings with the source or the author of the humourous account. These shared meanings are not necessarily local or regional, though they are often facilitated by local or communal connections. There may be also shared meanings of a more general nature, which may explain, for example, how many people today can still appreciate and enjoy the jokes and the humourous plots of Shakespearean drama, despite the cultural changes that have produced a different mentality, since the Elizabethan era.

Yet, not every joke or humourous account is equally enlightening. The most remarkable and successful jokes are probably those which reveal an aspect of reality, usually defective, through a paradox. Hence, their quality relies on their capacity to uncover by a simple strategy, which is usually called wit, a characteristic that has not been perceived earlier with the same clarity. "Wit is a sword, it is meant to make people feel the point, as well as see it", as G. K. Chesterton wrote in an essay about Mark Twain.[3] Despite its reliance upon some incongruity which amounts to a logical paradox, this kind of wittiness is strictly intellectual and cognitive, and could convey a deeper psychological, moral or metaphysical meaning. Its effect may be either enjoyable or distressing, depending on how inconvenient for the receiver is the truth that it reveals.

To receive the full effect of such an enlightening joke, one should however foster a willing cognitive disposition towards the world, which allows for a new apprehension that somewhat broadens one's knowledge of the world or of oneself. In this way, his or her inner *cosmos* is intentionally open towards the external world, though the *I* achieves the task of the cognitive elaboration on the basis of the contents already present in one's conscience. For example, in reading Marx Twain's account of a good education, according to which a "good breeding consists in concealing how much we think of ourselves and how little we think of the other person",[4] one may almost instantly acknowledge the fact that many educated people he/she came across with seemed to express a polite, but rather overstated concern for his/her person, and that he/she has often done the same when politely addressing other people, as he/she wanted to leave a good impression. Such an enlightening piece of explanation is thus elaborated by the *I* by recalling the memory of social interactions and of one's own conduct in some social circumstances. Afterwards, the *I* to some extent reorders the inner *cosmos* of the soul according to the light shed by this witty explanation of "good behaviour".

The integration into the inner *cosmos* of a new meaning which enlightens a part of one's conscious experience may also have a restorative and invigorating effect upon the soul. This effect could be, arguably, seen as deriving from a delight and an enjoyment which are the result of understanding. The correlation between

[3] G. K. Chesterton, *A Handful of Authors*: *Essays on Books and Writers*, ed. Dorothy Collins (Lanham, MD: Sheed and Ward Publishers, 1953), p. 11.

[4] Quoted by D. Nyberg, *The varnished truth*: *Truth telling and deceiving in ordinary life* (Chicago: University of Chicago Press, 1993), p. 158.

understanding, delight and joy is convincingly explained by a classic account of Thomas Aquinas, in the first book of his *Summa contra Gentiles*.

Intelligere, *Delectatio*, and *Gaudium*

According to Thomas Aquinas, every act of understanding (*intelligere*), insofar as it is directed to some good, is accompanied by an act of the will, "since the understood good is the proper object of the will" (*Contra Gentiles* lib. 1, cap. 72, n. 2).[5] The will is located in the intellect, as a part of mind which makes decisions by aiming at some good. One would never grasp the good by his/her intellect if one would not be endowed with will. Furthermore, the good understood by the intellect is perceived as delightful, because the will, which is a part of the intellect, takes pleasure in it in the same way as the appetite of concupiscence takes pleasure in the object of sensible delight (*Contra Gentiles* lib. 1, cap. 72, n. 5).[6] The understanding is thus all the more delightful insofar as it is more perfect. Thomas Aquinas further explains that delight (*delectatio*) is always of a present good. Hence it arises from a really conjoined good, whereas joy (*gaudium*) does not need to be of a conjoined good, since it suffices that the will is resting in an external object in order to take joy in it (*Contra Gentiles* lib. 1, cap. 90, n. 7).[7]

Now, according to the Thomistic-Aristotelian epistemology the intellectual grasp of an object results in a certain unity, since understanding is only a potency of the intellect which becomes actual only when it seizes an object. Hence, Thomas Aquinas speaks of delight, and not simply of joy, in connection with understanding – although in the first book of *Summa contra Gentiles* he mainly refers to the delight taken by the divine intellect in the object understood in order to sustain that God has will and that delight and joy are present in God. Yet, since the understanding of a human intellect follows a virtually similar path, we may consistently argue that delight is an effect of understanding even in the case of the human mind. And so is joy, since the requirement for *gaudium* is less demanding in terms of the unity of a subject with an external object.

A provisional conclusion might be that the act of understanding, which is intrinsic to the reception of a witty or an enlightening humourous account brings about delight and joy. Yet, the humourous ingredient which is likely to invigorate the soul is not thoroughly explained by this interpretation. A humourous account usually conveys also an incongruous element of some kind. Therefore, we will need to supplement this Thomistic explanation by focusing also on the paradoxical side of an enlightening humourous account.

---

[5] St Thomas Aquinas, *Summa contra Gentiles. Book One: God*, trans. with an introd. and notes Anton C. Pegis (Notre Dame, IN: University of Notre Dame Press, 1975), p. 239.

[6] Op. cit., p. 241.

[7] Op. cit., pp. 276–277.

The Paradox of Wittiness

The fact that humour presupposes some incongruities which resemble the philosophical paradoxes has already been examined by some theoretical approaches, among which those of Aristotle, Immanuel Kant and Søren Kierkegaard are the most famous.[8] Usually the incongruities arise from the distortions provided by the revelation of a successful joke, so that the latter contradicts the expectation. For example, when Oscar Wilde wrote in a tale that "I often have long conversations all by myself, and I am so clever that sometimes I don't understand a single word of what I am saying",[9] the paradoxical structure which apparently causes the humourous effect is created by the nonsensical dividing of the self in two parts, one of which is excused for being less intelligent through the other's alleged brightness, which in fact may amount to a simple incoherence.

Alternatively, the paradox may be stated from the very beginning, being further explained in an unexpected way, as in Chesterton's ingenious vindication of the morality of robbery: "Thiefs respect property. They merely wish the property to become their property that they may more perfectly respect it".[10] The humourous effect results now from the contradiction of the "normal" assumption that the respect for property also involves consideration for other people's possessions.

So, the structure which produces a humourous enlightening effect that is likely to refresh or invigorate the soul is somewhat inconsistent from a strictly logical standpoint. We should, however, add that in order to bring about a humourous effect that is also restorative this paradoxical structure has to be cheerful and somewhat gratifying for the intellect, so that its elaboration by the *I* could impress the most disinterested and spiritual part of the soul, as the gleam of sunshine, or the sight of blooming trees. Only such a joyful paradox that is not based upon a prejudice, or upon a codified mindset, could possibly possess a genuine spiritual quality that produces cheerfulness and joy, thus increasing the life of the soul.

In fact, in Stein's account of the powers of the soul, she explicitly refers to some sort of phenomena which may increase the life of the soul, as the splendour of the sun, the radiant blue of the sky, an encouraging word, an idyllic landscape, or the happy laughter of a baby.[11] We may also add to this list an enlightening humourous account of a paradoxical kind which may relieve the soul by providing it a new meaning (even if not by following a strict logical order) that is further integrated into the inner *cosmos*.

---

[8] See also Peter Cave, "Humour and Paradox Laid Bare", *Monist*, 88: 1 (Jan. 2005), p. 135.

[9] Oscar Wilde, *The Happy Prince and Other Stories* (Hertfordshire: Wordsworth Editions, 2003), p. 77.

[10] G. K. Chesterton, *The Man Who Was Thursday (A Nightmare)*, (Rockville, MD: Serenity Publishers, 2009), p. 38.

[11] Stein, op. cit., p. 450.

The Inward Elaboration of a Humourous Statement

The interaction of the inner *cosmos* with the outer world through the passive sense of humour, seen as a potentiality of the soul, thus brings not only a cognitive input, but also a restorative force that is likely to increase the life of the soul. However, the occurrence of this second effect seems to depend upon the cognitive disposition towards the external world that is willingly and somewhat responsibly assumed by the soul. The *I* could not properly receive the meaning of a humourous account without an intentional cognitive "openness" to the external world. A simple appetite for an entertaining account that is received in a whimsical way, as a mere fantasy, could hardly result in an adequate reception of its meaning. The most enlightening humourous account is usually received as conveying a certain truth that is recognized as such and is further elaborated by the *I*.

This process of inward elaboration presupposes a reflection that is based on a survey of one's own conscious memory. The memory contains of course not only personal recollections, but also some records of the experience of other people, real or fictional. This examination takes into account only the essential extracted from these records, which is quickly surveyed by the mind so as to corroborate the truth of the account. For example, when one reads an ironical remark about sincerity made by George Bernard Shaw in his *Maxims for Revolutionists* (1903), according to which "it is dangerous to be sincere unless you are also stupid", the first step of the elaboration is not to infer an analytical conclusion like "the sincerity of someone who possesses a keen mind is often undesirable", but to corroborate the remark with some conclusions drawn after taking a brief glance at the records stored in one's own memory, so as to testify if the statement is accurate. So, the elaboration is cognitive, in the sense that it aims at assessing the veracity of the remark, although it is not strictly formal or analytical. The process of elaboration is rather a synthetic one, since it considers a wider knowledge and only arrives at the conclusion after a more comprehensive glance on the conscious memory. After taking the first step of testifying the veracity of the statement, one may also draw an analytic conclusion, but only as a kind of cognitive supplement that is not necessary for the further stage of reacting to the humourous account in which tensions are released, usually by laughter.

The *I* also conserves the newly elaborated meaning by adapting it to the inner *cosmos*, that is to the inner order of experiential knowledge associated with some specific meanings. Yet, as I have argued, this process leads to an integration of a new meaning into the open inner *cosmos*, which increases the life of the soul, since the new understanding is accompanied by the delight and joy arisen by the humourous paradox.

Now, this delight plays a restorative function which provides a new vitality for the soul of the receiver, by giving it a new meaning which helps the *I* to reorder the inner *cosmos*. This function of a humourous account is especially welcome whenever the soul, given its fragility, passes through a more disconcerting event which produces a confusing or a painful effect that diminishes its life and causes tension or apprehension. The life of the soul is exposed to a continuous exchange of forces with the external world. Stein argues that this exchange of forces is meant to assist the development of a power which corresponds to the essential determination of the soul and is not present from the beginning, being acquired only in time.[12] She further explains that the forces which to the largest extent assist such development possess a certain spiritual nature, which could be absorbed by the soul and may boost its life. For example, the splendour of the sun in the morning does express something of a spiritual kind, like the triumph of a radiant and cheery light over darkness, and so does humour, especially when it is construed in an enlightening and paradoxical way.

Thus, the development of a force which conforms to the essential determination of the soul needs the assistance of other forces by which the *I* keeps the balance of the inner *cosmos*, and such a force is humour. This approach of the balancing role of humour with regard to the inner *cosmos* of the soul comes somewhat close to the play and relaxation explanation of humour provided by Thomas Aquinas, according to which entertainment and relaxation are an important part of a balanced human life (*Summa theologiae*, II–II, q. 168, a. 2).[13] It is also in some agreement with the relief theory of humour, since it also endorses the idea that the reaction to a humourous account plays the functional role of releasing tensions. Yet, it also plays an intellectual role, since it emphasizes also the cognitive shift produced by an englightening humorous account, whenever an input from the external world increases the clarity of understanding, thus boosting up the life of the soul.

---

[12] Ibid.

[13] St Thomas Aquinas, *Summa Theologica*: *Volume IV – Part III, First Section*, trans. Fathers of the English Dominican Province (New York: Cosimo Classics, 2007), p. 1872.

Program from The 62nd International Congress of Phenomenology: *The Forces of the Cosmos and the Ontopoietic Genesis of Life*, held in Paris, France, 2012

The World Institute for Advanced Phenomenological Research and Learning

1 Ivy Pointe Way, Hanover, New Hampshire 03755, United States,
Telephone: (802) 295–3487; Fax: (802) 295–5963;
Website: http://www.phenomenology.org

Anna-Teresa Tymieniecka, President; Daniela Verducci, Vice-President

The 62nd International Congress of Phenomenology

Location: Lucernaire Centre National d'Art et d'Essai, Paris, France, August 8–10, 2012

August 8–10, 2012

PROGRAM

Topic: ***THE FORCES OF THE COSMOS AND THE ONTOPOIETIC GENESIS OF LIFE***

The Congress begins at 9:00 a.m., August 8, 2012 at Lucernaire Centre National d'Art et d'Essai, Paris, France

Conference Director: Daniela Verducci, Vice-President, World Phenomenology Institute

Program Presided by: Anna-Teresa Tymieniecka, President World Phenomenology Institute

Local Organization Committee: Carmen Cozma, Claire Hill, Konrad Rokstad, Leszek Pyra, Daniela Verducci

A.-T. Tymieniecka (ed.), *Phenomenology of Space and Time: The Forces of the Cosmos and the Ontopoietic Genesis of Life: Book One*, Analecta Husserliana 116,
DOI 10.1007/978-3-319-02015-0,

International Scientific Committee, Directed by Angele Kremer-Marietti (FRANCE):

AZERBAIJAN: Salahaddin Khalilov; FRANCE: Claire Hill; ITALY: Angela Ales Bello, Francesco Totaro, Daniela Verducci; NORWAY: Konrad Rokstad, POLAND: Maria Bielawka, Mariola Sulkowska-Janowska; TURKEY: Klymet Selvi, Erkut Sezgin; UNITED STATES: Ogla Louchakova-Schwartz, Patricia Trutty-Coohill, Anna-Teresa Tymieniecka

SCHOLARLY PROGRAM

WEDNESDAY, AUGUST 8, 2012
9:00 – 12:30, Theatre Noir

PRESIDENTIAL ADDRESS
Anna-Teresa Tymieniecka

PLENARY SESSION I
Chaired by: Angele Kremer-Marietti, Paris, France

Communicative Virtues of the Phenomenology of Life
Daniela Verducci, Universita degli Studi di Macerata, Italy

Toward a New Enlightenment: Metaphysics as Philosophy of Life
Nicoletta Ghigi, CIRF (Roma), Italy

Moral Excellence as Cosmicization Human Beingness in the *Ontopoietic* Perspective
Carmen Cozma, University "Al.I.Cuza", Iasi, Romania

The Inseparable Link between "Cosmology" and "Life World" in Anna-Teresa Tymieniecka's Philosophy: The Originality of a New Perspective of the "Real Individual and Autonomous Being", A Possible Comparison with Hedwig Conrad-Martius' "Phenomenological Realism"
Francesco Alfieri, OFM, Convento "Cristo Re", Martina Franca, Italy

The Forces of the Cosmos before Genesis and Before Life: Some Remarks on Eugen Fink's Philosophy of the World
Simona Bertolini, Universite de Bologna, Italy

12:30 – 13:30 LUNCH

WEDNESDAY, AUGUST 8, 2012
13:30 – 17:00, Theatre Noir

SESSION I
Chaired by: Carmen Cozma, University "Al.I.Cuza", Iasi, Romania

The Modern Concept of the Individual and the "Forces of Nature"
Oliver W. Holmes, Wesleyan University, Middletown, Connecticut, United States

Platonian Chora in the Light Of Husserl's Concept of Transcendental Constitution
Maria Bielawka, Krakow, Poland

Transformation in Phenomenology: Husserl and Tymieniecka
Anar Jafarov, The East–west Research Center, Baku, Azerbaijan

Ontopoietic Process of Life in Kierkegaard's Books: Zoe and Bios
Elodie Gontier, Universite de la Sorbonne, Paris, France

Edifices
Semiha Akinci, Anadolu University, Tepebasi-Eskisehir, Turkey

The Problem of Unconscious Phenomena in Phenomenological Practices
Nevia Dolcini, University of Macau, China

WEDNESDAY, AUGUST 8, 2012
13:30 – 17:00, Paradis

SESSION II
Chaired by: Detlev Quintern, Istanbul, Turkey

The Cave, the Lifeworld and the Tradition: The Transcendence-Immanence Contrast Perspective
Abdul Rahim Afaki, University of Karachi, Pakistan

Wahdat al-Wujud and Logos of Life: The Philosophical Comparison
Konul Bunyadzade, The East–west Research Center, Baku, Azerbaijan

The Forces of the Cosmos and the Sense of Life in Sufism: A Philosophical Reading of the Sufi Way Towards the Creator
Benedetto Cortellesi, Pontifica Universtas Lateranensis, Italy

Consciousness of the Cosmos: A Thought Experiment Through Philosophy and Science Fiction
Sibel Oktar, Ozyegin University, Istanbul, Turkey

Intentionality of Time and Space and Quantum-Phenomenological Sense of Space
Mamuka G. Dolidze, Institute of Philosophy of Georgia, Tblisi, Georgia

Duality and the 2011 Nobel Prize in Physics
Tsung-I Dow, Emeritus: Florida Atlantic University, United States

WEDNESDAY, AUGUST 8, 2012
13:30 – 17:00, Theatre Rouge

SESSION III
Chaired by: Francesco Totaro, Universiata degli Studi de Macerata, Italy

Anthropological Regression in the Modern World Versus Anna-Teresa Tymieniecka's Metaphysics of Ontopoiesis of Life
Jan Szmyd, University of Krakow, Poland

Biologically Organized Quantum Vacuum and the Cosmic Origin of Cellular Life
Attila Grandpierre, Konkoly Observatory, Zebegeny, Hungary

The Question Concerning Technology in Merleau-Ponty's Critique of Objective Thought
Claus Halberg, University of Bergen, Norway

The Open Void
Marguerite Harris, Indiana University, Bloomington, Indiana, United States

Ontopoiesis of Eidolon and Transcendental Schematism in Cassirer and the Concept of Ontology in Meinong and Quine
Giuseppina Sgueglia, La Pontifica Universitas Lateranenensis, Italy

Logos Omnia, Cosmos and the Limits of Perception: Tymieniecka's Phenomenological Ontology in Light of Merleau-Ponty's Insights
Olga Louchakova-Schwartz, Institute of Transpersonal Psychology, Palo Alto, California, United States

THURSDAY, AUGUST 9, 2012
9:00 – 12:30, Theatre Noir

PLENARY SESSION II
Chaired by: Daniela Verducci, Universita degli Studi di Macerata, Italy

Human Soul and Life Horizons
Maija Kule, University of Latvia

Towards a Phenomenology of Life and Invisible
Giovanna Costanzo, Universita degli Studi di Messina, Italy

Intentionality, Telos, Transcendentality as Ontopoietic Forces of the Cosmos
Francesco Totaro, Universiata degli Studi de Macerata, Italy

Ontopoiesis in Ben Okri's Poetic Oeurve and *A Time for New Dreams* (2011)
Rosemary Gray, University of Pretoria, South Africa

Pythagoras at Chartres
Patricia Trutty-Coohill, Siena College, Siena, Loudonville, United States

Ciphering, Novum, Intuition and Imaginatio Creatrix: An Approach to Zoroaster's Gathas
Martin Schwartz, University of California, Berkeley, United States

12:30 – 13:30 LUNCH

THURSDAY, AUGUST 9, 2012
13:30 – 17:00, Theatre Noir

SESSION IV
Chaired by: Leszek Pyra, Krakow, Poland

The Cosmos and Bodily Life on Earth
Konrad Rokstad, University of Bergen, Norway

Jolanthe's Cosmos: Knowing Without Seeing
Detlev Quintern, Istanbul, Turkey

Philosophical Hermeneutics Confronted by that which is Different
Aleksandra Pawliszyn, Uniwersytet Gdanski, Poland

Evolution of Matter and Spirit, Rediscovering Slowackis' Mysticism and de Chardin's Theology
Piotr Popiolek, Pontifical University of John Paul II in Cracow, Poland

Anna-Teresa Tymieniecka's and Max Scheler's Phenomenology as the *Ontopoietic Genesis of a Manager's Life*
Bronislaw Bombala, University of Warmia and Masuria, Poland

The Eschatology of Ontopoiesis: Kanzi
Johannes Servan, University of Bergen, Norway

THURSDAY, AUGUST 9, 2012
13:30 – 17:00, Paradis

SESSION V
Chaired by: Maija Kūle, University of Latvia

Novel as Path: Mamardashvili's "Lectures on Proust"
Mara Stafecka, Rockford, Illinois, United States

Comparative Phenomenology of Singing and Dance as Artistic "Instruments" Incorporated into the Body of their Performer
Alessia Rita Vitale, University de la Sorbonne Paris IV, France

Phenomenology and Archeology: Methodical Insights and Thematic Inspirations
Jaroslava Vydrova, Filozoficky ustav SAV, Bratislava, Slovakia

Plotinus' "Enneads" and Self-Creation
Ineta Kivle, J. Vitols Latvian Academy of Music, Riga, Latvia

Directing Anatoly Vasilyev, from Individual Creative Manner to the Method
Valery Kolenova, Vladimir State University, Russia

Kant and Starry Sky
Rihards Kūlis, University of Latvia

THURSDAY, AUGUST 9, 2012
13:30 – 17:00, Theatre Rouge

SESSION VI
Chaired by: Rosemary Gray, University of Pretoria, South Africa

Teleology in Nature and Life-Transforming Art
Vladimir L. Marchenkov, Ohio University, Athens, United States

The Creative Potential of Humor
Anna Malecka, AGH - University of Science and Technology in Krakow, Poland

Artist's Personal Cosmogony, Andre Gide and Jaroslaw Iwaszkiewicz's Concept of Cosmos, Genesis of Life and Origin of Art
Daria Gosek, Jagiellonian University, Krakow, Poland

Educational Paradigm Shifting Towards Phenomenological Pedagogy
Klymet Selvi, Anadolu University, Tepebasi-Eskisehir, Turkey

The Call of Philosophizing by Writing and Reading Characters by Significations Unfolding in Internal Connections at the Moment of Spacing as Letters, Words, Voices of Speech
Erkut Sezgin, Istanbul Kultur University, Turkey

Notes for a Phenomenological-Existential and Hermeneutic Psychotherapy: an Introduction
Paulo Roberto R. Machado, Universidade de Evora – Portugal, Vlla Mariana - Sao Paulo, Brazil

FRIDAY, AUGUST 10, 2012
9:00 – 12:30, Theatre Noir

PLENARY SESSION III
Chaired by: Konrad Rokstad, University of Bergen, Norway

Cosmic Order and Exoneration of the Beautiful: Visions of the Problem in Contemporary Philosophy
Ella Buceniece, University of Latvia

The Law of Opposites in the Ontopoiesis of Life and in Language
Zaiga Ikere, Daugavpils University, Latvia

"One, Two, Three, but where is the Fourth...?" Plato's Conception Tripartite/ Quadripartite Nature of Being and Becoming and Its Evolvement in Contemporary Philosophy
Velga Vevere, University of Latvia

The Forces of Darkness and the Powers of Goodness: Jerzy Nowosielski's Concept
Katarzyna Stark, Krakow, Poland

Le Chaos du Monde Sensible et la Creation du sens Rudimentaire (à partir de Plotin)
Robert Karul, Institut de Philosophie, Bratislava, Slovakia

Cosmic Forces and Human Existence
Manjulika Ghosh, University of North Bengal, Darjeeling, India

12:30–13:30 LUNCH

FRIDAY, AUGUST 10, 2012
13:30–17:00, Theatre Noir

SESSION VII
Chaired by: Patricia Trutty-Coohill, Siena College, Siena, Loudonville, United States

The Unity of Eastern and Western Thought Traditions in A-T. Tymieniecka's Phenomenology of Life
Salahaddin Khalilov, The East–west Research Center, Baku, Azerbaijan

Transcendental Morphology: A Phenomenological Interpretation of Cosmos of Subject
Bence Peter Marosan, Eotvos Lorand University, Budapest, Hungary

The Outside's Inside: The Phenomenology of the External World in Hedwig Conrad-Martius' Thought
Ronny Miron, Bar Ilan University, Ramat Gan, Israel

Time and the Mystery of Existence
Ashok Kumar Malhotra, State University of New York, Oneonta, United States

The Idea of Contemporary Dance in the Context of Ontopoiesis of Life
Olga Polisadova, Vladimir State University, Russia

Templars and the Holy Grail: A Cross-Reference Relationship in an Ontopoietic Perspective
Mina Sehdev, Macerata, Italy

FRIDAY, AUGUST 10, 2012
13:30–17:00, Paradis

SESSION VIII
Chaired by: Ella Buceniece, University of Latvia

How to Approach Heideggerian Gods?
Jani Vanhala, University of Helsinki, Finland

The Forthcoming Sciences of Life and Life of Sciences: From Nietzsche and Husserl and Varela and Beyond
Ammar Zeifa, Universite Paris I, France

Motion in Crisis: Why the Analytic Principles of Thought Destroy Motion and Life in the World
Ion Soteropoulous, Apeiron Centre, Paris, France

"Heraclitus/Nietzsche/Heidegger in Πόλεμος": "τὰ δὲ Πάντα οἰακίζει Κεραυνός". Heraclitus
Kiymo Murata-Soraci, Tama University, Japan

***Lebenswelt* and Operational Methodology in the Philosophical and the Epistemological Reflections of Hugo Dingler**
Dario Sacchi, Universita Cattolica del Sacro Cuore, Milano, Italy

The Permanent Creativity of the Self
Stefano Polenta, Universita de Macerata, Italy

Somatic Unity in Cosmic Flux
Paul Martin Ryan, Czech Republic

FRIDAY, AUGUST 10, 2012
13:30–17:00, Theatre Rouge

SESSION IX
Chaired by: Erkut Sezgin, Istanbul Kultur University, Turkey

Cognition and Emotion: From Dichotomy to Ambiguity
Simen Øyen *and* Claus Halberg, University of Bergen, Norway

The Meeting of Man with Man
Leszek Pyra, Krakow, Poland

Humour, an Enlightening and Restorative Force of the Inner *Cosmos*; A Phenomenological Approach
Tereza-Brindusa Palade, National School of Political Studies and Public Administration, Bucharest, Romania

Eidos and Simulacrum: Turn on Logo to Linguistic-Matezis
Rimma Kurenkova *and* O. Gubanov, Vladimir Pedagogical Institute, Russia

Hyper Klein Bottle Logophysics Ontopoiesis of the Cosmos and Life
Diego Rapoport, Universidad Nacional de Quilmes, Beunos Aires, Argentina

Comments on Max Scheler's Thought and Philosophical Counseling
Lucrezia Piraino, Universita degli Studi di Messina, Italy

Active and Passive Analysis of a Value
Susi Ferrarello, University of Rome, La Sapienza, Italy

The manufacturer's authorised representative in the EU is Springer Nature Customer Service Centre GmbH, Europaplatz 3, 69115 Heidelberg, Germany. If you have any concerns regarding our products, please contact ProductSafety@springernature.com

Printed and bound by CPI Group (UK) Ltd, Croydon, CR0 4YY
01/10/2026
02225865-0020